Surveys in Differential Geometry

Vol. 1: Lectures given in 1990
edited by S.-T. Yau and H. Blaine Lawson

Vol. 2: Lectures given in 1993
edited by C.C. Hsiung and S.-T. Yau

Vol. 3: Lectures given in 1996
edited by C.C. Hsiung and S.-T. Yau

Vol. 4: Integrable systems
edited by Chuu Lian Terng and Karen Uhlenbeck

Vol. 5: Differential geometry inspired by string theory
edited by S.-T. Yau

Vol. 6: Essay on Einstein manifolds
edited by Claude LeBrun and McKenzie Wang

Vol. 7: Papers dedicated to Atiyah, Bott, Hirzebruch, and Singer
edited by S.-T. Yau

Vol. 8: Papers in honor of Calabi, Lawson, Siu, and Uhlenbeck
edited by S.-T. Yau

Vol. 9: Eigenvalues of Laplacians and other geometric operators
edited by A. Grigor'yan and S-T. Yau

Vol. 10: Essays in geometry in memory of S.-S. Chern
edited by S.-T. Yau

Vol. 11: Metric and comparison geometry
edited by Jeffrey Cheeger and Karsten Grove

Vol. 12: Geometric flows
edited by Huai-Dong Cao and S.-T. Yau

Vol. 13: Geometry, analysis, and algebraic geometry
edited by Huai-Dong Cao and S.-T.Yau

Vol. 14: Geometry of Riemann surfaces and their moduli spaces
edited by Lizhen Ji, Scott A. Wolpert, and S.-T. Yau

Volume XII

Surveys in Differential Geometry

Geometric flows

edited by
Huai-Dong Cao and Shing-Tung Yau

International Press
www.intlpress.com

Surveys in Differential Geometry, Vol. 12

Editors:
Huai-Dong Cao, Lehigh University
Shing-Tung Yau, Harvard University

2010 Mathematics Subject Classification 53C44

ISBN 978-1-57146-182-7

Paperback reissue 2010. Previously published in 2008 under ISBN 1-57146-118-6 (clothbound).

Typeset using the LaTeX system.

Preface

Geometric flows are non-linear parabolic differential equations which describe the evolution of geometric structures. Inspired by Hamilton's Ricci flow, the field of geometric flows has seen tremendous progress in the past 25 years and yields important applications to geometry, topology, physics, nonlinear analysis, and so on. Of course, the most spectacular development is Hamilton's theory of Ricci flow and its application to three-manifold topology, including the Hamilton-Perelman proof of the Poincaré conjecture.

This twelfth volume of the annual Surveys in Differential Geometry examines recent developments on a number of geometric flows and related subjects, such as Hamilton's Ricci flow, formation of singularities in the mean curvature flow, the Kähler-Ricci flow and Yau's uniformization conjecture, long-time behavior of the Yamabe flow, the Lagrangian mean curvature flow, existence and regularity results for extrinsic curvature flows in semi-Riemannian manifolds, global regularity and singular development for wave maps, the Li-Yau-Hamilton type inequalities and related monotonicity formulas, the Hamilton-Jacobi equation and 2+1-dimensional Einstein equations.

While the works contained in the present collection cast but a partial picture of the field as a whole, they do reflect the type of problems that geometric flows deal with, and some of the techniques involved. It is our hope that the present volume will benefit geometric analysts and other researchers interested in the subject. Finally, we would like to express our gratitude to the authors, for their efforts which made this volume possible.

Huai-Dong Cao
Shing-Tung Yau

Contents

Preface .... *v*

On the Conformal Scalar Curvature Equation and Related Problems
Simon Brendle 1

A Survey on the Kähler-Ricci Flow and Yau's Uniformization Conjecture
Albert Chau and Luen-Fai Tam 21

Recent Developments on Hamilton's Ricci flow
Huai-Dong Cao, Bing-Long Chen, and Xi-Ping Zhu 47

Curvature Flows in Semi-Riemannian Manifolds
Claus Gerhardt 113

Global Regularity and Singularity Development for Wave Maps
J. Krieger 167

Relativistic Teichmüller Theory – A Hamilton-Jacobi Approach to 2 + 1-Dimensional Einstein Gravity
Vincent Moncrief 203

Monotonicity and Li-Yau-Hamilton Inequalities
Lei Ni 251

Singularities of Mean Curvature Flow and Flow with Surgeries
Carlo Sinestrari 303

Some Recent Developments in Lagrangian Mean Curvature Flows
Mu-Tao Wang 333

On the Conformal Scalar Curvature Equation and Related Problems

Simon Brendle

1. The Yamabe problem

A fundamental result in two-dimensional Riemannian geometry is the uniformization theorem, which asserts that any Riemannian metric on a compact surface is conformally related to a metric of constant curvature. There are two ways of generalizing this result to higher dimensions. One problem, proposed by E. Calabi, is concerned with the existence of Kähler-Einstein metrics (or, more generally, extremal Kähler metrics). The other one was proposed in 1960 by H. Yamabe [**38**]. Yamabe claimed the following theorem:

THEOREM 1.1. *Let M be a compact manifold of dimension $n \geq 3$ without boundary, and let g be a Riemannian metric on M. Then there exists a metric $\tilde{g}$ which is conformally related to g and has constant scalar curvature.*

However, Yamabe's proof contained a serious gap. Theorem 1.1 was first verified by N. Trudinger [**36**] under the additional assumption that g has non-positive scalar curvature. T. Aubin [**3**] proved Theorem 1.1 assuming that $n \geq 6$ and (M, g) is not locally conformally flat. The remaining cases were solved by R. Schoen [**27**] using the positive mass theorem (see [**31, 37**]). A. Bahri [**4**] gave an alternative proof in the locally conformally flat case.

It is well known that the Yamabe problem is equivalent to the solvability of a semilinear elliptic PDE. Indeed, if $\tilde{g}$ is conformally related to g, then we can write $\tilde{g} = u^{\frac{4}{n-2}}\, g$, where u is a positive function on M. The scalar curvature of $\tilde{g}$ is related to the scalar curvature of g by the formula

$$R_{\tilde{g}}\, u^{\frac{n+2}{n-2}} = R_g\, u - \frac{4(n-1)}{n-2}\, \Delta_g u. \tag{1.1}$$

Here, Δ_g denotes the Laplace operator associated with the metric g. In particular, the metric $\tilde{g} = u^{\frac{4}{n-2}}\, g$ has scalar curvature c if and only if u is a positive solution of the PDE

$$\frac{4(n-1)}{n-2}\,\Delta_g u - R_g\,u + c\,u^{\frac{n+2}{n-2}} = 0. \tag{1.2}$$

The Yamabe problem admits a natural variational characterization. A function u satisfies (1.2) for some $c \in \mathbb{R}$ if and only if u is a critical point of the Yamabe functional

$$E_g(u) = \frac{\int_M \left(\frac{4(n-1)}{n-2}\,|du|_g^2 + R_g\,u^2\right) dvol_g}{\left(\int_M u^{\frac{2n}{n-2}}\,dvol_g\right)^{\frac{n-2}{n}}}. \tag{1.3}$$

The functional E_g resembles the Sobolev quotient associated with the embedding of $W^{1,2}(M,g)$ into $L^{\frac{2n}{n-2}}(M,g)$. Moreover, the functional E_g is closely related to the total scalar curvature functional. Indeed, it follows from (1.1) that $E_g(u) = \mathcal{E}(u^{\frac{4}{n-2}}\,g)$, where $\mathcal{E}$ denotes the normalized Einstein-Hilbert action:

$$\mathcal{E}(g) = \frac{\int_M R_g\,dvol_g}{\mathrm{Vol}_g(M)^{\frac{n-2}{n}}}.$$

Finally, the Yamabe constant of g is defined as the infimum of the Yamabe functional $E_g(u)$ over all positive functions on M:

$$Y(M,g) = \inf_{0<u\in C^1(M)} E_g(u). \tag{1.4}$$

The key step in the proof of Theorem 1.1 is an upper bound, due to T. Aubin and R. Schoen, for the Yamabe constant $Y(M,g)$. Once this inequality is known, the existence of a minimizer of E_g follows from standard PDE arguments.

THEOREM 1.2. *Assume that (M,g) is not conformally equivalent to the round sphere S^n. Then $Y(M,g) < Y(S^n)$, where $Y(S^n)$ denotes the Yamabe energy of the round metric on S^n.*

The Yamabe PDE (1.2) has a unique solution if $Y(M,g) \leq 0$. However, uniqueness fails for $Y(M,g) > 0$. As an example, consider the sphere S^n equipped with the round metric. This example is very special, because the Yamabe PDE is invariant under the conformal transformations on S^n. By a result of M. Obata [**25**], every metric of constant scalar curvature which is conformal to the round metric on S^n is given by the pull-back of the round metric under a conformal diffeomorphism. Hence, if g is the round metric on S^n, then every solution to (1.2) is minimizing, and the space of solutions can be identified with the unit ball B^{n+1}.

As another example, let g be the product metric on $S^{n-1}(1) \times S^1(L)$. In this case, all solutions of the Yamabe PDE are rotationally symmetric. If the length L of the S^1-factor is sufficiently small, then the Yamabe PDE has a unique solution (which is constant). On the other hand, the Yamabe PDE has many non-minimizing solutions if L is large. We refer to [**28**] for a detailed discussion of this example.

In this paper, we will discuss two PDE problems that arise in connection with the Yamabe problem. The first problem is concerned with the set of all constant scalar curvature metrics in a given conformal class. In particular, we are interested in knowing whether or not this set is compact. A compactness result for solutions of (1.2) would be useful in developing a Morse theory for the Yamabe functional (see [**29**] and [**30**] for details).

The second problem deals with the gradient flow for the Yamabe functional. This leads to the following evolution equation for the Riemannian metric:

$$\frac{\partial}{\partial t} g(t) = -(R_{g(t)} - r_{g(t)})\, g(t), \qquad g(0) = g_0. \tag{1.5}$$

Here, $R_{g(t)}$ denotes the scalar curvature of $g(t)$ and $r_{g(t)}$ is the mean value of the scalar curvature of $g(t)$. We are interested in the longtime behavior of this evolution equation: since (1.5) is the gradient flow to the Yamabe functional, one might expect that the flow converges to a metric of constant scalar curvature as $t \to \infty$.

2. Compactness of the set of constant scalar curvature metrics in a given conformal class

Fix a compact Riemannian manifold (M, g) of dimension $n \geq 3$ with positive Yamabe constant. In this section, we will address the question of whether the set of solutions to the Yamabe PDE

$$\frac{4(n-1)}{n-2} \Delta_g u - R_g\, u + 4n(n-1)\, u^{\frac{n+2}{n-2}} = 0 \tag{$\star$}$$

is compact in the C^2-topology. In some applications, it is convenient to approximate ($\star$) by subcritical problems of the form

$$\frac{4(n-1)}{n-2} \Delta_g u - R_g\, u + 4n(n-1)\, u^{\frac{n+2}{n-2}-\delta} = 0, \tag{$\star_\delta$}$$

where $\delta \in [0, \frac{4}{n-2})$. Given any real number $\delta_0 \in [0, \frac{4}{n-2})$, we denote by $\Phi(M, g; \delta_0)$ the set of all positive functions that satisfy either the Yamabe PDE ($\star$) or one of the subcritical problems:

$$\begin{aligned} &\Phi(M, g; \delta_0) \\ &= \Big\{ 0 < u \in C^2(M) : u \text{ is a solution of } (\star_\delta) \text{ for some } \delta \in [0, \delta_0] \Big\}. \end{aligned}$$

Clearly, compactness fails if (M, g) is conformally equivalent to the round sphere. The following result provides sufficient conditions for the set $\Phi(M, g; \delta_0)$ to be compact.

THEOREM 2.1. *Assume that (M, g) is not conformally equivalent to the round sphere S^n. Moreover, assume that one of the following conditions holds:*

(i) *(M, g) is locally conformally flat*
(ii) *$3 \leq n \leq 7$*
(iii) *$n \geq 8$ and $|W_g(p)| + |\nabla W_g(p)| > 0$ for all $p \in M$*

Given any real number $\delta_0 \in [0, \frac{4}{n-2})$, there exists a constant Λ, depending only on δ_0 and the background metric g, such that $\|u\|_{C^2(M)} \leq \Lambda$ for all functions $u \in \Phi(M, g; \delta_0)$.

R. Schoen proved Theorem 2.1 in the locally conformally flat case (see [**28, 29**]). Moreover, Schoen outlined a strategy for proving the theorem in the non-locally conformally flat case. In [**23**], Y.Y. Li and M. Zhu followed this strategy to prove the theorem in dimension 3. Compactness in dimensions 4 and 5 was first established by O. Druet [**15**] (see also [**21**] and [**24**]).

The case $n \geq 6$ is more subtle, and requires a careful analysis of the local properties of the background metric g near a blow-up point. A well-known conjecture asserts that the Weyl tensor should vanish to an order greater than $[\frac{n-6}{2}]$ at a blow-up point (see [**30**]). This conjecture is known as the Weyl Vanishing Conjecture. It has been verified in dimensions 6 and 7 by F.C. Marques [**24**] and, independently, by Y.Y. Li and L. Zhang [**21**]. Using this result and the positive mass theorem, these authors were able to prove compactness for all $n \leq 7$. In dimension $n \geq 8$, Li and Zhang proved that any blow-up point p satisfies $W_g(p) = 0$ and $\nabla W_g(p) = 0$ (see [**21**]). In particular, this proves the Weyl Vanishing Conjecture in dimensions 8 and 9. Moreover, this result shows that blow-up cannot occur if $|W_g(p)| + |\nabla W_g(p)| > 0$ for all $p \in M$.

The preceeding discussion raises the question as to whether the Weyl Vanishing Conjecture holds in dimension $n \geq 10$. Y.Y. Li and L. Zhang [**22**] proved the Weyl Vanishing Conjecture in dimensions 10 and 11. Very recently, M. Khuri, F. Marques, and R. Schoen [**19**] were able to verify the Weyl Vanishing Conjecture up to dimension 24. This result, combined with the positive mass theorem, can be used to rule out blow-up for $n \leq 24$ (see [**19**], Theorem 1.1):

THEOREM 2.2. *Assume that (M, g) is not conformally equivalent to the round sphere S^n. Moreover, assume that $n \leq 24$ and $M \setminus \{p\}$ is a spin manifold. For every real number $\delta_0 \in [0, \frac{4}{n-2})$, there exists a constant Λ, depending only on δ_0 and the background metric g, such that $\|u\|_{C^2(M)} \leq \Lambda$ for all functions $u \in \Phi(M, g; \delta_0)$.*

A key ingredient in the proof of both Theorem 2.1 and Theorem 2.2 is the Pohozaev identity. Let Ω be an open subset of M with smooth boundary, and let ν be the outward-pointing unit normal vector field along $\partial\Omega$. Moreover, suppose that V is a vector field defined on Ω. Using the divergence theorem, we obtain

$$\begin{aligned}
&-\int_\Omega \frac{1}{2}\left[\nabla^i V^j + \nabla^j V^i - \frac{2}{n}\operatorname{div}_g V\, g^{ij}\right] \partial_i u\, \partial_j u\, dvol_g \\
&- \frac{n-2}{4(n-1)}\int_\Omega R_g\, u\left[\langle V, \nabla u\rangle + \frac{n-2}{2n}\, u \operatorname{div}_g V\right] dvol_g \\
&+ \int_{\partial\Omega} \langle V, \nabla u\rangle\, \langle \nabla u, \nu\rangle - \int_{\partial\Omega} \frac{1}{2}\, |\nabla u|^2\, \langle V, \nu\rangle \\
&+ \frac{n-2}{2n}\int_{\partial\Omega} u\, \langle \nabla u, \nu\rangle \operatorname{div}_g V + \frac{(n-2)^2}{2}\int_{\partial\Omega} u^{\frac{2n}{n-2}-\delta}\, \langle V, \nu\rangle \\
&= \frac{n-2}{2n}\int_\Omega u\, \langle \nabla u, \nabla(\operatorname{div}_g V)\rangle\, dvol_g \\
&- \frac{(n-2)^2}{2}\, \delta \int_\Omega u^{\frac{n+2}{n-2}-\delta}\, \langle V, \nabla u\rangle\, dvol_g
\end{aligned}$$

whenever u is a solution of $(\star_\delta)$. This relation can be viewed as a generalization of the classical Pohozaev identity in Euclidean space. The proof of Theorem 2.1 rests on a precise analysis of the various terms in the Pohozaev identity. For example, in low dimensions and in the locally conformally flat case, one can use the positive mass theorem to show that blow-up is inconsistent with the Pohozaev identity. We refer to [**17**] and [**30**] for excellent surveys on this subject.

The compactness results described above can be used to obtain information concerning the number of constant scalar curvature metrics in a given conformal class (cf. [**19**] or [**30**]). To that end, we fix a compact Riemannian manifold (M, g) and a real number $\delta_0 \in (0, \frac{4}{n-2})$. We assume that the set $\Phi(M, g; \delta_0)$ is compact in the C^2-topology. Moreover, we assume that all positive solutions to $(\star)$ are non-degenerate. This implies that $(\star)$ has a finite number of solutions. We denote by N_k the number of positive solutions to $(\star)$ with Morse index k. If $\delta > 0$ is sufficiently small, then each solution to $(\star_\delta)$ is close (in the C^2-topology) to a solution of $(\star)$. Conversely, suppose we are given a solution u of $(\star)$. By the implicit function theorem, there exists a unique solution to $(\star_\delta)$ near u, and this solution has the same Morse index as u. Hence, if $\delta > 0$ is sufficiently small, then the subcritical problem $(\star_\delta)$ has exactly N_k solutions with Morse index k. Since the functional associated with $(\star_\delta)$ satisfies the Palais-Smale condition, the Morse inequalities hold for the subcritical problem $(\star_\delta)$. This allows us to draw the following conclusion:

COROLLARY 2.3. *Assume that the assumptions of Theorem 2.1 are satisfied, and all solutions to $(\star)$ are non-degenerate. Then*

$$\sum_{k=0}^{l} (-1)^{l-k}\, N_k \geq (-1)^l$$

for $l = 0, 1, 2, \ldots$.

3. Non-compactness and non-uniqueness results

In the previous section, we have discussed various conditions that imply compactness of the set $\Phi(M, g; \delta_0)$. In this section, we will address the opposite question: is it possible to construct Riemannian manifolds (M, g) such that the set of constant scalar curvature metrics in the conformal class of g fails to be compact? Until recently, the only known examples where compactness fails involved non-smooth background metrics. The first result in this direction was established by A. Ambrosetti and A. Malchiodi [**2**]. This result was subsequently improved by M. Berti and A. Malchiodi [**7**]. The following theorem was established in [**7**]:

THEOREM 3.1. *Let $k \geq 2$ and $n \geq 4k+3$. Then there exists a Riemannian metric g of class C^k on S^n with the following properties:*

(i) *g is not conformally flat*
(ii) *There exists a sequence of positive functions v_ν such that v_ν is a solution of $(\star)$ for all $\nu \in \mathbb{N}$ and $\sup_{S^n} v_\nu \to \infty$ as $\nu \to \infty$.*

Moreover, we can choose the metric g such that g is close to the round metric in the C^k-norm. See [**1**] for an excellent survey of this and related results. A related blow-up result was obtained recently by O. Druet and E. Hebey [**16**]: they showed that blow-up can occur for problems of the form $Lu + c\, u^{\frac{n+2}{n-2}} = 0$, where L is a small perturbation of the conformal Laplacian on S^n.

In a recent paper [**12**], the author showed that in high dimensions blow-up can occur even if the background metric g is C^∞-smooth. More precisely, for every $n \geq 52$, there exists a smooth Riemannian metric g on S^n such that the set $\Phi(M, g; 0)$ fails to be compact.

THEOREM 3.2. *Assume that $n \geq 52$. Then there exists a smooth Riemannian metric g on S^n and a sequence of positive functions $v_\nu \in C^\infty(S^n)$ $(\nu \in \mathbb{N})$ with the following properties:*

(i) *g is not conformally flat*
(ii) *The function v_ν is a solution of $(\star)$ for all $\nu \in \mathbb{N}$*
(iii) *$E_g(v_\nu) < Y(S^n)$ for all $\nu \in \mathbb{N}$, and $E_g(v_\nu) \to Y(S^n)$ as $\nu \to \infty$*
(iv) *$\sup_{S^n} v_\nu \to \infty$ as $\nu \to \infty$*

In particular, the sequence v_ν forms a singularity consisting of exactly one bubble.

In order to construct a metric on S^n with these properties, we pick a multi-linear form $W : \mathbb{R}^n \times \mathbb{R}^n \times \mathbb{R}^n \times \mathbb{R}^n \to \mathbb{R}$ which satisfies all the algebraic properties of the Weyl tensor. We assume that some components W_{ijkl} are non-zero, so that

$$\sum_{i,j,k,l=1}^n (W_{ijkl} + W_{ilkj})^2 > 0.$$

We then define a trace-free symmetric two-tensor $H_{ik}(x)$ on $\mathbb{R}^n$ by

$$H_{ik}(x) = \sum_{p,q=1}^{n} W_{ipkq}\, x_p\, x_q.$$

The proof of Theorem 3.2 rests on the following result, which is of some interest in itself (see [**12**], Proposition 24).

PROPOSITION 3.3. *Assume that $n \geq 52$. Moreover, let g be a smooth metric on $\mathbb{R}^n$ of the form $g(x) = \exp(h(x))$, where $h(x)$ is a trace-free symmetric two-tensor on $\mathbb{R}^n$ such that $|h(x)| + |\partial h(x)| + |\partial^2 h(x)| \leq \alpha$ for all $x \in \mathbb{R}^n$, $h(x) = 0$ for $|x| \geq 1$, and $h_{ik}(x) = \mu\,(\lambda^2 - |x|^2)\, H_{ik}(x)$ for $|x| \leq \rho$. Assume that $\mu \leq 1$ and $\lambda \leq \rho \leq 1$. If α and $\rho^{2-n}\,\mu^{-2}\,\lambda^{n-10}$ are sufficiently small, then there exists a positive solution v of $(\star)$ satisfying*

$$\int_{\mathbb{R}^n} v^{\frac{2n}{n-2}} < \Big(\frac{Y(S^n)}{4n(n-1)}\Big)^{\frac{n}{2}}$$

and

$$\sup_{|x| \leq \lambda} v(x) \geq c\,\lambda^{\frac{2-n}{2}}.$$

Here, c is a positive constant that depends only on n.

In order to prove Proposition 3.3, we consider the Hilbert space

$$\mathcal{E} = \Big\{w \in L^{\frac{2n}{n-2}}(\mathbb{R}^n) \cap W^{1,2}_{loc}(\mathbb{R}^n) : dw \in L^2(\mathbb{R}^n)\Big\}$$

with norm $\|w\|_{\mathcal{E}} = \|dw\|_{L^2(\mathbb{R}^n)}$. For every pair $(\xi, \varepsilon) \in \mathbb{R}^n \times (0, \infty)$, we define a closed subspace $\mathcal{E}_{(\xi,\varepsilon)} \subset \mathcal{E}$ by

$$\begin{aligned}\mathcal{E}_{(\xi,\varepsilon)} = \Big\{w \in \mathcal{E} : &\int_{\mathbb{R}^n} \frac{\varepsilon^2 - |x-\xi|^2}{\varepsilon^2 + |x-\xi|^2}\, u_{(\xi,\varepsilon)}^{\frac{n+2}{n-2}}\, w = 0 \quad \text{and} \\ &\int_{\mathbb{R}^n} \frac{2\varepsilon\,(x_k - \xi_k)}{\varepsilon^2 + |x-\xi|^2}\, u_{(\xi,\varepsilon)}^{\frac{n+2}{n-2}}\, w = 0 \quad \text{for } k = 1, \ldots, n\Big\}.\end{aligned}$$

For every pair $(\xi, \varepsilon) \in \mathbb{R}^n \times (0, \infty)$, the function

$$u_{(\xi,\varepsilon)}(x) = \Big(\frac{\varepsilon}{\varepsilon^2 + |x-\xi|^2}\Big)^{\frac{n-2}{2}}$$

is a solution of the PDE

$$\Delta u_{(\xi,\varepsilon)} + n(n-2)\, u_{(\xi,\varepsilon)}^{\frac{n+2}{n-2}} = 0.$$

Using the implicit function theorem, we can construct a nearby function $v_{(\xi,\varepsilon)}$ such that $v_{(\xi,\varepsilon)} - u_{(\xi,\varepsilon)} \in \mathcal{E}_{(\xi,\varepsilon)}$ and

$$\int_{\mathbb{R}^n} \Big(\langle dv_{(\xi,\varepsilon)}, d\psi \rangle_g + \frac{n-2}{4(n-1)} \, R_g \, v_{(\xi,\varepsilon)} \, \psi - n(n-2) \, |v_{(\xi,\varepsilon)}|^{\frac{4}{n-2}} \, v_{(\xi,\varepsilon)} \, \psi \Big) = 0$$

for all test functions $\psi \in \mathcal{E}_{(\xi,\varepsilon)}$. The problem is then reduced to finding a critical point of the function

$$\begin{aligned} \mathcal{F}_g(\xi,\varepsilon) = \int_{\mathbb{R}^n} \Big(|dv_{(\xi,\varepsilon)}|_g^2 + \frac{n-2}{4(n-1)} \, R_g \, v_{(\xi,\varepsilon)}^2 - (n-2)^2 \, |v_{(\xi,\varepsilon)}|^{\frac{2n}{n-2}} \Big) \\ - 2(n-2) \Big(\frac{Y(S^n)}{4n(n-1)} \Big)^{\frac{n}{2}}. \end{aligned} \tag{3.1}$$

For abbreviation, let $\overline{H}_{ik}(x) = (1 - |x|^2) \, H_{ik}(x)$. We next define a function $F : \mathbb{R}^n \times (0,\infty) \to \mathbb{R}$ by

$$\begin{aligned} F(\xi,\varepsilon) = \int_{\mathbb{R}^n} \frac{1}{2} \sum_{i,k,l=1}^n \overline{H}_{il}(x) \, \overline{H}_{kl}(x) \, \partial_i u_{(\xi,\varepsilon)}(x) \, \partial_k u_{(\xi,\varepsilon)}(x) \\ - \int_{\mathbb{R}^n} \frac{n-2}{16(n-1)} \sum_{i,k,l=1}^n (\partial_l \overline{H}_{ik}(x))^2 \, u_{(\xi,\varepsilon)}(x)^2 \\ + \int_{\mathbb{R}^n} \sum_{i,k=1}^n \overline{H}_{ik}(x) \, \partial_i \partial_k u_{(\xi,\varepsilon)}(x) \, z_{(\xi,\varepsilon)}(x), \end{aligned} \tag{3.2}$$

where $z_{(\xi,\varepsilon)} \in \mathcal{E}_{(\xi,\varepsilon)}$ satisfies the relation

$$\begin{aligned} &\int_{\mathbb{R}^n} \Big(\langle dz_{(\xi,\varepsilon)}, d\psi \rangle - n(n+2) \, u_{(\xi,\varepsilon)}(x)^{\frac{4}{n-2}} \, z_{(\xi,\varepsilon)} \, \psi \Big) \\ &= - \int_{\mathbb{R}^n} \sum_{i,k=1}^n \overline{H}_{ik} \, \partial_i \partial_k u_{(\xi,\varepsilon)} \, \psi \end{aligned}$$

for all test functions $\psi \in \mathcal{E}_{(\xi,\varepsilon)}$. The function $F(\xi,\varepsilon)$ is an even function of ξ. A direct computation yields

$$\begin{aligned} F(0,\varepsilon) = - \frac{(n-2)(n+4)}{16n(n-1)(n+2)} \, |S^{n-1}| \sum_{i,j,k,l=1}^n (W_{ijkl} + W_{ilkj})^2 \\ \cdot \Big[\frac{n-8}{n+4} \, \varepsilon^4 - 2 \, \varepsilon^6 + \frac{n+8}{n-10} \, \varepsilon^8 \Big] \int_0^\infty (1+r^2)^{2-n} \, r^{n+3} \, dr. \end{aligned}$$

Hence, if $n \geq 52$, then the function $F(\xi,\varepsilon)$ has a critical point of the form $(0,\varepsilon_*)$, where ε_* is a positive real number that depends only on n.

Given any compact set $A \subset \mathbb{R}^n \times (0,\infty)$, there exists a constant C such that

$$
\begin{aligned}
&|\lambda^{-8}\,\mu^{-2}\,\mathcal{F}_g(\lambda\xi,\lambda\varepsilon) - F(\xi,\varepsilon)| \\
&\leq C\,\lambda^{\frac{16}{n-2}}\,\mu^{\frac{4}{n-2}} + C\,\rho^{\frac{2-n}{2}}\,\mu^{-1}\,\lambda^{\frac{n-10}{2}} + C\,\rho^{2-n}\,\mu^{-2}\,\lambda^{n-10}
\end{aligned}
\tag{3.3}
$$

for all $(\xi,\varepsilon) \in A$. Using (3.3) one can show that the function $\mathcal{F}_g$ has a critical point near $(0,\lambda\varepsilon_*)$. This proves the existence of a highly concentrated solution to $(\star)$.

The following result is an immediate consequence of Proposition 3.3:

COROLLARY 3.4. *Assume that $n \geq 52$. Then there exists a sequence of smooth Riemannian metrics g_ν on S^n and a sequence of positive functions $v_\nu \in C^\infty(S^n)$ $(\nu \in \mathbb{N})$ with the following properties:*

(i) $\frac{4(n-1)}{n-2}\,\Delta_{g_\nu} v_\nu - R_{g_\nu}\,v_\nu + 4n(n-1)\,v_\nu^{\frac{n+2}{n-2}} = 0$ *for all* $\nu \in \mathbb{N}$

(ii) $E_{g_\nu}(v_\nu) < Y(S^n)$ *for all* $\nu \in \mathbb{N}$

(iii) *The metrics g_ν converge smoothly to a metric g_∞ as $\nu \to \infty$.*

(iv) *There exists a sequence of points $p_\nu \in S^n$ $(\nu \in \mathbb{N})$ such that $v_\nu(p_\nu) = \sup_{S^n} v_\nu \to \infty$ as $\nu \to \infty$.*

(v) *The points p_ν converge to a point $p_\infty \in S^n$ as $\nu \to \infty$. The point p_∞ satisfies $W_{g_\infty}(p_\infty) = 0$, $\nabla W_{g_\infty}(p_\infty) = 0$, and $\nabla^2 W_{g_\infty}(p_\infty) \neq 0$.*

The question arises as to whether Theorem 3.2 holds in dimension less than 52. Recent work of the author and F. Marques [**13**] indicates that similar blow-up examples exist for all $n \geq 25$. Hence, the condition $n \leq 24$ in Theorem 2.2 is sharp.

D. Pollack [**26**] has shown that every Riemannian metric with positive Yamabe constant is C^0 close to a smooth metric for which the Yamabe PDE has an arbitrarily large number of solutions. The solutions constructed in [**26**] have high energy and high Morse index.

Using the techniques developed in [**12**], one can show that, for $n \geq 52$, every Riemannian metric with positive Yamabe constant can be approximated in $W^{2,s}$ by a metric which allows blow-up:

PROPOSITION 3.5. *Let M be a compact manifold of dimension $n \geq 52$, and let g_0 be a smooth Riemannian metric on M with $Y(M,g_0) > 0$. Moreover, let β and s be real numbers such that $\beta > 0$ and $s > \frac{n}{2}$. Then we can find a smooth Riemannian metric g on M with the following properties:*

(i) $\|g - g_0\|_{W^{2,s}(M,g_0)} < \beta$

(ii) *There exists a sequence of positive functions $v_\nu \in C^\infty(S^n)$ $(\nu \in \mathbb{N})$ such that* $\frac{4(n-1)}{n-2}\,\Delta_g v_\nu - R_g\,v_\nu + 4n(n-1)\,v_\nu^{\frac{n+2}{n-2}} = 0$ *for all* $\nu \in \mathbb{N}$, $E_g(v_\nu) < Y(S^n)$ *for all* $\nu \in \mathbb{N}$, *and* $\sup_M v_\nu \to \infty$ *as* $\nu \to \infty$.

By the results of Li and Zhang [**21**], this statement is no longer true if we replace the $W^{2,s}$-norm by the C^2-norm.

In order to prove Proposition 3.5, we pick an arbitrary point $p \in M$. In geodesic normal coordinates around p, the metric g_0 satisfies $g_{0,ik}(x) = \delta_{ik} + O(|x|^2)$. Let $\eta : \mathbb{R} \to \mathbb{R}$ be a smooth cut-off function such that $\eta(t) = 1$ for $t \leq 1$ and $\eta(t) = 0$ for $t \geq 2$. We define a trace-free symmetric two-tensor $h(x)$ by

$$h_{ik}(x) = \sum_{N=N_0}^{\infty} \eta(4N^2 \, |x - y_N|) \, 2^{-N} \, (2^{-N} - |x - y_N|^2) \, H_{ik}(x - y_N),$$

where $y_N = (\frac{1}{N}, 0, \ldots, 0) \in \mathbb{R}^n$. Clearly, $h(x)$ is smooth. We next define a Riemannian metric g by

$$g(x) = [1 - \eta(N_0 \, |x|)] \, g_0(x) + \eta(N_0 \, |x|) \, \exp(h(x)).$$

It is easy to see that

$$\|g - g_0\|_{W^{2,s}(M,g_0)} < \beta$$

if N_0 is sufficiently large. Using the arguments in [**12**], one can construct a sequence of functions v_ν with the desired properties.

4. Deformation of Riemannian metrics by their scalar curvature

We next discuss a natural parabolic PDE associated with the Yamabe problem. Let M be a compact manifold of dimension $n \geq 3$, and let g_0 be a Riemannian metric on M. We consider the following flow of Riemannian metrics:

$$\frac{\partial}{\partial t} g(t) = -(R_{g(t)} - r_{g(t)}) \, g(t), \qquad g(0) = g_0. \tag{4.1}$$

As usual, $R_{g(t)}$ denotes the scalar curvature of $g(t)$. Moreover,

$$r_{g(t)} = \frac{\int_M R_{g(t)} \, dvol_{g(t)}}{\text{Vol}_{g(t)}(M)} \tag{4.2}$$

denotes the mean value of the scalar curvature of $g(t)$. (This guarantees that the volume of M is constant along the flow.) The evolution equation (4.1) is equivalent to a parabolic PDE for the conformal factor. To see this, we write $g(t) = u(t)^{\frac{4}{n-2}} \, g_0$, where $u(t)$ is a positive function defined on M. If $g(t)$ is a solution of (4.1), then $u(t)$ satisfies the PDE

$$\frac{\partial}{\partial t} u(t)^{\frac{n+2}{n-2}} = \frac{n+2}{4} \Big(\frac{4(n-1)}{n-2} \, \Delta_{g_0} u(t) - R_{g_0} \, u(t) + r_{g(t)} \, u(t)^{\frac{n+2}{n-2}} \Big). \tag{4.3}$$

This equation is the gradient flow to the Yamabe functional $E_{g_0}(u)$. It can be viewed as the parabolic analogue of the Yamabe PDE (1.2). For that reason, (4.1) is called the Yamabe flow.

The Yamabe flow was first proposed by R. Hamilton. Hamilton showed that the Yamabe flow exists for all time (i.e. the flow never develops a singularity in finite time).

Once longtime existence is known, the next issue is the asymptotic behavior of the flow as $t \to \infty$. This is not a difficult problem if $Y(M, g_0) \leq 0$: in this case, the flow converges exponentially to a metric of constant scalar curvature. However, the problem is much more subtle if $Y(M, g_0) > 0$. B. Chow [**14**] proved that the Yamabe flow converges to a metric of constant scalar curvature provided that (M, g_0) is locally conformally flat and g_0 has positive Ricci curvature. The second assumption was subsequently removed by R. Ye [**39**]: Ye showed that the flow converges to a metric of constant scalar curvature whenever (M, g_0) is locally conformally flat. The proof in [**39**] rests on the injectivity of the developing map and a reflection argument. This method uses in a fundamental way the assumption that (M, g_0) is locally conformally flat.

H. Schwetlick and M. Struwe [**33**] introduced a new approach based on concentration-compactness arguments. In low dimensions, they were able to prove the convergence of the Yamabe flow assuming only a bound on the initial energy. More precisely, if $3 \leq n \leq 5$ and the Yamabe energy of the initial metric is less than $\left[Y(M, g_0)^{\frac{n}{2}} + Y(S^n)^{\frac{n}{2}}\right]^{\frac{2}{n}}$, then the Yamabe flow converges to a metric of constant scalar curvature as $t \to \infty$. In [**9**], we proved the convergence of the Yamabe flow for initial metrics with arbitrary energy:

THEOREM 4.1. *Let (M, g_0) be a compact Riemannian manifold of dimension n. If $3 \leq n \leq 5$, then the Yamabe flow converges to a metric of constant scalar curvature as $t \to \infty$.*

In the remainder of this section, we will sketch the main steps involved in the proof of Theorem 4.1. The argument is particularly simple if (M, g_0) is the round sphere; see [**10**].

Since the volume of M is preserved by the flow, we may assume that $\text{Vol}_{g(t)}(M) = 1$ for all $t \geq 0$. For abbreviation, let $r_\infty = \lim_{t \to \infty} r_{g(t)}$. (The limit exists since the function $t \mapsto r_{g(t)}$ is decreasing.) It follows from work of Schwetlick and Struwe [**33**] that

$$\lim_{t \to \infty} \int_M |R_{g(t)} - r_\infty|^s \, dvol_{g(t)} = \lim_{t \to \infty} \int_M |R_{g(t)} - r_{g(t)}|^s \, dvol_{g(t)} = 0$$

for $1 < s < \frac{n+2}{2}$. Hence, if t_ν $(\nu \in \mathbb{N})$ is a sequence of times such that $t_\nu \to \infty$ as $\nu \to \infty$, then the functions $u(t_\nu)$ form a Palais-Smale sequence. The properties of such sequences are described by the following important result:

PROPOSITION 4.2. *Let c be a positive real number, and let u_ν $(\nu \in \mathbb{N})$ be a sequence of positive functions satisfying*

$$\int_M u_\nu^{\frac{2n}{n-2}} \, dvol_{g_0} = 1$$

for all $\nu \in \mathbb{N}$ *and*

$$\frac{4(n-1)}{n-2}\,\Delta_{g_0} u_\nu - R_{g_0}\,u_\nu + c\,u_\nu^{\frac{n+2}{n-2}} \to 0$$

in $L^{\frac{2n}{n+2}}(M)$. *After passing to a subsequence if necessary, we can find a non-negative integer* m, *a family of points* $p_{k,\nu} \in M$ $(1 \leq k \leq m,\ \nu \in \mathbb{N})$, *and a family of positive real numbers* $\varepsilon_{k,\nu}$ $(1 \leq k \leq m,\ \nu \in \mathbb{N})$ *with the following properties:*

(i) *For each* k, *we have* $\varepsilon_{k,\nu} \to 0$ *as* $\nu \to \infty$.

(ii) *For all* $i \neq j$, *we have*

$$\frac{\varepsilon_{i,\nu}}{\varepsilon_{j,\nu}} + \frac{\varepsilon_{j,\nu}}{\varepsilon_{i,\nu}} + \frac{d(p_{i,\nu}, p_{j,\nu})^2}{\varepsilon_{i,\nu}\,\varepsilon_{j,\nu}} \to \infty$$

as $\nu \to \infty$.

(iii) *There exists a non-negative function* u_∞ *such that*

$$u_\nu - \sum_{k=1}^m \Big(\frac{4n(n-1)}{c}\Big)^{\frac{n-2}{4}} \Big(\frac{\varepsilon_{k,\nu}}{\varepsilon_{k,\nu}^2 + d(p_{k,\nu}, \cdot)^2}\Big)^{\frac{n-2}{2}} \to u_\infty$$

in $W^{1,2}(M, g_0)$.

(iv) *The function* u_∞ *is a smooth solution of the PDE*

$$\frac{4(n-1)}{n-2}\,\Delta_{g_0} u_\infty - R_{g_0}\,u_\infty + c\,u_\infty^{\frac{n+2}{n-2}} = 0.$$

(v) *If* u_∞ *does not vanish identically, then* u_∞ *is positive everywhere, and we have* $c = \big[E_{g_0}(u_\infty)^{\frac{n}{2}} + m\,Y(S^n)^{\frac{n}{2}}\big]^{\frac{2}{n}}$. *Otherwise, we have* $c = \big[m\,Y(S^n)^{\frac{n}{2}}\big]^{\frac{2}{n}}$.

Proposition 4.2 is a variant of a result of M. Struwe [**35**]. Property (ii) is due to A. Bahri and J.M. Coron [**5**].

Our goal is to prove that the Yamabe flow converges to a metric of constant scalar curvature as $t \to \infty$. To that end, we need to show that volume concentration does not occur (i.e. $m = 0$). The following result is the key ingredient in the proof of Theorem 4.1:

PROPOSITION 4.3. *Assume that* $3 \leq n \leq 5$. *Let* c *be a positive real number, and let* u_ν $(\nu \in \mathbb{N})$ *be a sequence of positive functions satisfying*

$$\int_M u_\nu^{\frac{2n}{n-2}}\,dvol_{g_0} = 1$$

for all $\nu \in \mathbb{N}$ *and*

$$\frac{4(n-1)}{n-2}\,\Delta_{g_0} u_\nu - R_{g_0}\,u_\nu + c\,u_\nu^{\frac{n+2}{n-2}} \to 0$$

in $L^{\frac{2n}{n+2}}(M)$. *Then we can find positive real numbers* γ *and* C *such that*

$$\begin{aligned} &E_{g_0}(u_\nu) - c \\ &\leq C \left(\int_M \left| \frac{4(n-1)}{n-2} \, \Delta_{g_0} u_\nu - R_{g_0} \, u_\nu + c \, u_\nu^{\frac{n+2}{n-2}} \right|^{\frac{2n}{n+2}} dvol_{g_0} \right)^{\frac{n+2}{2n}(1+\gamma)} \end{aligned}$$

for infinitely many $\nu \in \mathbb{N}$.

If u_∞ vanishes, then we can choose $\gamma = 1$. The same is true if $u_\infty > 0$ is a non-degenerate solution of the Yamabe PDE.

In [**9**], Proposition 4.3 was stated for sequences of the form $u_\nu = u(t_\nu)$, where $g(t) = u(t)^{\frac{4}{n-2}} \, g_0$ is a solution of the Yamabe flow (see [**9**], Proposition 3.3). However, the proof works for general Palais-Smale sequences. The following result is an immediate consequence of Proposition 4.3:

Corollary 4.4. *Assume that* $3 \leq n \leq 5$. *Suppose that* $g(t) = u(t)^{\frac{4}{n-2}} \, g_0$ *is a solution to the Yamabe flow such that*

$$\int_M u(t)^{\frac{2n}{n-2}} \, dvol_{g_0} = 1$$

for all $t \geq 0$. *Let* $r_\infty = \lim_{t \to \infty} r_{g(t)}$. *Then there exist positive real numbers* γ *and* t_0 *such that*

$$\begin{aligned} &E_{g_0}(u(t)) - r_\infty \\ &\leq \left(\int_M \left| \frac{4(n-1)}{n-2} \, \Delta_{g_0} u(t) - R_{g_0} \, u(t) + r_\infty \, u(t)^{\frac{n+2}{n-2}} \right|^{\frac{2n}{n+2}} dvol_{g_0} \right)^{\frac{n+2}{2n}(1+\gamma)} \end{aligned}$$

for all $t \geq t_0$.

Corollary 4.4 can be used to rule out volume concentration along the Yamabe flow. Indeed, it follows from Corollary 4.4 and (1.1) that

$$r_{g(t)} - r_\infty \leq \left(\int_M |R_{g(t)} - r_\infty|^{\frac{2n}{n+2}} \, dvol_{g(t)} \right)^{\frac{n+2}{2n}(1+\gamma)}$$

for $t \geq t_0$. This implies

$$r_{g(t)} - r_\infty \leq C \left(\int_M |R_{g(t)} - r_{g(t)}|^{\frac{2n}{n+2}} \, dvol_{g(t)} \right)^{\frac{n+2}{2n}(1+\gamma)}$$

if t is sufficiently large. On the other hand, we have

$$r_{g(t)} - r_\infty = \frac{n-2}{2} \int_t^\infty \int_M (R_{g(\tau)} - r_{g(\tau)})^2 \, dvol_{g(\tau)} \, d\tau$$

for all $t \geq 0$. Putting these facts together, one can show that

$$\int_0^\infty \left(\int_M (R_{g(\tau)} - r_{g(\tau)})^2 \, dvol_{g(\tau)} \right)^{\frac{1}{2}} d\tau < \infty.$$

Hence, given any $\eta > 0$, we can find a positive real number r such that

$$\int_{B_r(p)} u(t)^{\frac{2n}{n-2}} \, dvol_{g_0} \leq \eta$$

for all $p \in M$ and all $t \geq 0$. (Here, $B_r(p)$ is a geodesic ball with respect to the background metric g_0.) This shows that volume concentration does not occur. Hence, the flow converges to a metric of constant scalar curvature as $t \to \infty$.

Therefore, it suffices to prove Proposition 4.3. The proof of this result uses a well-known inequality for analytic functions due to Lojasiewicz (cf. [**34**]). More importantly, we need a family of functions $\overline{u}_{(p,\varepsilon)}$ ($p \in M$, $0 < \varepsilon \leq \varepsilon_0$) with the following properties:

- The Yamabe energy of $\overline{u}_{(p,\varepsilon)}$ is bounded by

$$E_{g_0}(\overline{u}_{(p,\varepsilon)}) \leq Y(S^n)$$

  for all $p \in M$ and $0 < \varepsilon \leq \varepsilon_0$.
- As $\varepsilon \to 0$, the functions $\overline{u}_{(p,\varepsilon)}$ converge to the "standard bubble" on $\mathbb{R}^n$ after rescaling. More precisely,

$$\lim_{\varepsilon \to 0} \varepsilon^{\frac{n-2}{2}} \, \overline{u}_{(\varepsilon,p)}(\exp_p(\varepsilon\xi)) = \left(\frac{1}{1 + |\xi|^2} \right)^{\frac{n-2}{2}}$$

  for all $p \in M$ and $\xi \in T_pM$.

For $3 \leq n \leq 5$, the existence of a family of functions with these properties follows work of R. Schoen [**27**].

5. Convergence of the Yamabe flow for $n \geq 6$

In this section, we provide sufficient conditions for the convergence of the Yamabe flow in higher dimensions. Throughout this section, we fix a compact Riemannian manifold (M, g) of dimension $n \geq 6$ with positive Yamabe constant.

DEFINITION. *Let $d = [\frac{n-2}{2}]$. We denote by $\mathcal{Z}$ the set of all points $p \in M$ such that*

$$\limsup_{q \to p} d(p,q)^{2-d} \, |W_g(q)| = 0,$$

where W_g denotes the Weyl tensor of g.

In other words, a point $p\in M$ belongs to $\mathcal{Z}$ if and only if the Weyl tensor vanishes to an order greater than $[\frac{n-6}{2}]$ at p. Clearly, $\mathcal{Z}$ is a closed subset of M. Moreover, the set $\mathcal{Z}$ depends only on the conformal class of g.

THEOREM 5.1. *Assume that $\mathcal{Z} = \emptyset$. Then, for every initial metric in the conformal class of g, the Yamabe flow converges to a metric of constant scalar curvature as $t\to\infty$.*

The condition $\mathcal{Z} = \emptyset$ can be removed if the positive mass theorem holds. To explain this, consider a point $p\in\mathcal{Z}$, and let G be the Green's function of the conformal Laplacian with pole at p. Then the manifold $(M\setminus\{p\}, G^{\frac{4}{n-2}}\, g)$ is scalar flat and satisfies the decay conditions of R. Bartnik [**6**]. Hence, this manifold has a well-defined ADM mass.

We can prove the convergence of the Yamabe flow provided that the ADM mass of $(M\setminus\{p\}, G^{\frac{4}{n-2}}\, g)$ is positive for all $p\in\mathcal{Z}$. If $M\setminus\{p\}$ is a spin manifold, then the positivity of the ADM mass follows from work of Bartnik (see [**6**], Theorem 6.3; see also [**31**],[**37**]).

THEOREM 5.2. *Assume that $M\setminus\{p\}$ is a spin manifold. Then, for every choice of the initial metric g_0, the Yamabe flow converges to a metric of constant scalar curvature as $t\to\infty$.*

In order to prove Theorem 5.1, we construct a family of test functions $\overline{u}_{(p,\varepsilon)}$ with Yamabe energy less than $Y(S^n)$. If the Weyl tensor of g is non-zero at each point on M, we can use a result of T. Aubin [**3**] to construct a family of test functions with the required properties. In the remaining part of this section, we sketch how this construction can be generalized to deal with points where the Weyl tensor is equal to 0. The details can be found in [**11**].

Consider a point $p\in M$. After changing the metric g conformally, we may assume that $\det g(x) = 1 + O(|x|^{2d+2})$ in geodesic normal coordinates around p (see [**20**], Theorem 5.1, or [**32**], Theorem 3.1). Hence, we may write $g(x) = \exp(h(x))$, where $h(x)$ is a symmetric two-tensor satisfying $\operatorname{tr} h(x) = O(|x|^{2d+2})$ and $\sum_{k=1}^n h_{ik}(x)\, x_k = 0$ for $i = 1,\ldots,n$. Let $H_{ik}(x)$ be the Taylor polynomial of order d associated with the function $h_{ik}(x)$. Clearly, $H_{ik}(x)$ is a trace-free symmetric two-tensor on $\mathbb{R}^n$ satisfying $\sum_{k=1}^n H_{ik}(x)\, x_k = 0$ for $i = 1,\ldots,n$. Note that $H_{ik}(x) \not\equiv 0$ since $\mathcal{Z} = \emptyset$.

We denote by G the Green's function of the conformal Laplacian with pole at p. Let $\chi : \mathbb{R}\to\mathbb{R}$ is a non-negative cutoff function satisfying $\chi(t) = 1$ for $t\leq\frac{4}{3}$ and $\chi(t) = 0$ for $t\geq\frac{5}{3}$. Moreover, suppose that δ and ε are positive real numbers such that $\delta\geq 2\varepsilon$. We define a function u_ε by

$$u_\varepsilon(x) = \Big(\frac{\varepsilon}{\varepsilon^2 + |x|^2}\Big)^{\frac{n-2}{2}}.$$

We can find a vector field $V^{(\varepsilon,\delta)}$ on $\mathbb{R}^n$ such that

$$\sum_{k=1}^n \partial_k \Big[u_\varepsilon^{\frac{2n}{n-2}} \, (\chi(|x|/\delta) \, H_{ik} - \partial_i V_k^{(\varepsilon,\delta)} - \partial_k V_i^{(\varepsilon,\delta)} + \frac{2}{n} \operatorname{div} V^{(\varepsilon,\delta)} \, \delta_{ik})\Big] = 0$$

for $i = 1, \ldots, n$. We then define a function $v_{(\varepsilon,\delta)} : M \to \mathbb{R}$ by

$$v_{(\varepsilon,\delta)} = \chi(|x|/\delta) \, (u_\varepsilon + w_{(\varepsilon,\delta)}) + (1 - \chi(|x|/\delta)) \, \varepsilon^{\frac{n-2}{2}} \, G,$$

where

$$w_{(\varepsilon,\delta)} = \sum_{l=1}^n \partial_l u_\varepsilon \, V_l^{(\varepsilon,\delta)} + \frac{n-2}{2n} \, u_\varepsilon \operatorname{div} V^{(\varepsilon,\delta)}.$$

Note that $w_{(\varepsilon,\delta)}$ is a solution of the linear elliptic PDE

$$\Delta w_{(\varepsilon,\delta)} + n(n+2) \, u_\varepsilon^{\frac{4}{n-2}} \, w_{(\varepsilon,\delta)} = \frac{n-2}{4(n-1)} \, u_\varepsilon \sum_{i,k=1}^n \partial_i \partial_k \big[\chi(|x|/\delta) \, H_{ik}\big].$$

The function $w_{(\varepsilon,\delta)}$ represents a correction term which is needed to push the Yamabe energy of $v_{(\varepsilon,\delta)}$ below $Y(S^n)$. The idea of adding a small correction term to u_ε goes back to work of E. Hebey and M. Vaugon [**18**].

Our goal is to show that $E_g(v_{(\varepsilon,\delta)}) < Y(S^n)$ if ε is sufficiently small. The leading term in the asymptotic expansion for $E_g(v_{(\varepsilon,\delta)}) - Y(S^n)$ is, up to an overall factor, given by

$$\begin{aligned}
&- \int_{B_\delta(0)} \frac{1}{4} \, u_\varepsilon^2 \sum_{i,k,l=1}^n \partial_l H_{ik} \, \partial_l H_{ik} + \int_{B_\delta(0)} \frac{1}{2} \, u_\varepsilon^2 \sum_{i,k,l=1}^n \partial_k H_{ik} \, \partial_l H_{il} \\
&+ \int_{B_\delta(0)} 2 \, u_\varepsilon \, w_{(\varepsilon,\delta)} \sum_{i,k=1}^n \partial_i \partial_k H_{ik} + \int_{B_\delta(0)} \frac{4(n-1)}{n-2} \, |dw_{(\varepsilon,\delta)}|^2 \\
&- \int_{B_\delta(0)} \frac{4(n-1)}{n-2} \, n(n+2) \, u_\varepsilon^{\frac{4}{n-2}} \, w_{(\varepsilon,\delta)}^2.
\end{aligned}$$

In order to estimate this term, we use the identity

$$\begin{aligned}
&- \frac{1}{4} \, u_\varepsilon^2 \sum_{i,k,l=1}^n \partial_l H_{ik} \, \partial_l H_{ik} + \frac{1}{2} \, u_\varepsilon^2 \sum_{i,k,l=1}^n \partial_k H_{ik} \, \partial_l H_{il} \\
&+ 2 \, u_\varepsilon \, w_{(\varepsilon,\delta)} \sum_{i,k=1}^n \partial_i \partial_k H_{ik} + \frac{4(n-1)}{n-2} \, |dw_{(\varepsilon,\delta)}|^2
\end{aligned}$$

$$
\begin{aligned}
&- \frac{4(n-1)}{n-2}\, n(n+2)\, u_\varepsilon^{\frac{4}{n-2}}\, w_{(\varepsilon,\delta)}^2 \\
(5.1) \qquad &= -\frac{1}{4} \sum_{i,k,l=1}^n |Q_{ik,l}^{(\varepsilon,\delta)}|^2 - 2\, u_\varepsilon^{\frac{2n}{n-2}} \sum_{i,k=1}^n |T_{ik}^{(\varepsilon,\delta)}|^2 \\
&+ \frac{1}{2} \sum_{i=1}^n \bigg| \sum_{k=1}^n \Big(u_\varepsilon\, \partial_k T_{ik}^{(\varepsilon,\delta)} + \frac{2n}{n-2}\, \partial_k u_\varepsilon\, T_{ik}^{(\varepsilon,\delta)} \Big) \bigg|^2 \\
&+ \text{divergence terms.}
\end{aligned}
$$

Here, $T_{ik}^{(\varepsilon,\delta)}$ and $Q_{ik,l}^{(\varepsilon,\delta)}$ are defined by

$$T_{ik}^{(\varepsilon,\delta)} = H_{ik} - \partial_i V_k^{(\varepsilon,\delta)} - \partial_k V_i^{(\varepsilon,\delta)} + \frac{2}{n} \operatorname{div} V^{(\varepsilon,\delta)}\, \delta_{ik}$$

and

$$
\begin{aligned}
Q_{ik,l}^{(\varepsilon,\delta)} &= u_\varepsilon\, \partial_l T_{ik}^{(\varepsilon,\delta)} - \frac{2}{n-2}\, \partial_i u_\varepsilon\, T_{kl}^{(\varepsilon,\delta)} - \frac{2}{n-2}\, \partial_k u_\varepsilon\, T_{il}^{(\varepsilon,\delta)} \\
&+ \frac{2}{n-2}\, \partial_p u_\varepsilon\, T_{ip}^{(\varepsilon,\delta)}\, \delta_{kl} + \frac{2}{n-2}\, \partial_p u_\varepsilon\, T_{kp}^{(\varepsilon,\delta)}\, \delta_{il}.
\end{aligned}
$$

The identity (5.1) is closely related to the second variation formula for the Einstein-Hilbert action near the round metric on S^n (see [**8**], Section 4G). The first two terms on the right hand side are non-positive. Moreover, the third term vanishes in $B_\delta(0)$. To see this, observe that

$$\sum_{k=1}^n \partial_k \big(u_\varepsilon^{\frac{2n}{n-2}}\, T_{ik}^{(\varepsilon,\delta)} \big) = 0$$

for $|x| \leq \delta$. This implies

$$\sum_{k=1}^n \Big(u_\varepsilon\, \partial_k T_{ik}^{(\varepsilon,\delta)} + \frac{2n}{n-2}\, \partial_k u_\varepsilon\, T_{ik}^{(\varepsilon,\delta)} \Big) = 0$$

for $|x| \leq \delta$.

References

[1] A. Ambrosetti, *Multiplicity results for the Yamabe problem on S^n*, Proc. Natl. Acad. Sci. USA 99, 15252–15256 (2002)

[2] A. Ambrosetti and A. Malchiodi, *A multiplicity result for the Yamabe problem on S^n*, J. Funct. Anal. 168, 529–561 (1999)

[3] T. Aubin, *Équations différentielles non linéaires et problème de Yamabe concernant la courbure scalaire,* J. Math. Pures Appl. 55, 269–296 (1976)

[4] A. Bahri, *Proof of the Yamabe conjecture, without the positive mass theorem, for locally conformally flat manifolds,* Einstein metrics and Yang-Mills connections (ed. by Toshiki Mabuchi and Shigeru Mukai), Lecture Notes in Pure and Applied Mathematics, vol. 145, Marcel Dekker, New York, 1993, 1–26

[5] A. Bahri and J.M. Coron, *On a nonlinear elliptic equation involving the critical Sobolev exponent: The effect of the topology of the domain,* Comm. Pure Appl. Math. 41, 253–290 (1988)
[6] R. Bartnik, *The mass of an asymptotically flat manifold,* Comm. Pure Appl. Math. 39, 661–693 (1986)
[7] M. Berti and A. Malchiodi, *Non-compactness and multiplicity results for the Yamabe problem on S^n,* J. Funct. Anal. 180, 210–241 (2001)
[8] A.L. Besse, *Einstein manifolds,* Springer-Verlag, Berlin, 1987
[9] S. Brendle, *Convergence of the Yamabe flow for arbitrary initial energy,* J. Diff. Geom. 69, 217–278 (2005)
[10] S. Brendle, *A short proof for the convergence of the Yamabe flow on S^n*, Pure Appl. Math. Q. 3, 499–512 (2007)
[11] S. Brendle, *Convergence of the Yamabe flow in dimension 6 and higher*, Invent. Math. 170, 541–576 (2007)
[12] S. Brendle, *Blow-up phenomena for the Yamabe equation,* J. Amer. Math. Soc. (to appear)
[13] S. Brendle and F.C. Marques, *Blow-up phenomena for the Yamabe equation II*, J. Diff. Geom. (to appear)
[14] B. Chow, *The Yamabe flow on locally conformally flat manifolds with positive Ricci curvature,* Comm. Pure. Appl. Math. 45, 1003–1014 (1992)
[15] O. Druet, *Compactness for Yamabe metrics in low dimensions,* Internat. Math. Res. Notices 23, 1143–1191 (2004)
[16] O. Druet and E. Hebey, *Blow-up examples for second order elliptic PDEs of critical Sobolev growth,* Trans. Amer. Math. Soc. 357, 1915–1929 (2004)
[17] O. Druet and E. Hebey, *Elliptic equations of Yamabe type,* International Mathematics Research Surveys 1, 1–113 (2005)
[18] E. Hebey and M. Vaugon, *Le problème de Yamabe équivariant,* Bull. Sci. Math. 117, 241–286 (1993)
[19] M. Khuri, F. Marques, and R. Schoen, *A compactness theorem for the Yamabe problem,* preprint (2007)
[20] J.M. Lee and M. Parker, *The Yamabe problem,* Bull. Amer. Math. Soc. 17, 37–91 (1987)
[21] Y.Y. Li and L. Zhang, *Compactness of solutions to the Yamabe problem II,* Calc. Var. PDE 24, 185–237 (2005)
[22] Y.Y. Li and L. Zhang, *Compactness of solutions to the Yamabe problem III,* J. Funct. Anal. 245, 438–474 (2007)
[23] Y.Y. Li and L. Zhu, *Yamabe type equations on three-dimensional Riemannian manifolds,* Commun. Contemp. Math. 1, 1–50 (1999)
[24] F.C. Marques, *A-priori estimates for the Yamabe problem in the non-locally conformally flat case,* J. Diff. Geom. 71, 315–346 (2005)
[25] M. Obata, *The conjectures on conformal transformations of Riemannian manifolds,* J. Diff. Geom. 6, 247–258 (1971)
[26] D. Pollack, *Nonuniqueness and high energy solutions for a conformally invariant scalar equation,* Comm. Anal. Geom. 1, 347–414 (1993)
[27] R.M. Schoen, *Conformal deformation of a Riemannian metric to constant scalar curvature,* J. Diff. Geom. 20, 479–495 (1984)
[28] R.M. Schoen, *Variational theory for the total scalar curvature functional for Riemannian metrics and related topics,* Topics in the calculus of variations (ed. by Mariano Giaquinta), Lecture Notes in Mathematics, vol. 1365, Springer Verlag 1989, 120–154
[29] R.M. Schoen, *On the number of constant scalar curvature metrics in a conformal class,* Differential geometry (ed. by H. Blaine Lawson, Jr., and Keti Tenenblat),

Pitman Monographs and Surveys in Pure and Applied Mathematics, vol. 52, Longman Scientific & Technical 1991, 311–320

[30] R.M. Schoen, *A report on some recent progress on nonlinear problems in geometry,* In: Surveys in differential geometry, Lehigh University, Bethlehem, PA, 1991, 201–241

[31] R.M. Schoen and S.T. Yau, *On the proof of the positive mass conjecture in general relativity,* Comm. Math. Phys. 65, 45–76 (1979)

[32] R.M. Schoen and S.T. Yau, *Lectures on differential geometry, Volume I,* International Press, Boston, 1994

[33] H. Schwetlick and M. Struwe, *Convergence of the Yamabe flow for large energies,* J. Reine Angew. Math. 562, 59–100 (2003)

[34] L. Simon, *Asymptotics for a class of non-linear evolution equations with applications to geometric problems,* Ann. of. Math. 118, 525–571 (1983)

[35] M. Struwe, *A global compactness result for elliptic boundary value problems involving limiting nonlinearities,* Math. Z. 187, 511–517 (1984)

[36] N. Trudinger, *Remarks concerning the conformal deformation of Riemannian structures on compact manifolds,* Annali Scuola Norm. Sup. Pisa 22, 265–274 (1968)

[37] E. Witten, *A new proof of the positive energy theorem,* Comm. Math. Phys. 80, 381–402 (1981)

[38] H. Yamabe, *On a deformation of Riemannian structures on compact manifolds,* Osaka Math J. 12, 21–37 (1960)

[39] R. Ye, *Global existence and convergence of the Yamabe flow,* J. Diff. Geom. 39, 35–50 (1994)

A Survey on the Kähler-Ricci Flow and Yau's Uniformization Conjecture

Albert Chau[1] and Luen-Fai Tam[2]

ABSTRACT. Yau's uniformization conjecture states: a complete noncompact Kähler manifold with positive holomorphic bisectional curvature is biholomorphic to $\mathbb{C}^n$. The Kähler-Ricci flow has provided a powerful tool in understanding the conjecture, and has been used to verify the conjecture in several important cases. In this article we present a survey of the Kähler-Ricci flow with focus on its application to uniformization. Other interesting methods and results related to the study of Yau's conjecture are also discussed.

1. Introduction

A fundamental problem in complex geometry is to generalize the classical uniformization theorems on Riemann surfaces to higher dimensions. In Kähler geometry, the problem is to determine how curvature affects the underlying holomorphic structure of a Kähler manifold. In one complex dimension it is well known that a complete simply connected Riemann surface (M, g) is biholomorphic to either the Riemann sphere (when M is compact) or the complex plane (when M is noncompact) if the curvature is positive, and it is biholomorphic to the open unit disc if the curvature is negative and bounded from above away from zero.

In higher dimensions, for the compact case, the famous conjecture of Frankel says that a compact Kähler manifold with positive holomorphic bisectional curvature is biholomorphic to $\mathbb{CP}^n$. Frankel's conjecture was proved by Siu-Yau **[52]**. The stronger Hartshorne conjecture was proved by Mori **[37]**. In case of compact Kähler manifolds with nonnegative holomorphic bisectional curvature the uniformization of such manifolds was determined by Bando **[2]** for complex dimension three and by Mok **[35]** for all dimensions.

[1]Research partially supported by NSERC grant no. # 327637-06.
[2]Research partially supported by Earmarked Grant of Hong Kong #CUHK403005.

In the complete noncompact case there is a long standing conjecture due to Yau [**58**] in 1974 predicting analogous results for positively curved non-compact Kähler manifolds:

Yau's Conjecture [**58**] *A complete noncompact Kähler manifold with positive holomorphic bisectional curvature is biholomorphic to* $\mathbb{C}^n$.

Or in Yau's words [**58**, p. 620]: *The question is to demonstrate that every noncompact Kähler manifold with positive bisectional curvature is biholomorphic to the complex eulcidean space.*

The conjecture continues to generate much research activity, and although the full conjecture remains unproved, this research has produced many nice results and useful techniques and methods. In this survey we will discuss the method of using evolution equations, in particular the Kähler-Ricci flow, to study the conjecture.

In [**49**], Shi (under the supervision of Yau) began a program to use the Kähler-Ricci flow equation

$$\frac{dg_{i\bar{j}}}{dt} = -R_{i\bar{j}} \tag{1.1}$$

to prove Yau's uniformization conjecture[1]. (1.1) is a geometric evolution equation which deforms an initial Kähler metric in the direction of its Ricci curvature (see §2). The equation is a strictly parabolic system for $g_{i\bar{j}}$, so one generally expects the geometry to improve under this evolution. The idea is to show that the geometry improves to the point that it determines the holomorphic structure of the manifold. This idea has been successfully applied in many cases now. In this article we survey the progress on Yau's conjecture and the Kähler-Ricci flow program. We will also survey some more general results relating to the conjecture. Since there are many works, the survey cannot be exhaustive.

An outline of the paper is as follows. In §2 we introduce the Kähler-Ricci flow and its basic theory on complete noncompact Kähler manifolds, and on nonnegatively curved Kähler manifolds in particular. In §3 we discuss the rate of curvature decay on a complete noncompact nonnegatively curved Kähler manifold, which is also related to volume growth. This turns out to be useful for latter application. The Steinness of non-negatively curved noncompact Kähler manifolds is discussed in §4, and this begins our actual survey of uniformization results. In §5 we introduce the Kähler-Ricci solitons and present our uniformization theorem for steady and expanding gradient Kähler-Ricci solitons. We place this section here because Kähler-Ricci solitons are canonical models for the cases in §6 and the analogy drawn there

[1]The Kähler-Ricci flow had been used in Bando's work [**2**] (under the supervision of Yau) and Mok's work [**35**] on the uniformization of compact Kähler manifolds with nonnegative holomorphic bisectional curvature. The use of the Kähler-Ricci flow in these works however was not very extensive and involves only short-time behavior of the solutions. The study of long time behavior of the Kähler-Ricci flow in the compact case was initiated by Cao in [**6**].

between the Kähler-Ricci flow and complex dynamical systems. In §6 the connection between the Kähler-Ricci flow and complex dynamical systems is developed and used to prove a uniformization theorem for eternal solutions to the Kähler-Ricci flow and normalized Kähler-Ricci flow. As a corollary, we state a uniformization theorem for nonnegatively curved Kähler manifolds of average quadratic curvature decay, in particular for manifolds with maximum volume growth. Finally, in §7 we present the gap theorems for nonnegatively curved Kähler manifolds of faster than quadratic curvature decay. Whenever it is possible, we will sketch the main ideas of the proofs of the results.

2. Kähler-Ricci flow

The Ricci flow on a complete Riemannian manifold (M, g) is the following evolution equation for the metric g:

$$\begin{cases} \dfrac{dg(t)}{dt} = -2Rc(t) \\ g(0) = g, \end{cases} \tag{2.1}$$

where $Rc(t)$ is the Ricci tensor of $g(t)$. The Ricci flow was introduced by Hamilton [**27**] to study the Poincaré conjecture. In this seminal work, Hamilton proved (2.1) has a short time solution on any smooth compact Riemannian manifold. Hamilton then showed that the solution exists for all time and converges after rescaling on any compact three manifold with positive Ricci curvature.

We are interested in complete noncompact Riemannian manifolds. The first fundamental result for the Ricci flow on noncompact manifolds is the following short time existence result of Shi [**48**]:

THEOREM 2.1. [**48**] *Let (M^n, g) be a complete noncompact Riemannian manifold with bounded sectional curvature. Then there exists $0 < T < \infty$, depending only on the initial curvature bound, such that (2.1) has a solution $g(t)$ on $M \times [0, T]$. Moreover, for all $t \in [0, T]$ we have*

(i) *$g(t)$ has bounded sectional curvature and is equivalent to g.*
(ii) *For any integer $m \geq 0$, there is a constant depending only on n, m and the bound of the curvature of the initial metric g such that*

$$\sup_{x \in M} |\nabla^m Rm|(x, t) \leq \frac{C}{t^m}$$

for all $0 \leq t \leq T$. Here ∇ is the covariant derivative with respect to $g(t)$.

The following theorem of Shi [**50**] is fundamental to the theory and application of the Ricci flow to complete noncompact Kähler manifolds.

THEOREM 2.2. [**50**] *Let (M, g) be as in Theorem 2.1. Suppose (M, g) is Kähler. Then*

(i) *the solution $g(t)$ in the Theorem 2.1 is Kähler for $t \in [0, T]$; and*
(ii) *$g(t)$ has non-negative holomorphic bisectional curvature if the same is true for g.*

Hence in the Kähler category, we refer to the Ricci flow as the Kähler-Ricci flow and the equation becomes:

$$\frac{dg_{i\bar{j}}}{dt} = -R_{i\bar{j}}. \tag{2.2}$$

For compact Kähler manifolds, part (i) of Theorem 2.2 was proved by Bando [**2**] and part (ii) was proved by Bando [**2**] for complex three manifolds then by Mok [**35**] for all dimensions. Shi's proof of (i) for the noncompact case is different from that in [**2**]. Shi's proof of (ii) is similar to that in [**2**] and [**35**] which uses a so-called null-vector condition introduced in [**2**]. The proofs of both (i) and (ii) use the maximum principle for noncompact Riemannian manifolds which relies on the following result which was proved by Shi [**50**] using an idea of Greene-Wu [**24**]: Let (M, g) be a complete noncompact manifold with bounded curvature. Then one can construct a smooth function with bounded gradient and bounded Hessian, which is uniformly equivalent to the distance function from a point.

We now focus on complete non-compact Kähler manifolds with non-negative holomorphic bisectional curvature. The main long time existence result for the Kähler-Ricci flow in this setting is the following theorem of Shi [**50**]. We will denote the scalar curvature by $\mathcal{R}$ and define the average scalar curvature $k(x, r)$ as:

$$k(x, r) := \frac{1}{V_x(r)} \int_{B_x(r)} \mathcal{R} dV.$$

THEOREM 2.3. [**50**] *Let (M^n, g) be a complete noncompact Kähler manifold with bounded and nonnegative holomorphic bisectional curvature. Suppose*

$$k(x, r) \le \frac{C}{(1+r)^\epsilon} \tag{2.3}$$

for some constants C and $0 < \epsilon \le 2$, and all x and r. Then (2.2) has a long time solution $g(t)$. Moreover,

(i)
$$\mathcal{R}(x, t) \le Ct^{\frac{2(1-\epsilon)}{\epsilon}},$$
for some constant C for all x and t.

(ii) *For any integer $m \geq 0$, there is a constant C such that*

$$|\nabla^m Rm|(x,t) \leq Ct^{\frac{2(1-\epsilon)}{\epsilon}(m+2)}$$

for all x and t.

For example, if $\epsilon = 2$, then $t\mathcal{R}$ will be uniformly bounded on spacetime. The proof of the theorem is based on estimating the volume element

$$F(x,t) = \log \frac{\det(g_{i\bar{j}}(x,t))}{\det(g_{i\bar{j}}(x,0))}.$$

Shi proved that F stays bounded from below on any finite time interval. Note that the bound $F \leq 0$ from above follows from Theorem 2.2. Using this and a parabolic version of the third order estimate for the complex Monge-Ampère equation, Shi then showed that the Kähler-Ricci flow cannot develop a singularity in finite time. In the case of $\epsilon > 1$, there is also a method by Ni-Tam [**43**] for proving the long time existence. Namely, by extending the method of Mok-Siu-Yau [**36**], Ni-Shi-Tam [**41**] were able to construct a potential u_0, for the Ricci form of the initial metric, having uniformly bounded gradient. It is readily seen that $u = u_0 - F$ is a potential of the Ricci form for $g(t)$. Using the maximum principle, one can show that u also has uniformly bounded gradient. One then shows that $|\nabla u|^2 + \mathcal{R}$ is a subsolution of the time dependent heat equation, and applying the maximum principle again, one concludes that $|\nabla u|^2 + \mathcal{R}$ is bounded by its maximum at $t = 0$. Since $g(t)$ has nonnegative holomorphic bisectional curvature, the Riemannian curvature can also be bounded from this.

On the other hand, the estimate of F depends on the fact that

$$\Delta_0 F \leq \mathcal{R}(0) + e^F \frac{\partial F}{\partial t}$$

and some mean value inequalities (see [**43**] for example). Here Δ_0 is the Laplacian of the initial metric. Once we estimate F we can then estimate $\mathcal{R}$ using the fact that

$$F(x,t) = -\int_0^t \mathcal{R}(x,\tau)d\tau$$

and the Li-Yau-Hamilton type Harnack inequality Theorem 6.7 of Cao [**7**] which implies that $t\mathcal{R}$ is nondecreasing in time. In any case, the estimates of the covariant derivatives of the curvature tensor follow from the general method developed by Shi [**48**].

Note that in the theorem, condition (2.3) is uniform in x. In many cases this condition is at least true at a point by [**42**]. Using this and the pseudolocality of Ricci flow by Perelman [**44**], which can be generalized to complete noncompact manifolds, Yu and the authors [**16**] obtained the following result on long time existence:

THEOREM 2.4. *Let (M^n, g) be a complete non-compact Kähler manifold with non-negative holomorphic bisectional curvature with injectivity radius bounded away from zero such that*

$$|Rm|(x) \to 0$$

as $x \to \infty$. Then the Kähler-Ricci flow with initial data g has a long time solution $g(t)$ on $M \times [0, \infty)$.

It would be nice if one could remove the condition on the injectivity radius.

3. Curvature decay rate

In order to apply Kähler-Ricci flow to Yau's conjecture, the following question becomes important in light of Theorem 2.3: What can we expect of the the curvature decay rate on complete noncompact Kähler manifolds with positive bisectional curvature? There are some results on the volume and curvature of manifolds with nonnegative holomorphic bisectional curvature in [**41**]. For example, it was proved that if the scalar curvature of a complete noncompact Kähler manifold with nonnegative holomorphic bisectional curvature decays quadratically in the pointwise and average sense, and if the Ricci curvature is positive at some point, then the manifold must have maximum volume growth. On the other hand, the following was proved by Chen-Zhu [**21**]:

THEOREM 3.1. [**21**] *Let (M^n, g) be a complete noncompact Kähler manifold with positive holomorphic bisectional curvature. Then for any $x \in M$ there is a constant C which may depend on x such that*

$$\frac{1}{V_x(r)} \int_{B_x(r)} \mathcal{R} dV \leq \frac{C}{1+r}. \tag{3.1}$$

Later Ni and Tam [**42**] obtained the following, which generalizes the above results:

THEOREM 3.2. [**42**] *Let (M^n, g) be a complete noncompact Kähler manifold with nonnegative holomorphic bisectional curvature.*

(i) *Suppose M is simply connected, then $M = N \times M'$ holomorphically and isometrically, where N is a compact simply connected Kähler manifold, M' is a complete noncompact Kähler manifold and both N and M' have nonnegative holomorphic bisectional curvature. Moreover, the scalar curvature $\mathcal{R}'$ of M' satisfies the linear decay condition (3.1).*
(ii) *If the holomorphic bisectional curvature of M is positive at some point, then M itself satisfies the linear decay condition (3.1).*

The idea of the proof is to construct a strictly plurisubharmonic function u of linear growth. If such a u exists, then one can use L^2 theory to construct nontrivial holomorphic section s of the canonical line bundle using u as a weight function. The growth rate of u will give an estimate of the growth rate of the L^2 norm of the length $||s||$ of s. Using the Bochner type differential inequality:

$$\Delta \log ||s||^2 \geq \mathcal{R} \tag{3.2}$$

at the points where $s \neq 0$, one can show that $\log ||s||$ is at most linear growth by the mean value inequality of Li and Schoen [**32**]. Then the growth rate of $\log ||s||$ will give an estimate of the average of $\mathcal{R}$ over geodesic balls because of (3.2).

If the holomorphic bisectional curvature is positive, one just considers the Busemann function which is strictly plurisubharmonic at a point by a well known result of Wu [**55**]. In general, one may solve the heat equation with the Buesmann function as initial data. Then by a careful study of the solution for $t > 0$, one obtains a suitable strictly plurisubharmonic function in the case of Theorem 3.2 (ii).

Hence in some sense, linear decay is the slowest decay rate for the curvature of complete noncompact Kähler manifolds with nonnegative holomorphic bisectional curvature.

In case that the manifold has maximum volume growth, one may expect the curvature will decay faster. In fact, it was conjectured by Yau [**58**]: If M has maximal volume growth in the sense that $V_p(r) \geq Cr^{2n}$ for some $C > 0$ for some $p \in M$ for all r, then the curvature must decay quadratically in the average sense. Assuming the curvature is bounded, this was confirmed by Chen-Tang-Zhu [**20**] for complex surfaces and Chen-Zhu [**21**] for higher dimension under a much stronger assumption that the manifold has nonnegative curvature operator. Finally, Ni [**39**] proved this conjecture of Yau in general. More precisely, Ni obtained the following:

THEOREM 3.3. [**39**] *Let (M^n, g) be a complete noncompact Kähler manifold with bounded nonnegative holomorphic bisectional curvature. Suppose M has maximum volume growth. Then the scalar curvature $\mathcal{R}$ decay quadratically. That is to say, there exists $C > 0$ such that*

$$\frac{1}{V_x(r)} \int_{B_x(r)} \mathcal{R}\, dV \leq \frac{C}{(1+r)^2} \tag{3.3}$$

for all $x \in M$ and for all $r > 0$.

The main steps of the proof are as follow: First, use the results in [**42**], in particular Theorem 3.2, to prove that a non-flat gradient shrinking Kähler-Ricci soliton with nonnegative bisectional curvature must have zero asymptotic volume ratio. A gradient shrinking Kähler-Ricci soliton is a Kähler metric $g_{i\bar{j}}$ satisfying

$$R_{i\bar{j}} - \rho g_{i\bar{j}} = f_{i\bar{j}}, \quad f_{ij} = 0$$

for some smooth real-valued function f and for some $\rho > 0$. For an n dimensional Riemannian manifold with nonnegative Ricci curvature, the asymptotic volume ratio is defined as:

$$\mathcal{V}(g) = \lim_{r\to\infty} \frac{V_x(r)}{r^n}.$$

The limit exists and is independent of the base point x by the Bishop volume comparison theorem. In the case of maximal volume growth, this limit is non-zero.

The second step is to prove that any non-flat complete ancient solution of the Kähler-Ricci flow on a Kähler manifold with bounded and nonnegative holomorphic bisectional curvature must also have zero asymptotic volume ratio for all t. This is accomplished by showing that if this is not true, then by a blow down argument as in Perelman [**44**] for Riemannian manifolds with nonnegative curvature operator, one can construct a non-flat gradient shrinking Kähler-Ricci soliton with nonnegative bisectional curvature which has *nonzero* asymptotic volume ratio.

Now suppose (M, g) satisfies the conditions in Theorem 3.3, then one can solve the Kähler-Ricci flow equation with solution $g(t)$ with initial condition g, and can prove that $\mathcal{V}(g(t)) = \mathcal{V}(g)$ for all t. One then uses the second step above to prove that $g(t)$ must exist for all time and the scalar curvature must satisfy

$$\mathcal{R}(x,t) \leq \frac{C}{1+t}$$

for some C for all x and t. This is proved by contradiction: if this were not so, one could construct an ancient solution as above having non-zero asymptotic volume ratio, which contradicts the previous assertion. Finally, one may use this asymptotic behavior of $\mathcal{R}$ as $t \to \infty$ to get information of $\mathcal{R}(x, 0)$ as $x \to \infty$ and obtain (3.3). Here an argument similar to that in Perelman [**44**] is also used.

In order to get more information on the question on curvature decay rate, it would be helpful to construct examples. This is considerably easier in the Riemannian setting: constructing complete noncompact Riemannian manidfolds having positive sectional curvature. It is not easy to construct complete Kähler metrics on $\mathbb{C}^n$ with positive holomorphic bisectional curvature. The first example is by Klembeck [**31**] which has positive holomorphic bisectional curvature. Klembeck's example has linear curvature decay with volume growth like $V(r) \sim r^n$. Later, Cao [**8, 9**] constructed examples of $U(n)$ invariant metrics on $\mathbb{C}^n$ having positive holomorphic bisectional curvature. The examples in [**8**] have linear curvature decay and volume growth like

$V(r) \sim r^n$ while those in **[9]** have quadratic curvature decay and maximum volume growth.

Complete $U(n)$ invariant Kähler metrics with positive holomorphic bisectional curvature on $\mathbb{C}^n$ have been classified and examples have been constructed by Wu and Zheng **[56]**. They have developed a systematic way to construct examples having positive holomorphic bisectional curvature such that the scalar curvature $\mathcal{R}$ satisfies various decay rates. In particular, for any $1 \leq \theta \leq 2$ there are examples such that

$$\frac{1}{V_x(r)} \int_{B_x(r)} \mathcal{R}\, dV \leq \frac{C}{1 + r^\theta} \tag{3.4}$$

for some C for all r and for all x.

Finally, there are interesting results by Chen-Zhu **[21]** on volume growth of positively curved noncompact Kähler manifolds. For example, they proved that the volume of geodesic balls $B_p(r)$ in a complete noncompact Kähler manifold with positive holomorphic bisectional curvature must grow at least like r^n, where n is the complex dimension of the manifold. One may compare this with the well-known result of Calabi and Yau (see **[57]**, for example) that in case of Riemannian manifold with nonnegative Ricci curvature, the growth rate of geodesic balls is at least linear.

4. Steinness of nonnegatively curved manifolds

In **[51]**, Siu asked whether a complete noncompact Kähler manifold with positive holomorphic bisectional is Stein. This is a very interesting problem, and an affirmative answer to this should strongly support Yau's conjecture. In fact, Greene and Wu **[24]** first proved the following result on the complex structure of complete noncompact Kähler manifolds with positive curvature, which is part of the motivation behind Yau's conjecture:

THEOREM 4.1. **[24]** *Let (M, g) be a complete noncompact Kähler manifold with positive sectional curvature. Then M is Stein.*

The idea is to produce a smooth strictly plurisubharmonic exhaustion function. Then by a well-known result of Grauert, M will be Stein. As in the study of Riemannian manifolds with positive or nonnegative sectional curvature by Cheeger and Gromoll **[17]**, the method of Greene-Wu uses the Busemann function $\mathcal{B}(x)$. Here the Busemann function is an exhaustion function because the sectional curvature is nonnegative **[17]**. It is also plurisubharmonic if the holomorphic bisectional curvature is nonnegative and strictly plurisubharmonic at the points where the holomorphic bisectional curvature is positive by Wu **[55]**. However, $\mathcal{B}(x)$ is only Lipschitz, and so one needs to approximate $\mathcal{B}(x)$ by a smooth strictly plurisubharmonic exhaustion function. This can be done using the fact that $\mathcal{B}(x)$ is strictly plurisubharmonic and the method developed by Greene-Wu **[24]**.

Because of these, Theorem 4.1 can be improved. For example, the theorem is still true if we only assume that M has nonnegative sectional curvature and positive holomorphic bisectional curvature.

In the case that $\mathcal{B}$ is only plurisubharmonic, it is more efficient to use another method to approximate $\mathcal{B}(x)$. Namely, we solve the heat equation with initial data $\mathcal{B}(x)$. Ni and Tam [**42**] proved that if $u(x,t)$ is the solution to this, then $u(x,t)$ is still plurisubharmonic for $t > 0$. Moreover, they showed that the kernel $\mathcal{K}(x,t)$ of the complex Hessian of $u(x,t)$ is a parallel distribution. Using this one can obtain the following:

THEOREM 4.2. [**42**] *Let (M,g) be a complete noncompact Kähler manifold with nonnegative holomorphic bisectional curvature. Suppose either M has maximum volume growth or M has a pole. Then M is Stein.*

Here one uses the fact that $\mathcal{B}(x)$ is still an exhaustion function by [**47**] and so $u(x,t)$ is also an exhaustion function. As in the proof of Theorem 3.2, one can conclude that u is also strictly plurisubharmonic for $t > 0$. In the proof, one needs the following fact [**42**]: If (M,g) is a complete Kähler manifold with nonnegative holomorphic bisectional curvature which supports a nontrivial linear growth harmonic function, then the universal cover of M is a holomorphically isometric to the product of the complex $\mathbb{C}$ and another complete Kähler manifold with nonnegative holomorphic bisectional curvature.

Without the assumption that the manifold has maximum volume growth, if the Busemann function is an exhaustion function, then one can conclude that the universal cover is a product of a compact Hermitian symmetric manifold and a Stein manifold [**42**]. Using this, Fangyang Zheng (see [**42**]) obtained the following:

THEOREM 4.3. *Let (M,g) be a complete noncompact Kähler manifold with nonnegative sectional curvature. Then its universal cover is of the form $\widetilde{M} = \mathbb{C}^k \times \widetilde{N} \times \widetilde{L}$ where $\widetilde{N}$ is a compact Hermitian symmetric manifold, $\widetilde{L}$ is Stein and $\widetilde{L}$ contains no Euclidean factor. Moreover, M is a holomorphic and Riemannian fiber bundle with fiber $\widetilde{N} \times \widetilde{L}$ over a flat Kähler manifold $\mathbb{C}^k/\Gamma$. If in addition, the Ricci curvature is positive at some point, then M is simply connected and $M = \widetilde{M} = \widetilde{N} \times \widetilde{L}$ where $\widetilde{L}$ is diffeomorphic to the Euclidean space.*

So far the above results have made no assumptions on the boundedness of curvature. When the curvature is bounded, one can use the more powerful Kähler-Ricci flow (2.2), which can be solved for a short time by Theorem 2.1. The results in §2 suggests the study of $g(t)$ in connection to Yau's conjecture or more generally, the Steinness of M. When $g(t)$ exists for all time $0 \le t < \infty$ we expect the asymptotics of $g(t)$ to be particularly useful. In [**15**] the authors proved the following:

THEOREM 4.4. [**15**] *Let* (M^n, g_0) *be a complete non-compact Kähler manifold with bounded non-negative holomorphic bisectional curvature. Suppose the scalar curvature of* g_0 *is such that* $k(x,r) \le k(r)$ *for some function* $k(r)$ *satisfying*

$$k(r) \le \frac{C}{r} \tag{4.1}$$

as $r \to \infty$ *for some* $C > 0$. *Then* M *is holomorphically covered by a pseudoconvex domain in* $\mathbb{C}^n$ *which is homeomorphic to* $\mathbb{R}^{2n}$. *Moreover, if* M *has positive bisectional curvature and is simply connected at infinity, then* M *is biholomorphic to a pseudoconvex domain in* $\mathbb{C}^n$ *which is homeomorphic to* $\mathbb{R}^{2n}$, *and in particular,* M *is Stein.*

When $k(r) = \frac{C}{1+r^{1+\epsilon}}$, for $\epsilon > 0$, the result that M is biholomorphic to a pseudoconvex domain was proved by Shi [**50**] under the additional assumption that (M, g) has positive sectional curvature. Under the same decay condition and assuming maximum volume growth, similar results were obtained by Chen-Zhu [**19**]. The condition of positive sectional curvature in [**50**] was used to produce a convex compact exhaustion of M. The maximum volume growth condition in [**19**] was used to control the injectivity radius for $g(t)$. Note that the decay condition in Theorem 4.4 is almost optimal because of Theorem 3.2.

The idea of proof of Theorem 4.4 is as follows. Let g_0 be as in the theorem. Then one can find $g(t)$ which solves the Kähler-Ricci flow equation (2.2) with initial data $g(0) = g_0$ satisfying the conclusion in Theorem 2.3. By taking the universal cover and by using the result of Cao [**10**], we may assume that $g(t)$ has positive Ricci curvature. Let $g_i = g(i)$. Then for any fixed $p \in M$,

- **(a1)** $cg_i \le g_{i+1} \le g_i$ for some $1 > c > 0$ for all i.
- **(a2)** $|\nabla Rm(g_i)| + |Rm(g_i)| \le c'$ for some c' on $B_i(p, r_0)$, and for some $r_0 > 0$ for all i where $B_i(p, r_0)$ is the geodesic ball around p with respect to g_i.
- **(a3)** g_i is contracting in the following sense: For any ϵ, for any i, there $i' > i$ with
$$g_{i'} \le \epsilon g_i$$
in $B_i(p, r_0)$.

(a1)–**(a2)** follow from Theorem 2.3, and **(a3)** follows from Theorem 6.8 in §6.

Hence Theorem 4.4 is a consequence of the following:

THEOREM 4.5. [**15**] *Let* M^n *be a complex noncompact manifold. Suppose there exist a sequence of complete Kähler metrics* g_i, $i \ge 1$ *on* M *with properties* ***(a1)–(a3)***. *Then* M *is covered by a pseudoconvex domain in* $\mathbb{C}^n$ *which is homeomorphic to* $\mathbb{R}^{2n}$.

We now sketch some key ideas in the proof Theorem 4.5 . Fix some point $p \in M$. By **(a2)** we may lift the metric $g(t)$ to the tangent space via the exponential map at p and use L^2 theory to construct a local biholomorphism $\Phi_i : D(r) \to M$ for each i so that $\Phi_i^*(g_i)$ is equivalent to the Euclidean metric in $D(r)$ and equal to the Euclidean metric at 0, where $D(r)$ is the Euclidean ball of radius r with center at the origin (see Proposition 2.1 in [**14**]). Thus Φ_i provides a holomorphic normal coordinate at p, and one would like to consider an appropriate change of coordinate map from Φ_i and Φ_{i+1}. It is not clear one can do this however as Φ_i is generally not injective and may not even be a covering. Nevertheless, using **(a1)** and **(a2)**, we can find $F_{i+1} : D(r) \to \mathbb{C}^n$ such that $\Phi_i = \Phi_{i+1} \circ F_{i+1}$ for a smaller r if necessary yet independent of i. Note that this means that $F_{i+1}(D(r))$ is in the domain of Φ_{i+1}. Also, we may choose this r so that F_{i+1} is a biholomorphism onto its image. By **(a3)** one can show that these maps are essentially contracting in the sense that there are $n_i \uparrow \infty$ such that

$$F_{n_{i+1}} \circ \cdots F_{n_i+2} \circ F_{n_i+1}(D(r)) \subset D\left(\frac{r}{2}\right) \tag{4.2}$$

for every i. Now let us suppose each F_i can be extended to a biholomorphism of $\mathbb{C}^n$, which we still denote as F_i, on $\mathbb{C}^n$. We may then let

$$S_i = (F_{n_i} \circ \cdots \circ F_2)^{-1}(D(r))$$

and $\Omega = \bigcup_i S_i$. Then Ω will be a pseudoconvex domain in $\mathbb{C}^n$ and it will be homeomorphic to $\mathbb{R}^{2n}$ because $S_i \subset S_{i+1}$. Let Ψ be defined as

$$\Psi(z) = \Phi_{n_i} \circ F_{n_i} \circ \cdots \circ F_2(z)$$

for $z \in S_i$. Note that

$$\Phi_{n_{i+1}} \circ F_{n_{i+1}} \circ \cdots \circ F_{n_i} \circ \cdots \circ F_2(z) = \Phi_{n_i} \circ F_{n_i} \circ \cdots \circ F_2(z)$$

if $z \in S_i$ and thus Ψ is a well-defined nondegenerate map from Ω to M. Moreover, using **(a3)** and that g_i is 'shrinking', one can prove that Ψ is surjective thus proving Theorem 4.5. In case M is simply connected, one can also prove that Ψ is injective.

Now in general, F_i can not be extended to a biholomorphism of $\mathbb{C}^n$. The key is to show that the maps $F_i : D(r) \to \mathbb{C}^n$ can be approximated well enough by biholomorphisms of $\mathbb{C}^n$. For this one uses a theorem of Anderson-Lempert [**1**] which states that if F is a biholomorphism from a star-shape domain in $\mathbb{C}^n$ onto a Runge domain in $\mathbb{C}^n$, then F can be uniformly approximated by biholomorphisms of $\mathbb{C}^n$ on compact subsets of the domain. To use the result it is thus sufficient to show that the image of $F_i : D(r) \to \mathbb{C}^n$ is Runge. Once it is established that F_i can be approximated in this way, one uses the approximations to construct a map Ψ as above having the desired properties.

From the above proof, one expects to obtain stronger results depending on how much more can be said of the maps F_i. We will see this to be the case in the following sections §5 and §6.

5. Gradient Kähler-Ricci solitons

In very special cases, the maps F_i from the previous section will all be equal to a single map F. This is the case when $g(t)$ is a gradient Kähler-Ricci soliton.

Let us recall that Kähler-Ricci solitons are solutions to the Kähler-Ricci flow for which the metric evolves only by dilation and pull back along a one parameter family of biholomorphisms. More specifically, $(M, g_{i\bar{j}})$ is said to be a Kähler-Ricci soliton if there is a family of biholomorphisms ϕ_t on M, given by a holomorphic vector field V, such that $g_{ij}(x,t) = \phi_t^*(g_{ij}(x))$ is a solution of the normalized Kähler-Ricci flow:

$$\begin{aligned} \frac{\partial}{\partial t} g_{i\bar{j}}(x,t) &= -R_{i\bar{j}}(x,t) - \kappa g_{i\bar{j}}(x,t) \\ g_{i\bar{j}}(x,0) &= g_{i\bar{j}}(x) \end{aligned} \tag{5.1}$$

for some constant κ. The soliton is said to be *steady* if $\kappa = 0$, and *expanding* if $\kappa > 0$ (which will be normalized to be 1). In particular, note that $\phi_t : (M, g(t)) \to (M, g)$ is an isometry for every t, and thus Kähler-Ricci solitons can be viewed as generalized fixed points for the Kähler-Ricci flow. If in addition, the holomorphic vector field V is the gradient of a real valued function f, then the soliton is a gradient Kähler-Ricci soliton and the metric $g_{i\bar{j}}$ satisfies

$$\begin{aligned} f_{i\bar{j}} &= R_{i\bar{j}} + \kappa g_{i\bar{j}}, \\ f_{ij} &= 0. \end{aligned} \tag{5.2}$$

Conversely, if these equations are satisfied with bounded curvature, then the corresponding solution to (2.2) is a gradient Kähler-Ricci soliton by the uniqueness theorem of Chen-Zhu [**22**]. These equations are thus known as gradient Kähler-Ricci soliton equations.

Solitons are extremely important in studying the formation of singularities under the flow and hence the underlying structure of the manifolds. As described in [**29**] by Hamilton, solitons typically arise as limit solutions when one takes a dilation limit around a singularity forming under the flow (see [**28**] on limit solutions to the Ricci flow). In [**7**], Cao proved the following classification of limit solutions to the Kähler-Ricci flow.

THEOREM 5.1. [**7**] *Let $(M, g(t))$ be a family of complete Kähler metrics on a noncompact complex manifold M which form a solution to (5.1) for $t \in [0, \infty)$ if $\kappa > 0$ and for $t \in (-\infty, \infty)$ if $\kappa = 0$ such that the holomorphic bisectional curvature is nonnegative and the Ricci curvature is positive. Assume that the scalar curvature of $g(t)$ assumes its maximum in space time.*

Then $(M, g(t))$ is a gradient Kähler-Ricci soliton which is steady if $\kappa = 0$, and expanding if $\kappa = 1$.

The proof of Theorem 5.1 relies on the LYH type Harnack inequality Theorem 6.7 of Cao [**7**]. The theorem suggests that studying the uniformization of gradient Kähler-Ricci solitons should give insight into the uniformization of more general manifolds as in §6.

Let ϕ_t be the family of biholomorphisms defining a Kähler-Ricci soliton so that $g(t) = \phi_t^*(g(0))$. Assume the holomorphic bisectional curvature is bounded and nonnegative. Suppose p is a fixed point of the flow ϕ_t and let $F = \phi_1 = \phi_t|_{t=1}$. Then the injectivity radii of p with respect to $g(t)$ are constant. Hence one can construct biholomorphism Φ_i for $g(i)$ as in the proof of Theorem 4.5. In this case, we can actually take Φ_i such that $\Phi_{i+1}^{-1} = \Phi_1^{-1} \circ (\phi_1)^i$ and thus $\Phi_{i+1}^{-1} \circ \Phi_i = \Phi_1^{-1} \circ \phi \circ \Phi_1 = F_{i+1}$. In other words, the F_i's from §4 can in this case be taken as the single map. And studying the F_i's reduces here to studying the discrete complex dynamical system generated by F. Recall that p is a fixed point for F. In [**46**] Rosay-Rudin proved the following result for *attractive basins* of a fixed points for biholomorphisms of $\mathbb{C}^n$, which was pointed out to hold on general complex M by Varolin [**54**]:

THEOREM 5.2. [**46**] *Let F be a biholomorphism from a complex manifold M^n to itself and let $p \in M^n$ be a fixed point for F. Fix a complete Riemannian metric g on M and define the basin of attraction*

$$\Omega := \{x \in M : \lim_{k\to\infty} dist_g(F^k(x), p) = 0\}$$

where $F^k = F \circ F^{k-1}, F^1 = F$.

Then Ω is biholomorphic to $\mathbb{C}^n$ provided Ω contains an open neighborhood around p.

Now in many cases, the dynamical system generated by $F = \phi_1$ above can be shown to have a unique attractive fixed point with the whole manifold as a basin of attraction. By Theorem 5.2, such a soliton must then be biholomorphic to $\mathbb{C}^n$. This was observed by the authors in [**11**] where they proved:

THEOREM 5.3. [**11**] *If $(M, g_{i\bar{j}})$ is a complete non-compact gradient Kähler-Ricci soliton which is either steady with positive Ricci curvature so that the scalar curvature attains its maximum at some point, or expanding with non-negative Ricci curvature, then M is biholomorphic to $\mathbb{C}^n$.*

This result was obtained independently by Bryant [**5**] for the steady case.

To prove the theorem one only needs to check that there is a unique fixed point of the biholomorphisms ϕ_t and that ϕ_t is *contracting* on M. These are easily verified using the condition on the positivity or nonnegativity of the Ricci curvature.

The proof of Theorem 5.2 relies on the fact that F can be transformed to have a normal form around p. This fact is due to Sternberg [**53**] for real systems, and was later independently proved by Rosay-Rudin [**45**] for complex systems. We sketch the proof here for the case $M = \mathbb{C}^n$. Let $F : \mathbb{C}^n \to \mathbb{C}^n$ be a biholomorphism such that $F(0) = 0$. One then modifies F by a biholomorphism T near the origin, so that (i) $T \circ F \circ T^{-1}$ is close to an *upper triangular* map G; and (ii) $T'(0) = I$, the identity map. Here $G = (g_1, \dots, g_n)$ is an upper triangular map, if $g_1(z) = c_1 z_1$, $g_2(z) = c_2 z_2 + h(z_1)$,..., $g_n(z) = c_n z_n + h(z_1, \dots, z_{n-1})$ for some constants $c_1, \dots, c_n$. By (i) we mean that for any m, we can choose T and G so that $G^{-1} \circ T \circ F - T = O(|z|^m)$, with G being independent of m when m is large enough. It can then be shown that $\Psi = \lim_{k\to\infty} G^{-k} \circ T \circ F^k$ exists and is a biholomorphism from the basin onto $\mathbb{C}^n$. For any z in the basin, $F^k(z)$ is defined if k is large enough. On the other hand, F is shrinking so G is expanding. Hence one may expect the image will be the whole $\mathbb{C}^n$.

In special cases, the Kähler-Ricci flow actually performs this uniformization and the soliton metric converges in the re-scaled subsequence sense to a complete flat Kähler metric under the flow. The following result was obtained in [**12**].

THEOREM 5.4. [**12**] *Let $(M, g_{i\bar{j}})$ be a complete non-compact gradient Kähler-Ricci soliton as in Theorem 5.3 with smooth potential f and equilibrium point p. let $g_{i\bar{j}}(x,t)$ be the corresponding solution to (5.1) and let $\mathbf{v}_p \in T_p^{1,0}(M)$ be a fixed nonzero vector with $|\mathbf{v}_p|_0 = 1$. Then for any sequence of times $t_k \to \infty$, the sequence of complete Kähler metrics $\frac{1}{|\mathbf{v}_p|^2_{t_k}} g_{i\bar{j}}(x, t_k)$ subconverges on compact subsets of M to a complete flat Kähler metric $h_{i\bar{j}}$ on M if and only if $R_{i\bar{j}}(p) = \beta g_{i\bar{j}}(p)$ at $t = 0$ for some constant β. In particular,if this condition is satisfied then M is biholomorphic to $\mathbb{C}^n$.*

Note that the Theorem suggests in general, we do not expect to prove uniformization by rescaling a solution $g(t)$ to the Kähler-Ricci flow to obtain a complete Kähler flat metric as a limit.

6. Eternal solutions to the normalized Kähler-Ricci flow

In this section we generalize Theorem 5.3 for Kähler-Ricci solitons to *eternal* solutions to the normalized Kähler-Ricci flow. As a corollary of this we will present a uniformization theorem for the case of average quadratic curvature decay. This is a critical case for uniformization in light of the gap phenomenon for manifolds with faster than quadratic curvature decay discussed in §7. The first major result in this case was the following theorem of Mok [**34**].

THEOREM 6.1. [**34**] *Let (M^n, g) be a complete noncompact Kähler manifold with positive holomorphic bisectional curvature. Suppose that the* *following conditions are satisfied for some $p \in M^n$,*

(i) $Vol(B_p(r)) \geq C_1 r^{2n}$, *for all* $r \geq 0$
(ii) $\mathcal{R}(x) \leq \frac{C_2}{(d(p,x)+1)^2}$. *for some* $C_1, C_2, \epsilon > 0$.

Then M^n is biholomorphic to an affine algebraic variety.

It is well known that a complete noncompact Riemannian manifold with positive sectional curvature is diffeomorphic to the Euclidean space [**26**]. Using a result of Ramanujam [**45**] which states that an algebraic surface which is homeomorphic to $\mathbb{R}^4$ must be biholomorphic to $\mathbb{C}^2$, Mok [**34**] concluded:

Corollary 6.2. *Let (M^2, g) be a complete noncompact Kähler surface with positive sectional curvature satisfying conditions (i) and (ii) in the above theorem. Then M is biholomorphic to $\mathbb{C}^2$.*

The method of Mok is to construct enough polynomial growth holomorphic functions to embed M into some $\mathbb{C}^N$ so that the image will be an affine algebraic variety. The proof uses algebraic geometric methods. Later, in separate works, Chen-Tang-Zhu, Chen-Zhu and Ni used the Kähler-Ricci flow to improve Mok's result in Theorem 6.1. In particular in the case of $n = 2$, Chen-Zhu [**20, 21**] obtained the same result as in the Corollary 6.2 by only assuming positive and bounded bisectional curvature and maximal volume growth. Ni [**39**] further weakened the condition of positive holomorphic bisectional curvature to nonnegative holomorphic bisectional curvature. The main idea is to show that maximum volume growth still implies quadratic curvature decay condition, as mentioned in Theorem 3.3. Then one can still prove that M is homeomorphic to $\mathbb{R}^4$ and produce enough polynomial growth holomorphic functions to carry over Mok's method and to use the result of Ramanujam. Chen-Zhu [**18**] also proved that if a Kähler surface with bounded and positive sectional curvature is such that the integral of $(Ric)^2$ is finite, then the surface is biholomorphic to $\mathbb{C}^2$, using Ramanujam's result again. Ramanujam's theorem however is only true for complex surfaces, and for higher dimensions we need other methods.

We would like to use the results on Kähler-Ricci flow by Shi in §2 to generalize Theorem 5.3 for Kähler-Ricci solitons to general solutions $g(t)$ to the Kähler-Ricci flow. It is natural here to consider *eternal* solutions to (5.1), in other words solutions defined for $t \in (-\infty, \infty)$. It is readily seen that a steady or expanding gradient Kähler-Ricci soliton is indeed an eternal solution to (5.1). In light of this, one may expect that Theorem 5.3 is still true when $g(t)$ is an eternal solution to (5.1) with nonnegative uniformly bounded holomorphic bisectional curvature. This expectation is confirmed in Theorem 6.3 and was proved by the authors in [**14**]. Before stating the theorem, we first discuss the case of quadratic curvature decay in relation to eternal solutions.

Suppose (M, g) is complete noncompact with bounded and nonnegative holomorphic bisectional curvature so that its scalar curvature satisfies the quadratic decay condition (3.3). Then by the results in §2, (6.1) has a long time solution $\tilde{g}_{i\bar{j}}$ with initial data g, and the scalar curvature $\widetilde{\mathcal{R}}$ will satisfy $t\widetilde{R} \leq C$ for some constant C uniform on spacetime. Now if we let $g(t) = e^{-t}\tilde{g}_{i\bar{j}}(e^t)$, then $g(t)$ will be an eternal solution to the normalized Kähler-Ricci flow (5.1) for $\kappa = 1$. Moreover, it is easy to see that $g(t)$ has uniformly bounded nonnegative holomorphic bisectional curvature. Conversely, given such an eternal solution $g(t)$ one sees that $\widetilde{g}(t) = tg(\log t)$ solves the unnormalized Kähler-Ricci flow

$$\frac{\partial}{\partial t}\widetilde{g}_{i\bar{j}} = -\widetilde{R}_{i\bar{j}}. \tag{6.1}$$

for $t \geq 1$, and that $t\widetilde{R} \leq C$ for some uniform constant C. This in turns implies that the scalar curvature of the initial metric $g(0)$ satisfies the quadratic decay condition (3.3) see [**43, 39**].

Now we state our theorem on eternal solution:

THEOREM 6.3. [**14**] *Let M^n be a noncompact complex manifold. Suppose there is a smooth family of complete Kähler metrics $g(t)$ on M such that for $\kappa = 0$ or 1, $g(t)$ satisfies*

$$\frac{\partial}{\partial t}g_{i\bar{j}}(x,t) = -R_{i\bar{j}}(x,t) - \kappa g_{i\bar{j}}(x,t) \tag{6.2}$$

for all $t \in (-\infty, \infty)$ such that for every t, $g(t)$ has uniformly bounded nonnegative holomorphic bisectional curvature on M independent of t. Then M is holomorphically covered by $\mathbb{C}^n$.

By the remarks preceding the theorem, we have the following result by the authors:

THEOREM 6.4. [**14**] *Suppose (M^n, g) has holomorphic bisectional curvature which is bounded, non-negative and has average quadratic curvature decay. Then M is holomorphically covered by $\mathbb{C}^n$.*

Combining this with Theorem 3.3, we conclude:

COROLLARY 6.5. *Let (M^n, g) be a complete noncompact Kähler manifold with bounded and nonnegative holomorphic bisectional curvature such that M has maximum volume growth then M is biholomorphic to $\mathbb{C}^n$.*

REMARK 1. Corollary 6.5 was proved before Theorem 6.4 by the authors in [**13**].

As noted earlier, if we assume the holomorphic bisectional curvature is bounded and nonnegative, then Theorem 6.3 is basically a direct generalization of Theorem 5.3 for gradient Kähler-Ricci solitons. However, the proof of Theorem 6.3 is much more complicated. Beginning with a solution $g(t)$ to the Kähler-Ricci flow as in Theorem 6.3, fix some point $p \in M$ and construct maps Φ_i as in the proof of Theorem 4.5 in §4. For simplicity, we will assume that Φ_t is injective for every t (thus M is simply connected). In other words, we assume the injectivity radius of $g(t)$ at p is bounded from below independently of t. Such a bound exists in the case of [**13**], where maximum volume growth is assumed and removing the dependence on this bound is the essential generalization made in [**14**]. We may also assume that $g(t)$ has positive Ricci curvature because of a dimension reduction result of Cao [**10**].

Now for $N > 0$ sufficiently large, as in §4, we can find a sequences of biholomorphisms F_i from $D(r)$ onto its image which is inside $D(r)$:

$$F_{i+1} = \Phi^{-1}_{(i+1)N} \circ \Phi_{iN} : D(r) \to D(r) \tag{6.3}$$

for $i \geq 1$. These F_i's are basically the same as those in the proof of Theorem 4.5, which as noted in §5, can be chosen to be a single map F when $g(t)$ is gradient Kähler-Ricci soliton. One would now like to imitate the proof of Rosay-Rudin's Theorem 5.2. A key step in their proof was to transform F into a particularly nice form. Now the main difficulty here is simultaneously transforming the sequence $\{F_i\}$ into a likewise nice form. In [**30**], Johnsson-Varolin showed that this can be done provided that asymptotically they behave close enough to a single map F. This closeness is essentially in terms of the Lyapunov regularity of the F_i's (see [**3**] for the terminology). In terms of the Kähler-Ricci flow, the authors proved [**14**] that this transformability is possible due to the Lyapunov regularity of $g(t)$ as described in the following:

THEOREM 6.6. [**14**] *Let $M^n, g(t)$ be as in Theorem 6.3 such that the Ricci curvature of $g(t)$ is positive. Let $p \in M$ be fixed and let $\lambda_1(t) \geq \ldots \geq \lambda_n(t) > 0$ be the eigenvalues of $R_{i\bar{j}}(p,t)$ relative to $g_{i\bar{j}}(p,t)$. Then $\lim_{t\to\infty} \lambda_i(t)$ exists for $1 \leq i \leq n$ and there is a constant $C > 0$ such that $\lambda_n(t) \geq C$ for all $t \geq 0$. Moreover, if we let $\mu_1 > \ldots > \mu_l > 0$ be the distinct limits with multiplicities $d_1, \ldots, d_l$, then $V = T^{(1,0)}_p(M)$ can be decomposed orthogonally with respect to $g(0)$ as $V_1 \oplus \cdots \oplus V_l$ so that the following are true:*

(i) *If v is a nonzero vector in V_i for some $1 \leq i \leq l$, then $\lim_{t\to\infty} |v_i(t)| = 1$ and thus $\lim_{t\to\infty} Rc(v(t), \bar{v}(t)) = \mu_i$ and*

$$\lim_{t\to\infty} \frac{1}{t} \log \frac{|v|_t^2}{|v|_0^2} = -\mu_i - 1.$$

Moreover, the convergences are uniform over all $v \in V_i \setminus \{0\}$.

(ii) *For $1 \leq i, j \leq l$ and for nonzero vectors $v \in V_i$ and $w \in V_j$ where $i \neq j$, $\lim_{t\to\infty} \langle v(t), w(t) \rangle_t = 0$ and the convergence is uniform over all such nonzero vectors v, w.*

(iii) $\dim_{\mathbb{C}}(V_i) = d_i$ *for each* i.
(iv)

$$\sum_{i=1}^{l}(-\mu_i - 1)\dim_{\mathbb{C}} V_i = \lim_{t\to\infty}\frac{1}{t}\log\frac{\det(g_{i\bar{j}}(t))}{\det(g_{i\bar{j}}(0))}.$$

The theorem basically says that the eigenvalues of the Ricci tensor have limits and the eigenspaces are almost the same.

The proof of the theorem relies on an important differential Li-Yau-Hamilton(LYH) inequality for the Kähler-Ricci flow by Cao [**7**]:

THEOREM 6.7. [**7**] *Let $g(t)$ be a complete solution to the Kähler-Ricci flow (2.2) on $M \times [0,T)$ with bounded and nonnegative holomorphic bisectional curvature. For any $p \in M$ and holomorphic vector V at p, let*

$$Z_{i\bar{j}} = \frac{\partial}{\partial t}R_{i\bar{j}} + R_{i\bar{k}}Z_{k\bar{j}} + R_{i\bar{j},k}V_{\bar{k}} + R_{i\bar{j},\bar{k}}V_k + R_{i\bar{j}k\bar{l}}V_{\bar{k}}V_l + \frac{1}{t}R_{i\bar{j}}.$$

Then

$$Z_{i\bar{j}}W^iW^{\bar{j}} \ge 0$$

for all holomorphic vectors W at p.

Using this differential inequality the authors proved that

THEOREM 6.8. [**15**] *Let $g(t)$ be a complete solution to (2.2) with nonnegative holomorphic bisectional curvature such that for any $T > 0$, $g(t)$ has bounded curvature for all $t \in [0,T]$. Fix some $p \in M$ and let $\lambda_i(t)$ be the eigenvalues of $Rc(p,t)$ arranged in increasing order. Then*

$$t\lambda_k(t)$$

is nondecreasing in t for all $1 \le k \le n$.

Now under the condition of Theorem 6.6, the proof of Theorem 6.8 directly implies that $\lambda_i(t)$ is nondecreasing in t for every $1 \le i \le n$. This will imply that $\lim_{t\to\infty}\lambda_i(t)$ exists for all i. From this, one argues as in the proof of Theorem 5.1 in [**9**] to prove that $g(t)$ behaves like gradient Kähler-Ricci soliton with fixed point at p as $t \to \infty$ in the following sense: For any $t_k \to \infty$, there is a subsequence of $g(t+t_k)$ such that $(M, g(t+t_k))$ converge to a gradient Kähler-Ricci soliton. To prove this one actually only needs the convergence of the scalar curvature $R(p,t)$. In case the manifold has maximum volume growth, a more general result similar to this was obtained by Ni [**40**] independently. Now it is easy to see that if $g(t)$ is a gradient Kähler-Ricci soliton with fixed point p then Theorem 6.6 is true, and that in this case we do not even have to take limits. Observing this, one then argues carefully to obtain the results in Theorem 6.6.

We now return to our sketch of proof of Theorem 6.3. The $F_i's$ define a randomly iterated complex dynamical system on $D(r)$ with fixed point at the origin. Moreover, using that the Ricci curvature is bounded away from zero at p for all t by Theorem 6.6, one can show that the maps F_i are uniformly contracting at the origin and that $\lim_{i\to\infty} F_i \circ \cdots \circ F_1(D(r)) = 0$. This is one of the biggest differences between the F_i's here and those in §4. Although the maps there were eventually contracting, they are in general not uniformly contracting.

Theorem 6.6 can now be translated into Lyapunov regularity of the system $\{F_i\}$ which can roughly be described as follows. Let $A_i = F_i'(0)$. Then the system is Lyapunov regular at 0 if one can decomposed $\mathbb{C}^n$ orthogonally with respect the Euclidean metric as $E_1, \ldots, E_l$ such that if $E_k^{(i+1)} = A_{i+1}(E_k^{(i)})$, then $E_k^{(i)}$ are asymptotically orthogonal and for each k, A_i is asymptotically contracting at a constant rate on $E_k^{(i)}$.

Once this is established, one follows the construction in [**30**] to construct a sequence of biholomorphisms $G_i : \mathbb{C}^n \to \mathbb{C}^n$, and $T_i : D(r) \to D(r)$ (it may be necessary to take r smaller here, but independently of i). Here, the G_i's represent approximations of the F_i's in $\mathrm{Aut}(\mathbb{C}^n)$ which are lower triangular in a certain sense while the T_i's represent a sequence of holomorphic coordinate changes of $D(r)$. The following Lemma describes the extent of this approximation ([**13**] Lemma 5.7).

LEMMA 6.9. *Let $k \geq 0$ be an integer. Then*

$$\Psi_k = \lim_{l\to\infty} G_{k+1}^{-1} \circ G_{k+2}^{-1} \circ \cdots \circ G_{k+l}^{-1} \circ T_{k+l} \circ F_{k+l} \circ \cdots \circ F_{k+2} \circ F_{k+1}$$

exists and is a nondegenerate holomorphic map from $D(r)$ into $\mathbb{C}^n$. Moreover, there is a constant $\gamma > 0$ which is independent of k such that

$$\gamma^{-1} D(r) \subset \Psi_k(D(r)) \subset \gamma D(r).$$

Since F_i is uniformly contracting, one can show that the sets $\Omega_i = \Phi_{iT}(D(r))$ exhaust M as $i \to \infty$. This together with Lemma 6.9 and the definition of the $F_i's$ tell us that the sequence of maps

$$S_i = G_1^{-1} \circ \cdots G_i^{-1} \circ T_i \circ \Phi_{iN}^{-1} : \Omega_i \to \mathbb{C}^n$$

converges to a biholomorphic map Ψ from M into $\mathbb{C}^n$. It is shown in ([**13**], §5) that Ψ is onto, and thus M is biholomorphic to $\mathbb{C}^n$.

Now in case that the injectivity radius of p with respect to $g(t)$ is not bounded away from zero, one works on the pullback metrics $\hat{g}(t)$ of $g(t)$ under the exponential maps. In this setting the injectivity radius will be bounded away from zero and one can show that $\hat{g}(t)$ still behaves like gradient Kähler-Ricci soliton locally near 0 as $t \to \infty$. One then constructs maps F_i as in the proof of Theorem 4.5, which one then shows to be Lyapunov regular, and proceeds as above to obtain Theorem 6.3.

The fact that $\hat{g}$ behave like gradient Kähler-Ricci solitons locally near 0 as $t \to \infty$ can be used to prove the following corollary to the Theorem 6.3.

COROLLARY 6.10. *Let $M, g(t)$ as in Theorem 6.3. Suppose the Ricci curvature is positive with respect to $g(0)$ at some point p. Then M is simply connected and is biholomorphic to $\mathbb{C}^n$. In particular, if (M,g) is a complete noncompact Kähler manifold with bounded positive holomorphic bisectional curvature which satisfies the quadratic decay condition (3.3), then M is biholomorphic to $\mathbb{C}^n$.*

One might want to compare the last assertion of the corollary with the statement of Yau's conjecture and the result of Zheng's Theorem 4.3. The main point of the proof of the corollary is to show that the first fundamental group of M is finite. Since M is covered by $\mathbb{C}^n$, M must then be simply connected by a well-known result, see [**4**].

7. A Theorem of Mok-Siu-Yau and its generalizations

One may expect stronger results when the curvature decays faster than quadratic. In fact, there are gap theorems which tell us that curvature of a nonflat complete noncompact Kähler manifold (M,g) with nonnegative holomorphic bisectional curvature cannot decay too fast. These can be viewed as converses to the curvature decay Theorem 3.1. The following classic gap theorem of Mok-Siu-Yau [**36**] in 1981 was the first result supporting Yau's conjecture.

THEOREM 7.1. [**36**] *Let M be a complete noncompact Kähler manifold with nonnegative holomorphic bisectional curvature. Suppose that the following conditions are satisfied for some $p \in M$ and some $\epsilon > 0$:*

(i) $Vol(B_p(r)) \geq C_1 r^{2n}$, *for all* $r \geq 0$.

(ii) $\mathcal{R}(x) \leq \frac{C_2}{(d(p,x)+1)^{2+\epsilon}}$. *for some* $C_1, C_2, \epsilon > 0$.

(iii) *Either M is Stein or M has nonnegative sectional curvature.*

Then M is isometrically biholomorphic to $\mathbb{C}^n$.

By Theorem 4.2, condition (iii) is superfluous because of (i). The condition of maximum volume growth (i) however seems rather strong. In [**19**], Chen-Zhu proved the following:

THEOREM 7.2. [**19**] *Let M be a complete noncompact Kähler manifold with nonnegative and bounded holomorphic bisectional curvature. Suppose that for some positive real function $\epsilon(r)$ satisfying $\lim_{r\to\infty} \epsilon(r) = 0$ we have*

$$\frac{1}{Vol(B_x(r))} \int_{B_x(r)} \mathcal{R} dV \leq \frac{\epsilon(r)}{(1+r)^2} \tag{7.1}$$

for all $x \in M$ and for all $r > 0$. Then the universal cover of M is isometrically biholomorphic to $\mathbb{C}^n$.

The theorem says that if M has nonnegative holomorphic bisectional curvature such that the curvature is bounded and decays faster than quadratic on average *uniformly* at all points in M, then M is flat. It is not hard to see these hypothesis are implied by (i) and (ii) in Theorem 7.1. The proof of Theorem 7.2 uses the Kähler-Ricci flow. In fact, one can see from the proof of Theorem 2.3 that if we consider the normalized flow (5.1) with $\kappa = 1$, then the scalar curvature will tend to zero at infinity. Hence it must be zero initially by the monotonicity derived from Theorem 6.7. Note that in order to use the Kähler-Ricci flow, one needs to assume that the curvature is bounded. On the other hand, there is a way to prove a stronger result in this case without using the Kähler-Ricci flow by modifying the method of [**36**].

Let us go back to the proof of Mok-Siu-Yau [**36**]. Their method is as follows: First one solves the Poisson equation $\frac{1}{2}\Delta u = \mathcal{R}$ with good estimates. This can be done because assumptions (i) and (ii) in Theorem 7.1 give a good estimate of the Green's function. Secondly, one can show that $||\sqrt{-1}\partial\bar{\partial}u - \mathrm{Ric}||$ is subharmonic using the fact that M has nonnegative holomorphic bisectional curvature. Using the estimate of u, one may get an integral estimate of $||\partial\bar{\partial}u||^2$ on geodesic balls. Then by a mean value inequality, we conclude that $\sqrt{-1}\partial\bar{\partial}u = \mathrm{Ric}$. In particular, u is plurisubharmonic. Finally, one proves that u is constant implying that M is flat. One proves this by contradiction: assuming u is not constant, one produces a function v which at a point is strictly plurisubharmonic and satisfies $(\sqrt{-1}\partial\bar{\partial}v)^n \equiv 0$. If the manifold is Stein one embeds M in $\mathbb{C}^N$ for some N and proceeds to use the coordinate functions of $\mathbb{C}^N$ to construct such a v. If M has nonnegative sectional curvature one uses the Busemann function, together with u, to construct v.

Following this line of argument, Ni-Shi-Tam in [**41**] obtained a general result on the existence of a solution of the Poisson equation on complete noncompact Riemannian manifold with nonnegative Ricci curvature without any volume growth condition. In particular, it was shown there that if (7.1) holds at some point p, then one can still solve:

$$\frac{1}{2}\Delta u = \mathcal{R}$$

with

$$\limsup_{r\to\infty} \frac{u(x)}{\log d(p,x)} \leq 0. \tag{7.2}$$

These follow from the classic results of Li-Yau [**33**] on the heat kernel estimates on manifolds with nonnegative Ricci curvature, which provide an estimate for the Green's function without assuming maximum volume growth.

In order to prove that $\sqrt{-1}\partial\bar{\partial}u = \text{Ric}$, one needs the condition on the L^2 norm of $\mathcal{R}$:

$$\liminf_{r\to\infty}\left[\exp(-ar^2)\int_{B_p(r)}\mathcal{R}^2 dV\right] < \infty \tag{7.3}$$

for some $a > 0$. Note that $\mathcal{R}$ may be allowed to grow like $\exp(a'r^2)$ for some $a' > 0$. From this, one may then solve the heat equation with initial data u and use the maximum principle and the mean value inequality as developed by Li-Schoen [**32**] to conclude that u is indeed a potential of the Ric form. Here (7.3) is used to apply the maximum principle.

Finally, one can prove that u is constant by (7.2) and the following Liouville property for pluirsubharmonic functions of Ni-Tam [**42**]:

THEOREM 7.3. [**42**] *Let (M, g) be a complete noncompact Kähler manifold with nonnegative holomorphic bisectional curvature. Suppose u is a continuous plurisubharmonic function satisfying (7.2), then u is constant.*

We may assume here that u is bounded from below. The idea of proof of Theorem 7.3 uses methods as in the proof of Theorem 3.2 and uses the following result of Ni [**38**] to conclude that u is actually harmonic:

THEOREM 7.4. [**38**] *Let (M^n, g) be a complete noncompact Kähler manifold with nonnegative Ricci curvature. Suppose u is a plurisubharmonic function on M satisfying (7.2). Then $(\partial\bar{\partial}u)^n \equiv 0$.*

Hence u must be harmonic and therefore constant by a classical result of Cheng-Yau [**23**] on harmonic function on complete manifolds with nonnegative Ricci curvature. From these results Ni-Tam [**42**] proved the following, which is the best result up to now in the generalization of Theorem 7.1:

THEOREM 7.5. [**42**] *Let (M, g) be a complete noncompact Kähler manifold with nonnegative holomorphic bisectional curvature. Suppose the scalar curvature $\mathcal{R}$ satisfies (7.1) and (7.3) for some p. Then the universal cover of M is isometrically biholomorphic to $\mathbb{C}^n$.*

It is still unknown whether the condition (7.3) can be removed.

References

[1] Anderson, E. and Lempert, L., *On the group of holomorphic automorphisms of* $\mathbb{C}^n$, Invent. Math. **110** (1992), no. 2, 371–388.

[2] Bando, S., *On the classification of three-dimensional compact Kähler manifolds of nonnegative bisectional curvature*, J. Differential Geom. **19** (1984), 283–297.

[3] Barreira, L. and Pesin, Y. B., *Lyapunov Exponents and Smooth Ergodic Theory*, University Lecture Series v. 23, American Mathematical Society, 2001.

[4] Brown, K. S., *Cohomology of groups*, Springer-Verlag, 1982.

[5] Bryant, R., *Gradient Kähler Ricci solitons*, arXiv eprint 20024. arXiv: math.DG/0407453.

[6] Cao, H.-D., *Deformation of Kähler metrics to Kähler–Einstein metrics on compact Kähler manifolds*, Invent. Math. **81** (1985), no. 2, 359–372.
[7] Cao, H.-D., *On Harnack's inequality for the Kähler-Ricci flow*, Invent. Math. **109** (1992), 247–263.
[8] Cao, H.-D., *Existence of gradient Kähler-Ricci solitons*, Elliptic and parabolic methods in geometry (Minneapolis, MN, 1994), 1–16.
[9] Cao, H.-D., *Limits of solutions to the Kähler-Ricci flow*, J. Differential Geom. **45** (1997), 257–272.
[10] Cao, H.-D., *On Dimension reduction in the Kähler-Ricci flow*, Comm. Anal. Geom. **12** (2004), 305–320.
[11] Chau, A. and Tam, L.-F., *A note on the uniformization of gradient Kähler-Ricci solitons*, Math. Res. Lett. **12** (2005), no. 1, 19–21.
[12] Chau, A. and Tam, L.-F., *Gradient Kähler-Ricci solitons and complex dynamical systems*, in *Recent progress on some problems in several complex variables and partial differential equations*, 43–52, Contemp. Math., **400**, Amer. Math. Soc., Providence, RI, 2006.
[13] Chau, A. and Tam, L.-F., *On the complex structure of Kähler manifolds with non-negative curvature*, J. Differential Geom. **73** (2006), 491–530.
[14] Chau, A. and Tam, L.-F., *Non-negatively curved Kähler manifolds with average quadratic curvature decay*, to appear in Communications in Analysis and Geometry.
[15] Chau, A. and Tam, L.-F., *On the Steinness of a class of Kähler manifolds*, arXiv eprint 2006. arXiv: math.DG/0610535.
[16] Chau, A., Tam, L.-F. and Yu, C.-J., *Pseudolocality for the Ricci flow and applications*, arXiv eprint 2007. arXiv: math.DG/0701153.
[17] Cheeger, J., Gromoll, D., *On the structure of complete manifolds of nonnegative curvature*, Ann. of Math., **46** (1972), 413–433.
[18] Chen, B. L. and Zhu, X. P., *Positively curved complete noncompact Kähler manifolds*, arXiv eprint 2002. arXiv:math.DG/0211373.
[19] Chen, B. L. and Zhu, X. P., *On complete noncompact Kähler manifolds with positive bisectional curvature*, Math. Ann. **327** (2003), 1–23.
[20] Chen, B. L., Tang, S. H. and Zhu, X. P., *A uniformization theorem of complete noncompact Kähler surfaces with positive bisectional curvature*, J. Differential Geom. **67** (2004), 519–570.
[21] Chen, B. L. and Zhu, X. P., *Volume growth and curvature decay of positively curved Kähler manifolds*, Q. J. Pure Appl. Math. **1** (2005), no. 1, 68–108.
[22] Chen, B. L. and Zhu, X. P., *Uniqueness of the Ricci flow on complete noncompact manifolds*, J. Differential Geom. **74** (2006), 119–154.
[23] Cheng, S. Y. and Yau, S.-T., *Differential equations on Riemannian manifolds and their geometric applications*, Comm. Pure Appl. Math. **28** (1975), 333–354.
[24] Greene, R. E. and Wu, H., C^∞ *convex functions and manifolds of positive curvature*, Acta. Math., **137** (1976), 209–245.
[25] Greene, R. E. and Wu, H., C^∞ *approximations of convex, subharmonic, and plurisubharmonic functions*, Ann. Scient. Éc. Norm. Sup. **12** (1979), 47–84.
[26] Gromoll, D. and Meyer, W., *On complete open manifolds of positive curvature*, Ann. of Math. **90** (1969), 75–90.
[27] Hamilton, R. S., *Three manifolds with positive Ricci curvature*, J. of Differential Geometry. **17** (1982), 255-306.
[28] Hamilton, R. S., *A compactness property for solutions of the Ricci flow*, Amer. J. Math. **117** (1995), 545–572.
[29] Hamilton, R. S., *Formation of Singularities in the Ricci Flow*, Surveys in differential geometry, Vol. II (1995), 7–136.
[30] Jonsson, M. and Varolin, D., *Stable manifolds of holomorphic diffeomorphisms*, Invent. Math. **149** (2002), 409–430.

[31] Klembeck, P. F., *A complete Kähler metric of positive curvature on C^n*, Proc. Amer. Math. Soc. **64** (1977), 313–316.
[32] P. Li and R. Schoen, *L^p and mean value properties of subharmonic functions on Riemannian manifolds*, Acta. Math. **153** (1984), 279–301.
[33] P. Li and S.-T. Yau, *On the parabolic kernel of the Schrödinger operator*, Acta. Math. **156** (1986), 139–168.
[34] Mok, N., *An embedding theorem of complete Kähler manifolds of positive bisectional curvature onto affine algebraic varieties*, Bull. Soc. Math. France. **112** (1984), 179–258.
[35] Mok, N., *The uniformization theorem for compact Kähler manifolds of nonnegative holomorphic bisectional curvature*, J. Differential Geom. **27** (1988), 179–214.
[36] Mok, N., Siu, Y.-T. and Yau, S.-T., *The Poincaré-Lelong equation on complete Kähler manifolds*, Comp. Math. **44** (1981), 183–218.
[37] Mori, S., *Projective manifolds with ample tangent bundles*, Ann. of Math. (2) **110** (1979), 593–606.
[38] Ni, L., *Vanishing theorems on complete Kähler manifolds and their applications*, J. Differential Geom. **50** (1998), no 89–122.
[39] Ni, L., *Ancient solutions to Kähler-Ricci flow*, Math. Res. Lett. **12** (2005), 633–653.
[40] Ni, L., *A new Li-Yau-Hamilton estimate for Kähler-Ricci flow*, arXiv eprint 2005. arXiv:math.DG/0502495.
[41] Ni, L., Shi, Y.-G. and Tam, L.-F., *Poisson equation, Poincaré-Lelong equation and curvature decay on complete Kähler manifolds*, J. Differential Geom. **57** (2001), 339–388.
[42] Ni, L. and Tam, L.-F., *Plurisubharmonic functions and the structure of complete Kähler manifolds with nonnegative curvature*, J. Differential Geom. **64** (2003), 457–524.
[43] Ni, L. and Tam, L.-F., *Kähler-Ricci flow and the Poincaré-Lelong equation*, Comm. Anal. Geom. **12** (2004), 111–141.
[44] Perelman, G., *The entropy formula for the Ricci flow and its geometric applications*, arXiv:math.DG/0211159.
[45] Ramanujam, C. P., *A topological characterisation of the affine plane as an algebraic variety*, Ann. of Math. **94** (1971), 69–88.
[46] Rosay, J. P. and Rudin, W., *Holomorphic maps from $\mathbb{C}^n$ to $\mathbb{C}^n$*, Trans. AMS **310** (1988), 47–86.
[47] Shen, Z., *Complete manifolds with nonnegative Ricci curvature and large volume growth*, Invent. Math. **125** (1996), 393–404.
[48] Shi, W.-X., *Ricci deformation of the metric on complete noncompact Riemannian manifolds*, J. of Differential Geometry, **30** (1989), 223–301.
[49] Shi, W.-X., *Ricci deformation of the metric on complete noncompact Kähler manifolds*, PhD thesis, Harvard University, 1990.
[50] Shi, W.-X., *Ricci Flow and the uniformization on complete non compact Kähler manifolds*, J. of Differential Geometry, **45** (1997), 94–220.
[51] Siu, Y.-T., *Pseudoconvexity and the problem of Levi*, Bull. Amer. Math. Soc. **84** (1978), 481–512.
[52] Siu, Y.-T. and Yau, S.-T., *Compact Kähler manifolds of positive bisectional curvature*, Invent. Math. **59** (1980), 189–204.
[53] Sternberg, S., *Local contractions and a theorem of Poincaré*, Amer. J. Math. **79** (1957), 809–824.
[54] Varolin, D., *The density property for complex manifolds and geometric structures II*, Internat. J. Math. **11** (2000), no. 6, 837–847.
[55] Wu, H., *An elementary methods in the study of nonnegative curvature*, Acta. Math. **142** (1979), 57–78.

[56] Wu, H. and Zheng, F.-Y., *Examples of positively curved complete Kähler manifolds*, preprint.

[57] Yau, S.-T., *Some function-theoretic properties of complete Riemannian manifold and their applications to geometry*, Indiana University Mathematics Journal. Vol. 25, No. 7 (1976), 659–670.

[58] Yau, S.-T., *A review of complex differential geometry*, Proc. Sympos. Pure Math. **52** Part 2 (1991), 619–625.

Waterloo University, Department of Mathematics, The University of British Columbia, Vancouver, B.C., Canada V6T 1Z2
E-mail address: `chau@math.ubc.ca`

Department of Mathematics, The Chinese University of Hong Kong, Shatin, Hong Kong, China.
E-mail address: `lftam@math.cuhk.edu.hk`

Recent Developments on Hamilton's Ricci flow

Huai-Dong Cao, Bing-Long Chen, Xi-Ping Zhu

ABSTRACT. In 1982, Hamilton [**41**] introduced the Ricci flow to study compact three-manifolds with positive Ricci curvature. Through decades of works of many mathematicians, the Ricci flow has been widely used to study the topology, geometry and complex structure of manifolds. In particular, Hamilton's fundamental works (cf. [**12**]) in the past two decades and the recent breakthroughs of Perelman [**80, 81, 82**] have made the Ricci flow one of the most intricate and powerful tools in geometric analysis, and led to the resolutions of the famous Poincaré conjecture and Thurston's geometrization conjecture in three-dimensional topology.

In this survey, we will review the recent developments on the Ricci flow and give an outline of the Hamilton-Perelman proof of the Poincaré conjecture, as well as that of a proof of Thurston's geometrization conjecture.

1. Analytic Aspect

1.1. Short-time Existence and Uniqueness. Let M be an n-dimensional manifold without boundary. The Ricci flow

$$\partial_t g = -2Ric$$

introduced by Hamilton [**41**] is a degenerate parabolic evolution system on metrics. In his seminal paper [**41**], Hamilton used the Nash-Moser implicit function theorem to prove the following short-time existence and uniqueness theorem for the Ricci flow on compact manifolds.

THEOREM 1.1 (Hamilton [**41**]). *Let $(M, g_{ij}(x))$ be a compact Riemannian manifold. Then there exists a constant $T>0$ such that the Ricci flow $\partial_t g = -2Ric$, with $g_{ij}(x,0)=g_{ij}(x)$, admits a unique smooth solution $g_{ij}(x,t)$ for all $x \in M$ and $t \in [0,T)$.*

The degeneracy of the system is caused by the diffeomorphism invariance of the equation. By composing the Ricci flow with a family of suitably chosen diffeomorphisms, one can obtain a strictly parabolic system. This is the De Turck trick. The resulting system is called Ricci-De Turck flow. By using this trick, De Turck [**32**] gave a simpler proof of the above short-time existence and uniqueness result.

In 1989, Shi [**91**] generalized the above short-time existence result to complete noncompact manifolds with bounded curvature.

THEOREM 1.2 (Shi [**91**]). *Let $(M, g_{ij}(x))$ be a complete noncompact Riemannian manifold of dimension n with bounded curvature. Then there exists a constant $T>0$ such that the initial value problem*

$$\begin{cases} \dfrac{\partial}{\partial t} g_{ij}(x,t) = -2R_{ij}(x,t), \text{ on } M, t>0, \\ g_{ij}(x,0) = g_{ij}(x), \text{ on } M, \end{cases}$$

admits a smooth solution $g_{ij}(x,t)$, $t \in [0,T]$, with bounded curvature.

For the uniqueness of Ricci flow on complete noncompact manifolds, the situation is a little subtle. It is well known that, without extra growth conditions on the solutions, the uniqueness for the standard heat equation does not always hold. For example, even the simplest linear heat equation on $\mathbb{R}$ with zero as initial data has a nontrivial solution which grows faster than $e^{a|x|^2}$ for any $a>0$ whenever $t>0$. The bounded curvature condition for the Ricci flow in some sense resembles the growth assumption $e^{a|x|^2}$ for the heat equation. Heuristically, it is natural to ask the uniqueness of Ricci flow in the class of bounded curvature solutions.

Recently, the last two authors proved the following uniqueness theorem.

THEOREM 1.3 (Chen-Zhu [**24**]). *Let $(M, g_{ij}(x))$ be a complete noncompact Riemannian manifold of dimension n with bounded curvature. Let $g_{ij}(x,t)$ and $\bar{g}_{ij}(x,t)$ be two solutions to the Ricci flow on $M \times [0,T]$ with $g_{ij}(x)$ as the initial data and with bounded curvatures. Then $g_{ij}(x,t) = \bar{g}_{ij}(x,t)$ for all $(x,t) \in M \times [0,T]$.*

We remark that Perelman [**81**] sketched a different proof of the above uniqueness result for a special rotationally symmetric initial metric on $\mathbb{R}^3$. The detailed exposition of Perelman's uniqueness result was given by Lu-Tian [**63**].

1.2. Shi's Local Derivative Estimates. In the course of proving his short-time existence theorem in the noncompact case, Shi also obtained the following very useful local derivative estimates.

THEOREM 1.4 (Shi [**91**]). *There exist positive constants θ, $C_k, k=1,2,\ldots$, depending only on the dimension with the following property. Suppose that the curvature of a solution to the Ricci flow is bounded*

$$|Rm| \leq A, \quad on\ B_t(x_0, r_0) \times \left[0, \frac{\theta}{A}\right],$$

where $B_t(x_0,r_0)$ is compactly contained in the manifold, then we have

$$|\nabla^k Rm(p,t)|^2 \leq C_k A^2 \left(\frac{1}{r^{2k}} + \frac{1}{t^k} + A^k\right),$$

on $B_t\left(x_0, \frac{r_0}{2}\right)$, $t \in \left[0, \frac{\theta}{A}\right]$, for $k=1,2,\ldots$.

1.3. Advanced Maximum Principles. Maximum principle is a fundamental and powerful tool for studying heat equations. For Ricci flow, this principle was established by Hamilton [**41, 42**]. Roughly speaking, Hamilton's maximum principle states that if solutions to the corresponding ODE system always persist in some convex set C when they start from C, then the solutions to the original PDE system will also remain so as long as they stay in C at $t=0$. It turns out many key estimates, such as the Hamilton-Ivey pinching estimate, the Li-Yau-Hamilton estimate, are all proved by using this principle.

To introduce Hamilton's maximum principle, let us start with some basic set-up. We assume $(M, g_{ij}(x,t))$, $t \in [0,T]$, is a smooth complete solution to the Ricci flow with bounded curvature. Let V be an abstract vector bundle over M with a metric $h_{\alpha\beta}$, and connection $\nabla = \Gamma^{\alpha}_{i\beta}$ compatible with h. Now we may form the Laplace $\Delta\sigma = g^{ij}\nabla_i\nabla_j\sigma$ which acts on the sections $\sigma \in \Gamma(V)$ of V. Suppose $M_{\alpha\beta}(x,t)$ is a family of bounded symmetric bilinear forms on V satisfying the equation

$$\frac{\partial}{\partial t} M_{\alpha\beta} = \Delta M_{\alpha\beta} + u^i \nabla_i M_{\alpha\beta} + N_{\alpha\beta}, \tag{1.1}$$

where $u^i(t)$ is a time-dependent uniform bounded vector field on the manifold M, and $N_{\alpha\beta} = \mathcal{P}(M_{\alpha\beta}, h_{\alpha\beta})$ is a polynomial in $M_{\alpha\beta}$ formed by contracting products of $M_{\alpha\beta}$ with itself using the metric $h = \{h_{\alpha\beta}\}$. In [**41**], Hamilton established the following weak maximum principle: *Let $M_{\alpha\beta}$ be a bounded solution to (1.1) and suppose $N_{\alpha\beta}$ satisfies the condition that*

$$N_{\alpha\beta} v^\alpha v^\beta \geq 0 \qquad whenever \qquad M_{\alpha\beta} v^\beta = 0,$$

for $0 \leq t \leq T$. If $M_{\alpha\beta} \geq 0$ at $t=0$, then it remains so for $0 \leq t \leq T$.

Hamilton [**42**] also established a strong maximum principle for solutions to equation (1.1): *Let $M_{\alpha\beta}$ be a bounded solution to (1.1) with $u^i = 0$, and $N_{\alpha\beta}$ satisfies*

$$N_{\alpha\beta} \geq 0$$

whenever $M_{\alpha\beta}\geq 0$. *Suppose* $M_{\alpha\beta}\geq 0$ *at* $t=0$. *Then there exists an interval* $0<t<\delta$ *on which the rank of* $M_{\alpha\beta}$ *is constant and the null space of* $M_{\alpha\beta}$ *is invariant under parallel translation and invariant in time and also lies in the null space of* $N_{\alpha\beta}$.

The evolution equation of the curvature operator $M_{\alpha\beta}$ of the Ricci flow satisfies

$$\frac{\partial M_{\alpha\beta}}{\partial t}=\Delta M_{\alpha\beta}+M^2_{\alpha\beta}+M^{\#}_{\alpha\beta} \tag{1.2}$$

where $M^{\#}_{\alpha\beta}=C^{\xi\gamma}_{\alpha}C^{\eta\theta}_{\beta}M_{\xi\eta}M_{\gamma\theta}$ and $C^{\beta\gamma}_{\alpha}=\langle[\phi^{\beta},\phi^{\gamma}],\phi^{\alpha}\rangle$ are Lie structural constants in a standard basis of the Lie algebra consisting of two-forms. Choosing an orthonormal frame such that $M_{\alpha\beta}$ is diagonal with eigenvalues $\lambda_1\leq\lambda_2\leq\cdots$, then

$$M^2_{11}+M^{\#}_{11}=\lambda_1^2+\sum_{\xi,\eta\geq 2}(C^{\xi\eta}_{1})^2\lambda_{\xi}\lambda_{\eta}.$$

So $N_{\alpha\beta}=M^2_{\alpha\beta}+M^{\#}_{\alpha\beta}$ satisfies the assumption in Hamilton's strong maximum principle. Note also that if $M_{\alpha\alpha}=0$ for $\alpha\leq k$, and $M_{\alpha\alpha}>0$ for $\alpha>k$, then the condition $M^{\#}_{\alpha\alpha}=0$ for $\alpha\leq k$ implies

$$C^{\xi\gamma}_{\alpha}=<\phi^{\alpha},[\phi^{\xi},\phi^{\gamma}]>=0, \quad \text{if } \alpha\leq k \text{ and } \xi,\gamma>k.$$

This says that the image of $M_{\alpha\beta}$ is a Lie subalgebra. So Hamilton's strong maximum principle implies

THEOREM 1.5 (Hamilton [**42**]). *Suppose the curvature operator* $M_{\alpha\beta}$ *of the initial metric is nonnegative. Then, under the Ricci flow, for some interval* $0<t<\delta$ *the image of* $M_{\alpha\beta}$ *is a Lie subalgebra of* $so(n)$ *which has constant rank and is invariant under parallel translation and invariant in time.*

This implies that under the Ricci flow, any compact manifold with nonnegative curvature operator which admits no strictly positive curvature operator has special holonomy group. This theorem may lead to complete topological classification of compact manifolds with nonnegative curvature operators.

Note that the nonnegative curvature operators form a convex cone. For general convex set, Hamilton [**42**] developed an advanced maximum principle as follows.

Let $V\to M$ be a vector bundle with a fixed bundle metric h_{ab} and

$$\nabla_t:\quad \Gamma(V)\to\Gamma(V\otimes T^*M),\quad t\in[0,T]$$

be a smooth family of time-dependent connection compatible with h_{ab}. We may form the Laplacian

$$\Delta_t\sigma=g^{ij}(x,t)(\nabla_t)_i(\nabla_t)_j\sigma,$$

for $\sigma\in\Gamma(V)$.

Let $N : V \times [0,T] \to V$ be a fiber preserving map, i.e., $N(x,\sigma,t)$ is a time-dependent vector field defined on the bundle V and tangent to the fibers. Let K be a closed subset of V satisfying

(H1) K is invariant under parallel translation defined by the connection ∇_t for each $t \in [0,T]$;

(H2) the set $K_x \triangleq V_x \cap K$ is closed and convex in each fiber V_x.

Consider the following heat type equation

$$\frac{\partial}{\partial t}\sigma(x,t) = \Delta_t \sigma(x,t) + u^i(\nabla_t)_i \sigma(x,t) + N(x,\sigma(x,t),t) \tag{1.3}$$

where $u^i = u^i(t)$ is a time-dependent vector field on M which is uniformly bounded on $M \times [0,T]$, and $N(x,\sigma,t)$ is continuous in x, t and satisfies

$$|N(x,\sigma_1,t) - N(x,\sigma_2,t)| \le C_B|\sigma_1 - \sigma_2|$$

for all $x \in M$, $t \in [0,T]$ and $|\sigma_1| \le B$, $|\sigma_2| \le B$, where C_B is a positive constant depending only on B. We will also consider the corresponding ODE system

$$\frac{d\sigma_x}{dt} = N(x,\sigma_x,t) \tag{1.4}$$

for $\sigma_x = \sigma_x(t)$ in each fiber V_x. Hamilton's advanced maximum principle is the following:

THEOREM 1.6 (Hamilton [**42**]). *Let K be a closed subset of V satisfying (H1) and (H2). Suppose that for any $x \in M$ and any initial time $t_0 \in [0,T)$, and any solution $\sigma_x(t)$ of the ODE (1.4) which starts in K_x at t_0, the solution $\sigma_x(t)$ will remain in K_x for all later times. Then for any initial time $t_0 \in [0,T)$ the solution $\sigma(x,t)$ of the PDE (1.3) will remain in K for all later times if $\sigma(x,t)$ starts in K at time t_0 and the solution $\sigma(x,t)$ is uniformly bounded with respect to the bundle metric h_{ab} on $M \times [t_0,T]$.*

1.4. Hamilton-Ivey Pinching Estimate. The advanced maximum principle has several significant applications in Ricci flow. One of them is the following Hamilton-Ivey pinching estimate in dimension three.

THEOREM 1.7 ([**47, 52, 49**]). *Suppose we have a solution to the Ricci flow on a three-manifold which is complete with bounded curvature for each $t \ge 0$. Assume at $t = 0$ the eigenvalues $\lambda \ge \mu \ge \nu$ of the curvature operator at each point is bounded below by $\nu \ge -1$ and the scalar curvature at each point is bounded below by $R = (\lambda + \mu + \nu) \ge -1$. Then at all points and all times $t \ge 0$ we have the pinching estimate*

$$R \ge (-\nu)[\log(-\nu) + \log(1+t) - 3] \tag{1.5}$$

whenever $\nu < 0$.

This pinching estimate roughly says that if a solution to the Ricci flow on a three-manifold becomes singular at a time T, then the most negative sectional curvature will be small compared to the most positive sectional curvature. Thus after rescaling around the singularity, one will obtain a nonnegatively curved limit. This fact will play a crucial role in the classification of singularities of the Ricci flow on three-dimensional manifolds.

1.5. Li-Yau-Hamilton inequalities. The Harnack inequality, comparing values of a positive solution at different points in space-time, is a very useful property of parabolic equations. In their seminal paper [**61**], Li-Yau studied the heat equation and found a fundamental important inequality for the gradient of positive solutions. Integrating this inequality along suitable paths, they obtained the usual Harnack inequality.

In 1993, Hamilton [**44**] discovered a highly nontrivial matrix version of Li-Yau type inequality for the Ricci flow on complete manifolds with positive curvature operator. This inequality is called the Li-Yau-Hamilton inequality. We now describe these inequalities in detail.

Let us begin with the original Li-Yau inequality for the heat equation

$$\left(\frac{\partial}{\partial t}-\triangle\right)u=0. \tag{1.6}$$

THEOREM 1.8 (Li-Yau [**61**]). *Let (M,g_{ij}) be an n-dimensional complete Riemannian manifold with nonnegative Ricci curvature. Let $u(x,t)$ be any positive solution to the heat equation (1.6) for $t\in[0,\infty)$. Then we have*

$$\frac{\partial u}{\partial t}-\frac{|\nabla u|^2}{u}+\frac{n}{2t}u\geq 0 \text{ on } M\times(0,\infty). \tag{1.7}$$

The proof is a computation of $(\partial_t-\triangle)\left(\frac{\partial}{\partial t}\log u-|\nabla\log u|^2\right)$ and an application of the maximum principle.

Next recall that under the Ricci flow on a Riemann surface the scalar curvature satisfies the following heat type equation

$$\left(\frac{\partial}{\partial t}-\triangle\right)R=R^2.$$

By the maximum principle, the positivity of the curvature is preserved by the Ricci flow. Hamilton considered the quantity $Q=\frac{\partial}{\partial t}\log R-|\nabla\log R|^2$ and computed

$$\frac{\partial}{\partial t}\left(Q+\frac{1}{t}\right)\geq\triangle\left(Q+\frac{1}{t}\right)+2\nabla\log R\cdot\nabla\left(Q+\frac{1}{t}\right)+\left(Q-\frac{1}{t}\right)\left(Q+\frac{1}{t}\right).$$

From the maximum principle, it follows

THEOREM 1.9 (Hamilton [**43**]). *Let $g_{ij}(x,t)$ be a complete solution to the Ricci flow with bounded curvature on a surface M. Assume the scalar* *curvature of the initial metric is positive. Then*

$$\frac{\partial R}{\partial t} - \frac{|\nabla R|^2}{R} + \frac{R}{t} \geq 0.$$

For higher dimensions, the curvature operator is a matrix and satisfies a quite complicated evolution equation. Apparently, the first important thing is to find out the correct expression of the quantity we want to estimate. To this end, Hamilton observed a very useful fact: the Li-Yau inequality becomes an equality on the heat kernel $h(x,t) = (4\pi t)^{-\frac{n}{2}} e^{-\frac{|x|^2}{4t}}$ on $\mathbb{R}^n$ which can be considered as an "expanding soliton." In fact, the Li-Yau inequality is equivalent to the nonnegativity of the following quadratic form:

$$\frac{\partial u}{\partial t} + 2\langle \nabla u, V\rangle + u|V|^2 + \frac{n}{2t} u \geq 0.$$

ubstituting the optimal vector field $V = -\frac{\nabla u}{u}$, we recover (1.7). To illustrate the idea of forming the quantity (1.8), let us check the heat kernel $u(x,t) = (4\pi t)^{-\frac{n}{2}} e^{-\frac{|x|^2}{4t}}$. Differentiating the function u, we get

$$\nabla_j u + u V_j = 0 \tag{1.9}$$

where $V_j = \frac{x_j}{2t}$. Differentiating (1.9) again, we have

$$\nabla_i \nabla_j u + \nabla_i u V_j + \frac{u}{2t}\delta_{ij} = 0. \tag{1.10}$$

To make the expression in (1.10) symmetric in i, j, we multiply V_i to (1.9) and add it to (1.10)

$$\nabla_i \nabla_j u + \nabla_i u V_j + \nabla_j u V_i + u V_i V_j + \frac{u}{2t}\delta_{ij} = 0. \tag{1.11}$$

Taking the trace in (1.11), we obtain the Li-Yau expression

$$\frac{\partial u}{\partial t} + 2\nabla u \cdot V + u|V|^2 + \frac{n}{2t} u = 0$$

for the heat kernel u. Moreover, the above formulation suggests the matrix Li-Yau type inequality (1.11) discussed in [**50**].

Based on similar considerations, Hamilton found the matrix Li-Yau type expression for Ricci flow which vanishes on expanding gradient Ricci solitons, and established the following fundamental important inequality.

THEOREM 1.10 (Hamilton [**44**]). *Let $g_{ij}(x,t)$ be a complete solution with bounded curvature to the Ricci flow on a manifold M^n for $t \in (0,T)$ and suppose the curvature operator of $g_{ij}(x,t)$ is nonnegative. Then for any one-form W_a and any two-form U_{ab} we have*

$$M_{ab} W_a W_b + 2P_{abc} U_{ab} W_c + R_{abcd} U_{ab} U_{cd} \geq 0$$

for $x \in M$ and $t \in (0,T)$, where

$$M_{ab} = \Delta R_{ab} - \frac{1}{2}\nabla_a\nabla_b R + 2R_{acbd}R_{cd} - R_{ac}R_{bc} + \frac{1}{2t}R_{ab},$$
$$P_{abc} = \nabla_a R_{bc} - \nabla_b R_{ac}.$$

Consequently, for any one-form V_a, we have

$$\frac{\partial R}{\partial t} + \frac{R}{t} + 2\nabla_a R \cdot V_a + 2R_{ab}V_aV_b \geq 0. \tag{1.12}$$

Integrating (1.12) along suitable space-time paths, we obtain

COROLLARY 1.11 (Hamilton [**44**]). *Under the assumption of Theorem 1.10, for any $x_1, x_2 \in M$, $t_1 < t_2$, we have*

$$R(x_1,t_1) \leq \frac{t_2}{t_1} e^{d_{t_1}(x_1,x_2)^2/2(t_2-t_1)} \cdot R(x_2,t_2). \tag{1.13}$$

We remark that Brendle [**5**] has extended the Li-Yau-Hamilton inequality under certain curvature assumption which is weaker than nonnegative curvature operator but stronger than nonnegative sectional curvature. Also in the Kähler case, the above Li-Yau-Hamilton inequality has been generalized by the first author in [**9**] under the weaker assumption of nonnegative bisectional curvature.

Define the Lie bracket on $\Lambda^2 M \oplus \Lambda^1 M$ by

$$[U \oplus W, V \oplus X] = [U,V] \oplus (U \rfloor X - V \rfloor W),$$

and a degenerate inner product

$$\langle U \oplus V, W \oplus X\rangle = \langle U, W\rangle.$$

Hamilton [**47**] observed that the equation satisfied by the quantity Q can be formally written as

$$\frac{\partial}{\partial t}Q - \Delta Q = Q^2 + Q^{\#} \tag{1.14}$$

under appropriate space-time extensions of U and W. This fascinating structure led Hamilton to write in his survey [**47**]:

"The geometry would seem to suggest that the Harnack inequality is some sort of jet extension of positive curvature operator on some bundle including translations as well as rotation, this is somehow all related to solitons where the solution moves by translation. It would be very helpful to have a proper understanding of this suggestion."

In [**27**], Chow and Chu verified this geometric interpretation of Hamilton, by showing that the Li-Yau-Hamilton quantity is in fact the curvature of a torsion free connection compatible with a degenerate metric on space-time.

2. Sphere Theorems and Uniformization Conjectures

2.1. Differential Sphere Theorems. One of the basic problems in Riemannian geometry is the classification of positively curved manifolds. The classical (Topological) Sphere Theorem due to Rauch [**87**], Berger [**4**], Klingenberg [**57**] (cf. [**18**]) states that *a simply connected Riemannian manifold with* $1/4$*-pinched sectional curvature, in the sense that sectional curvatures at each point varying in the interval* $(1,4]$*, is homeomorphic to* $\mathbb{S}^n$. In 1951, Rauch [**87**] actually conjectured that such a Riemannian manifold is diffeomorphic to $\mathbb{S}^n$. This question is known as the *Differential Sphere Theorem.* The classical result of differential sphere theorem under δ-pinched assumption for δ close to 1 (with the best $\delta = 0.87$) was obtained by Gromoll [**36**], Calabi, Sugimoto-Shiohama [**94**], Karcher [**55**], Ruh [**88**] etc (cf. [**18**]). Note that the well-known theorem of Cheeger-Gromoll-Meyer (cf. [**18**]) asserts that any complete *noncompact* Riemannian manifold of positive sectional curvature is diffeomorphic to Euclidean space $\mathbb{R}^n$.

The Ricci flow has profound application in proving various differential sphere theorems. In his 1982 seminal paper [**41**], Hamilton proved the following famous sphere theorem.

THEOREM 2.1 (Hamilton [**41**]). *A compact* 3*-manifold with positive Ricci curvature is diffeomorphic to a spherical space form, i.e., the three-sphere* $\mathbb{S}^3$ *or a quotient of it by a finite group of fixed point free isometries in the standard metric.*

The idea of the proof is to study the long-time behavior of the Ricci flow with the given metric of positive Ricci curvature as the initial data and to obtain spherical space forms as its asymptotic limit. A sketch of the proof, as shown in [**46**], can be described as follows. First of all, if we diagonalize the 3×3 curvature operator matrix $M_{\alpha\beta}$ with eigenvalues $\lambda \geq \mu \geq \nu$, the corresponding ODE system to the evolution equation of the curvature operator is given by

$$\begin{cases} \dfrac{d}{dt}\lambda = \lambda^2 + \mu\nu \\ \dfrac{d}{dt}\mu = \mu^2 + \lambda\nu \\ \dfrac{d}{dt}\nu = \nu^2 + \mu\lambda \end{cases}$$

It is easy to see $\lambda \geq \mu \geq \nu$ is preserved by this system. For any $0 < \delta \leq 2$, we compute

$$\frac{d}{dt}(\mu + \nu - \delta\lambda) \geq 0, \tag{2.1}$$

if $\mu + \nu = \delta\lambda$. This implies $\mu + \nu \geq \frac{\delta}{1+\delta}(\lambda + \mu + \nu)$ is preserved by the above ODE system. That is equivalent to say, for any $0 < \delta' \leq \frac{1}{3}$, the Ricci pinching

$Ric \geq \delta' Rg$ is preserved by the Ricci flow. Moreover, considering the convex set

$$K = \{M_{\alpha\beta} \mid \lambda - \nu \leq C(\lambda + \mu + \nu)^{1-\eta}\},$$

it is not hard to see

$$\frac{d}{dt}\log(\lambda - \nu) = \lambda + \nu - \mu,$$

and

$$\frac{d}{dt}\log(\lambda + \mu + \nu) \geq \left(1 + \left(\frac{\delta}{2(1+\delta)}\right)^2\right)(\lambda + \nu - \mu)$$

if $\mu + \nu \geq \frac{\delta}{1+\delta}(\lambda + \mu + \nu) > 0$. So $\frac{d}{dt}\log\frac{\lambda-\nu}{(\lambda+\mu+\nu)^{1-\eta}} \leq 0$ for some $\eta > 0$. Thus Hamilton's advanced maximum principle implies

$$\frac{\left|Ric - \dfrac{R}{3}g\right|}{R^{1-\eta}} \leq C.$$

where C is some positive constant. By a blow up argument and using the second Bianchi identity, we then obtain the gradient estimate with small coefficient, i.e., for any $\varepsilon > 0$ there is $C_\varepsilon > 0$ such that

$$\max_{t\leq T}\max_{x\in M}|\nabla Rm(x,t)| \leq \varepsilon \max_{t\leq T}\max_{x\in M}|Rm(x,t)|^{\frac{3}{2}} + C_\varepsilon.$$

On the other hand, we know that the solution to the Ricci flow exists only for a finite time and curvatures become unbounded. Now dilate the metrics around maximum curvature points so that the maximum curvature becomes one. Combining the gradient estimate, the pinching estimate, and the Bonnet-Meyers theorem, we know the diameter is bounded for the rescaled solution. Furthermore, by Klingenberg's injectivity radius estimate for $\frac{1}{4}$-pinched manifold and Shi's derivative estimate, we may take a smooth convergent subsequence, whose limit is a round sphere $\mathbb{S}^3$.

The combination of the above Hamilton's sphere theorem and Hamilton's strong maximum principle gives a complete classification of 3-dimensional compact manifolds with nonnegative Ricci curvature, see [**42**]. Actually, if the Ricci curvature can't be deformed to strictly positive, then the kernel of Ricci tensor gives rise to a parallel distribution of the tangent bundle. By De Rham splitting theorem, the universal cover is either flat or splits. Consequently, *a compact three-manifold with nonnegative Ricci curvature is diffeomorphic to* $\mathbb{S}^3$ *or a quotient of one of the spaces* $\mathbb{S}^3$ *or* $\mathbb{S}^2 \times \mathbb{R}^1$ *or* $\mathbb{R}^3$ *by a group of fixed point free isometries in the standard metrics.*

By using his advanced maximum principle in a similar way, Hamilton [**42**] also proved a 4-dimensional differentiable sphere theorem:

THEOREM 2.2 (Hamilton [**42**]). *A compact 4-manifold with positive curvature operator is diffeomorphic to the 4-sphere* $\mathbb{S}^4$ *or the real projective space* $\mathbb{R}P^4$.

One of the key steps in the proof is to show the existence of pinching sets associated to the cone $C=\{Rm>0\}$ of positive curvature operators. Here a pinching set Z is a closed convex subset in the space of curvature operators that is invariant under the flow of ODE $\frac{dRm}{dt}=Rm^2+Rm^{\#}$ (cf. equation (1.2)) and such that $|\mathring{Rm}|\leq c|Rm|^{1-\delta}$, for some constants $\delta>0$ and $c>0$ and all $Rm\in Z$, where $|\mathring{Rm}|$ denotes the traceless part of Rm. The latter condition implies that when $|Rm|$ is large, the rescaled curvature operator with maximal norm one becomes almost constant curvature. When $n=4$, 2-forms can be written as the direct sum of self-dual 2-forms and anti self-dual forms, hence curvature operators admit block decompositions. Using this fact and by an elaborate argument, Hamilton [**42**] was able to construct pinching sets associated to the cone $C=\{Rm>0\}$ such that any compact subset K of C is contained in some pinching set Z. Once this is established, it follows (as in the case of $n=3$) every initial metric of positive curvature will evolve under the normalized Ricci flow to a round metric in the limit.

In [**42**], Hamilton also obtained the following classification theorem for four-manifolds with nonnegative curvature operator.

THEOREM 2.3 (Hamilton [**42**]). *A compact four-manifold with nonnegative curvature operator is diffeomorphic to one of the spaces* $\mathbb{S}^4$ *or* $\mathbb{CP}^2$ *or* $\mathbb{S}^2\times\mathbb{S}^2$ *or a quotient of one of the spaces* $\mathbb{S}^4$ *or* $\mathbb{CP}^2$ *or* $\mathbb{S}^3\times\mathbb{R}^1$ *or* $\mathbb{S}^2\times\mathbb{S}^2$ *or* $\mathbb{S}^2\times\mathbb{R}^2$ *or* $\mathbb{R}^4$ *by a group of fixed point free isometries in the standard metrics.*

We note that H. Chen [**26**] extended Theorem 2.2 to 2-positive curvature operator. Here *2-positive curvature operator* means the sum of the least two eigenvalues of the curvature operator is positive. Later, by using the Ricci flow, differential sphere theorems for higher dimensions under some suitable pointwise pinching conditions were obtained by Huisken [**51**] (see also Margerin [**65, 66**] and Nishikawa [**79**]).

Naturally, one would ask if a compact Riemannian manifold M^n, with $n\geq 5$, of positive curvature operator (or 2-positive curvature operator) is diffeomorphic to a space form. This was in fact conjectured so by Hamilton and proved only very recently by Böhm-Wilking [**8**].

THEOREM 2.4 (Böhm-Wilking [**8**]). *A compact Riemannian manifold of dimension* $n\geq 5$ *with a two-positive curvature operator is diffeomorphic to a spherical space form.*

In [**8**], Böhm-Wilking developed a powerful new method to construct closed convex sets, which are invariant under the Ricci flow, in the space of curvature operators. They introduced a linear transformation $l_{a,b}$ $(a, b > 0)$ on the space of curvature operators defined by $l_{a,b}(Rm) = Rm + aRm_I + bRm_0$ so that it increases the scalar curvature part Rm_I and the traceless Ricci part Rm_0 of Rm by factors of a and b respectively. A crucial property they found is that the associated transformation $D_{a,b}(Rm) = l_{a,b}^{-1}((l_{a,b}Rm)^2 + (l_{a,b}Rm)^{\#}) - Rm^2 - Rm^{\#}$ is independent of the Weyl curvature part of Rm. Based on this, they can construct new invariant cones from old ones. By choosing appropriate constants $a's$ and $b's$, this construction gives rise to a (continuous) pinching family $C(s)$, $s \in [0, 1)$, of invariant closed convex cones such that $C(0)$ is the cone of 2-nonnegative curvature operators and, as $s \to 1$, $C(s)$ approaches $\{cI : c \in \mathbb{R}_+\}$, the set of constant curvature operators. From this pinching family $C(s)$, one can then construct a generalized pinching set F which is a certain special invariant convex set in the space of curvature operators, so that F contains the initial data and $F \backslash C(s)$ is compact for every $s \in [0, 1)$. Since the curvature operator of the evolving metric under the Ricci flow has to diverge to infinity in finite time, it must be contained in every $C(s)$ when the time is large after rescaling. Thus, the solution to the normalized Ricci flow converges to a round metric in the limit.

We have seen that the Ricci flow preserves the *positive curvature operator* condition in all dimensions and preserves the positive *Ricci curvature* condition in dimension 3. On the other hand, in [**48**] Hamilton also proved that the *positive isotropic curvature* (PIC) condition is preserved by the Ricci flow in dimension 4. We remark that in 1988, by using minimal surface theory, Micallef and Moore [**67**] were able to prove that any compact simply connected n-dimensional manifold with positive isotropic curvature is homeomorphic to the n-sphere $\mathbb{S}^n$, and the condition of positive isotropic curvature is weaker than both positive curvature operator and 1/4-pinched. Very recently Brendle-Schoen [**6**], and independently H. Nguyen [**73**], proved that the PIC condition is preserved by the Ricci flow in all dimensions $n \geq 4$.[1] More excitingly, Brendle and Scheon [**6**] showed that when the initial metric has (pointwise) 1/4-pinched sectional curvature (in fact under the weaker curvature condition that $M \times \mathbb{R}^2$ has PIC, see [**6**]), the Ricci flow will converge to a spherical space form. As a corollary, they proved the long-standing Differential Sphere Theorem.

THEOREM 2.5 (Brendle-Schoen [**6**]). *Let M be a compact manifold with (pointwise) 1/4-pinched sectional curvature. Then M is diffeomorphic to $\mathbb{S}^n$ or a quotient of $\mathbb{S}^n$ by a group of fixed point free isometries in the standard metrics.*

[1]See also Andrews-Nguyen [**1**] for a proof that 1/4-pinched flag curvature is preserved for $n = 4$ which has some common features.

By generalizing the usual strong maximum principle to a powerful version, Brendle and Schoen [**7**] even obtained the following rigidity result, which extends the well-known rigidity result of Berger (cf. [**18**]).

THEOREM 2.6 (Brendle-Schoen [**7**]). *Let M be a compact manifold with (pointwise) weakly 1/4-pinched sectional curvature in the sense that* $0 < sect(P_1) \leq 4sect(P_2)$ *for all two-planes* $P_1, P_2 \in T_pM$. *If M is not diffeomorphic to a spherical space form, then it is isometric to a locally symmetric space.*

2.2. Kähler Manifolds with Nonnegative Holomorphic Bisectional Curvature. The classical uniformization theorem for Riemann surfaces implies that a complete simply connected Riemann surface with positive curvature is biholomorphic to either the Riemann sphere or the complex plane. The classification (in holomorphic category) of positively curved Kähler manifolds in higher dimensions is one of the most important problems in complex differential geometry. Corresponding to positive sectional curvature condition in Riemannian geometry, one usually considers the positive holomorphic bisectional curvature in complex differential geometry.

Let M^n be a complex n-dimensional compact Kähler manifold. The famous *Frankel conjecture* states that: *if M^n has positive holomorphic bisectional curvature, then it is biholomorphic to the complex projective space* $\mathbb{CP}^n$. This was independently proved by Mori [**71**] and Siu-Yau [**100**] by using different methods. After the work of Mori and Siu-Yau, it is natural to ask the similar question for the semi-positive case. This is often called the *generalized Frankel conjecture*. The complex three-dimensional case was first obtained by Bando [**3**]. When the curvature operator of M^n is assumed to be nonnegative, the result was proved by the first author and Chow [**11**]. The general case of the generalized Frankel conjecture is proved by Mok [**69**].

THEOREM 2.7 (Mok [**69**]). *Let (M^n, h) be a compact complex n-dimensional Kähler manifold of nonnegative holomorphic bisectional curvature and let $(\tilde{M}^n, \tilde{h})$ be its universal covering space. Then there exists nonnegative integers $k, N_1, \ldots, N_l, p$ and irreducible compact Hermitian symmetric spaces $M_1, \ldots, M_p$ of rank ≥ 2 such that $(\tilde{M}^n, \tilde{h})$ is isometrically biholomorphic to*

$$(\mathbb{C}^k, g_0) \times (\mathbb{CP}^{N_1}, \theta_1) \times \cdots \times (\mathbb{CP}^{N_l}, \theta_l) \times (M_1, g_1) \times \cdots \times (M_p, g_p)$$

where g_0 denotes the Euclidean metric on $\mathbb{C}^k$, $g_1, \ldots, g_p$ are canonical metrics on $M_1, \ldots, M_p$, and $\theta_i, 1 \leq i \leq l$, is a Kähler metric on $\mathbb{CP}^{N_i}$ carrying nonnegative holomorphic bisectional curvature.

Mok's method of proving the generalized Frankel conjecture in [**69**] depends on Mori's theory of rational curves on Fano manifolds, so his proof is not completely transcendental. Recently, by using the strong maximum principle of Brendle-Schoen in [**7**], H. L. Gu [**39**] gave a simpler and completely transcendental proof of the generalized Frankel conjecture. The above Mok's theorem on generalized Frankel conjecture is indeed a factorization theorem for compact case. Based on the arguments in [**39**], we now formulate a new factorization theorem for noncompact cases as follows.

THEOREM 2.8. *Let (M,h) be a complete noncompact Kähler manifold with bounded and nonnegative holomorphic bisectional curvature. Then one of the following holds:*

(i) M admits a Kähler metric with bounded and positive bisectional curvature;

(ii) The universal cover $\hat{M}$ of M splits holomorphically, isometrically and nontrivially as

$$\hat{M}=\mathbb{C}^k\times M_1\times\cdots\times M_{l_1}\times N_1\times\cdots\times N_{l_2}$$

where k,l_1,l_2 are nonnegative integers, $\mathbb{C}^k$ is the complex Euclidean space with flat metric, $M_i, 1\leq i\leq l_1$, are complete (compact or noncompact) Kähler manifolds with bounded and nonnegative bisectional curvature admitting a Kähler metric with bounded and positive bisectional curvature, $N_j, 1\leq j\leq l_2$, are irreducible compact Hermitian symmetric spaces of rank ≥ 2 with the canonical metrics.

PROOF. We evolve the metric h by the Kähler Ricci flow:

$$\begin{cases}\dfrac{\partial}{\partial t}g_{i\bar{j}}(x,t)=-R_{i\bar{j}}(x,t),\\ g_{i\bar{j}}(x,0)=h_{i\bar{j}}(x).\end{cases}$$

Then by Shi's short-time existence theorem, we know that there is a $T>0$ such that the Ricci flow has a smooth solution $g_{i\bar{j}}(t)$ with bounded curvature for $t\in[0,T)$. It is well-known (from [**69**] and [**91**]) that the solution $g_{i\bar{j}}(t)$ still has nonnegative holomorphic bisectional curvature. By lifting the solution to the universal cover $\hat{M}$ of M, then the pull back evolving metric $\hat{g}_{i\bar{j}}(t)$ is a solution to the Ricci flow on $\hat{M}$. Clearly we may assume that the holomorphic bisectional curvature of the solution $\hat{g}_{i\bar{j}}(t)$ vanishes somewhere at each time $t\in[0,T)$; otherwise we will have case (i).

By applying the standard De Rham decomposition theorem, we know that the universal cover $(\hat{M},\hat{g}_{i\bar{j}}(t)), t\in[0,\delta)$, can be isometrically and holomorphically splitted as

$$(\hat{M},\hat{g}_{i\bar{j}}(t))=(\mathbb{C}^k,\hat{g}^0)\times(\hat{M}_1,\hat{g}^1_{i\bar{j}}(t))\times\cdots(\hat{M}_p,\hat{g}^p_{i\bar{j}}(t))$$

for some $\delta\in(0,T)$, where each $(\hat{M}_\alpha,\hat{g}^\alpha_{i\bar{j}}(t)), 1\leq\alpha\leq p$, is irreducible and non-flat, $\hat{g}^0$ is the standard flat metric.

Consider each irreducible and non-flat factor $(\hat{M}_\alpha, \hat{g}^\alpha_{i\bar{j}}(t)), 1 \leq \alpha \leq p$. Suppose $(\hat{M}_\alpha, \hat{g}^\alpha_{i\bar{j}}(0))$ is not locally symmetric. We shall show that $(\hat{M}_\alpha, \hat{g}^\alpha_{i\bar{j}}(t)), t \in (0, \delta')$, has positive holomorphic bisectional curvature everywhere on $(0, \delta')$ for some $0 < \delta' < \delta$.

Since the smooth limit of locally symmetric space is also locally symmetric, we obtain that there exists $\delta' \in (0, T)$ such that $(\hat{M}_\alpha, \hat{g}^\alpha_{i\bar{j}}(t))$ is not locally symmetric for $t \in (0, \delta')$. Combining the Kählerity of $\hat{g}^\alpha_{i\bar{j}}(t)$ and Berger's holonomy theorem, we know that the holonomy group $\mathrm{Hol}(\hat{g}^\alpha(t))$ of $(\hat{M}_\alpha, \hat{g}^\alpha(t))$ is $U(n_\alpha)$, where $n_\alpha = \dim_C \hat{M}_\alpha$.

Recall the evolution equation of holomorphic bisectional curvature under an evolving orthnormal frame $\{e_i\}$ according to Hamilton [**42**]

$$\frac{\partial}{\partial t}\hat{R}^\alpha_{i\bar{i}j\bar{j}} = \triangle \hat{R}^\alpha_{i\bar{i}j\bar{j}} + \sum_{p,q}(\hat{R}^\alpha_{i\bar{i}p\bar{q}}\hat{R}^\alpha_{q\bar{p}j\bar{j}} - |\hat{R}^\alpha_{i\bar{p}j\bar{q}}|^2 + |\hat{R}^\alpha_{i\bar{j}p\bar{q}}|^2).$$

Let P be the fiber bundle with the fixed metric h and the fiber P_x over $x \in \hat{M}^\alpha$ consisting of all 2-vectors $\{X, Y\} \subset T^{1,0}_x(\hat{M}^\alpha)$. Now define a function u on $P \times (0, \delta')$ by

$$u(\{X, Y\}, t) = \hat{R}^\alpha(X, \overline{X}, Y, \overline{Y}),$$

where $\hat{R}^\alpha$ denotes the pull-back of the curvature tensor of $\hat{g}^\alpha_{i\bar{j}}(t)$. For simplicity, we denote $R = \hat{R}^\alpha$. Since $(\hat{M}^\alpha, \hat{g}^\alpha_{i\bar{j}}(t))$ has nonnegative holomorphic bisectional curvature, we have $u \geq 0$. Let $N = \{(\{X, Y\}, t) | u(\{X, Y\}, t) = 0, X \neq 0, Y \neq 0\} \subset P \times (0, \delta')$. We will show in the following that if N is not empty then it is invariant under the parallel translation.

For fixed e_i, consider the Hermitian form $H_i(X, Y) = R(e_i, \overline{e_i}, X, \overline{Y})$ and let $\{E_p\}$ be an orthonormal basis associated to eigenvectors of H_i. In these basis we have

$$\sum_{p,q} R_{i\bar{i}p\bar{q}}R_{q\bar{p}j\bar{j}} = \sum_p R(e_i, \overline{e_i}, E_p, \overline{E_p})R(E_p, \overline{E_p}, e_j, \overline{e_j}),$$

and

$$\sum_{p,q} |R_{i\bar{p}j\bar{q}}|^2 = \sum_{p,q} |R(e_i, \overline{E_p}, e_j, \overline{E_q})|^2.$$

Moreover, we claim

$$\sum_{p,q} R_{i\bar{i}p\bar{q}}R_{q\bar{p}j\bar{j}} - \sum_{p,q} |R_{i\bar{p}j\bar{q}}|^2 \geq c_1 \cdot \min\{0, \inf_{|\xi| = 1, \xi \in V} D^2u(\{e_i, e_j\}, t)(\xi, \xi)\}$$

for some constant $c_1 > 0$, where V denotes the vertical spaces of the fiber bundle P.

Indeed, inspired by Mok [**69**], for any given $\varepsilon_0>0$ and each fixed $q\in\{1,2,\ldots,n\}$, we consider the function

$$\widetilde{G}_q(\varepsilon)=(R+\varepsilon_0R_0)(e_i+\varepsilon E_q,\overline{e_i+\varepsilon E_q},e_j+\varepsilon\sum_p C_pE_p,\overline{e_j+\varepsilon\sum_p C_pE_p}),$$

where R_0 is a curvature operator defined by $(R_0)_{i\bar{j}k\bar{l}}=g_{i\bar{j}}g_{k\bar{l}}+g_{i\bar{l}}g_{k\bar{j}}$ and C_p are complex constants to be determined later. For simplicity, we denote $\widetilde{R}=R+\varepsilon_0R_0$, then

$$\widetilde{G}_q(\varepsilon)=\widetilde{R}(e_i+\varepsilon E_q,\overline{e_i+\varepsilon E_q},e_j+\varepsilon\sum_p C_pE_p,\overline{e_j+\varepsilon\sum_p C_pE_p}).$$

Then a direct computation gives

$$\begin{aligned}\frac{1}{2}\cdot\frac{d^2\widetilde{G}_q(\varepsilon)}{d\varepsilon^2}\bigg|_{\varepsilon=0}&=\widetilde{R}(E_q,\overline{E_q},e_j,\overline{e_j})+\sum_p|C_p|^2\widetilde{R}(e_i,\overline{e_i},E_p,\overline{E_p})\\&\quad+2Re\sum_p\overline{C_p}\widetilde{R}(e_i,\overline{E_q},e_i,\overline{E_p})+2Re\sum_p C_p\widetilde{R}(e_i,\overline{e_i},E_p,\overline{E_q}).\end{aligned}$$

Writing $C_p=x_pe^{i\theta_p}$, $(p\geq1)$ for some real numbers x_p,θ_p to be determined later, the above identity becomes:

$$\begin{aligned}\frac{1}{2}\cdot\frac{d^2\widetilde{G}_q(\varepsilon)}{d\varepsilon^2}\bigg|_{\varepsilon=0}&=\widetilde{R}(E_q,\overline{E_q},e_j,\overline{e_j})+\sum_p|x_p|^2\widetilde{R}(e_i,\overline{e_i},E_p,\overline{E_p})\\&\quad+2\sum_p x_p\cdot Re(e^{-i\theta_p}\widetilde{R}(e_i,\overline{E_q},e_j,\overline{E_p})+e^{i\theta_p}\widetilde{R}(e_i,\overline{e_j},E_p,\overline{E_q})).\end{aligned}$$

Following Mok [**69**], by setting $A_p=\widetilde{R}(e_i,\overline{e_j},E_p,\overline{E_q})$, $B_p=\widetilde{R}(e_i,\overline{E_q},e_j,\overline{E_p})$, we have:

$$\begin{aligned}\frac{1}{2}\cdot\frac{d^2\widetilde{G}_q(\varepsilon)}{d\varepsilon^2}\bigg|_{\varepsilon=0}&=\widetilde{R}(E_q,\overline{E_q},e_j,\overline{e_j})+\sum_p|x_p|^2\widetilde{R}(e_i,\overline{e_i},E_p,\overline{E_p})\\&\quad+\sum_p x_p(e^{-i\theta_p}B_p+e^{i\theta_p}\overline{B_p}+e^{i\theta_p}A_p+e^{-i\theta_p}\overline{A_p})\\&=\widetilde{R}(E_q,\overline{E_q},e_j,\overline{e_j})+\sum_p|x_p|^2\widetilde{R}(e_i,\overline{e_i},E_p,\overline{E_p})\\&\quad+\sum_p x_p\cdot(\overline{e^{i\theta_p}(A_p+\overline{B_p})}+e^{i\theta_p}(A_p+\overline{B_p}))\end{aligned}$$

By choosing θ_p such that $e^{i\theta_p}(A_p+\overline{B_p})$ is real and positive, the identity becomes:

$$\begin{aligned}\frac{1}{2}\cdot\frac{d^2\widetilde{G}_q(\varepsilon)}{d\varepsilon^2}\bigg|_{\varepsilon=0}&=\widetilde{R}(E_q,\overline{E_q},e_j,\overline{e_j})+\sum_p|x_p|^2\widetilde{R}(e_i,\overline{e_i},E_p,\overline{E_p})\\&\quad+2\sum_p x_p\cdot|A_p+\overline{B_p}|.\end{aligned}$$

If we change e_i by $e^{i\varphi}e_i$, then $A_p = \widetilde{R}(e_i, \overline{e_j}, E_p, \overline{E_q})$ is replaced by $e^{i\varphi}A_p$, and $\overline{B_p} = \overline{\widetilde{R}(e_i, \overline{E_q}, e_j, \overline{E_p})}$ is replaced by $e^{-i\varphi}\overline{B_p}$. Then we have:

$$\begin{aligned}\frac{1}{2}\cdot\frac{d^2\widetilde{F}_q(\varepsilon)}{d\varepsilon^2}\bigg|_{\varepsilon=0} &= \widetilde{R}(E_q,\overline{E_q},e_j,\overline{e_j}) + \sum\nolimits_p |x_p|^2\widetilde{R}(e_i,\overline{e_i},E_p,\overline{E_p}) \\ &\quad + 2\sum\nolimits_p x_p\cdot|e^{i\varphi}A_p + e^{-i\varphi}\overline{B_p}|,\end{aligned}$$

where

$$\widetilde{F}_q(\varepsilon) = \widetilde{R}\left(e^{i\varphi}e_i + \varepsilon E_q, \overline{e^{i\varphi}e_i + \varepsilon E_q}, e_j + \varepsilon\sum\nolimits_p C_pE_p, \overline{e_j + \varepsilon\sum\nolimits_p C_pE_p}\right).$$

Since the curvature operators R and R_0 have nonnegative and positive holomorphic bisectional curvature respectively, we know that the operator $\widetilde{R} = R_0 + \varepsilon_0 R_0$ has positive holomorphic bisectional curvature. Now by choosing $x_p = -\frac{|e^{i\varphi}A_p + e^{-i\varphi}\overline{B_p}|}{\widetilde{R}(e_i,\overline{e_i},E_p,\overline{E_p})}$, for $p \geq 1$, it follows that

$$\frac{1}{2\pi}\int_0^{2\pi}\left(\frac{1}{2}\cdot\frac{d^2\widetilde{F}_q(\varepsilon)}{d\varepsilon^2}\bigg|_{\varepsilon=0}\right)d\varphi = \widetilde{R}(E_q,\overline{E_q},e_j,\overline{e_j}) - \sum\nolimits_p\frac{|A_p|^2+|B_p|^2}{\widetilde{R}(e_i,\overline{e_i},E_p,\overline{E_p})}$$

and then

$$\begin{aligned}&\widetilde{R}(e_i,\overline{e_i},E_q,\overline{E_q})\cdot\frac{1}{2\pi}\int_0^{2\pi}\left(\frac{1}{2}\cdot\frac{d^2\widetilde{F}_q(\varepsilon)}{d\varepsilon^2}\bigg|_{\varepsilon=0}\right)d\varphi \\ &= \widetilde{R}(e_i,\overline{e_i},E_q,\overline{E_q})\widetilde{R}(E_q,\overline{E_q},e_j,\overline{e_j}) \\ &\quad - \sum\nolimits_p\frac{|A_p|^2+|B_p|^2}{\widetilde{R}(e_i,\overline{e_i},E_p,\overline{E_p})}\widetilde{R}(e_i,\overline{e_i},E_q,\overline{E_q}).\end{aligned}$$

Note that

$$\begin{aligned}\widetilde{F}_q(\varepsilon) &= \widetilde{R}(e^{i\varphi}e_i + \varepsilon E_q, \overline{e^{i\varphi}e_i + \varepsilon E_q}, e_j + \varepsilon\sum\nolimits_p C_pE_p, \overline{e_j + \varepsilon\sum\nolimits_p C_pE_p}) \\ &= \widetilde{R}(e_i + \varepsilon e^{-i\varphi}E_q, \overline{e_i + \varepsilon e^{-i\varphi}E_q}, e_j + \varepsilon\sum\nolimits_p C_pE_p, \overline{e_j + \varepsilon\sum\nolimits_p C_pE_p}).\end{aligned}$$

Interchanging the roles of E_q and E_p, and then taking summation, we have

$$\begin{aligned}&\sum\nolimits_q 2\widetilde{R}(e_i,\overline{e_i},E_q,\overline{E_q})\widetilde{R}(E_q,\overline{E_q},e_j,\overline{e_j}) \\ &\geq c_1\cdot\min\{0, \inf_{|\xi|=1,\xi\in V} D^2\widetilde{u}(\{e_i,e_j\},t)(\xi,\xi)\} \\ &\quad + \sum\nolimits_{p,q}(|A_p|^2+|B_p|^2)\left(\frac{\widetilde{R}(e_i,\overline{e_i},E_q,\overline{E_q})}{R(e_i,\overline{e_i},E_p,\overline{E_i})} + \frac{\widetilde{R}(e_i,\overline{e_i},E_p,\overline{E_p})}{\widetilde{R}(e_i,\overline{e_i},E_q,\overline{E_q})}\right) \\ &\geq c_1\cdot\min\{0, \inf_{|\xi|=1,\xi\in V} D^2\widetilde{u}(\{e_i,e_j\},t)(\xi,\xi)\} + 2\sum\nolimits_{p,q}|\widetilde{R}(e_i,\overline{E_q},e_j,\overline{E_p})|^2,\end{aligned}$$

where $\widetilde{u}(\{X,Y\},t)=\widetilde{R}(X,\overline{X},Y,\overline{Y})=R(X,\overline{X},Y,\overline{Y})+\varepsilon_0 R_0(X,\overline{X},Y,\overline{Y})$ and c_1 is a positive constant which depends on the bound of the curvature R, but does not depend on ε_0.

Hence

$$\begin{aligned}&\textstyle\sum_p \widetilde{R}(e_i,\overline{e_i},E_p,\overline{E_p})\widetilde{R}(E_p,\overline{E_p},e_j\overline{e_j})-\sum_{p,q}|\widetilde{R}(e_i,\overline{E_p},e_j,\overline{E_q})|^2\\&\geq c_1\cdot\min\{0,\inf_{|\xi|=1,\xi\in V}D^2\widetilde{u}(\{e_i,e_j\},t)(\xi,\xi)\}.\end{aligned}$$

Since $\varepsilon_0>0$ is arbitrary, we can let $\varepsilon_0\to 0$ and it follows that:

$$\sum_{p,q}R_{i\bar{i}p\bar{q}}R_{q\bar{p}j\bar{j}}-\sum_{p,q}|R_{i\bar{p}j\bar{q}}|^2\geq c_1\cdot\min\{0,\inf_{|\xi|=1,\xi\in V}D^2u(\{e_i,e_j\},t)(\xi,\xi)\},$$

for some constant $c_1>0$. Therefore we proved our claim.

By the definition of u and the evolution equation of the holomorphic bisectional curvature, we know that

$$\begin{aligned}\tfrac{\partial}{\partial t}u(\{X,Y\},t)&=\triangle u(\{X,Y\},t)+\textstyle\sum_{p,q}R(X,\overline{X},e_p,\overline{e_q})R(e_q,\overline{e_p},Y,\overline{Y})\\&\quad-\textstyle\sum_{p,q}|R(X,\overline{e_p},Y,\overline{e_q})|^2+\sum_{p,q}|R(X,\overline{Y},e_p,\overline{e_q})|^2.\end{aligned}$$

Therefore, from the above inequality, we obtain that:

$$\frac{\partial u}{\partial t}\geq Lu+c_1\cdot\min\{0,\inf_{|\xi|=1,\xi\in V}D^2u(\xi,\xi)\},$$

where L is the horizontal Laplacian on P, V denotes the vertical subspaces. By Proposition 2 in [**7**] and note that the curvature is nonnegative and bounded, we know that the set

$$N=\{(\{X,Y\},t)|u(\{X,Y\},t)=0,X\neq 0,Y\neq 0\}\subset P\times(0,\delta')$$

is invariant under parallel transport.

Next, we claim that $R_{i\bar{i}j\bar{j}}>0$ for all $t\in(0,\delta')$. Indeed, suppose not. Then $R_{i\bar{i}j\bar{j}}=0$ for some $t\in(0,\delta')$. Therefore

$$(\{e_i,e_j\},t)\in N.$$

Combining $R_{i\bar{i}j\bar{j}}=0$ with the evolution equation of the curvature operator and the first variation, we can obtain that

$$\begin{cases}\sum_{p,q}(R_{i\bar{i}p\bar{q}}R_{q\bar{p}j\bar{j}}-|R_{i\bar{p}j\bar{q}}|^2)=0,\\ R_{i\bar{j}p\bar{q}}=0,\quad\forall p,q,\\ R_{i\bar{i}p\bar{j}}=R_{j\bar{j}p\bar{i}}=0,\quad\forall p.\end{cases}$$

We define an orthonormal 2-frames $\{\widetilde{e_i}, \widetilde{e_j}\} \subset T_x^{1,0}(\hat{M}^\alpha)$ by

$$\widetilde{e_i} = \sin\theta \cdot e_i - \cos\theta \cdot e_j,$$

$$\widetilde{e_j} = \cos\theta \cdot e_i + \sin\theta \cdot e_j.$$

Then

$$\overline{\widetilde{e_i}} = \sin\theta \cdot \overline{e_i} - \cos\theta \cdot \overline{e_j},$$
$$\overline{\widetilde{e_j}} = \cos\theta \cdot \overline{e_i} + \sin\theta \cdot \overline{e_j}.$$

Since N is invariant under parallel transport and $(\hat{M}^\alpha, \hat{g}^\alpha_{i\bar{j}}(t))$ has holonomy group $U(n_\alpha)$, we obtain that

$$(\{\widetilde{e_i}, \widetilde{e_j}\}, t) \in N,$$

that is,

$$R(\widetilde{e_i}, \overline{\widetilde{e_i}}, \widetilde{e_j}, \overline{\widetilde{e_j}}) = 0.$$

On the other hand,

$$\begin{aligned}
R(\widetilde{e_i}, \overline{\widetilde{e_i}}, \widetilde{e_j}, \overline{\widetilde{e_j}}) &= \sin^2\theta\cos^2\theta R_{i\bar{i}i\bar{i}} + \sin^3\theta\cos\theta R_{i\bar{i}i\bar{j}} + \sin^3\theta\cos\theta R_{i\bar{i}j\bar{i}} \\
&\quad + \sin^4\theta R_{i\bar{i}j\bar{j}} - \sin\theta\cos^3\theta R_{i\bar{j}i\bar{i}} - \sin^2\theta\cos^2\theta R_{i\bar{j}i\bar{j}} \\
&\quad - \sin^2\theta\cos^2\theta R_{i\bar{j}j\bar{i}} - \sin^3\theta\cos\theta R_{i\bar{j}j\bar{j}} - \cos^3\theta\sin\theta R_{j\bar{i}i\bar{i}} \\
&\quad - \sin^2\theta\cos^2\theta R_{j\bar{i}i\bar{j}} - \sin^2\theta\cos^2\theta R_{j\bar{i}j\bar{i}} - \cos\theta\sin^3\theta R_{j\bar{i}j\bar{j}} \\
&\quad + \cos^4\theta R_{j\bar{j}i\bar{i}} + \cos^3\theta\sin\theta R_{j\bar{j}i\bar{j}} + \cos^3\theta\sin\theta R_{j\bar{j}j\bar{i}} \\
&\quad + \cos^2\theta\sin^2\theta R_{j\bar{j}j\bar{j}} \\
&= \cos^2\theta\sin^2\theta(R_{i\bar{i}i\bar{i}} + R_{j\bar{j}j\bar{j}}).
\end{aligned}$$

So we have $R_{j\bar{j}j\bar{j}} + R_{i\bar{i}i\bar{i}} = 0$, if we choose θ such that $\cos^2\theta\sin^2\theta \neq 0$. But this contradicts with the fact that $(\hat{M}^\alpha, \hat{g}^\alpha_{i\bar{j}}(t))$ has positive holomorphic sectional curvature. Hence we proved that $R_{i\bar{i}j\bar{j}} > 0$, for all $t \in (0, \delta')$.

This completes the proof of Theorem 2.8. □

We remark that a (rough) factorization theorem, according to whether the manifold supports a strictly plurisubharmonic function, was obtained earlier by Ni and Tam [**76**] without assuming the curvature to be bounded.

Finally, by combining with the resolution of the Frankel conjecture, our (more precise) factorization Theorem 2.8 can reduce the classification of complete noncompact Kähler manifolds with bounded and nonnegative bisectional curvature to the case of strictly positive bisectional curvature. In the latter case there is a long standing conjecture due to Yau (Problem 34 in [**101**]):

Yau's Conjecture (Yau [**101**]) *A complete noncompact Kähler manifold of positive holomorphic bisectional curvature is biholomorphic to a complex Euclidean space.*

In recent years, there have been many research activities in studying this conjecture of Yau. The Ricci flow has been found to be a useful tool to approach it. The following partial affirmative answer, due to Chen-Tang-Zhu [**20**] in complex dimension $n=2$ and Chau-Tam [**16**] for all dimensions, was obtained via the Ricci flow.

THEOREM 2.9. *Let M be a complete noncompact n-dimensional Kähler manifold of positive and bounded holomorphic bisectional curvature. Suppose there exists a positive constant C_1 such that for a fixed base point x_0, we have*

$$Vol(B(x_0,r)) \geq C_1 r^{2n} \qquad 0 \leq r < +\infty,$$

then M is biholomorphic to $\mathbb{C}^n$.

We refer the readers to the survey article of A. Chau and L. F. Tam [**17**] in this volume for more information on works related to the Kähler-Ricci flow and Yau's uniformization conjecture.

3. Perelman's Noncollapsing Result

In the celebrated work [**80**], Perelman proved a remarkable (local) noncollapsing result for the Ricci flow on compact manifolds in all dimensions. This (local) noncollapsing result had been conjectured by Hamilton in his survey paper [**47**] and is crucial in applying Hamilton's compactness theorem to understand the structure of singularities of the Ricci flow. Below, we follow Perelman [**80**] to give two approaches for deriving his noncollapsing result.

3.1. Perelman's Conjugate Heat Equation Approach. For the Ricci flow on a compact manifold, Perelman [**80**] introduced a new functional

$$\mathcal{W}(g_{ij},f,\tau)=\int_M [\tau(R+|\nabla f|^2)+f-n](4\pi\tau)^{-\frac{n}{2}}e^{-f}dV. \tag{3.1}$$

This functional has played a very important role in the Ricci flow; see also the more recent works by Feldman-Ilmanen-Ni [**34**], Cao-Hamilton-Ilmanen [**13**], Ma [**64**], Li [**60**], Zhang [**107**], Ye [**102**]–[**105**], X. Cao [**15**], Ling [**62**], etc.

Perelman proved that the $\mathcal{W}$-functional is monotone in time when the metric g evolves under the Ricci flow, the function f evolves under the backward heat equation

$$\frac{\partial f}{\partial \tau}=\Delta f-|\nabla f|^2+R-\frac{n}{2\tau},$$

and $\frac{d\tau}{dt} = -1$. This entropy monotonicity can be interpreted as a Li-Yau type estimate for the conjugate heat equation

$$\Box u := \frac{\partial u}{\partial \tau} - \Delta u + Ru = 0, \tag{3.2}$$

where $\tau = T - t$, and $g_{ij}(x,t)$, $0 \le t < T$, is a solution to the Ricci flow. Note that $u = (4\pi\tau)^{-\frac{n}{2}} e^{-f}$ satisfies the conjugate heat equation if and only if f satisfies the above backward heat equation.

By considering the shrinking Ricci solitons, one can find the analogous Li-Yau expression for the conjugate heat equation to be

$$H = 2\Delta f - |\nabla f|^2 + R + \frac{f-n}{\tau}.$$

(We learned this argument from Hamilton. The details can be found in [**14**].) By direct computations, one has

$$\frac{\partial H}{\partial \tau} = \Delta H - 2\nabla f \cdot \nabla H - \frac{1}{\tau} H - 2\left| R_{ij} + \nabla_i \nabla_j f - \frac{1}{2\tau} g_{ij} \right|^2.$$

Set

$$v = \tau H u = \big(\tau(R + 2\Delta f - |\nabla f|^2) + f - n\big)u, \tag{3.3}$$

then

$$\frac{\partial v}{\partial \tau} = \Delta v - Rv - 2\tau u \left| R_{ij} + \nabla_i \nabla_j f - \frac{1}{2\tau} g_{ij} \right|^2. \tag{3.4}$$

If u is a fundamental solution to (3.2), one can show $\lim_{\tau \to 0+} \tau H \le 0$ (see [**75**]). Then the maximum principle implies Perelman's Li-Yau type estimate for the conjugate heat equation:

$$H \le 0$$

for all $\tau \in (0, T]$. Along any space-time path $(\gamma(\tau), \tau)$, $\tau \in [0, \bar{\tau}]$ with $\gamma(0) = p$, $\gamma(\bar{\tau}) = q$, there holds

$$\frac{d}{d\tau}\big(2\sqrt{\tau} f(\gamma(\tau), \tau)\big) \le \sqrt{\tau}\big(R + |\dot{\gamma}(\tau)|^2_{g_{ij}(\tau)}\big).$$

If one defines

$$\mathcal{L}(\gamma) \triangleq \int_0^{\bar{\tau}} \sqrt{\tau}\big(R + |\dot{\gamma}(\tau)|^2_{g_{ij}(\tau)}\big) d\tau, \tag{3.5}$$

and

$$l(q, \bar{\tau}) \triangleq \inf_{\gamma} \frac{1}{2\sqrt{\bar{\tau}}} \mathcal{L}(\gamma), \tag{3.6}$$

where the inf is taken over all space curves $\gamma(\tau), 0 \le \tau \le \bar{\tau}$, joining p and q, then

$$f(q, \bar{\tau}) \le l(q, \bar{\tau}).$$

This leads to a lower estimate for the fundamental solution u of the conjugate heat equation,

$$u(q, \bar{\tau}) \ge (4\pi\bar{\tau})^{-\frac{n}{2}} e^{-l(q,\bar{\tau})}. \tag{3.7}$$

Now since v happens to be the integrand of the $\mathcal{W}$-functional, by integrating (3.4), one obtains

(3.8)

$$\frac{d}{dt}\mathcal{W}(g_{ij}(t), f(t), \tau(t)) = \int_M 2\tau \left| R_{ij} + \nabla_i \nabla_j f - \frac{1}{2\tau} g_{ij} \right|^2 (4\pi\tau)^{-\frac{n}{2}} e^{-f} dV \ge 0.$$

Let

$$\mu(M, g, \tau) = \inf \left\{ \mathcal{W}(g, f, \tau) \middle| \int (4\pi\tau)^{-\frac{n}{2}} e^{-f} dv = 1 \right\}, \tag{3.9}$$

then we have the monotonicity of Perelman's entropy:

LEMMA 3.1 (Perelman [**80**]). *$\mu(M, g(t), T - t)$ is nondecreasing along compact Ricci flow; moreover, the monotonicity is strict unless we are on a shrinking gradient soliton.*

A direct consequence is the following important noncollapsing theorem of Perelman.

THEOREM 3.2 (Perelman [**80**]). *Let $g_{ij}(x, t)$, $0 \le t \le T$, be a smooth solution to the Ricci flow on an n-dimensional compact manifold M. Then there exists a constant $\kappa > 0$ depending only on T and the initial metric such that the following holds: if $r_0 \le \sqrt{T}$ and $|Rm|(x, t_0) \le r_0^{-2}$ on $B_{t_0}(x_0, r_0)$, then*

$$vol_{t_0}(B_{t_0}(x_0, r_0)) \ge \kappa r_0^n.$$

Indeed, let ξ be a smooth nonnegative non-increasing function, which is 1 on $(-\infty, \frac{1}{2}]$ and 0 on $[\frac{3}{4}, \infty)$. Substituting

$$u = (4\pi r_0^2)^{-\frac{n}{2}} e^{-f} = \frac{\xi\left(\frac{d_{t_0}(x_0,x)}{r_0}\right)}{\int_M \xi\left(\frac{d_{t_0}(x_0,x)}{r_0}\right) dv_{t_0}}$$

into (3.1), we have

$$W(g_{ij}(t_0), f, r_0^2) \le C(n) + \log \frac{vol_{t_0}(B_{t_0}(x_0, r_0))}{r_0^n}.$$

By Lemma 3.1, we have

$$W(g_{ij}(t_0), f, r_0^2) \geq \mu(M, g_{ij}(0), r_0^2 + t_0).$$

Note that the right hand side is controlled by the log Sobolev constant of the initial metric (on the scales $\leq \sqrt{2T}$). This proves the noncollapsing theorem.

3.2. Perelman's Reduced Volume Approach. There is another way to get the noncollapsing result by establishing the comparison geometry to the $\mathfrak{L}$-length introduced in (3.5). Moreover, this comparison geometric approach could be adapted to get the noncollapsing for *surgical* solutions. We now discuss this approach.

In [**80**], Perelman introduced the $\mathfrak{L}$-length defined in (3.5) as a suitable renormalized distance function on potentially infinite dimensional space-time manifold, where the Ricci flow was embedded there. More explicitly, let $\frac{\partial g_{ij}}{\partial \tau} = 2Ric$ be a solution to the Ricci flow on M with $\tau = T - t$. Consider

$$\tilde{M} = M \times \mathbb{S}^N \times \mathbb{R}^+$$

with the following metric

$$\begin{aligned}
\tilde{g}_{ij} &= g_{ij}, \\
\tilde{g}_{\alpha\beta} &= \tau g_{\alpha\beta}, \\
\tilde{g}_{oo} &= \frac{N}{2\tau} + R, \\
\tilde{g}_{i\alpha} &= \tilde{g}_{io} = \tilde{g}_{\alpha o} = 0,
\end{aligned}$$

where i, j are coordinate indices on M, α, β are coordinate indices on $\mathbb{S}^N$ and the coordinate τ on R^+ has index o. The metric $g_{\alpha\beta}$ on $\mathbb{S}^N$ has constant sectional curvature $\frac{1}{2N}$. This construction may be viewed as a "regularization" of what Chow-Chu did in [**27**]. Perelman twisted the sign of the time and coupled the space-time with a solution to the Ricci flow with positive curvatures on manifolds of very big dimensions. As we mentioned before, Chu and Chow [**27**] found a geometric interpretation of Li-Yau-Hamilton inequality. So the following proposition of Perelman is not surprising.

PROPOSITION 3.3. *The components of the curvature tensor of the metric $\tilde{g}$ coincide (module N^{-1}) with the components of the Li-Yau-Hamilton quadratic.*

The key observation due to Perelman is the following

PROPOSITION 3.4 (Perelman [**80**]). *The Ricci curvature of the metric $\tilde{g}$ is flat (module N^{-1}), i.e. $|\tilde{Ric}|_{\tilde{g}} = O\left(\frac{1}{N}\right)$.*

Recall that we have Bishop-Gromov volume comparison theorem on manifolds with Ricci curvature bounded from below. The above proposition 3.4 motivated an important monotonicity formula: the reduced volume.

Actually, by looking at the $\tilde{g}$ length of a space time curve $\gamma(\tau)$, $0 \le \tau \le \bar{\tau}$,

$$\begin{aligned}&\int_0^{\bar{\tau}} \sqrt{\left(\frac{N}{2\tau}+R\right)+|\dot{\gamma}(\tau)|^2_{g_{ij}(\tau)}}d\tau\\ &\quad = \sqrt{2N\bar{\tau}} + \frac{1}{\sqrt{2N}}\int_0^{\bar{\tau}} \sqrt{\tau}(R+|\dot{\gamma}(\tau)|^2_{g_{ij}})d\tau + O(N^{-\frac{3}{2}}),\end{aligned}$$

we find the expression of $\mathcal{L}$ distance $\int_0^{\bar{\tau}} \sqrt{\tau}(R+|\dot{\gamma}(\tau)|^2_{g_{ij}})d\tau$. By computing the volumes of geodesic spheres of radii $\sqrt{2N\bar{\tau}}$ on $\tilde{M}$ and $\mathbb{R}^{n+N+1}$, we have

$$\frac{Vol(S_{\tilde{M}}(\sqrt{2N\bar{\tau}}))}{Vol(S_{R^{n+N+1}}(\sqrt{2N\bar{\tau}}))} \approx const \cdot N^{-\frac{n}{2}} \cdot \int_M (\bar{\tau})^{-\frac{n}{2}} \exp\left\{-\frac{1}{2\sqrt{\bar{\tau}}}L(x,\bar{\tau})\right\} dV_M.$$

Proposition 3.4 indicates that the quantity

$$\int_M (\bar{\tau})^{-\frac{n}{2}} \exp\{-\frac{1}{2\sqrt{\bar{\tau}}}L(x,\bar{\tau})\}dV_M$$

should be non-increasing in $\bar{\tau}$. This quantity is called *Perelman's reduced volume* and we denote it by $V(\bar{\tau})$.

The rigorous proof of this monotonicity can be obtained in the following way. One computes the first and second variation for the $\mathcal{L}$-length (3.5) to get

LEMMA 3.5 (Perelman [**80**]). *For the reduced distance $l(q,\bar{\tau})$ defined in (3.6), there hold*

$$\frac{\partial l}{\partial \bar{\tau}} = -\frac{l}{\bar{\tau}} + R + \frac{1}{2\bar{\tau}^{3/2}}K \tag{3.10}$$

$$|\nabla l|^2 = -R + \frac{l}{\bar{\tau}} - \frac{1}{\bar{\tau}^{3/2}}K \tag{3.11}$$

$$\Delta l \le -R + \frac{n}{2\bar{\tau}} - \frac{1}{2\bar{\tau}^{3/2}}K. \tag{3.12}$$

where

$$K = \int_0^{\bar{\tau}} \tau^{\frac{3}{2}} Q(X) d\tau,$$

and

$$Q(X) = -R_\tau - \frac{R}{\tau} - 2<\nabla R, X> + 2Ric(X,X)$$

is the trace Li-Yau-Hamilton quadratic. Moreover, the equality in (3.12) holds if and only if the solution along the $\mathfrak{L}$ minimal geodesic γ satisfies the gradient soliton equation

$$R_{ij} + \frac{1}{2\sqrt{\bar{\tau}}}\nabla_i\nabla_j L = \frac{1}{2\bar{\tau}} g_{ij}.$$

The combination of (3.10)-(3.12) gives

$$\left(\frac{\partial}{\partial\tau} - \triangle + R\right)((4\pi\bar{\tau})^{-\frac{n}{2}} e^{-l}) \leq 0. \tag{3.13}$$

If the manifold is compact, then by integrating, the reduced volume satisfies

$$\frac{d}{d\bar{\tau}} V(\bar{\tau}) = \frac{d}{d\bar{\tau}} \int_M (4\pi\bar{\tau})^{-\frac{n}{2}} e^{-l} dV_{\bar{\tau}}(q) \leq 0, \tag{3.14}$$

and equality holds if and only if we are on a shrinking gradient soliton.

To carry the monotonicity to noncompact manifolds, Perelman [**80**] established a Jacobian comparison for the exponential map associated to the $\mathfrak{L}$-length. From the $\mathfrak{L}$-length, one defines an $\mathfrak{L}$-exponential map (with parameter $\bar{\tau}$) $\mathfrak{L}exp(\bar{\tau}) : T_pM \to M$ as follows: for any $X \in T_pM$, set $\mathfrak{L}exp_X(\bar{\tau}) = \gamma(\bar{\tau})$, where γ is an $\mathfrak{L}$-geodesic satisfying $\gamma(0) = p$ and $\lim_{\tau\to 0}\sqrt{\tau}\dot{\gamma}(\tau) = X$. Let $\mathfrak{J}(\tau)$ be the Jacobian of the $\mathfrak{L}$-exponential map along $\gamma(\tau)$, $0 \leq \tau \leq \bar{\tau}$. Then by the standard computation of Jacobi fields, we obtain

$$\frac{d}{d\tau}\log \mathfrak{J}(\tau) = \Delta l + R$$

along any minimal $\mathfrak{L}$-geodesic γ. Combining with equations (3.10)-(3.12) in the Lemma (3.5), this gives

THEOREM 3.6 (Perelman's Jacobian comparison [**80**]). *Along any minimal $\mathfrak{L}$-geodesic γ, we have*

$$\frac{d}{d\tau}\{(4\pi\tau)^{-\frac{n}{2}} exp(-l(\tau))\mathfrak{J}(\tau)\} \leq 0. \tag{3.15}$$

Consequently, we obtain

THEOREM 3.7 (Monotonicity of the Perelman's reduced volume). *Let g_{ij} be a family of complete metrics evolving by the Ricci flow $\frac{\partial}{\partial\tau} g_{ij} = 2R_{ij}$ on a manifold M with bounded curvature. Fix a point p in M and let $l(q,\tau)$ be the reduced distance from $(p, 0)$. Then*

(i) *Perelman's reduced volume*

$$\tilde{V}(\tau) = \int_M (4\pi\tau)^{-\frac{n}{2}} exp(-l(q,\tau)) dV_\tau(q)$$

is finite and nonincreasing in τ;

(ii) *the monotonicity is strict unless we are on a gradient shrinking soliton.*

Now we are going to use the reduced volume to derive a slightly weaker version of Theorem 3.2. The advantage of this new method is that it allows to be adapted to the case that the solutions are only locally defined. This will be extremely important in the analysis of surgical solutions.

DEFINITION 3.8. *We say a solution to the Ricci flow is κ-noncollapsed at (x_0, t_0) on the scale r for positive constants κ and r if it satisfies the following property: if $|Rm|(x,t) \leq r^{-2}$ for all $x \in B_{t_0}(x_0, r)$ and $t \in [t_0 - r^2, t_0]$, then we have*

$$vol_{t_0}(B_{t_0}(x_0, r)) \geq \kappa r^n.$$

THEOREM 3.9 (Perelman [**80**]). *Let (M^n, g_{ij}) be a complete Riemannian manifold with bounded curvature $|Rm| \leq k_0$ and with injectivity radius bounded from below by $inj(M, g_{ij}) \geq i_0$. Let $g_{ij}(x,t)$, $t \in [0,T)$ be a smooth solution to the Ricci flow with bounded curvature for each $t \in [0,T)$ and $g_{ij}(x,0) = g_{ij}(x)$. Then there is a $\kappa > 0$ depending only on k_0, i_0 and T such that the solution is κ-noncollapsed on scales $\leq \sqrt{T}$.*

A sketch of the proof is given as follows. Argue by contradiction. Suppose $|Rm|(x,t) \leq r^{-2}$ for all $x \in B_{t_0}(x_0, r)$ and $t \in [t_0 - r^2, t_0]$, but

$$\frac{vol_{t_0}(B_{t_0}(x_0, r))}{r^n} = \varepsilon^n$$

is very small. Write

(3.16)

$$\begin{aligned}
\tilde{V}(\varepsilon r^2) &= \int_M (4\pi\varepsilon r^2)^{-\frac{n}{2}} \exp(-l(q, \varepsilon r^2)) dV_{t_0 - \varepsilon r^2}(q) \\
&\leq \int_{\mathcal{L}exp_{\{|v| \leq \frac{1}{4}\varepsilon^{-1/2}\}}(\varepsilon r^2)} (4\pi\varepsilon r^2)^{-\frac{n}{2}} \exp(-l(q, \varepsilon r^2)) dV_{t_0 - \varepsilon r^2}(q) \\
&\quad + \int_{\mathcal{L}exp_{\{|v| > \frac{1}{4}\varepsilon\varepsilon^{-1/2}\}}(\varepsilon r^2)} (4\pi\varepsilon r^2)^{-\frac{n}{2}} \exp(-l(q, \varepsilon r^2)) dV_{t_0 - \varepsilon r^2}(q) \\
&\leq I + II.
\end{aligned}$$

First of all, it can be shown $\mathcal{L}exp_{\{|X| \leq \frac{1}{4}\varepsilon^{-\frac{1}{2}}\}}(\varepsilon r^2) \subset B_{t_0}(x_0, r)$, and $l(q, \varepsilon r^2) \geq -C(n)\varepsilon$ on $B_{t_0}(x_0, r)$. This implies $I \leq C(n)\varepsilon^{-\frac{n}{2}}\varepsilon^n = C(n)\varepsilon^{\frac{n}{2}}$. By the monotonicity (3.15) of $\mathcal{L}$-Jacobian, one has

$$II \leq \int_{\{|X| \geq \frac{1}{4}\varepsilon^{-\frac{1}{2}}\}} (4\pi)^{-\frac{n}{2}} exp(-|X|^2) dX \leq \varepsilon^{\frac{n}{2}},$$

hence

$$\tilde{V}(\varepsilon r^2) \leq C(n)\varepsilon^{\frac{n}{2}}. \tag{3.17}$$

On the other hand, by Lemma (3.5),

$$\frac{\partial \bar{L}}{\partial \tau} + \triangle \bar{L} \leq 2n \tag{3.18}$$

where $\bar{L}(q,\tau) = 4\tau l(q,\tau)$. It follows from maximum principle that there is a point $q_0 \in M$ such that $l(q_0, t_0 - \frac{1}{C(n)k_0}) \leq \frac{n}{2}$. Since the geometry is controlled on $B_0(q_0,1) \times [0, \frac{1}{C(n)k_0}]$, one then has $l(q,t_0) \leq Const.$ on $B_0(q_0,1)$, which implies

$$\tilde{V}(t_0) \geq \int_{B_0(P_0,1)} (4\pi t_0)^{-\frac{n}{2}} e^{-l} dv \geq Const. > 0.$$

This contradicts with the monotonicity of the reduced volume when ε is small enough.

4. The Formation of Singularities

Given a compact Riemannian manifold (M,g), we evolve the metric by the Ricci flow

$$\frac{\partial g_{ij}}{\partial t} = -2R_{ij}.$$

We say a solution $g_{ij}(x,t)$, $t \in [0,T)$ is a *maximal solution* to the Ricci flow with $g_{ij}(x,0) = g_{ij}(x)$, if either $T = \infty$, or $T < \infty$ and the curvature becomes unbounded as $t \to T$. If T is finite, we say the solution develops singularities at the time T. In the early 90's, Hamilton initiated the program to investigate the formation of singularities.

4.1. Hamilton's Compactness Theorem. To understand the structure of a singularity, similar to the study of minimal surface theory and harmonic map theory, one tries to dilate the solution around the singularity and then take a limit of the rescaled sequence of solutions. In order to do this, a compactness theorem for solutions to the Ricci flow is needed.

The standard compactness theorems for Riemannian manifolds in $C^{1,\alpha}$ norm are available by the works of Gromov [**38**], Peters [**83**], Greene-Wu [**35**], etc. Thanks to Shi's derivative estimates (Theorem 1.4), we know that all the derivatives of curvature are guaranteed to be bounded once curvature is bounded for any solution to the Ricci flow. Based on this fact, Hamilton [**46**] established a C^∞ compactness theorem for solutions of the Ricci flow. A slight generalization of Hamilton's compactness theorem is given by below.

THEOREM 4.1 (Hamilton [**46, 47**]; see also [**14**]). *Let $(M_k, g_k(t), p_k)$, $t \in (A,\Omega]$ with $A < 0 \leq \Omega$, be a sequence of evolving marked complete Riemannian manifolds. Consider a sequence of geodesic balls $B_0(p_k, s_k) \subset M_k$ of* *radii* $s_k (0 < s_k \leq +\infty)$*, with* $s_k \to s_\infty (\leq +\infty)$*, around the base points* p_k *in the metrics* $g_k(0)$*. Suppose each* $g_k(t)$ *is a solution to the Ricci flow on* $B_0(p_k, s_k) \times (A, \Omega]$*. Suppose also*

(i) *for every radius* $r < s_\infty$ *there exist positive constants* $C(r)$ *and* $k(r)$ *independent of* k *such that the curvature tensors* $Rm(g_k)$ *of the evolving metrics* $g_k(t)$ *satisfy the bound*

$$|Rm(g_k)| \leq C(r)$$

on $B_0(p_k, r) \times (A, \Omega]$ *for all* $k \geq k(r)$*, and*

(ii) *there exists a constant* $\delta > 0$ *such that the injectivity radii of* M_k *at* p_k *in the metric* $g_k(0)$ *satisfy the bound*

$$inj(M_k, p_k, g_k(0)) \geq \delta > 0$$

for all $k = 1, 2, \ldots$.

Then there exists a subsequence of $(B_0(p_k, s_k), g_k(t), p_k)$ *over* $t \in (A, \Omega]$ *which converges in* C^∞_{loc} *topology to a solution* $(B_\infty, g_\infty(t), p_\infty)$ *over* $t \in (A, \Omega]$ *to the Ricci flow, where, at the time* $t = 0$*,* B_∞ *is a geodesic open ball centered at* $p_\infty \in B_\infty$ *with the radius* s_∞*. Moreover the limiting solution is complete if* $s_\infty = +\infty$*.*

4.2. Hamilton's Classification of Singularities. In [**47**], Hamilton divided all maximal solutions, according to the blow-up rate of maximal curvatures $K_{\max}(t) := \sup_{x \in M} |Rm|(x, t)$, into three types:

Type I: $T < +\infty$ and $\sup_{t \in [0,T)} (T - t) K_{max}(t) < +\infty$;

Type II: (a) $T < +\infty$ but $\sup_{t \in [0,T)} (T - t) K_{max}(t) = +\infty$;
(b) $T = +\infty$ but $\sup_{t \in [0,T)} t K_{max}(t) = +\infty$;

Type III: (a) $T = +\infty, \sup_{t \in [0,T)} t K_{max}(t) < +\infty$, and

$$\limsup_{t \to +\infty} t K_{max}(t) > 0;$$

(b) $T = +\infty, \sup_{t \in [0,T)} t K_{max}(t) < +\infty$, and

$$\limsup_{t \to +\infty} t K_{max}(t) = 0.$$

To understand the structure of a singularity, one can follow Hamilton in [**47**] by first picking a sequence of space-time points (x_k, t_k) which approach the singularity, then rescaling the solution around these points so that the norm of the curvature of each rescaled solution in the sequence is bounded by 2 everywhere and equal to 1 at the chosen points. (Such space-time points (x_k, t_k) are called *almost maximal points* to the maximal solution). The noncollapsing theorem of Perelman in the previous section gives the desired injectivity radius estimate (ii) for the rescaled sequence of solutions.

Thus one can apply Hamilton's compactness theorem to take a limit and conclude that any rescaling limit must be one of the singularity models in the following sense.

DEFINITION 4.2 (Hamilton [**47**]). *A solution $g_{ij}(x,t)$ to the Ricci flow on the manifold M, where either M is compact or at each time t the metric $g_{ij}(\cdot,t)$ is complete and has bounded curvature, is called a* **singularity model** *if it is not flat and of one of the following three types:*

***Type I**: The solution exists for $t \in (-\infty, \Omega)$ for some constant Ω with $0 < \Omega < +\infty$ and*

$$|Rm| \leq \Omega/(\Omega - t)$$

everywhere with equality somewhere at $t = 0$;

***Type II**: The solution exists for $t \in (-\infty, +\infty)$ and*

$$|Rm| \leq 1$$

everywhere with equality somewhere at $t = 0$;

***Type III**: The solution exists for $t \in (-A, +\infty)$ for some constant A with $0 < A < +\infty$ and*

$$|Rm| \leq A/(A + t)$$

everywhere with equality somewhere at $t = 0$.

In the special case of nonnegative curvatures, the singularity models of Type II and III can be further characterized as Ricci solitons.

THEOREM 4.3.

*(i) (Hamilton [**45**]) Any Type II singularity model with nonnegative curvature operator and positive Ricci curvature must be a (steady) Ricci soliton.*

*(ii) (Chen-Zhu [**21**]) Any Type III singularity model with nonnegative curvature operator and positive Ricci curvature must be a homothetically expanding Ricci soliton.*

*(iii) (Cao [**10**]) Any Type II or III singularity model on a Kähler manifold with nonnegative holomorphic bisectional curvature and positive Ricci curvature must be a steady Kähler-Ricci soliton or an expanding Kähler-Ricci soliton.*

For Type I singularity models, N. Sesum [**90**] obtained the following characterization in the compact case.

THEOREM 4.4. *Let $(M, g_{ij}(x,t))$ be a compact Type I singularity model obtained as a rescaling limit of Type I maximal solution. Then $(M, g_{ij}(x,t))$ must be a (non-flat) gradient shrinking Ricci soliton.*

Very recently Naber [**72**] showed that a suitable rescaling limit of any Type I maximal solution is a gradient shrinking soliton. However, it is still an interesting question when the rescaling limit must be non-flat.

In recent years, there have been some research activities on the question what kind of singularity models can be realized by the Ricci flow. We have seen from the Differential Sphere Theorems obtained in [**41, 42, 8**], and [**6**] that manifolds with positive curvatures (positive Ricci curvature in dimension 3, positive or two-positive curvature operator and 1/4-pinch in dimensions greater than 3) always develop *spherical Type I* singularities in the sense the singularity model is the round sphere. Apart from the spherical Type I singularities, there should exist a *necklike Type I* singularity in the sense the singularity models are the round cylinders. The existence of necklike Type I singularities was first demonstrated by M. Simon [**93**] on noncompact warped product $\mathbb{R}\times_f \mathbb{S}^n$. Later, Feldman-Ilmanen-Knopf [**33**] also found such necklike Type I singularities on some noncompact Kähler manifold, the total space of certain holomorphic line bundle L^{-k} over the complex projective space $\mathbb{CP}^n$. The existence of neckpinch Type I singularities on compact manifolds was recently proved by S. Angenent and D. Knopf [**2**] on $\mathbb{S}^{n+1}$ with suitable rotationally symmetric metrics. It is also interesting to see if a Type II singularity could be really formed in the Ricci flow. In [**31**], Daskalopoulos and Hamilton showed that a Type II singularity can be developed by the Ricci flow on the noncompact $\mathbb{R}^2$. The intuition of forming a Type II singularity on compact manifolds was described by Hamilton [**47**] (see also [**28**] and [**99**]) and the existence of a Type II singularity on compact manifolds was also proposed as an open question in the introduction of the book of Chow-Lu-Ni [**29**]. Most recently, Hui-Ling Gu and the last author [**40**] extended some arguments of Perelman to higher dimensions so as to show that a Type II singularity can be formed by the Ricci flow on $\mathbb{S}^n$ with suitable rotationally symmetric metric for all $n\geq 3$.

4.3. Ancient κ-solutions. Once we have a basic understanding for those singularities developed by almost maximum points, we now want to consider those singularities which might not come from almost maximum points. If we are considering a general singularity developed by the Ricci flow in a finite time on a compact manifold, then any rescaling limit around the singularity will define at least on $(-\infty,0)$, called an *ancient solution*. Moreover, by Perelman's noncollapsing result, there is some positive constant κ so that the rescaling limit is κ-noncollapsing for all scales. So any rescaling limit for singularities developed by the Ricci flow on compact manifolds is κ-noncollapsing and defined at least on the time interval $(-\infty,0)$. Up to now, all understandings to these rescaling limits are restricted on the class that have nonnegative curvature operators. That is, according to Perelman [**80**], we only consider *ancient κ-solutions*, i.e., each of them is defined on $(-\infty,0)$, has bounded and nonnegative curvature operators and is κ-noncollapsing for all scales for some $\kappa>0$. Notice, by Hamilton-Ivey pinching estimate and Perelman's noncollapsing result, that any rescaling limit of the Ricci flow on compact three-manifolds is an ancient κ-solution for some $\kappa > 0$.

The main purpose of this section is to review the properties of ancient κ-solutions and eventually to get a rather complete understanding to the ancient κ-solutions in certain low dimension cases. Firstly, two-dimensional ancient κ-solutions have been completely classified by Hamilton [**47**].

THEOREM 4.5. *Any two-dimensional κ-noncollapsing non-flat ancient solution must be either the round sphere $\mathbb{S}^2$ or the round real projective space $\mathbb{RP}^2$*

In fact, Hamilton [**47**] proved a somewhat stronger result: any two-dimensional complete non-flat ancient soltion of bounded curvature must be the round sphere $\mathbb{S}^2$, the round real project space $\mathbb{RP}^2$, or the cigar soliton. Note that the cigar soliton does not satisfy the κ-noncollapsing property for large scales.

Three-dimensional ancient κ-solutions have not yet been completely classified. Nevertheless, Perelman obtained a complete classification to a special class of three-dimensional ancient κ-solutions – the shrinking gradient solitons.

LEMMA 4.6 (Perelman [**81**]). *Let $(M, g_{ij}(t))$ be a nonflat gradient shrinking soliton to the Ricci flow on a three-manifold. Suppose $(M, g_{ij}(t))$ has bounded and nonnegative sectional curvature and is κ-noncollapsed on all scales for some $\kappa > 0$. Then $(M, g_{ij}(t))$ is one of the followings:*

(i) the round three-sphere $\mathbb{S}^3$, or its metric quotients;
(ii) the round infinite cylinder $\mathbb{S}^2 \times \mathbb{R}$, or its $\mathbb{Z}_2$ quotients.

Perelman's proof is based on the investigation of the shrinking soliton equation

$$R_{ij} + f_{ij} + \frac{g_{ij}}{2t} = 0, \qquad t < 0.$$

By applying Hamilton's strong maximum principle, one can easily characterize the shrinking soltions as either the round three-sphere $\mathbb{S}^3$, or the round infinite cylinder $\mathbb{S}^2 \times \mathbb{R}$ or a metric quotient of them, except the case when the soliton is noncompact and has positive sectional curvature everywhere. We now briefly describe Perelman's arguments in excluding the possibility of such noncompact 3-dimensional solitons with positive sectional curvature.

Consider the metric at $t = -1$. By investigating the soliton equation and the second variation formula, we find that $f(x) \approx \frac{1}{4}d^2(x, x_0)$ and $|\nabla f|^2 \approx \frac{1}{4}d^2(x, x_0)$. From $\nabla_i R = 2R_{ij}\nabla_j f$, we know R is increasing along the integral curves of the potential function f. It is not hard to see that the solution at infinity splits off a line $\mathbb{R}$. By comparing the existence time of the Ricci flow on standard $\mathbb{S}^2$, we find $\bar{R} = \limsup_{d_{-1}(x,x_0)} R(x, -1) \leq 1$. Consider the area of level sets of f, we have

$$\begin{aligned}\frac{d}{dt}area\{f=a\} &= \int_{\{f=a\}} div\left(\frac{\nabla f}{|\nabla f|}\right)\\ &\geq \int_{\{f=a\}} \frac{1}{|\nabla f|}(1-R) \geq \frac{1-\bar{R}}{2\sqrt{a}} area\{f=a\}.\end{aligned}$$

This forces $\bar{R}=1$ and the area of $\{f=a\}$ is increasing to the area of $\mathbb{S}^2$ with constant curvature $\frac{1}{2}$. On the other hand, by the Gauss equation and the soliton equation, the intrinsic curvature of $\{f=a\}(a \gg 1)$ can be computed as

$$K = R_{1212} + \frac{det(Hess(f))}{|\nabla f|^2} < \frac{1}{2},$$

which is a contradiction with the Gauss-Bonnet formula.

Remark: The above Perelman's result has been improved by Ni-Wallach [**77**] and Naber [**72**] in which they dropped the assumption on κ-noncollapsing condition and replaced nonnegative sectional curvature by nonnegative Ricci curvature. In addition, Ni-Wallach [**77**] can allow the curvature to grow as fast as $e^{ar^2(x)}$, where $r(x)$ is the distance function and a is a suitable small positive constant. In particular, Ni-Wallach's result implies that any 3-dimensional noncompact non-flat gradient shrinking soliton with nonnegative Ricci curvature and with curvature not growing faster than $e^{ar^2(x)}$ must be a quotient of the round infinite cylinder $\mathbb{S}^2 \times \mathbb{R}$. Now using the work of the second author in [**19**], we can further improve this latter result of Ni-Wallach as follows.

PROPOSITION 4.7. *Let (M^3, g_{ij}) be a 3-dimensional complete noncompact non-flat shrinking gradient soliton. Then (M^3, g_{ij}) is a quotient of the round neck $\mathbb{S}^2 \times \mathbb{R}$.*

PROOF. In view of the result of Ni-Wallach mentioned above, it suffices to show that our shrinking gradient soliton in fact has nonnegative Ricci curvature and satisfies the growth restriction on curvature.

First of all, by the work of the second author (see Corollary 2.4 of [**19**]), we know that the sectional curvature of g_{ij} must be nonnegative.

Next we claim that the scalar curvature, hence the curvature tensor, of g_{ij} grows at most quadratically in distance. Indeed, from the shrinking soliton equation

$$R_{ij} + f_{ij} - \frac{1}{2}g_{ij} = 0,$$

it is not hard to see

$$R_{jl}\nabla_l f = \frac{1}{2}\nabla_j R$$

and

$$R + |\nabla f|^2 - f = Const.$$

It then follows that $|\nabla f|^2 \leq f + Const$, because R is nonnegative. Thus, we obtain

$$|\nabla \sqrt{|f|+1}| \leq Const,$$

and hence

$$|f|(x) \leq C(d(x, x_0)^2 + 1).$$

Therefore,

$$R(x) \leq C(d(x, x_0)^2 + 1).$$

This completes the proof of the proposition. □

Clearly Perelman's argument using the Gauss-Bonnet formula imposes a restriction on the dimension. Thus an interesting open question is whether a similar classification of non-negatively curved shrinking solitons holds in higher dimensions. For $n=4$, Ni and Wallach [**78**] showed that any 4-dimensional complete gradient shrinking soliton with nonnegative curvature operator and positive isotropic curvature, satisfying certain additional assumptions, is either a quotient of $\mathbb{S}^4$ or a quotient of $\mathbb{S}^3 \times \mathbb{R}$. Based on this result of Ni-Wallach, Naber [**72**] proved that

PROPOSITION 4.8 (Naber [**72**]). *Any 4-dimensional complete noncompact shrinking Ricci soliton with bounded nonnegative curvature operator is isometric to either* $\mathbb{R}^4$, *or a finite quotient of* $\mathbb{S}^3 \times \mathbb{R}$ *or* $\mathbb{S}^2 \times \mathbb{R}^2$.

For higher dimensions, Gu and the last author [**40**] proved that any complete, *rotationally symmetric*, non-flat, n-dimensional ($n \geq 3$) shrinking Ricci soliton with κ-noncollapsing on all scales and with bounded and nonnegative sectional curvature must be the round sphere $\mathbb{S}^n$ or the round cylinder $\mathbb{S}^{n-1} \times \mathbb{R}$. Subsequently, Kotschwar [**59**] proved a more general result that the only complete shrinking Ricci solitons (without curvature sign and bound assumptions) of rotationally symmetric metrics (on S^n, $\mathbb{R}^n$ and $\mathbb{R} \times S^{n-1}$) are, respectively, the round, flat, and standard cylindrical metrics. Ni-Wallach [**77**] and Petersen-Wylie [**86**] also proved a classification result on gradient shrinking solitons with vanishing Weyl curvature tensor which includes all the rotationally symmetric ones. For additional recent results on shrinking or expanding Ricci solitons, see the works of Petersen and Wylie [**84, 85**].

Let us come back to the discussion on general ancient κ-solutions. Given a three-dimensional ancient κ-solution, one can pick a suitable sequence of space-time points (x_k, t_k) with $t_k \to -\infty$ as in [**81**] and take a rescaling limit, usually called a blow-down limit. By using the monotonicity of the reduced volume, Perelman [**80**] showed that the blow-down limit is necessarily a shrinking Ricci soliton. Then, based on the above classification lemma (Lemma 4.6) for three-dimensional shrinking Ricci solitons and imitating the argument as in proving his noncollapsing result, Perelman [**81**] obtained the following important universal noncollapsing property for all three-dimensional ancient κ-solutions.

PROPOSITION 4.9 (Universal Noncollapsing [**81**]). *There exists a positive constant κ_0 with the following property. Suppose we have a non-flat three-dimensional ancient κ-solution for some $\kappa > 0$. Then either the solution is κ_0-noncollapsed on all scales, or it is a metric quotient of the round three-sphere.*

This universal noncollapsing property for three-dimensional ancient κ-solutions has been used indispensably by Perelman in [**81**] to prove the noncollapsing of surgical solutions to the Ricci flow with surgery. When extending the Hamilton-Perelman theory of three-dimensional Ricci flow with surgery to higher dimensions, one must meet the question how to verify the universal property to ancient κ-solutions. Due to the lack of complete classification of higher dimensional positively curved shrinking Ricci solitons, it is desirable to find an alternative way, without using the classification of shrinking Ricci solitons, to prove the universal noncollapsing property. Indeed, an alternative approach had been given by the last two authors in [**25**] to handle a class $\mathcal{C}$ of ancient κ-solutions without any knowledge of classification of gradient shrinking solitons. Roughly speaking, the class $\mathcal{C}$ contains all ancient κ-solutions where each of them at infinity splits as $\mathbb{S}^{n-1} \times \mathbb{R}$. In particular, all ancient three-dimensional κ-solutions and four-dimensional ancient κ-solutions with restrictive isotropic curvature pinching belong to this class $\mathcal{C}$. Here we say a four-dimensional ancient κ-solution satisfies *restricted isotropic curvature pinching* if there is some fixed $\Lambda > 0$ such that

$$a_3 \leq \Lambda a_1, \; c_3 \leq \Lambda c_1, \; b_3^2 \leq a_1 c_1,$$

where $Rm = \begin{pmatrix} A & B \\ {}^tB & C \end{pmatrix}$ is the usual block decomposition of curvature operator in dimension 4 and a_i, b_i, c_i are eigenvalues of the corresponding matrixes A, B, C. By Hamilton's pinching estimate in [**48**], such four-dimensional ancient κ-solutions with restricted isotropic curvature pinching appears naturally as the singularity models of Ricci flow on compact four-manifolds with positive isotropic curvature.

Dimension reduction is a useful approach to understand the structure of singularities in the theory of minimal surfaces or harmonic maps. In his survey paper [**47**], Hamilton systemically developed the dimension reduction method for the Ricci flow. From Hamilton's classification to two-dimensional ancient solutions, one observes that any two-dimensional complete ancient solution of bounded curvature cannot be of maximal volume growth. Based on this observation and by applying a dimension reduction argument, Perelman [**80**] proved

PROPOSITION 4.10 (Non-maximal Volume Growth). *Let M be an n-dimensional complete noncompact Riemannian manifold. Suppose $g_{ij}(x,t)$, $x \in M$ and $t \in (-\infty, T)$ with $T > 0$, is a nonflat ancient solution of the Ricci flow with a nonnegative curvature operator and bounded* *curvature. Then the asymptotic volume ratio of the solution metric satisfies*

$$\nu_M(t) = \lim_{r \to +\infty} \frac{Vol_t(B_t(O,r))}{r^n} = 0$$

for each $t \in (-\infty, T)$.

The same result for the Ricci flow on Kähler manifolds has been independently discovered by the last two authors in [**23**]. Moreover, for the Ricci flow on Kähler manifolds, it is proved by the last two authors and Tang [**20**] in complex dimension two and by Ni [**74**] for all dimensions that the nonnegative curvature operator condition can be replaced by the weaker condition of nonnegative holomorphic bisectional curvature.

By a standard rescaling argument, using the above non-maximal volume growth property, Perelman [**80**] got a local curvature bound of solutions in terms of local volume lower bound. Conversely, the noncollapsing estimate of Perelman says that local curvature bound can control the local volume lower bound. Hence the combination of these two facts would imply an elliptic type estimate, which allows one to compare the values of the curvatures at different points at the same time. Such an estimate was first implicitly given by Perelman in [**80**]. The following version is taken from [**14**] and [**25**].

PROPOSITION 4.11 (Elliptic Type Estimate). *There exist a positive constant* η *and a positive function* $\omega : [0, +\infty) \to (0, +\infty)$ *with the following properties. Suppose that* $(M, g_{ij}(t))$, $-\infty < t \leq 0$, *is a 3-dimensional ancient* κ*-solution or a 4-dimensional ancient* κ*-solution with restricted isotropic curvature pinching, for some* $\kappa > 0$. *Then*

(i) *for every* $x, y \in M$ *and* $t \in (-\infty, 0]$, *there holds*

$$R(x,t) \leq R(y,t) \cdot \omega(R(y,t)d_t^2(x,y));$$

(ii) *for all* $x \in M$ *and* $t \in (-\infty, 0]$, *there hold*

$$|\nabla R|(x,t) \leq \eta R^{\frac{3}{2}}(x,t) \text{ and } |R_t|(x,t) \leq \eta R^2(x,t).$$

Let us come back to consider three-dimensional ancient κ-solutions. In view of Hamilton's dimension reduction, each noncompact three-dimensional ancient κ-solution splits off a line at infinity. Then by combining the classification of two-dimensional ancient κ-solutions, we see that each noncompact non-flat three-dimensional ancient κ-solution is asymptotic to a round cylinder at infinity. On the other hand, by applying the universal noncollapsing Proposition 4.9 and the above elliptic type estimate Proposition 4.11, we know that the space of non-flat three-dimensional ancient κ-solutions is compact modulo scalings and the quotients of the round sphere $\mathbb{S}^3$. This compactness property and asymptotically cylindric property allow us to use a standard rescaling argument to get a canonical neighborhood property, due to Perelman [**81**], for three-dimensional ancient κ-solutions.

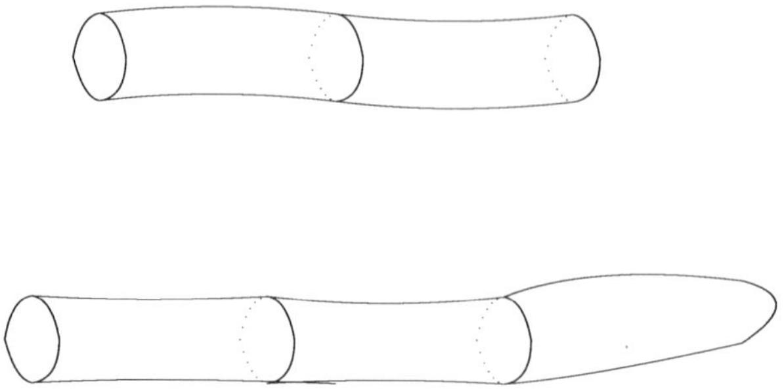

FIGURE 1. ε-neck and ε-cap.

Before stating the canonical neighborhood result, we introduce the terminologies of evolving ε-neck and ε-cap (see Figure 1).

Fix $\varepsilon > 0$. Let $g_{ij}(x,t)$ be a non-flat ancient κ-solution on a three-manifold M for some $\kappa > 0$. We say that a point $x_0 \in M$ is the *center of an evolving ε-neck* at $t=0$, if the solution $g_{ij}(x,t)$ in the set $\{(x,t)| -\varepsilon^{-2}Q^{-1} < t \leq 0, d_t^2(x,x_0) < \varepsilon^{-2}Q^{-1}\}$, where $Q = R(x_0,0)$, is, after scaling with factor Q, ε-close (in $C^{[\varepsilon^{-1}]}$ topology) to the corresponding set of the evolving round cylinder, having scalar curvature one at $t=0$. An evolving ε-cap is the time slice at the time t of an evolving metric on $\mathbb{B}^3$ or $\mathbb{RP}^3 \setminus \bar{\mathbb{B}}^3$ such that the region outside some suitable compact subset of $\mathbb{B}^3$ or $\mathbb{RP}^3 \setminus \bar{\mathbb{B}}^3$ is an evolving ε-neck.

THEOREM 4.12 (Canonical neighborhood theorem [**81**]). *For every sufficiently small $\varepsilon > 0$ one can find positive constants $C_1 = C_1(\varepsilon)$, $C_2 = C_2(\varepsilon)$ with the following property. Suppose we have a three-dimensional nonflat (compact or noncompact) ancient κ-solution $(M, g_{ij}(x,t))$. Then either the ancient solution is the round $\mathbb{RP}^2 \times \mathbb{R}$, or every point (x,t) has an open neighborhood B, with $B_t(x,r) \subset B \subset B_t(x,2r)$ for some $0 < r < C_1 R(x,t)^{-\frac{1}{2}}$, which falls into one of the following three categories:*

(a) B is an **evolving ε-neck**, *or*
(b) B is an **evolving ε-cap**, *or*
(c) B is a compact manifold (without boundary) with positive sectional curvature (thus it is diffeomorphic to the round three-sphere $\mathbb{S}^3$ or its metric quotients); furthermore, the scalar curvature of the ancient κ-solution in B at time t is between $C_2^{-1}R(x,t)$ and $C_2 R(x,t)$, and its volume in cases (a) and (b) satisfies

$$(C_2 R(x,t))^{-\frac{3}{2}} \leq Vol_t(B) \leq \varepsilon r^3.$$

Finally, we remark that this canonical neighborhood theorem has been extended by the last two authors **[25]** to all four-dimensional ancient κ-solutions with restrictive isotropic curvature pinching.

4.4. Singularity Structure Theorem. Let (M, g_{ij}) be a compact oriented three-manifold. Evolve the metric g_{ij} by the Ricci flow. Denote by $[0, T)$ the maximal time interval. Suppose $T < \infty$, then $\sup_{x\in M} |Rm|(x,t) \to \infty$ as $t \to T$. Let (x_k, t_k) be a sequence of almost maximal points, i.e. $\sup_{t\leq t_k} |Rm|(\cdot, t) \leq C|Rm|(x_k, t_k)$, $t_k \to T$, for some uniform constant C. Scale the solution around (x_k, t_k) with factor $Q_k = |Rm|(x_k, t_k)$ and shift the time t_k to 0. By applying Hamilton's compactness theorem, Perelman's local non-collapsing theorem, as well as Hamilton-Ivey pinching estimate, one can extract a convergent subsequence such that the limit is an oriented ancient κ-solution. Observe that $\mathbb{RP}^2 \times \mathbb{R}$ is excluded since it is not orientable. Consequently, for an arbitrarily given $\varepsilon > 0$, the solution around the points x_k and at times $t_k \to T$ have canonical neighborhoods which are either an ε-neck, or an ε-cap, or a compact positively curved manifold (without boundary). This gives the structure of singularities coming from a sequence of (almost) maximum points.

However the above argument does not work for singularities coming from a sequence of points (y_k, s_k) with $s_k \to T$ and $|Rm(y_k, s_k)| \to +\infty$ when $|Rm(y_k, s_k)|$ is not comparable with the maximum of the curvature at time s_k, since we cannot take a limit directly. To overcome this difficulty, Perelman **[80]** developed a refined blow up argument.

For convenience of stating the estimates, we may assume the initial data is normalized, namely, the norm of the curvature operator is less than $\frac{1}{10}$ and the volume of the unit ball is bigger than 1.

THEOREM 4.13 (Singularity structure theorem **[80]**). *Given $\varepsilon > 0$ and $T_0 > 0$, one can find $r_0 > 0$ with the following property. If $g_{ij}(x,t), x \in M$ and $t \in [0, T)$ with $1 < T \leq T_0$, is a solution to the Ricci flow on a compact oriented three-manifold M with normalized initial metric, then for any point (x_0, t_0) with $t_0 \geq 1$ and $Q = R(x_0, t_0) \geq r_0^{-2}$, the solution in $\{(x,t) \mid d_{t_0}^2(x, x_0) < \varepsilon^{-2}Q^{-1}, t_0 - \varepsilon^{-2}Q^{-1} \leq t \leq t_0\}$ is, after scaling by the factor Q, ε-close (in $C^{[\varepsilon^{-1}]}$-topology) to the corresponding subset of some oriented ancient κ-solution (for some $\kappa > 0$).*

We now would like to give a outline of the proof. The proof is divided into four steps. The first three steps are basically following the line given by Perelman in **[80]**; while the last step is an alternative argument which is taken from **[14]** or **[25]**.

The proof is an argument by contradiction. Suppose for some $\varepsilon > 0$, $T_0 > 1$, there exist a sequence of $r_k \to 0$, $1 < T_k \leq T_0$ and solutions $(M_k, g_k(\cdot, t))$, $t \in [0, T_k)$, satisfying the assumption of the theorem, but the conclusion of the theorem fails at some $x_k \in M_k$ and times $t_k \geq 1$

with $Q_k = R_k(x_k,t_k) \geq r_k^{-2}$. For each such solution, we adjust the point (x_k,t_k) so that the value of the curvature at (x_k,t_k) is as large as possible so that the conclusion of the theorem fails at (x_k,t_k), but holds for any $(x,t) \in M_k \times [t_k - H_k Q_k^{-1}, t_k]$ satisfying $R_k(x,t) \geq 2Q_k$, where $H_k = \frac{1}{4} r_k^{-2} \to +\infty$ as $k \to +\infty$.

Let $(M_k, \tilde{g}_k(\cdot,t), x_k)$ be the rescaled solutions obtained by rescaling $(M_k, g_k(\cdot,t))$ around x_k with the factors $Q_k = R_k(x_k,t_k)$ and shifting the time t_k to the new time zero. Denote by $\tilde{R}_k$ the rescaled scalar curvature. We will show that a subsequence of the rescaled solutions $(M_k, \tilde{g}_k(\cdot,t), x_k)$ converges in C^∞_{loc} topology to an ancient κ-solution. This will be a contradiction.

The argument is divided into four steps.

Step 1. First of all, we need a local bound on curvature.
For each $(\bar{x},\bar{t})$ with $t_k - \frac{1}{2}H_k Q_k^{-1} \leq \bar{t} \leq t_k$, we have

$$R_k(x,t) \leq 4\bar{Q}_k$$

whenever $\bar{t} - c\bar{Q}_k^{-1} \leq t \leq \bar{t}$ and $d^2_{\bar{t}}(x,\bar{x}) \leq c\bar{Q}_k^{-1}$, where $\bar{Q}_k = Q_k + R_k(\bar{x},\bar{t})$ and $c > 0$ is a small universal constant.

This result is a simple consequence of the gradient estimate (ii) in Proposition 4.11. Indeed, since any ancient κ-solution satisfies the gradient estimate

$$|\nabla R^{-\frac{1}{2}}| + |\frac{\partial}{\partial t} R^{-1}| \leq 2\eta, \tag{4.1}$$

the desired curvature bound follows directly from integrating the gradient estimate along a space-time path.

Step 2. This step is to show that the curvature of rescaled solution is bounded at bounded distance at time $t=0$. The detailed exposition to this step was first given by Kleiner-Lott in the first version of their notes [**56**]. The idea of the proof can be described as follows.

For all $\rho \geq 0$, set

$$M(\rho) = \sup\{\tilde{R}_k(x,0) \mid k \geq 1, x \in M_k \text{ with } d_0(x,x_k) \leq \rho\}$$

and

$$\rho_0 = \sup\{\rho \geq 0 \mid M(\rho) < +\infty\}.$$

By Hamilton-Ivey's pinching estimate, it suffices to show $\rho_0 = +\infty$. Still argue by contradiction. Suppose there is a sequence of points y_k so that the rescaled $\tilde{R}(y_k,0) \to +\infty$ and $\tilde{d}_0(x_k,y_k) \to \rho_0 > 0$. Connecting x_k and y_k with a minimal geodesic γ_k. By Step 1, Hamilton's compactness theorem 4.1 and Perelman's noncollapsing theorem, there is a convergent subsequence such that the limit has nonnegative sectional curvature on the ball of radius ρ_0. The curvature still blows up along the limiting geodesic γ_∞ by the gradient estimate for ancient κ-solutions. Then by the choice of the points x_k, one can show each such point on the limiting geodesic has a neck-like neighborhood. So, by adding the end point q_∞ to the limit geodesic, the union of the limiting space and the added point q_∞ has nonnegative curvature in Alexandrov space sense. By blowing up the tangent cone at the q_∞, we get a non-flat solution to the Ricci flow on the cone, which is a contradiction to Hamilton's strong maximum principle.

Since the curvature is bounded at time 0, by gradient estimate (4.1) and Hamilton's compactness theorem, one can show the limit solution is actually defined on the space-time open subset $\{(y,\tilde{t}) : y \in M_\infty, t \in [-\frac{1}{4}\eta^{-1} \tilde{R}_\infty(y)^{-1}, 0]\}$ containing $M_\infty \times \{0\}$.

Step 3. This step is to show that the limit $(M_\infty, \tilde{g}_\infty(\cdot,0), x_\infty)$ at the time slice $\{t=0\}$ has bounded curvature.

If the curvature is unbounded, by the virtue of Hamilton's dimension reduction, we can choose a sequence of points $q_j \to \infty$, and take a rescaled limit around q_j to get infinite number of tiny ε-necks. But this is a contradiction with the following basic geometry lemma, which was written down by the last two authors in [**25**].

LEMMA 4.14. *There exists a constant $\varepsilon_0 = \varepsilon_0(n) > 0$ such that every complete noncompact Riemannian manifold (M^n, g_{ij}) of nonnegative sectional curvature has a positive constant r_0 such that any ε-neck of radius r on (M^n, g_{ij}) with $\varepsilon \leq \varepsilon_0$ must have $r \geq r_0$.*

Here we call an open subset $N \subset M^n$ to be an *ε-neck of radius r* if $(N, r^{-2}g_{ij})$ is ε-close, in $C^{[\varepsilon^{-1}]}$ topology, to a standard neck $\mathbb{S}^{n-1} \times (-\varepsilon^{-1}, \varepsilon^{-1})$ where $\mathbb{S}^{n-1}$ has the scalar curvature 1.

As a consequence, the limit can be extended backward to some uniform interval $(-C, 0]$ for some $C > 0$.

Step 4. This step is to show the limit can be extended backward to $-\infty$.

Denote by

$$t' = \inf\{ \ \tilde{t} \ \mid \ \text{we can take a smooth limit on } (\tilde{t}, 0] \text{ from a subsequence of the rescaled solutions } \tilde{g}_k\}.$$

By the Li-Yau-Hamilton inequality, which must hold on the limit since the curvature is bounded by Step 1 and Step 3, and Hamilton's compactness theorem, one can show that there is a subsequence of the rescaled solutions $\tilde{g}_k$ which converges in C^∞_{loc} topology to a smooth limit $(M_\infty, \tilde{g}_\infty(\cdot,t))$ on the maximal time interval $(t', 0]$. We next claim that $t' = -\infty$.

Suppose not, then the curvature of the limit $(M_\infty, \tilde{g}_\infty(\cdot,t))$ becomes unbounded as $t \to t' > -\infty$. By applying the maximum principle, we see that the infimum of the scalar curvature is nondecreasing in time. Thus there exists some point $y_\infty \in M_\infty$ such that

$$\tilde{R}_\infty(y_\infty, t' + \frac{c}{3}) < \frac{3}{2}$$

where $c > 0$ is some universal small constant.

By using Step 1, we see that the limit $(M_\infty, \tilde{g}_\infty(\cdot, t))$ at a small neighborhood of the point $(y_\infty, t' + \frac{c}{3})$ extends backward to the time interval $[t' - \frac{c}{3}, t' + \frac{c}{3}]$. Moreover, one can show the distances at the time t and the time 0 are roughly equivalent in the following sense

$$d_t(x,y) \geq d_0(x,y) \geq d_t(x,y) - const. \tag{4.2}$$

This estimate ensures that the limit around the point y_∞ at any time $t \in (t', 0]$ is exactly the original limit around x_∞ at the time $t = 0$. By repeating the same arguments as in the above Step 2 and Step 3 to the solution around (y_k, t) for $t \in [t' - \frac{c}{3}, t' + \frac{c}{3}]$, we conclude the original limit $(M_\infty, \tilde{g}_\infty(\cdot, t))$ is actually well defined on the time slice $M_\infty \times \{t'\}$ and also has uniformly bounded curvature for all $t \in [t', 0]$. This is a contradiction.

Therefore the proof of the theorem is completed.

We remark that this singularity structure theorem had been extended by the last two authors in **[25]** to the Ricci flow on compact four-manifolds with positive isotropic curvature.

5. Ricci Flow with Surgery

In this section, we will discuss the surgery theory of the Ricci flow on three-dimensional manifolds. We also mention its extension to four-dimensional manifolds with positive isotropic curvature.

5.1. The Solution at the First Singular Time. Given any compact three-manifold M with an arbitrary Riemannian metric. By dilation, we may always assume that the metric is normalized so that the absolute values of the eigenvalues of its curvature operator at each point are bounded by $1/10$ and every geodesic ball of radius one has a volume of at least one. Let us evolve the normalized metric by the Ricci flow

$$\frac{\partial g_{ij}}{\partial t} = -2R_{ij},$$

and let $g(t)$, $t \in [0, T)$ be the maximal solution. If $T < \infty$, then curvature becomes unbounded as t tends to T, we say the maximal solution develops singularities as t tends to T and T is a singular time.

After obtaining the structure of points with suitably large curvature before the first singular time as in Theorem 4.13, we can give a clear picture of the solution near the singular time T as follows.

For the given $\varepsilon > 0$ and the maximal solution $(M, g_{ij}(\cdot, t))$ on $[0, T)$, with $T < \infty$, we can find $r_0 > 0$ depending only on T and ε such that each point (x, t), with $R(x, t) \geq r_0^{-2}$, admits a canonical neighborhood which is either an ε-neck, or an ε-cap, or a compact positively curved manifold (without boundary). In the last case the solution, by the well-known theorem of Hamilton in [**41**] (see also Theorem 2.1), becomes extinct at time T and the manifold M is diffeomorphic to the round three-sphere $\mathbb{S}^3$ or a metric quotient of $\mathbb{S}^3$.

Let Ω denote the set of all points in M where the curvature stays bounded as $t \to T$. If Ω is empty, then the solution becomes extinct at time T. In this case, either the manifold M is compact and positively curved, or it is entirely covered by ε-necks and ε-caps shortly before the maximal time T. So the manifold M is diffeomorphic to either $\mathbb{S}^3$, or a metric quotient of the round $\mathbb{S}^3$, or $\mathbb{S}^2 \times \mathbb{S}^1$, or $\mathbb{RP}^3 \# \mathbb{RP}^3$.

We now consider the case when Ω is nonempty. By using the local derivative estimates of Shi (Theorem 1.4), we see that as $t \to T$ the solution metric $g(t)$ has a smooth limit $\bar{g}$ on Ω. Let $\bar{R}$ denote the scalar curvature of $\bar{g}$. For any $0 < \rho < r_0$, let us consider the set

$$\Omega_\rho = \{x \in \Omega \mid \bar{R}(x) \leq \rho^{-2}\}.$$

First, we need some terminologies:

A metric on $\mathbb{S}^2 \times \mathbb{I}$, such that each point is contained in some ε-neck, is called an **ε-tube**, or an **ε-horn**, or a **double ε-horn**, if the scalar curvature stays bounded on both ends, or stays bounded on one end and tends to infinity on the other end, or tends to infinity on both ends, respectively (see Figure 2);

A metric on $\mathbb{B}^3$ or $\mathbb{RP}^3 \backslash \bar{\mathbb{B}}^3$ is called an **capped ε-horn** if each point outside some compact subset is contained in an ε-neck and the scalar curvature tends to infinity on the end (see Figure 3).

Now take any ε-neck in $(\Omega, \bar{g})$ and consider a point x on one of its boundary components. If $x \in \Omega \backslash \Omega_\rho$, then there is either an ε-cap or an ε-neck, adjacent to the initial ε-neck. In the latter case we can take a point on the boundary of the second ε-neck and continue. This procedure can either terminate when we get into Ω_ρ or an ε-cap, or go on indefinitely, producing an ε-horn. The same procedure can be repeated for the other boundary component of the initial ε-neck. Therefore, taking into account that Ω has no compact components, we conclude that each ε-neck of $(\Omega, \bar{g})$ is contained in a subset of Ω of one of the following types:

(a) an ε-tube with boundary components in Ω_ρ, or
(b) an ε-cap with boundary in Ω_ρ, or
(c) an ε-horn with boundary in Ω_ρ, or
(d) a capped ε-horn, or
(e) a double ε-horn.

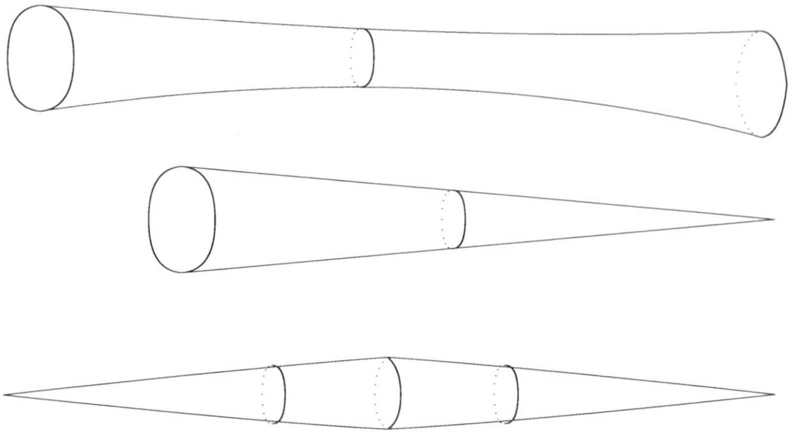

FIGURE 2. ε-tube, ε-horn and double ε-horn.

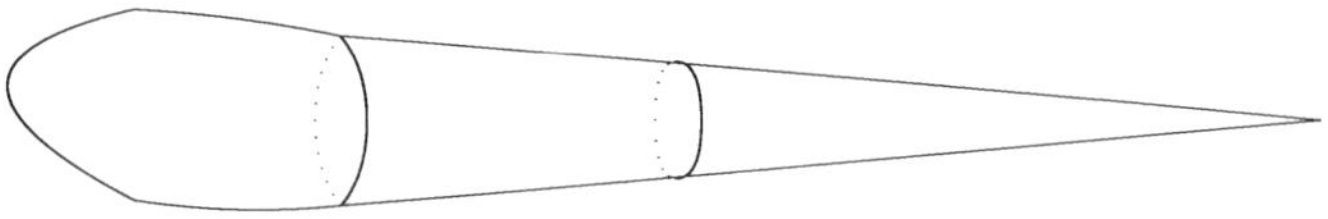

FIGURE 3. Capped ε-horn.

Similarly, each ε-cap of $(\Omega, \bar{g}_{ij})$ is contained in a subset of Ω of either type (b) or type (d).

It is clear that there is a definite lower bound (depending on ρ) on the volume of subsets of type (a), (b), and (c). So there can be only a finite number of them. Thus we conclude that there is only a finite number of components of Ω, containing points of Ω_ρ, and every such component has a finite number of ends, each being an ε-horn. On the other hand, every component of Ω containing no points of Ω_ρ is either a capped ε-horn, or a double ε-horn. If we look at the solution $g(t)$ at a slightly earlier time, the above argument shows that each ε-neck or ε-cap of $(M, g(t))$ is contained in a subset of type (a) or (b), while the ε-horns, capped ε-horns and double ε-horns (at the maximal time T) are connected together to form ε-tubes and ε-caps at any time t shortly before T (see Figure 4).

Let us denote by Ω_j, $1 \leq j \leq m$, the connected components of Ω which contain points of Ω_ρ. Then *the initial three-manifold M is diffeomorphic to a connected sum of $\bar{\Omega}_j$, $1 \leq j \leq m$, with a finite number of copies of $\mathbb{S}^2 \times \mathbb{S}^1$*

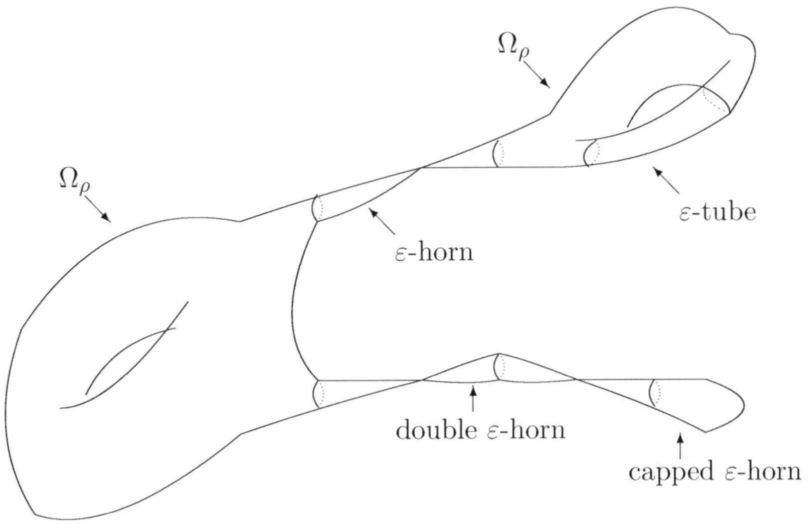

FIGURE 4. Solution at a maximal time.

(which correspond to gluing a tube to two boundary components of the same Ω_j*), and a finite number of copies of* $\mathbb{RP}^3$. Here $\bar{\Omega}_j$, $j=1,2,\ldots,m$, is the compact manifold (without boundary) obtained from Ω_j by taking an ε-neck in every ε-horn of Ω_j, cutting it along the middle two-sphere, removing the horn-shaped end, and gluing back a cap (or more precisely, a differentiable three-ball).

5.2. Definition of Surgical Solutions. We have seen that when the Ricci flow develops singularities, it gives a natural way to split the underlying manifold M into pieces $\Omega_1,\ldots,\Omega_m$ – the components of Ω containing points of Ω_ρ. Thus to capture the topology of M, one only needs to understand the topologies of the compact orientable three-manifolds $\bar{\Omega}_j$, $1 \le j \le m$, described above.

Let us evolve each $\bar{\Omega}_j$ by the Ricci flow again and, when the solution develops singularities, perform the above surgeries to get new compact orientable three-manifolds. By repeating this procedure, we will obtain a "weak" solution to the Ricci flow, called a solution to the *Ricci flow with surgery* or a *surgically modified solution* to the Ricci flow.

To get the topological information of the initial manifold M from the Ricci flow with surgery, we have to construct a surgically modified solution so that it has at most a finite number of surgeries at each finite interval and admits a well-understood long-time behavior. In this section, we only consider the question of how to construct a surgically modified solution with at most a finite number of surgeries at each finite interval.

Let us look at the above construction for surgical solutions in more detail. Arbitrarily fix a small positive constant ε. On the given compact orientable three-manifold M with a normalized Riemannian metric, we evolve the normalized metric by the Ricci flow to obtain a maximal solution defined on the maximal time interval $[0, t_1)$ with $t_1 < +\infty$. By the theorem on the structure of singularity, there exists a small positive constant r_0 such that any point (x, t) at which (the norm of) the curvature is greater than r_0^{-2} has a canonical neighborhood. Then, according to the above discussions, we can cut off canonical neighborhoods to get a new compact orientable (not necessarily connected) three-manifold M_1. Clearly, there are still some points, in the remaining parts near the surgery region, on M_1 at which the curvature are not less than r_0^{-2} and then we cannot expect that the metric of M_1 is still normalized. After evolving M_1 on a maximal time interval $[t_1, t_2)$ with $t_2 < +\infty$, we can only find canonical neighborhoods on the region where the curvature is at least r_0^{-4} (since, to apply the theorem on the structure of singularity, we have to dilate M_1 with a factor at least r_0^{-2}). By performing the surgery again, we get a compact orientable (not necessarily connected) three-manifold M_2 and there are still some points on M_2 with curvature not less than r_0^{-4}. By repeating this process, we will get a surgically modified solution on some time interval $[0, T_{\max})$ with the surgery times $0 < t_1 < \cdots < t_k < \cdots < T_{\max}$ such that at each $t_k, k = 1, 2, \ldots,$ the curvature is at least r_0^{-2k} somewhere. Intuitively, under this kind of surgery procedures, the curvatures would become higher and higher and the time intervals (t_{k-1}, t_k) become shorter and shorter. So, the surgery times of such constructed surgically modified solution are likely to accumulate in finite time. The trouble is basically caused by the inability to recognize the canonical neighborhoods on some fixed size of (high) curvature.

If one can improve the above surgery procedures so that there exists a *uniform* size on curvature to recognize canonical neighborhoods, then one will be able to cut down the solution so that its curvature never exceeds such a designed uniform size and hence each surgery will drop at least a fixed amount of volume. This, in turn, will prevent the surgery times from accumulating since one can easily show that the volume of the surgically modified solution can grow (in time) at most exponentially. So, what one really needs is to design a surgery procedure such that one can find a uniform positive function $r(t)$ on $[0, +\infty)$ so that any point (x, t) on the surgically modified solution at which the curvature is greater than $r(t)^{-2}$ has a canonical neighborhood. The theorem on the structure of singularity precisely ensures the existence of such a uniform function $r(t)$ for *smooth* solutions.

Thus, to prevent the accumulation of surgery times, we are led to construct surgically modified solutions which satisfy the following canonical neighborhood assumption (we refer the readers to Section 7.3 in [**14**] for precise definitions):

Canonical neighborhood assumption (with accuracy ε): There exists a nonincreasing positive function $r : [0, +\infty) \to (0, +\infty)$ such that at each time t, each point x, where the scalar curvature $R(x, t)$ is at least $r^{-2}(t)$, has a neighborhood B falling into one of the three categories:

(a) B is a strong ε-neck, or
(b) B is an ε-cap, or
(c) B is a compact manifold (without boundary) of positive curvature.

The Hamilton-Ivey curvature pinching estimate is a special feature on three-dimension. It plays an important role in the proof of the theorem of structure of singularity. Thus one should also require the surgical solutions to satisfy the following Hamilton-Ivey pinching condition:
Pinching assumption: The eigenvalues $\lambda \geq \mu \geq \nu$ of the curvature operator Rm of the surgical solution at each point and each time satisfy

$$R \geq (-\nu)[\log(-\nu) + \log(1 + t) - 3]$$

whenever $\nu < 0$.

5.3. Long-Time Existence of Surgical Solutions. Let ε be an arbitrarily given small positive constant. We now describe how to use an inductive argument to construct a long time surgically modified solution satisfying the pinching assumption and the canonical neighborhood assumption (with accuracy ε).

Start with a (smooth) maximal solution $g(t)$, $t \in [0, T)$, to the Ricci flow on the compact, oriented three-manifold M with normalized initial metric. By the Hamilton-Ivey pinching estimate and Theorem 4.13 on the structure of singularity, we see that the maximal solution $g(t)$ satisfies the pinching assumption and the canonical neighborhood assumption on the maximal time interval $[0, T)$. If $T = +\infty$, we have the desired long time solution. Thus, without loss of generality, we may assume $T < +\infty$ and hence the solution goes singular at time T.

Suppose that we have a surgically modified solution on $[0, T)$ (with $T < +\infty$ and with the normalized metric as initial data) which satisfies the pinching assumption and the canonical neighborhood assumption (with accuracy ε), becomes singular at time T, and has only a finite number of surgery times on $[0, T)$. Let Ω denote the set of all points in M where the curvature stays bounded as $t \to T$. Then the solution $g(t)$ has a smooth limit $\bar{g}$, defined on Ω, as $t \to T$.

For some $\delta > 0$ to be chosen much smaller than ε, we let $\rho = \delta r(T)$, where $r(t)$ is the positive nonincreasing function in the definition of the canonical neighborhood assumption. We then consider the corresponding compact set

$$\Omega_\rho = \{x \in \Omega \mid \bar{R}(x) \leq \rho^{-2}\},$$

where $\bar{R}$ is the scalar curvature of $\bar{g}$.

If Ω_ρ is empty, then the manifold (near the maximal time T) is entirely covered by ε-necks, ε-caps and compact components with positive curvature. As a consequence, the manifold is diffeomorphic to the union of a finite number of copies of $\mathbb{S}^3$, or metric quotients of the round $\mathbb{S}^3$, or $\mathbb{S}^2\times\mathbb{S}^1$, or a connected sum of them. Thus when Ω_ρ is empty, the procedure stops here, and we say the **solution becomes extinct**.

We now assume Ω_ρ is not empty. As was explained before, we only need to consider those components Ω_j, $1\le j\le m$, of Ω which contain points of Ω_ρ. We will perform surgical procedures, which have been roughly described before, by finding an ε-neck in all horns of Ω_j, $1\le j\le m$, and then cutting it along the middle two-sphere, removing the horn-shaped end, and gluing back a cap.

However, in order to maintain the pinching assumption and the canonical neighborhood assumption *with the same accuracy* after surgery, we will need to find sufficiently "fine" necks in the ε-horns and to glue sufficiently "fine" caps. Note that $\delta>0$ is to be chosen much smaller than $\varepsilon>0$.

Actually, one can show (due to Perelman [**81**], see Lemma 7.3.2 [**14**]) that in every ε-horn of Ω_j, $1\le j\le m$, there exists a δ-neck with its radius depending only on δ and $r(T)$. This gives us the "fine" necks in the ε-horns.

To construct "fine" caps, we consider the semi-infinite standard round cylinder $N_0=\mathbb{S}^2\times(-\infty,4)$ with the metric g_0 of scalar curvature 1. Denote by z the coordinate of the second factor $(-\infty,4)$. Let f be a smooth nondecreasing convex function on $(-\infty,4)$ defined by

$$\begin{cases} f(z)=0, \quad z\le 0,\\ f(z)=ce^{-\frac{P}{z}}, \quad z\in(0,3],\\ f(z) \text{ is strictly convex on } z\in[3,3.9],\\ f(z)=-\frac{1}{2}\log(16-z^2), \quad z\in[3.9,4),\end{cases}$$

where the (small) constant $c>0$ and (big) constant $P>0$ will be determined later (see Figure 5).

Let us replace the standard metric g_0 on the portion $\mathbb{S}^2\times[0,4)$ of the semi-infinite cylinder by the conformal change $e^{-2f}g_0$. Then the resulting metric $\hat{g}$ is smoothly defined on $\mathbb{R}^3$ obtained by adding a point to $\mathbb{S}^2\times(-\infty,4)$ at $z=4$. We denote by $C(c,P)=(\mathbb{R}^3,\hat{g})$, and call it a **standard capped infinite cylinder** (see Figure 6). Clearly $C(c,P)$ has nonnegative sectional curvature and positive scalar curvature everywhere.

As a side remark, one might wonder whether we should also cut off all those ε-tubes and ε-caps in the surgery procedure. However, in general one may not be able to find a "fine" neck in an ε-tube or an ε-cap, and surgeries at "rough" ε-necks will certainly lose some accuracy. If one performs the surgeries at the necks with some fixed accuracy ε on the high curvature

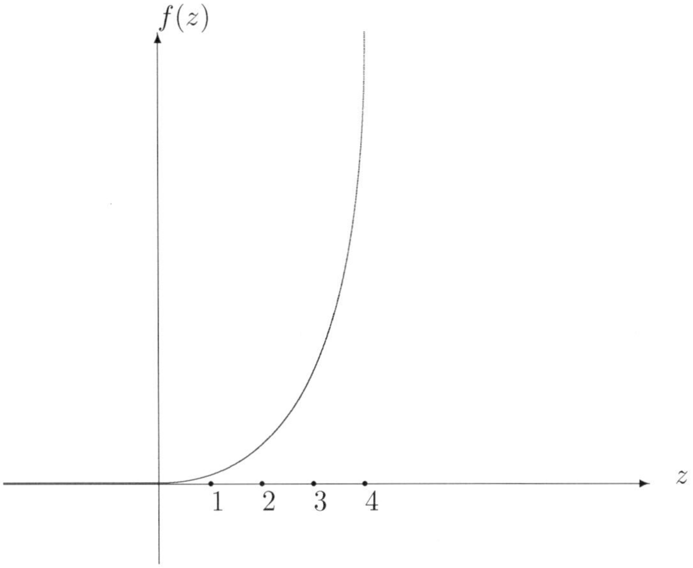

FIGURE 5. The function $f(z)$.

FIGURE 6. Standard capped infinite cylinder.

region at each surgery time, then it is possible that the errors of surgeries may accumulate to a certain amount so that at some later time one cannot recognize the structure of very high curvature region. This prevents us to carry out the whole process in finite time with finite steps. This is the reason why we will only perform the surgeries at ε-horns.

We can now perform Hamilton's geometric surgery procedure as follows. Take an ε-horn with boundary in Ω_ρ and take a δ-neck N of radius h, $0 < h < \delta\rho$, in the ε-horn. By definition, $(N, h^{-2}\bar{g})$ is δ-close (in the $C^{[\delta^{-1}]}$ topology) to the standard round neck $\mathbb{S}^2 \times \mathbb{I}$ of scalar curvature 1 with $\mathbb{I} = (-\delta^{-1}, \delta^{-1})$. The parameter $z \in \mathbb{I}$ induces a function on the δ-neck N.

Let us cut the δ-neck N along the middle (topological) two-sphere $N \bigcap \{z = 0\}$. Without loss of generality, we may assume that the right hand half portion $N \bigcap \{z \geq 0\}$ is contained in the horn-shaped end. Let φ be a smooth bump function with $\varphi = 1$ for $z \leq 2$, and $\varphi = 0$ for $z \geq 3$. Construct a new metric $\tilde{g}$ on a (topological) three-ball $\mathbb{B}^3$ as

$$\tilde{g}=\begin{cases} \bar{g}, & z=0, \\ e^{-2f}\bar{g}, & z\in[0,2], \\ \varphi e^{-2f}\bar{g}+(1-\varphi)e^{-2f}h^2g_0, & z\in[2,3], \\ h^2e^{-2f}g_0, & z\in[3,4]. \end{cases}$$

The surgery, called a δ-**cutoff surgery**, is to replace the horn-shaped end by the cap $(\mathbb{B}^3,\tilde{g})$ (see Figure 7).

We remark that this type of surgery is topologically trivial. But it is geometrically significant: after suitable adjusting the parameters c, P and δ, the pinching assumption will survive under the surgeries. Indeed, we can prove

THEOREM 5.1 (see Lemma 7.3.4 in [**14**]). *There are universal positive constants δ_0, c_0 and P_0 such that if one takes a δ-cutoff surgery at a δ-neck of radius h at time T with $\delta\le\delta_0$ and $h^{-2}\ge 2e^2\log(1+T)$, then one can choose $c=c_0$ and $P=P_0$ in the definition of $f(z)$ such that after the surgery, the pinching condition*

$$\tilde{R}\ge(-\tilde{\nu})[\log(-\tilde{\nu})+\log(1+T)-3]$$

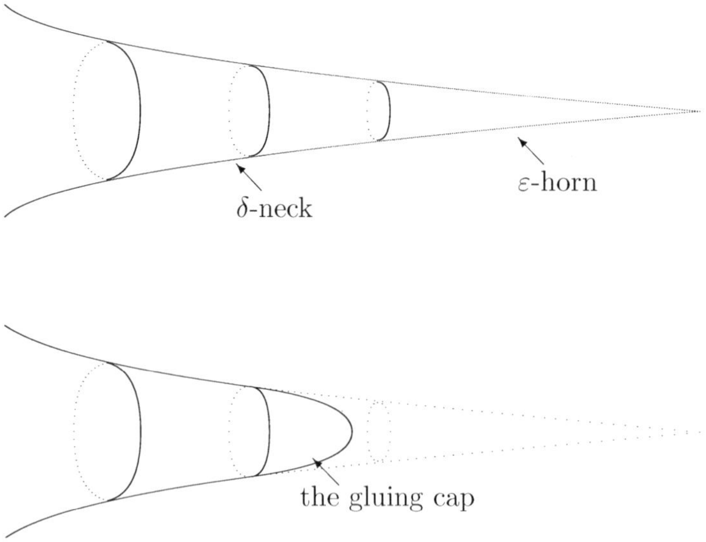

FIGURE 7. δ-cutoff surgery.

still holds whenever $\tilde{\nu} < 0$. *Here* $\tilde{R}$ *is the scalar curvature of* $\tilde{g}$ *and* $\tilde{\nu}$ *is the least eigenvalue of the curvature operator of* $\tilde{g}$.

Define the positive function $\bar{\delta}(t)$ on $[0, +\infty)$ by

$$\bar{\delta}(t) = \min\left\{\frac{1}{2e^2\log(1+t)}, \delta_0\right\}.$$

From now on we will always assume $0 < \delta < \bar{\delta}(t)$ for any δ-cutoff surgery at a time $t > 0$ and take $c = c_0$ and $P = P_0$ so that the pinching assumption is preserved under the surgeries at T. After performing the δ-cutoff surgeries for all Ω_j, $1 \leq j \leq m$, we obtain the compact (without boundary), orientable three-manifolds $\bar{\Omega}_j$, $1 \leq j \leq m$. With these new compact manifolds as initial data, we can continue the solution under the Ricci flow until it becomes singular again at some later time $T' > T$. By the Hamilton-Ivey estimate (Theorem 1.7), we see that the solution still satisfies the pinching assumption on the extended time interval $[0, T')$.

By dilation and Theorem 4.13 on the structure of singularity, there always exists a nonincreasing positive function $r = r'(t)$, defined on $[0, +\infty)$, such that the canonical neighborhood assumption (with accuracy ε) holds on the extended time interval $[0, T')$ with the positive function $r = r'(t)$. Nevertheless, in order to prevent the surgery times from accumulating, the key is to choose the nonincreasing positive functions $r(t)$ uniformly. By a further restriction on the positive function $\bar{\delta}(t)$ we can verify the canonical neighborhood assumption with a uniform $r(t)$.

THEOREM 5.2 (Justification of the canonical neighborhood assumption [**81**]). *Given any small* $\varepsilon > 0$, *there exist decreasing sequences* $0 < \widetilde{r}_j < \varepsilon$ *and* $0 < \widetilde{\delta}_j < \varepsilon^2$, $j = 1, 2, \ldots$, *with the following property. Define the positive function* $\widetilde{\delta}(t)$ *on* $[0, +\infty)$ *by* $\widetilde{\delta}(t) = \widetilde{\delta}_j$ *for* $t \in [(j-1)\varepsilon^2, j\varepsilon^2)$. *Suppose there is a surgically modified solution, defined on* $[0, T)$ *with* $T < +\infty$, *to the Ricci flow which satisfies the following:*

(1) it starts on a compact orientable three-manifold with normalized initial metric, and
(2) it has only a finite number of surgeries such that each surgery at a time $t \in (0, T)$ *is a* $\delta(t)$*-cutoff surgery with*

$$0 < \delta(t) \leq \min\{\widetilde{\delta}(t), \bar{\delta}(t)\}.$$

Then on each time interval $[(j-1)\varepsilon^2, j\varepsilon^2] \bigcap [0, T)$, $(j = 1, 2, \ldots)$, *the solution satisfies the canonical neighborhood assumption (with accuracy* ε*) with* $r = \widetilde{r}_j$.

This result was first given by Perelman in [**81**]. It extends the singularity structure theorem (Theorem 4.13) for smooth solutions to surgically modified solutions. However, when one tries to adapt the arguments of the smooth case to the surgical case, they will encounter several difficulties: how to generalize the non local collapsing theorem of Perelman to surgical solutions to get the local injectivity radius bound; how to apply Hamilton's compactness theorem to surgically modified solutions; how to extend the rescaling limits backward in time without touching the surgical regions. Below we will give a brief description of the proof.

The proof is by induction: having constructed our sequences for $1 \leq j \leq m$, we make one more step, defining $\widetilde{r}_{m+1}$ and $\widetilde{\delta}_{m+1}$. We follow a very clever idea of Perelman [**81**] by redefining $\widetilde{\delta}_m = \widetilde{\delta}_{m+1}$ in order to push the surgical regions to infinity in space.

We argue by contradiction. Suppose for some sequences of positive numbers $r^\alpha \to 0$ and $\widetilde{\delta}^{\alpha\beta} \to 0$, there exist sequences of solutions $g_{ij}^{\alpha\beta}$ to the Ricci flow with surgery, with a compact orientable normalized three-manifold as initial data, so that

(i) each δ-cutoff at time $t \in [(m-1)\varepsilon^2, (m+1)\varepsilon^2]$ satisfies $\delta \leq \widetilde{\delta}^{\alpha\beta}$; and
(ii) the solutions satisfy the statement of the proposition on $[0, m\varepsilon^2]$, but violate the canonical neighborhood assumption (with accuracy ε) with $r = r^\alpha$ on $[m\varepsilon^2, (m+1)\varepsilon^2]$.

For each solution $g_{ij}^{\alpha\beta}$, we choose $\bar{t}^{\alpha\beta}$ to be the nearly first time for which the canonical neighborhood assumption (with accuracy ε) is violated at some $(\bar{x}^{\alpha\beta}, \bar{t}^{\alpha\beta})$ but the canonical neighborhood assumption with accuracy parameter 2ε does hold on $t \in [m\varepsilon^2, \bar{t}^{\alpha\beta}]$.

Let $\widetilde{g}_{ij}^{\alpha\beta}$ be the rescaled solutions around $(\bar{x}^{\alpha\beta}, \bar{t}^{\alpha\beta})$ with factors $R(\bar{x}^{\alpha\beta}, \bar{t}^{\alpha\beta})$ $(\geq (r^\alpha)^{-2} \to +\infty$ as $\alpha \to +\infty)$ and shift $\bar{t}^{\alpha\beta}$ to zero. We hope to take a subsequential limit of the rescaled solutions as $\alpha, \beta \to \infty$ and show that the limit is an orientable ancient κ-solution, which will give the desired contradiction.

To do so, we first need to get a uniform lower bound for the injectivity radii of the rescaled sequence $\widetilde{g}_{ij}^{\alpha\beta}$ at the marking points $(\bar{x}^{\alpha\beta}, \bar{t}^{\alpha\beta})$. Based on the fact that the canonical neighborhood assumption with accuracy parameter 2ε holds for $t \in [m\varepsilon^2, \bar{t}^{\alpha\beta}]$, we appeal the following lemma to show that the (unscaled) sequence $g_{ij}^{\alpha\beta}$ is κ-noncollapsed for some $\kappa > 0$ independent of α, β.

LEMMA 5.3 (Perelman [**81**], see also Lemma 7.4.2 in [**14**]). *Given $\varepsilon > 0$, suppose we have constructed the sequences satisfying the proposition for $1 \leq j \leq l$ (for some positive integer l). Then there exists $\kappa > 0$, such that for any r, $0 < r < \varepsilon$, one can find $\widetilde{\delta} = \widetilde{\delta}(r, \varepsilon)$, $0 < \widetilde{\delta} < \varepsilon^2$, which may also depend on the already constructed sequences, with the following property. Suppose we have a solution with a compact oriented normalized three-manifold as initial data, to the Ricci flow with finite number of surgeries on a time interval $[0, \bar{T}]$ with $l\varepsilon^2 \leq \bar{T} < (l+1)\varepsilon^2$, satisfying the assumptions and the conclusions of Proposition 7.4.1 on $[0, l\varepsilon^2)$, and the canonical neighborhood* *assumption (with accuracy ε) with r on $[l\varepsilon^2, \bar{T}]$, as well as $0 < \delta(t) \leq \widetilde{\delta}$ for any δ-cutoff surgery with $\delta = \delta(t)$ at a time $t \in [(l-1)\varepsilon^2, \bar{T}]$. Then the solution is κ-noncollapsing on $[0, \bar{T}]$ for all scales less than ε.*

The major observation in the proof of the lemma is that the space-time curves near the region of surgery carry large reduced distance, so one can find suitable surgically unaffected cone-like regions and to apply Perelman's Jacobian comparison theorem (formula (3.15)) there as in the proof Theorem 3.9. The universal noncollapsing property of ancient κ-solutions is substantially used in this proof. We have to mention that the noncollapsing constant κ obtained in such a way does not depend on the canonical neighborhood parameter r in $[(l-1)\varepsilon^2, \bar{T}]$. This is the key point of the lemma[2].

The uniform noncollapsing estimate guarantees the desired injectivity radius bound for the rescaled sequence $\widetilde{g}_{ij}^{\alpha\beta}$.

Next we need to get a uniform curvature bound for the rescaled sequence $\widetilde{g}_{ij}^{\alpha\beta}$ on compact subsets around the marked points $\bar{x}^{\alpha\beta}$. Since the (unscaled) solutions satisfy the canonical neighborhood assumption with accuracy parameter 2ε on $[m\varepsilon^2, \bar{t}^{\alpha\beta}]$, we can use the gradient estimates in the canonical neighborhood assumption to get a uniform curvature estimate for the rescaled solutions $\widetilde{g}_{ij}^{\alpha\beta}$ in some small space-time neighborhoods of $(\bar{x}^{\alpha\beta}, \bar{t}^{\alpha\beta})$. The sizes of these neighborhoods in space (at the new time zero) are uniform, but the time interval may vary due to the surgeries. If the time interval for the solution is too short, one can not apply Shi's derivative estimates to obtain uniform higher derivatives estimates. This prevent us from applying Hamilton's compactness theorem, which requires a uniform time interval for all the solutions in the sequence, to take a limit for the surgically modified rescaled solutions. To overcome this difficulty, in [**14**] and [**25**], the authors established three time-extension results:

The first assertion says that if we have curvature estimates for the renormalized solutions on a box $\widetilde{B}_0(\bar{x}, A) \times [-b, 0]$, then the solution can be extended to a larger time interval.

Assertion 1. *For arbitrarily fixed α, $0 < A < +\infty$, $1 \leq C < +\infty$ and $0 \leq B < \frac{1}{2}\varepsilon^2 (r^\alpha)^{-2} - \frac{1}{8}\eta^{-1}C^{-1}$, there is a $\beta_0 = \beta_0(\varepsilon, A, B, C)$ (independent of α) such that if $\beta \geq \beta_0$ and the rescaled solution $\widetilde{g}_{ij}^{\alpha\beta}$ on the ball $\widetilde{B}_0(\bar{x}, A)$ is defined on a time interval $[-b, 0]$ with $0 \leq b \leq B$ and the scalar curvature satisfies*

$$\widetilde{R}(x,t) \leq C, \quad \text{on } \widetilde{B}_0(\bar{x}, A) \times [-b, 0],$$

then the rescaled solution $\widetilde{g}_{ij}^{\alpha\beta}$ on the ball $\widetilde{B}_0(\bar{x}, A)$ is also defined on the extended time interval $[-b - \frac{1}{8}\eta^{-1}C^{-1}, 0]$.

---

[2] We have also learned the very recent works of Ye [**105**] and Zhang [**108**] on how to obtain a uniform Sobolev inequality, which is independent of the number of surgeries, and use it to derive κ-noncollapsing for surgically modified Ricci flow.

If the solution cannot be defined on the larger interval $[-b-\frac{1}{8}\eta^{-1}C^{-1},0]$, it must hit the surgery region. Since the surgery was done on the δ-neck, the solution on the surgery region is close to a standard solution by the uniqueness theorem of Ricci flow [**24**]. For standard solutions, we have canonical neighborhood decompositions. Note that once the solution is defined on $\widetilde{B}_0(\bar{x},A)$ for a subinterval $[-b-\nu,0]$ of $[-b-\frac{1}{8}\eta^{-1}C^{-1},0]$, the curvature bound on this region follows directly from the gradient estimate on canonical neighborhoods with accuracy parameter 2ε. This curvature bound guarantees that the solution near the point $\bar{x}$ is close to a standard solution until the time 0. Since the canonical neighborhood assumption with accuracy parameter ε is violated at $(\bar{x},\bar{t})$ by assumption, this gives a contradiction.

Assertion 2. *For arbitrarily fixed α, $0<A<+\infty$, $1\leq C<+\infty$ and $0\leq B<\frac{1}{2}\varepsilon^2(r^\alpha)^{-2}-\frac{1}{50}\eta^{-1}$, there is a $\beta_0=\beta_0(\varepsilon,A,B,C)$ (independent of α) such that if $\beta\geq\beta_0$ and the rescaled solution $\tilde{g}_{ij}^{\alpha\beta}$ on the ball $\widetilde{B}_0(\bar{x},A)$ is defined on a time interval $[-b+\epsilon',0]$ with $0<b\leq B$ and $0<\epsilon'<\frac{1}{50}\eta^{-1}$ and the scalar curvature satisfies*

$$\widetilde{R}(x,t)\leq C \quad \text{on } \widetilde{B}_0(\bar{x},A)\times[-b+\epsilon',0],$$

and there is a point $y\in\widetilde{B}_0(\bar{x},A)$ such that $\widetilde{R}(y,-b+\epsilon')\leq\frac{3}{2}$, then the rescaled solution $\tilde{g}_{ij}^{\alpha\beta}$ at y is also defined on the extended time interval $[-b-\frac{1}{50}\eta^{-1},0]$ and satisfies the estimate

$$\widetilde{R}(y,t)\leq 15$$

for $t\in[-b-\frac{1}{50}\eta^{-1},-b+\epsilon']$.

Assertion 2 follows the same philosophy as in Assertion 1. Once we have the curvature estimates, and the solution hits the surgery, it must maintain the shape of standard solution until the time 0 by the uniqueness theorem in [**24**]. In practice, the point y will come from an almost minimal value of the scalar curvature, so its curvature is uniformly bounded.

The following Assertion 3 is based on the observation that the standard solution satisfies $R(x_1,t)\leq D''R(x_2,t)$ for any $t\in[0,\frac{1}{2}]$ and any two points x_1,x_2, where D'' is a universal constant.

Assertion 3. *For arbitrarily fixed α, $0<A<+\infty$, $1\leq C<+\infty$, there is a $\beta_0=\beta_0(\varepsilon,AC^{\frac{1}{2}})$ such that if any point (y_0,t_0) with $0\leq -t_0<\frac{1}{2}\varepsilon^2(r^\alpha)^{-2}-\frac{1}{8}\eta^{-1}C^{-1}$ of the rescaled solution $\tilde{g}_{ij}^{\alpha\beta}$ for $\beta\geq\beta_0$ satisfies $\widetilde{R}(y_0,t_0)\leq C$, then either y_0 can be defined at least on $[t_0-\frac{1}{16}\eta^{-1}C^{-1},t_0]$ and the scalar curvature satisfies*

$$\widetilde{R}(y_0,t)\leq 10C \quad \text{for } t\in[t_0-\frac{1}{16}\eta^{-1}C^{-1},t_0],$$

or we have

$$\widetilde{R}(x_1,t_0)\leq 2D''\widetilde{R}(x_2,t_0)$$

for any two points $x_1,x_2\in\widetilde{B}_{t_0}(y_0,A)$, where D'' is the above universal constant.

Based on these three time-extension results, we can adapt the arguments in the proof of Theorem 4.13 on the structure of singularity to the surgically modified solutions.

We next argue as in the second step of the proof of Theorem 4.13 to show that the curvatures of the rescaled solutions $\tilde{g}^{\alpha_m\beta_m}$ at new time zero (after shifting) stay uniformly bounded at bounded distances from $\bar{x}$ as $m\to\infty$. More precisely, we will prove the following assertion:

Assertion 4. *For the rescaled solutions $\tilde{g}_{ij}^{\alpha_m\beta_m}$, we have that for any $L>0$, there are constants $C(L)>0$ and $m(L)$ such that for all $m\geq m(L)$ the rescaled solutions $\tilde{g}_{ij}^{\alpha_m\beta_m}$ satisfy*

(i) $\tilde{R}(x,0)\leq C(L)$ for all points x with $\tilde{d}_0(x,\bar{x})\leq L$;

(ii) the ball $\tilde{B}_0(\bar{x},L)$ is defined at least on the time interval $[-\frac{1}{16}\eta^{-1}C(L)^{-1},0]$.

For all $\rho>0$, set

$$M(\rho)=\sup\{\tilde{R}(x,0)\mid \tilde{d}_0(x,\bar{x})\leq\rho \text{ in the rescaled solutions } \tilde{g}_{ij}^{\alpha_m\beta_m}\}$$

and

$$\rho_0=\sup\{\rho>0\ |M(\rho)<+\infty\}.$$

Note that the gradient estimate implies that $\rho_0>0$. For (i), it suffices to prove $\rho_0=+\infty$.

Suppose $\rho_0<+\infty$. By Assertion 3 or Assertion 1, we have for any $0<\rho<\rho_0$, the rescaled solutions on the balls $\tilde{B}_0(\bar{x},\rho)$ are defined on the time interval $[-\frac{1}{16}\eta^{-1}M(\rho)^{-1},0]$ for all large m. Once the solution is defined on this time interval, by gradient estimate and Shi's derivative estimate, we know that the covariant derivatives of the curvatures of all order on $\tilde{B}_0(\bar{x},\rho-\frac{1}{m})\times[-\frac{1}{32}\eta^{-1}M(\rho)^{-1},0]$ are also uniformly bounded. Hence Hamilton's compactness theorem is applicable now. Then we can apply the similar argument as in Step 2 of the proof of Theorem 4.13 to prove Assertion 4.

For any subsequence (α_m,β_m) of (α,β) with $r^{\alpha_m}\to 0$ and $\widetilde{\delta}^{\alpha_m\beta_m}\to 0$ as $m\to\infty$, by Assertion 4, the κ-noncollapsing and Hamilton's compactness theorem, we can extract a C^∞_{loc} convergent subsequence of $\tilde{g}_{ij}^{\alpha_m\beta_m}$ over some space-time open subsets containing $t=0$. As in the proof of Singularity Structure Theorem 4.13, we can use Lemma 4.14 to show *any* such limit has bounded curvature at $t=0$.

Choose $\alpha_m,\beta_m\to\infty$ so that $r^{\alpha_m}\to 0$, $\widetilde{\delta}^{\alpha_m\beta_m}\to 0$, and Assertions 1-3 hold with $\alpha=\alpha_m,\beta=\beta_m$ for all $A\in\{p/q\mid p,q=1,2,\dots,m\}$, and $B,C\in\{1,2,\dots,m\}$. By Assertion 4, we may assume the rescaled solutions $\widetilde{g}_{ij}^{\alpha_m\beta_m}$ converge in C^∞_{loc} topology at the time $t=0$. Since the curvature of the limit at $t=0$ is bounded, it follows from Assertion 1 and the choice of the sequence (α_m,β_m) that the limiting $(M_\infty,\widetilde{g}_{ij}^\infty(\cdot,t))$ is defined at least on a backward time interval $[-a,0]$ for some positive constant a and is a smooth solution to the Ricci flow there.

We can further extend the limit backward in time to infinity to get an ancient κ-solution. We omit the details here and refer the reader to consult [**14**] and [**25**]. The idea of time extension is used throughout our proof. We emphasize that, comparing to the no surgery case, this is a crucial point.

Summing up, for any $\varepsilon > 0$, there exist nonincreasing positive functions $\widetilde{\delta}(t)$ and $\widetilde{r}(t)$, defined on $[0, +\infty)$, such that for an arbitrarily given positive function $\delta(t)$ with $\delta(t) < \widetilde{\delta}(t)$ on $[0, +\infty)$, the Ricci flow with surgery has a solution on $[0, T_{\max})$ obtained by evolving the Ricci flow and by performing δ-cutoff surgeries at a sequence of times $0 < t_1 < t_2 < \cdots < t_i < \cdots < T_{\max}$, with $\delta(t_i) \leq \delta \leq \widetilde{\delta}(t_i)$ at each time t_i, so that the pinching assumption and the canonical neighborhood assumption (with accuracy ε) with $r = \widetilde{r}(t)$ are satisfied. (At this moment we still do not know whether the surgery times t_i are discrete).

Each δ-cutoff surgery at time t_i cuts down the volume at least at an amount depending only on $\delta(t_i)$ and $\widetilde{r}(t_i)$, while the volume of the surgically modified solution can be bounded by

$$V(t) \leq V(0)e^{-R_{min}(0)t}.$$

Thus the surgery times t_i cannot accumulate in any finite interval. When the solution becomes extinct at some finite time $T_{\max}$, the solution at a time slightly before $T_{\max}$ is entirely covered by canonical neighborhoods and then the initial manifold is diffeomorphic to a connected sum of a finite copies of $\mathbb{S}^2 \times \mathbb{S}^1$ and $\mathbb{S}^3/\Gamma$ (the metric quotients of round three-sphere). So we have the following long-time existence result, which was proposed by Perelman in [**81**].

THEOREM 5.4 (Long-time Existence Theorem). *For any given small constant $\varepsilon > 0$, there exist nonincreasing (continuous) positive functions $\widetilde{\delta}(t)$ and $\widetilde{r}(t)$, defined on $[0, +\infty)$, such that for any arbitrarily given (continuous) positive function $\delta(t)$ with $\delta(t) \leq \widetilde{\delta}(t)$ on $[0, +\infty)$, the Ricci flow with surgery, with an arbitrarily given compact orientable normalized three-manifold as initial data, has the following property: either*

(i) it is defined on a finite interval $[0, T_{\max})$ and obtained by evolving the Ricci flow and by performing a finite number of cutoff surgeries, with each δ-cutoff at a time $t \in (0, T_{\max})$ having $\delta = \delta(t)$, so that the solution becomes extinct at $T_{\max}$, and the initial manifold is diffeomorphic to a connected sum of a finite copies of $\mathbb{S}^2 \times \mathbb{S}^1$ and $\mathbb{S}^3/\Gamma$ (the metric quotients of round three-sphere) ; or

(ii) it is defined on $[0, +\infty)$ and obtained by evolving the Ricci flow and by performing at most a countably many cutoff surgeries, with each δ-cutoff at a time $t \in [0, +\infty)$ having $\delta = \delta(t)$, so that the pinching assumption and the canonical neighborhood assumption (with accuracy ε) with $r = \widetilde{r}(t)$ are satisfied, and there exist at most a finite number of surgeries on every finite time interval.

5.4. Topological Implications. As the first consequence of the above long-time existence result, one can obtain a complete classification to compact three-manifolds with nonnegative scalar curvature. Indeed, if the initial manifold has positive scalar curvature, then the solution becomes extinct in finite time and the manifold is either flat or diffeomorphic to a connected sum of a finite number of copies of $\mathbb{S}^2 \times \mathbb{S}^1$ and $\mathbb{S}^3/\Gamma$ (the metric quotients of round three-spheres). This improves the well-known topological classification of Schoen-Yau [**89**] for compact three-manifolds with nonnegative scalar curvature.

The famous **Poincar´e conjecture** states that *any simply connected compact three-manifold (without boundary) is homeomorphic to the three-sphere.* Recent works of Perelman [**82**] (see also detailed exposition given in Morgan-Tian [**70**]) and Colding-Minicozzi [**30**] proved that any surgically modified solution to the Ricci flow on a simply connected compact three-manifold must be extinct in finite time. Thus *The combination of the assertion (i) of the above long-time theorem 5.4 and the finite extinction result of Perelman and Colding-Minicozzi gives a complete proof to the Poincaré conjecture.*

The idea of proving the extinction result is adapted from Hamilton [**49**] where the argument was used to show the incompressibility of the boundary tori of hyperbolic pieces. This argument for the finite time extinction can be roughly described as follows. Suppose there would exist a surgical solution $g_{ij}(t)$ on the infinite time interval $[0, +\infty)$. Since the manifold is simply connected, a well-known result of J. P. Serre implies some higher homotopic group of the manifold is nontrivial. Then one can use the nontrivial homotopic group to construct a minimal surface $\Sigma(t)$ for each $t \in [0, +\infty)$. Denote by $A(t)$ the area of $\Sigma(t)$. By an argument of Schoen-Yau [**89**] of using the Gauss-Bonnet formula, one can bound the the time derivative of the area function

$$\frac{dA(t)}{dt} \le -f(t)$$

for some positive function $f(t)$ which is nonintegrable on $[0, +\infty)$. Then it gives the desired contradiction.

To conclude this section, we mention the application of the Ricci flow with surgery to the classification of four-manifolds with positive isotropic curvature. Recall that a Riemannian four-manifold is said to have *positive isotropic curvature* if for every orthonormal four-frame the curvature tensor satisfies

$$R_{1313} + R_{1414} + R_{2323} + R_{2424} > 2R_{1234}.$$

An *incompressible space form* N^3 in a four-manifold M^4 is a three-dimensional submanifold diffeomorphic to $\mathbb{S}^3/\Gamma$ (the quotient of the three-sphere by a group of isometries without fixed point) such that the fundamental group $\pi_1(N^3)$ injects into $\pi_1(M^4)$. The space form is said to be *essential* unless $\Gamma = \{1\}$, or $\Gamma = \mathbb{Z}_2$ and the normal bundle is non-orientable. In [**25**], the last two authors obtained the following long-time existence result for the Ricci flow with surgery for a class of four-manifolds.

THEOREM 5.5 (Chen-Zhu [**25**]). *Let M^4 be a compact four-manifold with no essential incompressible space-form and with a metric g_{ij} of positive isotropic curvature. Then we have a finite collection of smooth solutions $g_{ij}^{(k)}(t)$, $k=0,1,\ldots,m$, to the Ricci flow, defined on $M_k^4 \times [t_k, t_{k+1})$, $(0=t_0<\cdots<t_{m+1})$ with $M_0^4=M^4$ and $g_{ij}^{(0)}(t_0)=g_{ij}$, which go singular as $t \to t_{k+1}$, such that the following properties hold:*

(i) *for each $k=0,1,\ldots,m-1$, the compact (possible disconnected) four-manifold M_k^4 contains an open set Ω_k such that the solution $g_{ij}^{(k)}(t)$ can be smoothly extended to $t=t_{k+1}$ over Ω_k;*
(ii) *for each $k=0,1,\ldots,m-1$, $(\Omega_k, g_{ij}^{(k)}(t_{k+1}))$ and $(M_{k+1}^4, g_{ij}^{(k+1)}(t_{k+1}))$ contain compact (possible disconnected) four-dimensional submanifolds with smooth boundary, which are isometric and then can be denoted by N_k^4;*
(iii) *for each $k=0,1,\ldots,m-1$, $M_k^4 \setminus N_k^4$ consists of a finite number of disjoint pieces diffeomorphic to $\mathbb{S}^3 \times \mathbb{I}$, $\mathbb{B}^4$ or $\mathbb{RP}^4 \setminus \mathbb{B}^4$, while $M_{k+1}^4 \setminus N_k^4$ consists of a finite number of disjoint pieces diffeomophic to $\mathbb{B}^4$;*
(iv) *for $k=m$, M_m^4 is diffeomorphic to the disjoint union of a finite number of $\mathbb{S}^4$, or $\mathbb{RP}^4$, or $\mathbb{S}^3 \times \mathbb{S}^1$, or $\mathbb{S}^3 \tilde{\times} \mathbb{S}^1$, or $\mathbb{RP}^4 \# \mathbb{RP}^4$.*

As a direct consequence, it gives a complete proof to the following classification result of Hamilton [**48**].

COROLLARY 5.6. *A compact four-manifold with no essential incompressible space-form and with a metric of positive isotropic curvature is diffeomorphic to $\mathbb{S}^4$, or $\mathbb{RP}^4$, or $\mathbb{S}^3 \times \mathbb{S}^1$, or $\mathbb{S}^3 \tilde{\times} \mathbb{S}^1$,or a connected sum of them.*

6. Geometrization of Three-manifolds

In the late 70's and early 80's, Thurston [**95, 96, 97**] proved a number of remarkable results on the existence of geometric structures on three-manifolds, especially the celebrated Haken manifold theorem. These results motivated him to formulate a profound conjecture

Thurston's Geometrization Conjecture *Let M be a compact, orientable and prime three-manifold. Then there is an embedding of a finite number of disjoint unions, possibly empty, of incompressible two-tori $\coprod_i T_i^2 \subset M$ such that every component of the complement admits a locally homogeneous Riemannian metric of finite volume.*

In dimension three, every locally homogeneous manifold with finite volume is modeled on one of the following eight homogeneous manifolds (see for example Theorem 3.8.4 of [**98**]):

(1) $\mathbb{S}^3$, the round three-sphere;
(2) $\mathbb{R}^3$, the Euclidean space;
(3) $\mathbb{H}^3$, the standard hyperbolic space;
(4) $\mathbb{S}^2 \times \mathbb{R}$;
(5) $\mathbb{H}^2 \times \mathbb{R}$;
(6) Nil, the three-dimensional nilpotent Heisenberg group (consisting of upper triangular 3×3 matrices with diagonal entries 1);
(7) $\widetilde{PSL}(2,\mathbb{R})$, the universal cover of the unit sphere bundle of $\mathbb{H}^2$;
(8) Sol, the three-dimensional solvable Lie group.

According to Kneser [**58**] and Milnor [**68**], every compact orientable three-manifold admits a unique decomposition as a finite connected sum of orientable prime three-manifolds. Thus the geometrization conjecture is a complete classification to three-dimensional manifolds. In particular, the Poincaré conjecture can be deduced from Thurston's geometrization conjecture. Indeed, suppose that we have a compact simply connected three-manifold that satisfies the conclusion of the geometrization conjecture. If it were not diffeomorphic to the three-sphere $\mathbb{S}^3$, there would be a prime factor in the prime decomposition of the manifold. Since the prime factor still has a vanishing fundamental group, the torus decomposition (by Jaco-Shalen [**53**] and Johannsen [**54**]) of the prime factor in the geometrization conjecture must be trivial. Thus the prime factor is a compact homogeneous manifold model. From the list of above eight models, we see that the only compact three-dimensional model is $\mathbb{S}^3$. This is a contradiction. Consequently, the compact simply connected three-manifold is diffeomorphic to $\mathbb{S}^3$.

The approach to prove the geometrization conjecture via the Ricci flow is to analyze long time behavior of surgically modified solutions. The argument of Perelman in [**81**] for the long-time behavior of surgical solutions is basically along the line given by Hamilton [**49**], in which Hamilton obtained the geometrization for a special class of solutions to the Ricci flow on three-manifolds, the so called nonsingular solutions. In the following, we present the long-time behavior analysis and sketch the proof of Thurston's geometrization conjecture.

Since we already have a complete (topological) classification to compact three-manifolds with nonnegative scalar curvature, we now assume that our initial manifold does not admit any metric with nonnegative scalar curvature and that *once we get a compact component with nonnegative scalar curvature, it is immediately removed.* Also by Theorem 5.4 (i), we only need to consider those solutions to the Ricci flow with surgery which exist for all time $t \geq 0$.

Let $g_{ij}(t)$, $0 \leq t < +\infty$, be a solution to the Ricci flow with δ-cutoff surgeries, constructed by the above long-time existence theorem (Theorem

5.2) with normalized initial data. Let $0=t_0<t_1<t_2<\cdots<t_k<\cdots$ be the surgery times. On each time interval (t_{k-1},t_k), the minimum of the scalar curvature $R_{\min}(t)$ at time t satisfies the differential inequality

$$\frac{d}{dt}R_{\min}(t)\geq\frac{2}{3}R^2_{\min}(t)$$

for $t\in(t_{k-1},t_k)$, $k=1,2,\ldots$. Since the surgeries occur only on regions of very large scalar curvature, it follows that

$$R_{\min}(t)\geq-\frac{3}{2}\cdot\frac{1}{t+\frac{3}{2}}$$

for all $t\geq 0$. Meanwhile, on each time interval (t_{k-1},t_k), the volume V satisfies the evolution equation

$$\frac{d}{dt}V=-\int RdV$$

and hence

$$\frac{d}{dt}V\leq\frac{3}{2}\cdot\frac{1}{(t+\frac{3}{2})}V.$$

Since the cutoff surgeries do not increase volume, the function $V(t)(t+\frac{3}{2})^{-\frac{3}{2}}$ is nonincreasing on $[0,+\infty)$. and there holds

$$\frac{V(t)}{(t+\frac{3}{2})^{\frac{3}{2}}}\leq\frac{V(0)}{(\frac{3}{2})^{\frac{3}{2}}}\exp\left\{-\int_0^t\left(R_{\min}(t)+\frac{3}{2(t+\frac{3}{2})}\right)dt\right.$$
$$\left.-\int_0^t\frac{1}{V}\int_M(R-R_{\min}(t))dVdt\right\}$$

for all $t>0$. This inequality implies that whenever we have a rescaling limit along a sequence times $t^\alpha\to\infty$ and with factors $(t^\alpha)^{-1}$, the limit must be a hyperbolic manifold. Then by extending the elliptic type estimate in Proposition 4.11 to surgical solutions, one will be able to obtain the following important thick-thin decomposition (see Figure 8) for surgically modified solutions.

Theorem 6.1 (Thick-thin decomposition theorem). *For any $w>0$ and $0<\varepsilon\leq\frac{1}{2}w$, there exists a positive constant $\rho=\rho(w,\varepsilon)\leq 1$ with the following property. Suppose $g_{ij}(t)$, $t\in[0,+\infty)$, is a surgically modified solution constructed by the above long-time existence theorem. Then for any arbitrarily fixed $\xi>0$, for t large enough, the manifold M_t at time t admits a decomposition*

$$M_t=M_{thin}(w,t)\cup M_{thick}(w,t)$$

with the following properties:

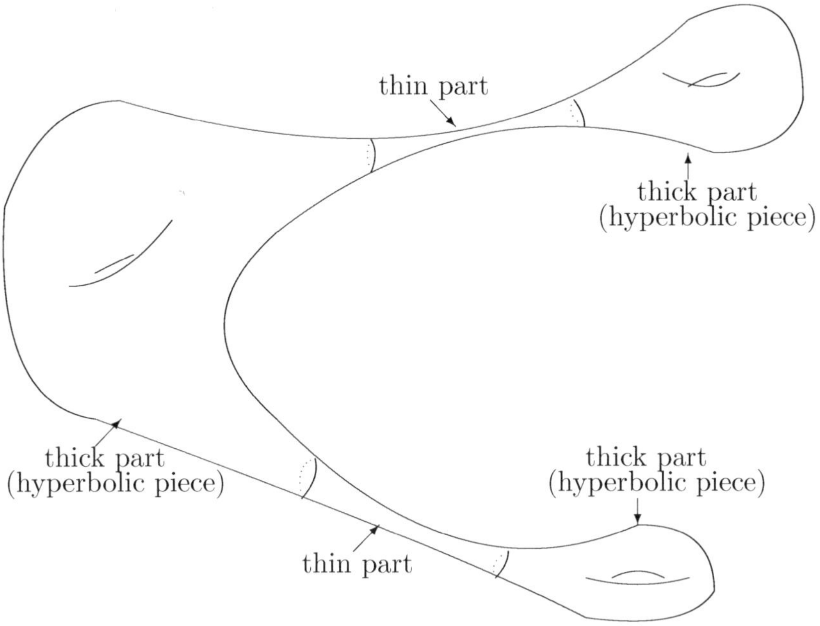

FIGURE 8. Thick-thin decomposition.

(a) For every $x \in M_{thin}(w,t)$, there exists some $r = r(x,t) > 0$, with $0 < r\sqrt{t} < \rho\sqrt{t}$, such that

$$Rm \geq -(r\sqrt{t})^{-2} \quad on \quad B_t(x, r\sqrt{t}), \quad and$$

$$Vol_t(B_t(x, r\sqrt{t})) < w(r\sqrt{t})^3;$$

(b) For every $x \in M_{thick}(w,t)$, we have

$$|2tR_{ij} + g_{ij}| < \xi \quad on \quad B_t(x, \rho\sqrt{t}), \quad and$$

$$Vol_t(B_t(x, \rho\sqrt{t})) \geq \frac{1}{10} w(\rho\sqrt{t})^3.$$

Moreover, if we take any sequence $t^\alpha \to +\infty$ and points $x^\alpha \in M_{thick}(w, t^\alpha)$, then a subsequence of the rescaled metrics of $g_{ij}(t^\alpha)$ around x^α with factor $(t^\alpha)^{-1}$ converge smoothly to a complete hyperbolic manifold of finite volume with constant sectional curvature $-\frac{1}{4}$.

This thick-thin decomposition theorem was implicitly given by Perelman in **[81]** without any restriction on the parameters ε and w. The above weaker version with the restriction is taken from **[14]**. The difference is because of our difficulty in understanding the original argument of Perelman **[81]**.

Fortunately, this weaker version still allows us to complete the proof of the geometrization conjecture.

Without loss of generality, we may further assume that the initial manifold M is irreducible. By our surgery procedures, the solution manifold M_t at each time t consists of a finite number of components where one of the components, called the *essential component* and denoted by $M_t^{(1)}$, is diffeomorphic to the initial manifold M while the rest are diffeomorphic to the 3-sphere $\mathbb{S}^3$.

Based on the thick-thin decomposition theorem, we can modify the arguments of Hamilton in [**49**] to obtain the following long-time behavior result.

THEOREM 6.2 (Long-time Behavior Theorem). *Let $w>0$ and $0<\varepsilon\leq\frac{1}{2}w$ be any small positive constants and let $(M_t,g_{ij}(t)), 0<t<+\infty$, be a solution to the Ricci flow with surgery constructed by the long-time existence theorem. Then one of the following holds: either*

(i) for all sufficiently large t, we have $M_t=M_{thin}(w,t)$; or

(ii) there exists a sequence of times $t^\alpha\to+\infty$ such that the scalings of the essential component $(M_{t^\alpha}^{(1)},g_{ij}(t^\alpha))$, with factor $(t^\alpha)^{-1}$, converge in the C^∞ topology to a hyperbolic metric on the initial compact manifold M with constant sectional curvature $-\frac{1}{4}$; or

(iii) we can find a finite collection of complete noncompact hyperbolic three-manifolds $\mathcal{H}_1,\ldots,\mathcal{H}_m$, of finite volume, and compact subsets $K_1,\ldots,K_m$ of $\mathcal{H}_1,\ldots,\mathcal{H}_m$ respectively obtained by truncating each cusp of the hyperbolic manifolds along constant mean curvature torus of small area, and for all t beyond some time $T<+\infty$ we can find diffeomorphisms $\varphi_l, 1\leq l\leq m$, of K_l into M_t so that as long as t is sufficiently large, the metric $t^{-1}\varphi_l^*(t)g_{ij}(t)$ is as close to the hyperbolic metric as we like on the compact sets $K_1,\ldots,K_m$; moreover, the complement $M_t\backslash(\varphi_1(K_1)\cup\cdots\cup\varphi_m(K_m))$ is contained in the thin part $M_{thin}(w,t)$, and the boundary tori of each K_l are incompressible in the sense that each φ_l injects $\pi_1(\partial K_l)$ into $\pi_1(M_t)$.

If case (ii) holds, then it is clear that the initial manifold M is geometrizable. While if case (iii) holds, then it follows from Thurston's theorem on Haken manifolds that the initial manifold M is also geometrizable. Thus it remains to consider case (i). For case (i), we will appeal to the following collapsing result, which was first announced (in a more general version) by Perelman [**81**] and proved by Shioya-Yamaguchi [**92**].

THEOREM 6.3 (Collapsing Theorem). *Suppose (M^α,g_{ij}^α) is a sequence of compact orientable three-manifolds without boundary, and $w^\alpha\to0$. Assume that for each point $x\in M^\alpha$ there exists a radius $\rho=\rho^\alpha(x)$, not exceeding the diameter of M^α, such that the volume of the ball $B(x,\rho)$ in the metric g_{ij}^α is* *at most* $w^\alpha \rho^3$ *and the sectional curvature on* $B(x,\rho)$ *is at least* $-\rho^{-2}$. *Then* M^α, *for sufficiently large* α, *are diffeomorphic to graph manifolds.*

Now the proof of the Thurston geometrization conjecture can be completed in the following way.

Let M be a compact, orientable and irreducible three-manifold. Arbitrarily given a (normalized) Riemannian metric on the manifold M, we use it as initial data for the Ricci flow. Take an arbitrarily sequence of small positive constants $w^\alpha \to 0$ as $\alpha \to +\infty$. For each fixed α, we set $\varepsilon = w^\alpha/2 > 0$. Then we apply the long-time existence theorem (Theorem 5.2) to get a sequence of surgically modified solutions $(M_t^\alpha, g_{ij}^\alpha(t))$ on maximal time intervals $[0, T_{\max}^\alpha)$ satisfying the pinching assumption and the canonical assumption (with the accuracy parameter $\varepsilon = w^\alpha/2$). We may assume that the maximal time $T_{\max}^\alpha = +\infty$ for all α and the surgical solutions $(M_t^\alpha, g_{ij}^\alpha(t))$ always satisfy assertion (i) of the long-time behavior theorem. That is, for each α, $M_t^\alpha = M_{thin}(w^\alpha, t)$ when t is sufficiently large.

Clearly we only need to consider the essential component $(M_t^\alpha)^{(1)}$. We divide the discussion into two cases:

(1) there is a positive constant $1 < C < +\infty$ such that for each α there is a sufficiently large time $t_\alpha > 0$ such that

$$r(x, t_\alpha)\sqrt{t_\alpha} < C \cdot diam((M_{t_\alpha}^\alpha)^{(1)})$$

for *all* $x \in (M_{t_\alpha}^\alpha)^{(1)} \subset M_{thin}(w^\alpha, t_\alpha)$;

(2) there are a subsequence α_k and sequences of positive constants $C_k \to +\infty$ and $T_k < +\infty$ such that for each $t \geq T_k$, we have

$$r(x(t), t)\sqrt{t} \geq C_k \cdot diam((M_t^{\alpha_k})^{(1)})$$

for *some* $x(t) \in (M_t^{\alpha_k})^{(1)}$, $k = 1, 2, \ldots$.

In case (1), we apply the collapsing theorem of Shioya-Yamaguchi to conclude that $(M_{t_\alpha}^\alpha)^{(1)}$ are graph manifolds. So the initial manifold M is geometrizable.

For case (2), if there are subsequences α_k (still denoted by α_k) and $t_k \in (T_k, +\infty)$ such that

$$Vol_{t_k}((M_{t_k}^{\alpha_k})^{(1)}) < w_k'(diam((M_{t_k}^{\alpha_k})^{(1)}))^3$$

for some sequence $w_k' \to 0$, then we can apply the collapsing theorem again to conclude that $(M_{t_k}^{\alpha_k})^{(1)}$ are diffeomorphic to graph manifolds. On the other hand, if there is a positive constant w' such that

$$Vol_t((M_t^{\alpha_k})^{(1)}) \geq w'(diam((M_t^{\alpha_k})^{(1)}))^3$$

for each k and all $t \geq T_k$, we can obtain the curvature estimates and take a rescaling limit to conclude that the initial manifold M is diffeomorphic to a flat manifold. So the initial manifold M is also geometrizable in case (2).

Therefore, we see that the Thurston geometrization conjecture is true.

References

[1] Andrews, B. and Nguyen, H., *Four-manifolds with 1/4-pinched flag curvatures,* Preprint, (2007).

[2] Angenent, S. and Knopf, D., *An example of neckpinching for Ricci flow on S^{n+1}*, Math. Res. Lett. **11** (2004), no. 4, 493–518.

[3] Bando, S., *On the classification of three-dimensional compact Kähler manifolds of nonnegative bisectional curvature,* J. Differential Geom. **19** (1984), no. 2, 283–297.

[4] Berger, M., *Les variétés Riemanniennes 1/4-pincées,* Ann. Scuola Norm. Sup. Pisa, **14** (1960), 161–170.

[5] Brendle, S., *A generalization of Hamilton's Harnack inequality for the Ricci flow,* preprint, arXiv:math.DG/0707.2192

[6] Brendle, S. and Schoen, R., *Manifolds with 1/4-pinched curvature are space forms,* preprint, arXiv:math.DG/0705.0766.

[7] Brendle, S. and Schoen, R., *Classification of manifolds with weakly 1/4-pinched curvatures,* preprint, arXiv:math.DG/0705.3963.

[8] Böhm, C., and Wilking, B., *Manifolds with positive curvature operator are space forms*, preprint, arxiv:math.DG/0606187.

[9] Cao, H.-D., *On Harnack's inequalities for the Kähler-Ricci flow,* Invent. Math. **109** (1992), no. 2, 247–263.

[10] Cao, H.-D., *Limits of Solutions to the Kähler-Ricci flow*, J. Differential Geom., **45** (1997), 257–272.

[11] Cao, H.-D. and Chow, B., *Compact Kähler manifolds with nonnegative curvature operator,* Invent. Math. **83** (1986), 553–556.

[12] Cao, H.-D., Chow, B., Chu, S.-C., and Yau, S.-T. (editors), *Collected Papers on Ricci Flow, Series in Geometry and Topology,* **37**. Cambridge, MA: International Press, 2003.

[13] Cao, H.-D., Hamilton, R. S., and Ilmanen, T., Gaussian densities and stability for some Ricci solitons, preprint, arXiv:math/0404165v1

[14] Cao, H.-D. and Zhu, X.-P., *A complete proof of Poincare and geometrization conjectures–application of Hamilton-Perelman theory of Ricci flow*, Asian J. math. **102** (2006), 165–492.

[15] Cao, X., *First Eigenvalues of Geometric Operators under the Ricci Flow*, preprint, arXiv:0710.3947.

[16] Chau, A. and Tam, L. F., *On the complex structure of Kähler manifolds with nonnegative curvture*, J. Differential Geom., **73** (2006), 491–530.

[17] Chau, A. and Tam, L. F., *A survey on the Kähler-Ricci flow and Yau's uniformization conjecture,* preprint, arXiv: math. DG/0702257.

[18] Cheeger, J. and Ebin, D., *Comparison theorems in Riemannian geometry,* North-Holland (1975).

[19] Chen, B. L., *Strong uniqueness of the Ricci flow,* preprint, arXiv:math.DG/0706.3081.

[20] Chen, B. L., Tang, S. H. and Zhu, X. P., *A uniformization theorem of complete noncompact Kähler surfaces with positive bisectional curvature*, J. Diff. Geometry **67** (2004) 519–570.

[21] Chen, B. L. and Zhu, X. P., *Complete Riemannian manifolds with pointwise pinched curvature*, Invent. Math. **140** (2000), no. 2, 423–452.

[22] Chen, B. L. and Zhu, X. P., *On complete noncompact Kähler manifolds with positive bisectional curvature*, Math. Ann. **327** (2003) 1–23.

[23] Chen, B. L. and Zhu, X. P., *Volume growth and curvature decay of positively curved Kähler manifolds*, Quarterly Journal of Pure and Applied Mathematics Vol. **1**. no. 1 68–108 (2005).

[24] Chen, B. L. and Zhu, X. P., *Uniqueness of the Ricci flow on complete noncompact manifolds*, J. Diff. Geom. **74** (2006), 119–154.
[25] Chen, B. L. and Zhu, X. P., *Ricci flow with Surgery on four-manifolds with positive isotropic curvature*, J. Diff. Geom. **74** (2006), 177–264.
[26] Chen, H., *Pointwise quater-pinched 4 manifolds*, Ann. Global Anal. Geom. **9** (1991), 161–176.
[27] Chow, B. and Chu, S. C., *A geometric interpretation of Hamilton's Harnack inequality for the Ricci flow,* Math. Res. Lett.**2** (1995) 701–718.
[28] Chow, B. and Knopf, D., *The Ricci flow: An introduction*, Mathematical Surveys and Monographs, Amer. Math. Soc., Providence, RI, 2004.
[29] Chow, B., Lu, P. and Ni, L., *Hamilton's Ricci Flow,* Lectures in Contemporary Mathematics, Amer. Math. Soc., Providence, RI, 2006.
[30] Colding, T. H. and Minicozzi, W., P.II, *Estimates for the extinction time for the Ricci flow on certain 3-manifolds and a question of Perelman,* J. Amer. Math. Soc. **18** (2005), no. 3, 561–569.
[31] Daskalopoulos, P. and Hamilton, R. S., *Geometric estimates for the logarithmic fast diffusion equation*, Communications in Analysis and Geometry, vol. 12, no. 1–2, pp. 143–164, 2004.
[32] De Turck, D., *Deforming metrics in the direction of their Ricci tensors* J.Diff. Geom. **18** (1983), 157–162.
[33] Feldman, M., Ilmanen, T., and Knopf, D., *Rotationally symmetric shrinking and expanding gradient Kähler-Ricci solitons*, J. Differential Geom. **65** (2003), 169–209.
[34] Feldman, M., Ilmanen, T., and Ni, L., *Entropy and reduced distance for Ricci expanders*, J. Geom. Anal. **15** (2005), no. 1, 49–62.
[35] Greene, R. E. and Wu, H., *Lipschitz convergence of Riemannian manifolds*, Pacific J. Math. **131** (1988), 119–141.
[36] Gromoll, D., *Differenzierbare Strukturen und Metriken positiver Krümmung auf Sphären,* Math. Ann., **164**, (1966), 353–371.
[37] Gromoll, D. and Meyer, W., *On complete open manifolds of positive curvature*, Ann. of Math., **90** (1969), 95–90.
[38] Gromov, M., *Metric Structures for Riemannian and Non-Riemannian Spaces,* Edited by J.LaFontaine and P.Pansu English translation by Sean Michael Bates, 1998, Birkäuser.
[39] Gu, H. L., *A simple proof for the generalized Frankel conjecture,* preprint, arXiv:math.DG/0707.0035.
[40] Gu, H. L. and Zhu, X. P., *The Existence of Type II Singularities for the Ricci Flow on S^{n+1}*, preprint, arXiv:math.DG/0707.0033.
[41] Hamilton, R. S., *Three manifolds with positive Ricci curvature* , J. Diff. Geom. **17** (1982), 255–306.
[42] Hamilton, R. S., *Four-manifolds with positive curvature operator*, J. Differential Geom. **24** (1986), 153–179.
[43] Hamilton, R. S., *The Ricci flow on surfaces*, Contemporary Mathematics **71** (1988) 237–261.
[44] Hamilton, R. S., *The Harnack estimate for the Ricci flow*, J. Diff. Geom. **37** (1993), 225–243.
[45] Hamilton, R. S., *Eternal solutions to the Ricci flow*, J. Diff. Geom. **38** (1993), no. 1, 1–11.
[46] Hamilton, R. S., *A compactness property for solution of the Ricci flow*, Amer. J. Math. **117** (1995), 545–572.
[47] Hamilton, R. S., *The formation of singularities in the Ricci flow*, Surveys in Differential Geometry (Cambridge, MA, 1993), **2**, 7–136, International Press, Combridge, MA, 1995.

[48] Hamilton, R. S., *Four manifolds with positive isotropic curvature*, Commu. in Analysis and Geometry, **5** (1997), 1–92.
[49] Hamilton, R. S., *Non-singular solutions to the Ricci flow on three manifolds*, Commu. Anal. Geom. **1** (1999), 695–729.
[50] Hamilton, R. S., *A matrix Harnack estimate for the heat equation*, Commu. Anal. Geom. **1** (1993), 88–99.
[51] Huisken, G., *Ricci deformation of the metric on a Riemanian mnifold* J. Differential Geom. **21** (1985), 47–62.
[52] Ivey, T., *Ricci solitons on compact three-manifolds*, Diff. Geom. Appl. **3** (1993), 301–307.
[53] Jaco, W. and Shalen, P. *Seifert fibered spaces in 3-manifolds*, Mem. Amer. Math. Soc., **21** (1979), no. 220, viii+192 pp.
[54] Johannson, K., *Homotopy equivalences of 3-manifolds with boundaries*, Lecture Notes in Mathematics, **761** (1979), Springer, Berlin, ii+303 pp.
[55] Karcher, H., *Pinching implies strong pinching,* Comment. Math. Hevl. **46** (1971), 124–126.
[56] Kleiner, B. and Lott, J., *Note on Perelman's paper,* http://www.math. lsa.umich.edu /research /ricciflow /perelman.html.
[57] Klingenberg, W., *Über Riemannsche Mannigfaltigkeiten mit positiver Krümmung,* Comment. Math. Helv. **35** (1961), 47–54.
[58] Kneser, H., *Geschlossene Flächen in dreidmensionalen mannigfaltigkeiten,* Jahresbericht der Deut. Math. Verein., **38** (1929), 248–260.
[59] Kotschwar, B., *On rotationally invariant shrinking gradient Ricci solitons*, preprint, arXiv:math/0702597.
[60] Li, J.-F., *Eigenvalues and energy functionals with monotonicity formulae under Ricci flow*, Math. Ann. **338** (2007), no. 4, x927–946.
[61] Li, P. and Yau, S. T., *On the parabolic kernel of the Schrödinger operator,* Acta Math. **156** (1986), no. 3–4, 153–201.
[62] Ling, J., *Some Asymptotic Behavior of the first Eigenvalue along the Ricci Flow*, arXiv:0710.4326v1.
[63] Lu, P. and Tian, G., *The uniqueness of standard solutions in the work of Perelman,* Available at www.math.lsa.umich.edu /lott/ricciflow /StanUniqWork2.pdf,2005.
[64] Ma, L., *Eigenvalue monotonicity for the Ricci-Hamilton flow*, Ann. Global Anal. Geom. **29** (2006), no. 3, 287–292.
[65] Margerin, C., *Une caractrisation optimale de la structure diffrentielle standard de la sphre en terme de courbure pour (presque) toutes les dimensions. I. Les noncs,* (French) [A sharp curvature-type characterization of the smooth sphere in all but a few dimensions. I. Statement of the theorems] C. R. Acad. Sci. Paris Sr. I Math. **319** (1994), no. 6, 605–607.
[66] Margerin, C., *A sharp characterization of the smooth 4-sphere in curvature terms*, Comm. Anal. Geom. **6** (1998), no. 1, 21–65.
[67] Micallef, M. and Moore, J. D., *Minimal two-spheres and the topology of manifolds with positive curvature on totally isotropic two-planes*, Ann. of Math. (2) **127** (1988) 199–227.
[68] Milnor, J., *A unique factorization theorem for 3-manifolds*, Amer. J. Math., **84** (1962), 1–7.
[69] Mok, N., *The uniformization theorem for compact Kähler manifolds of nonnegative holomorphic bisectional curvature*, J. Differential Geom. **27** (1988), 179–214.
[70] Morgan, J. and Tian, G.*Ricci flow and the Poincaré conjecture*, preprint, arXiv:math.DG/0605667.
[71] Mori, S., *Projective manifolds with ample tangent bundles*, Ann. of Math. **100** (1979), 593–606.

[72] Naber, A., *Noncompact Shrinking 4-Solitons with Nonnegative Curvature*, preprint, arXiv:math.DG/07105579.
[73] Nguyen, H., Ph.D. thesis, the Australian National University.
[74] Ni, L., *Ancient solution to Kahler-Ricci flow*, Math. Res. Lett. **12** (2005), 633–653.
[75] Ni, L., *A note on Perelman's LYH inequality,* Comm. Anal. Geom. **14** (2006), 883–905.
[76] Ni, L. and Tam, L. F., *Plurisubharmonic functions and the structure of complete Kähler manifolds with nonnegative curvature*, J. Differential Geom. **64** (2003), no. 3, 457–524.
[77] Ni, L. and Wallach, N., *On a classification of the gradient shrinking solitons*, preprint, arXiv:math.DG/07103194
[78] Ni, L. and Wallach, N., *On 4-dimensional gradient shrinking solitons*, arXiv:math.DG/07103195
[79] Nishikawa, S., *Deformation of Riemannian metrics and manifolds with bounded curvature ratios*, Geometric measure theory and the calculus of variations (Arcata, Calif., 1984), 343–352, Proc. Sympos. Pure Math., 44, Amer. Math. Soc., Providence, RI, 1986
[80] Perelman, G., *The entropy formula for the Ricci flow and its geometric applications*, arXiv:math.DG/0211159.
[81] Perelman, G., *Ricci flow with surgery on three manifolds,* arXiv: math. DG/0303109.
[82] Perelman, G., *Finite extinction time to the solutions to the Ricci flow on certain three manifolds,* arXiv:math.DG/0307245.
[83] Peters, S., *Convergence of Riemannian mainfolds*, Comp. Math. **62** (1987), 3–16.
[84] Petersen, P. and Wylie, P., *Rigidity of gradient Ricci Solitons*, arXiv:math.DG/07103174
[85] Petersen, P. and Wylie, P., *On gradient Ricci solitons with Symmetry*, arXiv:math.DG/07103595
[86] Petersen, P. and Wylie, P., *On the classification of gradient Ricci solitons*, arXiv: math.DG/07121298
[87] Rauch, H. E., *A contribution to differential geometry in the large,* Ann. of Math. **54**(1951), 38–55.
[88] Ruh, E., *Curvature and differentiable structure on spheres,* Comment. Math. Hevl. 46, (1971), 127–136.
[89] Schoen, R. and Yau, S. T., *Existence of incompressible minimal surfaces and the topology of three-dimensional manifolds with nonnegative scalar curvature,* Ann. of Math. (2) **110** (1979), no. 1, 127–142.
[90] Sesum, N., *Limiting behaviour of the Ricci flow,* arXiv.math.DG/0402194.
[91] Shi, W. X., *Deforming the metric on complete Riemannian manifold*, J. Differential Geometry **30** (1989), 223–301.
[92] Shioya, T. and Yamaguchi, T., *Volume collapsed three-manifolds with a lower curvature bound,* Math. Ann. **333** (2005), 131–155.
[93] Simon. M., *A class of Riemannian manifolds that pinch when evolved by Ricci flow,* Manuscripta Math. **101** (2000), no. 1, 89–114.
[94] Sugimoto. M. and Shiohama. K., *On the differentiable pinching problem,* Math. Ann. **195**, (1971), 1–16.
[95] Thurston, W. P., *Three-dimensional manifolds, Kleinian groups and hyperbolic geometry,* Bull. Amer. Math. Soc. (N.S.) **6** (1982), no. 3, 357–381.
[96] Thurston, W. P., *Hyperbolic structures on* 3*-manifolds. I. Deformation of acylindrical manifolds,* Ann. of Math. (2) **124** (1986), no. 2, 203–246.
[97] Thurston, W. P., *Hyperbolic structures on 3-manifolds II: Surface groups and 3-manifolds which fiber over the circle,* preprint, 1988.

[98] Thurston, W. P., *Three-dimensional geometry and topology,* Vol. 1. Edited by Silvio Levy. Princeton Mathematical Series, 35. Princeton University Press, Princeton, NJ, 1997.

[99] Topping, P., *Lectures on the Ricci flow*, L.M.S. Cambridge University Press, 2006.

[100] Siu, Y. T. and Yau, S. T., *Complex Kähler manifolds of positive bisectional curvature,* Invent. Math. **59** (1980), 189–204.

[101] Yau, S. T., *Problem Section* in *Seminar on Differential Geometry,* Edited by S. T. Yau, Princeton University Press, 1982.

[102] Ye, R., *The logarithmic Sobolev inequality along the Ricci flow,* arXiv:math.DG/0707.2424

[103] Ye, R., *The logarithmic Sobolev inequality along the Ricci flow: the case* $\lambda_0(g_0)=0$, arXiv:math.DG/0708.2005.

[104] Ye, R., *The Log Entropy Functional Along the Ricci Flow,* arXiv:math.DG/0708.2008.

[105] Ye, R., *Sobolev Inequalities, Riesz Transforms and the Ricci Flow,* arXiv:math.DG/0709.0512.

[106] Ye, R., *Entropy Functionals, Sobolev Inequalities and kappa-Noncollapsing Estimates along the Ricci Flow*, arXiv:math.DG/0709.2724

[107] Zhang, Q., *A uniform Sobolev inequality under Ricci flow,* arXiv: math.DG/07061594.

[108] Zhang, Q., *Strong non-collapsing and uniform Sobolev inequalities for Ricci flow with surgeries*, arXiv:math.DG/07121329

Curvature Flows in Semi-Riemannian Manifolds

Claus Gerhardt

ABSTRACT. We prove that the limit hypersurfaces of converging curvature flows are stable, if the initial velocity has a weak sign, and give a survey of the existence and regularity results.

CONTENTS

1. Introduction 114
2. Notations and preliminary results 114
3. Evolution equations for some geometric quantities 117
4. Essential parabolic flow equations 122
5. Stability of the limit hypersurfaces 129
6. Existence results 140
7. The inverse mean curvature flow 155
8. The IMCF in ARW spaces 157
References 164

2000 *Mathematics Subject Classification.* 35J60, 53C21, 53C44, 53C50, 58J05.
Key words and phrases. semi-Riemannian manifold, mass, stable solutions, cosmological spacetime, general relativity, curvature flows, ARW spacetime.
This research was supported by the Deutsche Forschungsgemeinschaft.
Date: April 3, 2007.

1. Introduction

In this paper we want to give a survey of the existence and regularity results for extrinsic curvature flows in semi-Riemannian manifolds, i.e., Riemannian or Lorentzian ambient spaces, with an emphasis on flows in Lorentzian spaces. In order to treat both cases simultaneously terminology like spacelike, timelike, etc., that only makes sense in a Lorentzian setting should be ignored in the Riemannian case.

The general stability result for the limit hypersurfaces of converging curvature flows in Section 5 is new. The regularity result in Theorem 6.5—especially the time independent $C^{m+2,\alpha}$-estimates—for converging curvature flows that are graphs is interesting too.

2. Notations and preliminary results

The main objective of this section is to state the equations of Gauß, Codazzi, and Weingarten for hypersurfaces. In view of the subtle but important difference that is to be seen in the *Gauß equation* depending on the nature of the ambient space—Riemannian or Lorentzian—which we already mentioned in the introduction, we shall formulate the governing equations of a hypersurface M in a semi-Riemannian (n+1)-dimensional space N, which is either Riemannian or Lorentzian. Geometric quantities in N will be denoted by $(\bar{g}_{\alpha\beta}), (\bar{R}_{\alpha\beta\gamma\delta})$, etc., and those in M by $(g_{ij}), (R_{ijkl})$, etc. Greek indices range from 0 to n and Latin from 1 to n; the summation convention is always used. Generic coordinate systems in N resp. M will be denoted by (x^α) resp. (ξ^i). Covariant differentiation will simply be indicated by indices, only in case of possible ambiguity they will be preceded by a semicolon, i.e., for a function u in N, (u_α) will be the gradient and $(u_{\alpha\beta})$ the Hessian, but e.g., the covariant derivative of the curvature tensor will be abbreviated by $\bar{R}_{\alpha\beta\gamma\delta;\epsilon}$. We also point out that

$$\bar{R}_{\alpha\beta\gamma\delta;i} = \bar{R}_{\alpha\beta\gamma\delta;\epsilon} x_i^\epsilon \tag{2.1}$$

with obvious generalizations to other quantities.

Let M be a *spacelike* hypersurface, i.e., the induced metric is Riemannian, with a differentiable normal ν. We define the signature of ν, $\sigma = \sigma(\nu)$, by

$$\sigma = \bar{g}_{\alpha\beta}\nu^\alpha\nu^\beta = \langle \nu, \nu \rangle. \tag{2.2}$$

In case N is Lorentzian, $\sigma = -1$, and ν is timelike.

In local coordinates, (x^α) and (ξ^i), the geometric quantities of the spacelike hypersurface M are connected through the following equations

$$x_{ij}^\alpha = -\sigma h_{ij}\nu^\alpha \tag{2.3}$$

the so-called *Gauß formula*. Here, and also in the sequel, a covariant derivative is always a *full* tensor, i.e.,

$$x^\alpha_{ij} = x^\alpha_{,ij} - \Gamma^k_{ij} x^\alpha_k + \bar{\Gamma}^\alpha_{\beta\gamma} x^\beta_i x^\gamma_j. \tag{2.4}$$

The comma indicates ordinary partial derivatives.

In this implicit definition the *second fundamental form* (h_{ij}) is taken with respect to $-\sigma\nu$.

The second equation is the *Weingarten equation*

$$\nu^\alpha_i = h^k_i x^\alpha_k, \tag{2.5}$$

where we remember that ν^α_i is a full tensor.

Finally, we have the *Codazzi equation*

$$h_{ij;k} - h_{ik;j} = \bar{R}_{\alpha\beta\gamma\delta} \nu^\alpha x^\beta_i x^\gamma_j x^\delta_k \tag{2.6}$$

and the *Gauß equation*

$$R_{ijkl} = \sigma\{h_{ik}h_{jl} - h_{il}h_{jk}\} + \bar{R}_{\alpha\beta\gamma\delta} x^\alpha_i x^\beta_j x^\gamma_k x^\delta_l. \tag{2.7}$$

Here, the signature of ν comes into play.

DEFINITION 2.1. (i) Let $F \in C^0(\bar{\Gamma}) \cap C^{2,\alpha}(\Gamma)$ be a strictly monotone curvature function, where $\Gamma \subset \mathbb{R}^n$ is a convex, open, symmetric cone containing the positive cone, such that

$$F_{|\partial\Gamma} = 0 \quad \wedge \quad F_{|\Gamma} > 0. \tag{2.8}$$

Let N be semi-Riemannian. A spacelike, orientable[1] hypersurface $M \subset N$ is called *admissible*, if its principal curvatures with respect to a chosen normal lie in Γ. This definition also applies to subsets of M.

(ii) Let M be an admissible hypersurface and f a function defined in a neighbourhood of M. M is said to be an *upper barrier* for the pair (F, f), if

$$F_{|M} \geq f \tag{2.9}$$

(iii) Similarly, a spacelike, orientable hypersurface M is called a *lower barrier* for the pair (F, f), if at the points $\Sigma \subset M$, where M is admissible, there holds

$$F_{|\Sigma} \leq f. \tag{2.10}$$

Σ may be empty.

(iv) If we consider the mean curvature function, $F = H$, then we suppose F to be defined in $\mathbb{R}^n$ and any spacelike, orientable hypersurface is admissible.

[1] A hypersurface is said to be orientable, if it has a continuous normal field.

One of the assumptions that are used when proving a priori estimates is that there exists a strictly convex function $\chi \in C^2(\bar{\Omega})$ in a given domain Ω. We shall state sufficient geometric conditions guaranteeing the existence of such a function. The lemma below will be valid in Lorentzian as well as Riemannian manifolds, but we formulate and prove it only for the Lorentzian case.

LEMMA 2.2. *Let N be globally hyperbolic, $\mathcal{S}_0$ a Cauchy hypersurface, (x^α) a special coordinate system associated with $\mathcal{S}_0$, and $\bar{\Omega} \subset N$ be compact. Then, there exists a strictly convex function $\chi \in C^2(\bar{\Omega})$ provided the level hypersurfaces $\{x^0 = \text{const}\}$ that intersect $\bar{\Omega}$ are strictly convex.*

PROOF. For greater clarity set $t = x^0$, i.e., t is a globally defined time function. Let $x = x(\xi)$ be a local representation for $\{t = \text{const}\}$, and t_i, t_{ij} be the covariant derivatives of t with respect to the induced metric, and $t_\alpha, t_{\alpha\beta}$ be the covariant derivatives in N, then

$$0 = t_{ij} = t_{\alpha\beta} x_i^\alpha x_j^\beta + t_\alpha x_{ij}^\alpha, \tag{2.11}$$

and therefore,

$$t_{\alpha\beta} x_i^\alpha x_j^\beta = -t_\alpha x_{ij}^\alpha = -\bar{h}_{ij} t_\alpha \nu^\alpha. \tag{2.12}$$

Here, (ν^α) is past directed, i.e., the right-hand side in (2.12) is positive definite in $\bar{\Omega}$, since (t_α) is also past directed.

Choose $\lambda > 0$ and define $\chi = e^{\lambda t}$, so that

$$\chi_{\alpha\beta} = \lambda^2 e^{\lambda t} t_\alpha t_\beta + \lambda e^{\lambda t} t_{\alpha\beta}. \tag{2.13}$$

Let $p \in \Omega$ be arbitrary, $\mathcal{S} = \{t = t(p)\}$ be the level hypersurface through p, and $(\eta^\alpha) \in T_p(N)$. Then, we conclude

$$e^{-\lambda t} \chi_{\alpha\beta} \eta^\alpha \eta^\beta = \lambda^2 |\eta^0|^2 + \lambda t_{ij} \eta^i \eta^j + 2\lambda t_{0j} \eta^0 \eta^i, \tag{2.14}$$

where t_{ij} now represents the left-hand side in (2.12), and we infer further

$$\begin{aligned} e^{-\lambda t} \chi_{\alpha\beta} \eta^\alpha \eta^\beta &\ge \tfrac{1}{2} \lambda^2 |\eta|^{0^2} + [\lambda \epsilon - c_\epsilon] \sigma_{ij} \eta^i \eta^j \\ &\ge \tfrac{\epsilon}{2} \lambda \{-|\eta^0|^2 + \sigma_{ij} \eta^i \eta^j\} \end{aligned} \tag{2.15}$$

for some $\epsilon > 0$, and where λ is supposed to be large. Therefore, we have in $\bar{\Omega}$

$$\chi_{\alpha\beta} \ge c \bar{g}_{\alpha\beta}, \quad c > 0, \tag{2.16}$$

i.e., χ is strictly convex. □

3. Evolution equations for some geometric quantities

Curvature flows are used for different purposes, they can be merely vehicles to approximate a stationary solution, in which case the flow is driven not only by a curvature function but also by the corresponding right-hand side, an external force, if you like, or the flow is a pure curvature flow driven only by a curvature function, and it is used to analyze the topology of the initial hypersurface, if the ambient space is Riemannian, or the singularities of the ambient space, in the Lorentzian case.

In this section we are treating very general curvature flows[2] in a semi-Riemannian manifold $N = N^{n+1}$, though we only have the Riemannian or Lorentzian case in mind, such that the flow can be either a pure curvature flow or may also be driven by an external force. The nature of the ambient space, i.e., the signature of its metric, is expressed by a parameter $\sigma = \pm 1$, such that $\sigma = 1$ corresponds to the Riemannian and $\sigma = -1$ the Lorentzian case. The parameter σ can also be viewed as the signature of the normal of the spacelike hypersurfaces, namely,

$$\sigma = \langle \nu, \nu \rangle. \tag{3.1}$$

Properties like spacelike, achronal, etc., however, only make sense, when N is Lorentzian and should be ignored otherwise.

We consider a strictly monotone, symmetric, and concave curvature $F \in C^{4,\alpha}(\Gamma)$, homogeneous of degree 1, a function $0 < f \in C^{4,\alpha}(\Omega)$, where $\Omega \subset N$ is an open set, and a real function $\Phi \in C^{4,\alpha}(\mathbb{R}_+)$ satisfying

$$\dot\Phi > 0 \quad \text{and} \quad \ddot\Phi \le 0. \tag{3.2}$$

For notational reasons, let us abbreviate

$$\tilde f = \Phi(f). \tag{3.3}$$

Important examples of functions Φ are

$$\Phi(r) = r, \quad \Phi(r) = \log r, \quad \Phi(r) = -r^{-1} \tag{3.4}$$

or

$$\Phi(r) = r^{\frac{1}{k}}, \quad \Phi(r) = -r^{-\frac{1}{k}}, \qquad k \ge 1. \tag{3.5}$$

REMARK 3.1. The latter choices are necessary, if the curvature function F is not homogeneous of degree 1 but of degree k, like the symmetric polynomials H_k. In this case we would sometimes like to define $F = H_k$ and not $H_k^{1/k}$, since

$$F^{ij} = \frac{\partial F}{\partial h_{ij}} \tag{3.6}$$

[2]We emphasize that we are only considering flows driven by the extrinsic curvature, not by the intrinsic curvature.

is then divergence free, if the ambient space is a spaceform, cf. Lemma 5.8 on page 139, though on the other hand we need a concave operator for technical reasons, hence we have to take the k-th root.

The curvature flow is given by the evolution problem

$$\begin{aligned} \dot{x} &= -\sigma(\Phi - \tilde{f})\nu, \\ x(0) &= x_0, \end{aligned} \tag{3.7}$$

where x_0 is an embedding of an initial compact, spacelike hypersurface $M_0 \subset \Omega$ of class $C^{6,\alpha}$, $\Phi = \Phi(F)$, and F is evaluated at the principal curvatures of the flow hypersurfaces $M(t)$, or, equivalently, we may assume that F depends on the second fundamental form (h_{ij}) and the metric (g_{ij}) of $M(t)$; $x(t)$ is the embedding of $M(t)$ and σ the signature of the normal $\nu = \nu(t)$, which is identical to the normal used in the Gaussian formula (2.3) on page 114.

The initial hypersurface should be *admissible*, i.e., its principal curvatures should belong to the convex, symmetric cone $\Gamma \subset \mathbb{R}^n$.

This is a parabolic problem, so short-time existence is guaranteed, cf. [**18**, Chapter 2.5]

There will be a slight ambiguity in the terminology, since we shall call the evolution parameter *time*, but this lapse shouldn't cause any misunderstandings, if the ambient space is Lorentzian.

At the moment we consider a sufficiently smooth solution of the initial value problem (3.7) and want to show how the metric, the second fundamental form, and the normal vector of the hypersurfaces $M(t)$ evolve. All time derivatives are *total* derivatives, i.e., covariant derivatives of tensor fields defined over the curve $x(t)$, cf. [**17**, Chapter 11.5]; t is the flow parameter, also referred to as *time*, and (ξ^i) are local coordinates of the initial embedding $x_0 = x_0(\xi)$ which will also serve as coordinates for the the flow hypersurfaces $M(t)$. The coordinates in N will be labelled (x^α), $0 \le \alpha \le n$.

LEMMA 3.2 (Evolution of the metric). *The metric g_{ij} of $M(t)$ satisfies the evolution equation*

$$\dot{g}_{ij} = -2\sigma(\Phi - \tilde{f})h_{ij}. \tag{3.8}$$

PROOF. Differentiating

$$g_{ij} = \langle x_i, x_j \rangle \tag{3.9}$$

covariantly with respect to t yields

$$\begin{aligned} \dot{g}_{ij} &= \langle \dot{x}_i, x_j \rangle + \langle x_i, \dot{x}_j \rangle \\ &= -2\sigma(\Phi - \tilde{f})\langle x_i, \nu_j \rangle = -2\sigma(\Phi - \tilde{f})h_{ij}, \end{aligned} \tag{3.10}$$

in view of the Codazzi equations. □

LEMMA 3.3 (Evolution of the normal). *The normal vector evolves according to*

$$\dot{\nu} = \nabla_M(\Phi - \tilde{f}) = g^{ij}(\Phi - \tilde{f})_i x_j. \tag{3.11}$$

PROOF. Since ν is unit normal vector we have $\dot{\nu} \in T(M)$. Furthermore, differentiating

$$0 = \langle \nu, x_i \rangle \tag{3.12}$$

with respect to t, we deduce

$$\langle \dot{\nu}, x_i \rangle = -\langle \nu, \dot{x}_i \rangle = (\Phi - \tilde{f})_i. \tag{3.13}$$

□

LEMMA 3.4 (Evolution of the second fundamental form). *The second fundamental form evolves according to*

$$\dot{h}_i^j = (\Phi - \tilde{f})_i^j + \sigma(\Phi - \tilde{f}) h_i^k h_k^j + \sigma(\Phi - \tilde{f}) \bar{R}_{\alpha\beta\gamma\delta} \nu^\alpha x_i^\beta \nu^\gamma x_k^\delta g^{kj} \tag{3.14}$$

and

$$\dot{h}_{ij} = (\Phi - \tilde{f})_{ij} - \sigma(\Phi - \tilde{f}) h_i^k h_{kj} + \sigma(\Phi - \tilde{f}) \bar{R}_{\alpha\beta\gamma\delta} \nu^\alpha x_i^\beta \nu^\gamma x_j^\delta. \tag{3.15}$$

PROOF. We use the *Ricci identities* to interchange the covariant derivatives of ν with respect to t and ξ^i

$$\begin{aligned} \tfrac{D}{dt}(\nu_i^\alpha) &= (\dot{\nu}^\alpha)_i - \bar{R}^\alpha{}_{\beta\gamma\delta} \nu^\beta x_i^\gamma \dot{x}^\delta \\ &= g^{kl}(\Phi - \tilde{f})_{ki} x_l^\alpha + g^{kl}(\Phi - \tilde{f})_k x_{li}^\alpha - \bar{R}^\alpha{}_{\beta\gamma\delta} \nu^\beta x_i^\gamma \dot{x}^\delta. \end{aligned} \tag{3.16}$$

For the second equality we used (3.11). On the other hand, in view of the Weingarten equation we obtain

$$\tfrac{D}{dt}(\nu_i^\alpha) = \tfrac{D}{dt}(h_i^k x_k^\alpha) = \dot{h}_i^k x_k^\alpha + h_i^k \dot{x}_k^\alpha. \tag{3.17}$$

Multiplying the resulting equation with $\bar{g}_{\alpha\beta} x_j^\beta$ we conclude

$$\dot{h}_i^k g_{kj} - \sigma(\Phi - \tilde{f}) h_i^k h_{kj} = (\Phi - \tilde{f})_{ij} + \sigma(\Phi - \tilde{f}) \bar{R}_{\alpha\beta\gamma\delta} \nu^\alpha x_i^\beta \nu^\gamma x_j^\delta \tag{3.18}$$

or equivalently (3.14).

To derive (3.15), we differentiate

$$h_{ij} = h_i^k g_{kj} \tag{3.19}$$

with respect to t and use (3.8). □

We emphasize that equation (3.14) describes the evolution of the second fundamental form more meaningfully than (3.15), since the mixed tensor is independent of the metric.

LEMMA 3.5 (Evolution of $(\Phi - \tilde f)$). *The term* $(\Phi - \tilde f)$ *evolves according to the equation*

$$(3.20)\qquad \begin{aligned}(\Phi - \tilde f)' - \dot\Phi F^{ij}(\Phi - \tilde f)_{ij} &= \sigma\dot\Phi F^{ij} h_{ik}h_j^k(\Phi - \tilde f) + \sigma\tilde f_\alpha\nu^\alpha(\Phi - \tilde f)\\ &\quad + \sigma\dot\Phi F^{ij}\bar R_{\alpha\beta\gamma\delta}\nu^\alpha x_i^\beta\nu^\gamma x_j^\delta(\Phi - \tilde f),\end{aligned}$$

where

$$(3.21)\qquad (\Phi - \tilde f)' = \frac{d}{dt}(\Phi - \tilde f)$$

and

$$(3.22)\qquad \dot\Phi = \frac{d}{dr}\Phi(r).$$

PROOF. When we differentiate F with respect to t we consider F to depend on the mixed tensor h_i^j and conclude

$$(3.23)\qquad (\Phi - \tilde f)' = \dot\Phi F_j^i \dot h_i^j - \tilde f_\alpha \dot x^\alpha;$$

The equation (3.20) then follows in view of (3.7) and (3.14). □

REMARK 3.6. The preceding conclusions, except Lemma 3.5, remain valid for flows which do not depend on the curvature, i.e., for flows

$$(3.24)\qquad \begin{aligned}\dot x &= -\sigma(-f)\nu = \sigma f\nu,\\ x(0) &= x_0,\end{aligned}$$

where $f = f(x)$ is defined in an open set Ω containing the initial spacelike hypersurface M_0. In the preceding equations we only have to set $\Phi = 0$ and $\tilde f = f$.

The evolution equation for the mean curvature then looks like

$$(3.25)\qquad \dot H = -\Delta f - \sigma\{|A|^2 + \bar R_{\alpha\beta}\nu^\alpha\nu^\beta\}f,$$

where the Laplacian is the Laplace operator on the hypersurface $M(t)$. This is exactly the derivative of the mean curvature operator with respect to normal variations as we shall see in a moment.

But first let us consider the following example.

EXAMPLE 3.7. Let (x^α) be a future directed Gaussian coordinate system in N, such that the metric can be expressed in the form

$$(3.26)\qquad d\bar s^2 = e^{2\psi}\{\sigma(dx^0)^2 + \sigma_{ij}dx^i dx^j\}.$$

Denote by $M(t)$ the coordinate slices $\{x^0 = t\}$, then $M(t)$ can be looked at as the flow hypersurfaces of the flow

$$(3.27)\qquad \dot x = -\sigma(-e^\psi)\bar\nu,$$

where we denote the geometric quantities of the slices by $\bar g_{ij}$, $\bar\nu$, $\bar h_{ij}$, etc.

Here x is the embedding

$$x = x(t, \xi^i) = (t, x^i). \tag{3.28}$$

Notice that, if N is Riemannian, the coordinate system and the normal are always chosen such that $\nu^0 > 0$, while, if N is Lorentzian, we always pick the past directed normal.

Hence the mean curvature of the slices evolves according to

$$\dot{\bar{H}} = -\Delta e^{\psi} - \sigma\{|\bar{A}|^2 + \bar{R}_{\alpha\beta}\bar{\nu}^\alpha\bar{\nu}^\beta\}e^{\psi}. \tag{3.29}$$

We can now derive the linearization of the mean curvature operator of a spacelike hypersurface, compact or non-compact.

3.8. Let $M_0 \subset N$ be a spacelike hypersurface of class C^4. We first assume that M_0 is compact; then there exists a tubular neighbourhood $\mathcal{U}$ and a corresponding *normal* Gaussian coordinate system (x^α) of class C^3 such that $\frac{\partial}{\partial x^0}$ is normal to M_0.

Let us consider in $\mathcal{U}$ of M_0 spacelike hypersurfaces M that can be written as graphs over M_0, $M = \operatorname{graph} u$, in the corresponding normal Gaussian coordinate system. Then the mean curvature of M can be expressed as

$$H = \{-\Delta u + \bar{H} - \sigma v^{-2} u^i u^j \bar{h}_{ij}\}v, \tag{3.30}$$

where $\sigma = \langle\nu, \nu\rangle$, and hence, choosing $u = \epsilon\varphi$, $\varphi \in C^2(M_0)$, we deduce

$$\begin{aligned} \frac{d}{d\epsilon} H_{|_{\epsilon=0}} &= -\Delta\varphi + \dot{\bar{H}}\varphi \\ &= -\Delta\varphi - \sigma(|\bar{A}|^2 + \bar{R}_{\alpha\beta}\nu^\alpha\nu^\beta)\varphi, \end{aligned} \tag{3.31}$$

in view of (3.29).

The right-hand side is the derivative of the mean curvature operator applied to φ.

If M_0 is non-compact, tubular neighbourhoods exist locally and the relation (3.31) will be valid for any $\varphi \in C_c^2(M_0)$ by using a partition of unity.

The preceding linearization can be immediately generalized to a hypersurface M_0 solving the equation

$$F_{|_{M_0}} = f, \tag{3.32}$$

where $f = f(x)$ is defined in a neighbourhood of M_0 and $F = F(h_{ij})$ is curvature operator.

LEMMA 3.9. *Let M_0 be of class $C^{m,\alpha}$, $m \geq 2$, $0 \leq \alpha \leq 1$, satisfy* (3.32). *Let $\mathcal{U}$ be a (local) tubular neighbourhood of M_0, then the linearization of the* *operator $F-f$ expressed in the normal Gaussian coordinate system (x^α) corresponding to $\mathcal{U}$ and evaluated at M_0 has the form*

$$-F^{ij}u_{ij}-\sigma\{F^{ij}h_i^kh_{kj}+F^{ij}\bar R_{\alpha\beta\gamma\delta}\nu^\alpha x_i^\beta\nu^\gamma x_j^\delta+f_\alpha\nu^\alpha\}u, \tag{3.33}$$

where u is a function defined in M_0, and all geometric quantities are those of M_0; the derivatives are covariant derivatives with respect to the induced metric of M_0. The operator will be self-adjoint, if F^{ij} is divergence free.

PROOF. For simplicity assume that M_0 is compact, and let $u\in C^2(M_0)$ be fixed. Then the hypersurfaces

$$M_\epsilon=\operatorname{graph}(\epsilon u) \tag{3.34}$$

stay in the tubular neighbourhood $\mathcal{U}$ for small ϵ, $|\epsilon|<\epsilon_0$, and their second fundamental forms (h_{ij}) can be expressed as

$$v^{-1}h_{ij}=-(\epsilon u)_{ij}+\bar h_{ij}, \tag{3.35}$$

where $\bar h_{ij}$ is the second fundamental form of the coordinate slices $\{x^0=\text{const}\}$.

We are interested in

$$\frac{d}{d\epsilon}(F-f)_{|\epsilon=0}. \tag{3.36}$$

To differentiate F with respect to ϵ it is best to consider the mixed form (h_i^j) of the second fundamental form to derive

$$\frac{d}{d\epsilon}(F-f)=F_j^i\dot h_i^j-\frac{\partial f}{\partial x^0}u=-F^{ij}u_{ij}+F_j^i\dot{\bar h}_i^j u-\frac{\partial f}{\partial x^0}u, \tag{3.37}$$

where the equation is evaluated at $\epsilon=0$ and $\dot{\bar h}_i^j$ is the derivative of $\bar h_i^j$ with respect to x^0.

The result then follows from the evolution equation (3.14) for the flow (3.27), i.e., we have to replace $(\Phi-\tilde f)$ in (3.14) by -1. □

4. Essential parabolic flow equations

From (3.14) on page 119 we deduce with the help of the Ricci identities a parabolic equation for the second fundamental form

LEMMA 4.1. *The mixed tensor h_i^j satisfies the parabolic equation*

$$\begin{aligned}\dot h_i^j-\dot\Phi F^{kl}h_{i;kl}^j&=\sigma\dot\Phi F^{kl}h_{rk}h_l^rh_i^j-\sigma\dot\Phi Fh_{ri}h^{rj}+\sigma(\Phi-\tilde f)h_i^kh_k^j\\&\quad-\tilde f_{\alpha\beta}x_i^\alpha x_k^\beta g^{kj}+\sigma\tilde f_\alpha\nu^\alpha h_i^j+\dot\Phi F^{kl,rs}h_{kl;i}h_{rs;}{}^j\\&\quad+\ddot\Phi F_iF^j+2\dot\Phi F^{kl}\bar R_{\alpha\beta\gamma\delta}x_m^\alpha x_i^\beta x_k^\gamma x_r^\delta h_l^m g^{rj}\end{aligned} \tag{4.1}$$

$$\begin{aligned}
&- \dot{\Phi} F^{kl} \bar{R}_{\alpha\beta\gamma\delta} x_m^\alpha x_k^\beta x_r^\gamma x_l^\delta h_i^m g^{rj} - \dot{\Phi} F^{kl} \bar{R}_{\alpha\beta\gamma\delta} x_m^\alpha x_k^\beta x_i^\gamma x_l^\delta h^{mj} \\
&+ \sigma \dot{\Phi} F^{kl} \bar{R}_{\alpha\beta\gamma\delta} \nu^\alpha x_k^\beta \nu^\gamma x_l^\delta h_i^j - \sigma \dot{\Phi} F \bar{R}_{\alpha\beta\gamma\delta} \nu^\alpha x_i^\beta \nu^\gamma x_m^\delta g^{mj} \\
&+ \sigma(\Phi - \tilde{f}) \bar{R}_{\alpha\beta\gamma\delta} \nu^\alpha x_i^\beta \nu^\gamma x_m^\delta g^{mj} \\
&+ \dot{\Phi} F^{kl} \bar{R}_{\alpha\beta\gamma\delta;\epsilon} \{\nu^\alpha x_k^\beta x_l^\gamma x_i^\delta x_m^\epsilon g^{mj} + \nu^\alpha x_i^\beta x_k^\gamma x_m^\delta x_l^\epsilon g^{mj}\}.
\end{aligned}$$

PROOF. We start with equation (3.14) on page 119 and shall evaluate the term

$$(\Phi - \tilde{f})_i^j; \tag{4.2}$$

since we are only working with covariant spatial derivatives in the subsequent proof, we may—and shall—consider the covariant form of the tensor

$$(\Phi - \tilde{f})_{ij}. \tag{4.3}$$

First we have

$$\Phi_i = \dot{\Phi} F_i = \dot{\Phi} F^{kl} h_{kl;i} \tag{4.4}$$

and

$$\Phi_{ij} = \dot{\Phi} F^{kl} h_{kl;ij} + \ddot{\Phi} F^{kl} h_{kl;i} F^{rs} h_{rs;j} + \dot{\Phi} F^{kl,rs} h_{kl,;i} h_{rs;j}. \tag{4.5}$$

Next, we want to replace $h_{kl;ij}$ by $h_{ij;kl}$. Differentiating the Codazzi equation

$$h_{kl;i} = h_{ik;l} + \bar{R}_{\alpha\beta\gamma\delta} \nu^\alpha x_k^\beta x_l^\gamma x_i^\delta, \tag{4.6}$$

where we also used the symmetry of h_{ik}, yields

$$\begin{aligned}
h_{kl;ij} = h_{ik;lj} &+ \bar{R}_{\alpha\beta\gamma\delta;\epsilon} \nu^\alpha x_k^\beta x_l^\gamma x_i^\delta x_j^\epsilon \\
&+ \bar{R}_{\alpha\beta\gamma\delta} \{\nu_j^\alpha x_k^\beta x_l^\gamma x_i^\delta + \nu^\alpha x_{kj}^\beta x_l^\gamma x_i^\delta + \nu^\alpha x_k^\beta x_{lj}^\gamma x_i^\delta + \nu^\alpha x_k^\beta x_l^\gamma x_{ij}^\delta\}.
\end{aligned} \tag{4.7}$$

To replace $h_{kl;ij}$ by $h_{ij;kl}$ we use the Ricci identities

$$h_{ik;lj} = h_{ik;jl} + h_{ak} R^a{}_{ilj} + h_{ai} R^a{}_{klj} \tag{4.8}$$

and differentiate once again the Codazzi equation

$$h_{ik;j} = h_{ij;k} + \bar{R}_{\alpha\beta\gamma\delta} \nu^\alpha x_i^\beta x_k^\gamma x_j^\delta. \tag{4.9}$$

To replace $\tilde{f}_{ij}$ we use the chain rule

$$\begin{aligned}
\tilde{f}_i &= \tilde{f}_\alpha x_i^\alpha, \\
\tilde{f}_{ij} &= \tilde{f}_{\alpha\beta} x_i^\alpha x_j^\beta + \tilde{f}_\alpha x_{ij}^\alpha.
\end{aligned} \tag{4.10}$$

Then, because of the Gauß equation, Gaussian formula, and Weingarten equation, the symmetry properties of the Riemann curvature tensor and the assumed homogeneity of F, i.e.,

$$F = F^{kl} h_{kl}, \tag{4.11}$$

we deduce (4.1) from (3.14) on page 119 after reverting to the mixed representation. □

REMARK 4.2. If we had assumed F to be homogeneous of degree d_0 instead of 1, then we would have to replace the explicit term F—occurring twice in the preceding lemma—by d_0F.

If the ambient semi-Riemannian manifold is a space of constant curvature, then the evolution equation of the second fundamental form simplifies considerably, as can be easily verified.

LEMMA 4.3. *Let N be a space of constant curvature K_N, then the second fundamental form of the curvature flow* (3.7) *on page* 118 *satisfies the parabolic equation*

$$\begin{aligned} \dot{h}_i^j - \dot{\Phi} F^{kl} h_{i;kl}^j &= \sigma \dot{\Phi} F^{kl} h_{rk} h_l^r h_i^j - \sigma \dot{\Phi} F h_{ri} h^{rj} + \sigma(\Phi - \tilde{f}) h_i^k h_k^j \\ &\quad - \tilde{f}_{\alpha\beta} x_i^\alpha x_k^\beta g^{kj} + \sigma \tilde{f}_\alpha \nu^\alpha h_i^j + \dot{\Phi} F^{kl,rs} h_{kl;i} h_{rs;}{}^j \\ &\quad + \ddot{\Phi} F_i F^j + K_N \{(\Phi - \tilde{f}) \delta_i^j + \dot{\Phi} F \delta_i^j - \dot{\Phi} F^{kl} g_{kl} h_i^j\}. \end{aligned} \tag{4.12}$$

Let us now assume that the open set $\Omega \subset N$ containing the flow hypersurfaces can be covered by a Gaussian coordinate system (x^α), i.e., Ω can be topologically viewed as a subset of $I \times \mathcal{S}_0$, where $\mathcal{S}_0$ is a compact Riemannian manifold and I an interval. We assume furthermore, that the flow hypersurfaces can be written as graphs over $\mathcal{S}_0$

$$M(t) = \{\, x^0 = u(x^i) \colon x = (x^i) \in \mathcal{S}_0 \,\}; \tag{4.13}$$

we use the symbol x ambiguously by denoting points $p = (x^\alpha) \in N$ as well as points $p = (x^i) \in \mathcal{S}_0$ simply by x, however, we are careful to avoid confusions.

Suppose that the flow hypersurfaces are given by an embedding $x = x(t, \xi)$, where $\xi = (\xi^i)$ are local coordinates of a compact manifold M_0, which then has to be homeomorphic to $\mathcal{S}_0$, then

$$\begin{aligned} x^0 &= u(t, \xi) = u(t, x(t, \xi)), \\ x^i &= x^i(t, \xi). \end{aligned} \tag{4.14}$$

The induced metric can be expressed as

$$g_{ij} = \langle x_i, x_j \rangle = \sigma u_i u_j + \sigma_{kl} x_i^k x_j^l, \tag{4.15}$$

where

$$u_i = u_k x_i^k, \tag{4.16}$$

i.e.,

$$g_{ij} = \{\sigma u_k u_l + \sigma_{kl}\} x_i^k x_j^l, \tag{4.17}$$

hence the (time dependent) *Jacobian* (x_i^k) is invertible, and the (ξ^i) can also be viewed as coordinates for $\mathcal{S}_0$.

Looking at the component $\alpha = 0$ of the flow equation (3.7) on page 118 we obtain a scalar flow equation

$$\dot{u} = -e^{-\psi} v^{-1} (\Phi - \tilde{f}), \tag{4.18}$$

which is the same in the Lorentzian as well as in the Riemannian case, where

$$v^2 = 1 - \sigma \sigma^{ij} u_i u_j, \tag{4.19}$$

and where

$$|Du|^2 = \sigma^{ij} u_i u_j \tag{4.20}$$

is of course a scalar, i.e., we obtain the same expression regardless, if we use the coordinates x^i or ξ^i.

The time derivative in (4.18) is a total time derivative, if we consider u to depend on $u = u(t, x(t, \xi))$. For the partial time derivative we obtain

$$\frac{\partial u}{\partial t} = \dot{u} - u_k \dot{x}_i^k = -e^{-\psi} v (\Phi - \tilde{f}), \tag{4.21}$$

in view of (3.7) on page 118 and our choice of normal $\nu = (\nu^\alpha)$

$$(\nu^\alpha) = \sigma e^{-\psi} v^{-1} (1, -\sigma u^i), \tag{4.22}$$

where $u^i = \sigma^{ij} u_j$.

Controlling the C^1-norm of the graphs $M(t)$ is tantamount to controlling v, if N is Riemannian, and $\tilde{v} = v^{-1}$, if N is Lorentzian. The evolution equations satisfied by these quantities are also very important, since they are used for the a priori estimates of the second fundamental form.

Let us start with the Lorentzian case.

LEMMA 4.4 (Evolution of $\tilde{v}$). *Consider the flow (3.7) in a Lorentzian space N such that the spacelike flow hypersurfaces can be written as graphs over $\mathcal{S}_0$. Then, $\tilde{v}$ satisfies the evolution equation*

$$\begin{aligned}\dot{\tilde{v}} - \dot{\Phi} F^{ij} \tilde{v}_{ij} = &-\dot{\Phi} F^{ij} h_{ik} h_j^k \tilde{v} + [(\Phi - \tilde{f}) - \dot{\Phi} F] \eta_{\alpha\beta} \nu^\alpha \nu^\beta \\ &- 2\dot{\Phi} F^{ij} h_j^k x_i^\alpha x_k^\beta \eta_{\alpha\beta} - \dot{\Phi} F^{ij} \eta_{\alpha\beta\gamma} x_i^\beta x_j^\gamma \nu^\alpha \\ &- \dot{\Phi} F^{ij} \bar{R}_{\alpha\beta\gamma\delta} \nu^\alpha x_i^\beta x_k^\gamma x_j^\delta \eta_\epsilon x_l^\epsilon g^{kl} \\ &- \tilde{f}_\beta x_i^\beta x_k^\alpha \eta_\alpha g^{ik},\end{aligned} \tag{4.23}$$

where η is the covariant vector field $(\eta_\alpha) = e^{\psi}(-1, 0, \dots, 0)$.

PROOF. We have $\tilde{v} = \langle \eta, \nu \rangle$. Let (ξ^i) be local coordinates for $M(t)$. Differentiating $\tilde{v}$ covariantly we deduce

$$\tilde{v}_i = \eta_{\alpha\beta} x_i^\beta \nu^\alpha + \eta_\alpha \nu_i^\alpha, \tag{4.24}$$

$$\tilde{v}_{ij} = \eta_{\alpha\beta\gamma} x_i^\beta x_j^\gamma \nu^\alpha + \eta_{\alpha\beta} x_{ij}^\beta \nu^\alpha + \eta_{\alpha\beta} x_i^\beta \nu_j^\alpha + \eta_{\alpha\beta} x_j^\beta \nu_i^\alpha + \eta_\alpha \nu_{ij}^\alpha \tag{4.25}$$

The time derivative of $\tilde{v}$ can be expressed as

$$\begin{aligned} \dot{\tilde{v}} &= \eta_{\alpha\beta} \dot{x}^\beta \nu^\alpha + \eta_\alpha \dot{\nu}^\alpha \\ &= \eta_{\alpha\beta} \nu^\alpha \nu^\beta (\Phi - \tilde{f}) + (\Phi - \tilde{f})^k x_k^\alpha \eta_\alpha \\ &= \eta_{\alpha\beta} \nu^\alpha \nu^\beta (\Phi - \tilde{f}) + \dot{\Phi} F^k x_k^\alpha \eta_\alpha - \tilde{f}_\beta x_i^\beta x_k^\alpha g^{ik} \eta_\alpha, \end{aligned} \tag{4.26}$$

where we have used (3.11) on page 119.

Substituting (4.25) and (4.26) in (4.23), and simplifying the resulting equation with the help of the Weingarten and Codazzi equations, we arrive at the desired conclusion. □

In the Riemannian case we consider a normal Gaussian coordinate system (x^α), for otherwise we won't obtain a priori estimates for v, at least not without additional strong assumptions. We also refer to $x^0 = r$ as the *radial distance function.*

LEMMA 4.5 (Evolution of v). *Consider the flow (3.7) in a normal Gaussian coordinate system where the $M(t)$ can be written as graphs of a function $u(t)$ over some compact Riemannian manifold $\mathcal{S}_0$. Then the quantity*

$$v = \sqrt{1 + |Du|^2} = (r_\alpha \nu^\alpha)^{-1} \tag{4.27}$$

satisfies the evolution equation

$$\begin{aligned} \dot{v} - \dot{\Phi} F^{ij} v_{ij} = &-\dot{\Phi} F^{ij} h_{ik} h_j^k v - 2v^{-1} \dot{\Phi} F^{ij} v_i v_j \\ &+ r_{\alpha\beta} \nu^\alpha \nu^\beta [(\Phi - \tilde{f}) - \dot{\Phi} F] v^2 + 2\dot{\Phi} F^{ij} h_i^k r_{\alpha\beta} x_k^\alpha x_j^\beta v^2 \\ &+ \dot{\Phi} F^{ij} \bar{R}_{\alpha\beta\gamma\delta} \nu^\alpha x_i^\beta x_j^\gamma x_k^\delta r_\epsilon x_m^\epsilon g^{mk} v^2 \\ &+ \dot{\Phi} F^{ij} r_{\alpha\beta\gamma} \nu^\alpha x_i^\beta x_j^\gamma v^2 + \tilde{f}_\alpha x_m^\alpha g^{mk} r_\beta x_k^\beta v^2. \end{aligned} \tag{4.28}$$

PROOF. Similar to the proof of the previous lemma. □

The previous problems can be generalized to the case when the right-hand side f is not only defined in N or in $\bar{\Omega}$ but in the *tangent bundle* $T(N)$ resp. $T(\bar{\Omega})$. Notice that the tangent bundle is a manifold of dimension $2(n+1)$, i.e., in a *local trivialization* of $T(N)$ f can be expressed in the form

$$f = f(x, \nu) \tag{4.29}$$

with $x \in N$ and $\nu \in T_x(N)$, cf. [**17**, Note 12.2.14]. Thus, the case $f = f(x)$ is included in this general set up. The symbol ν indicates that in an equation

$$F_{|_M} = f(x, \nu) \tag{4.30}$$

we want f to be evaluated at (x, ν), where $x \in M$ and ν is the normal of M in x.

The *Minkowski problem* or *Minkowski type problems* are also covered by the present setting, though the Minkowski problem has the additional property that the problem is transformed via the *Gauß map* to a different semi-Riemannian manifold as a *dual problem* and solved there. Minkowski type problems have been treated in [**5, 16, 23**] and [**21**].

REMARK 4.6. The equation (4.30) will be solved by the same methods as in the special case when $f = f(x)$, i.e., we consider the same curvature flow, the evolution equation (3.7) on page 118, as before.

The resulting evolution equations are identical with the natural exception, that, when f or $\tilde{f}$ has to be differentiated, the additional argument has to be considered, e.g.,

$$\tilde{f}_i = \tilde{f}_\alpha x_i^\alpha + \tilde{f}_{\nu^\beta} \nu_i^\beta = \tilde{f}_\alpha x_i^\alpha + \tilde{f}_{\nu^\beta} x_k^\beta h_i^k \tag{4.31}$$

and

$$\dot{\tilde{f}} = \tilde{f}_\alpha \dot{x}^\alpha + \tilde{f}_{\nu^\beta} \dot{\nu}^\beta = -\sigma(\Phi - \tilde{f})\tilde{f}_\alpha \nu^\alpha + \tilde{f}_{\nu^\beta} g^{ij} (\Phi - \tilde{f})_i x_j^\beta. \tag{4.32}$$

The most important evolution equations are explicitly stated below.

Let us first state the evolution equation for $(\Phi - \tilde{f})$.

LEMMA 4.7 (Evolution of $(\Phi - \tilde{f})$). *The term $(\Phi - \tilde{f})$ evolves according to the equation*

$$\begin{aligned}(\Phi - \tilde{f})' - \dot{\Phi} F^{ij} (\Phi - \tilde{f})_{ij} = {} & \sigma \dot{\Phi} F^{ij} h_{ik} h_j^k (\Phi - \tilde{f}) \\ & + \sigma \tilde{f}_\alpha \nu^\alpha (\Phi - \tilde{f}) - \tilde{f}_{\nu^\alpha} x_i^\alpha (\Phi - \tilde{f})_j g^{ij} \\ & + \sigma \dot{\Phi} F^{ij} \bar{R}_{\alpha\beta\gamma\delta} \nu^\alpha x_i^\beta \nu^\gamma x_j^\delta (\Phi - \tilde{f}),\end{aligned} \tag{4.33}$$

where

$$(\Phi - \tilde{f})' = \frac{d}{dt}(\Phi - \tilde{f}) \tag{4.34}$$

and

$$\dot{\Phi} = \frac{d}{dr}\Phi(r). \tag{4.35}$$

Here is the evolution equation for the second fundamental form.

LEMMA 4.8. *The mixed tensor h_i^j satisfies the parabolic equation*

$$
\begin{aligned}
\dot{h}_i^j - \dot{\Phi} F^{kl} h_{i;kl}^j = {} & \sigma \dot{\Phi} F^{kl} h_{rk} h_l^r h_i^j - \sigma \dot{\Phi} F h_{ri} h^{rj} + \sigma(\Phi - \tilde{f}) h_i^k h_k^j \\
& - \tilde{f}_{\alpha\beta} x_i^\alpha x_k^\beta g^{kj} + \sigma \tilde{f}_\alpha \nu^\alpha h_i^j - \tilde{f}_{\alpha\nu^\beta} (x_i^\alpha x_k^\beta h^{kj} + x_l^\alpha x_k^\beta h_i^k g^{lj}) \\
& - \tilde{f}_{\nu^\alpha \nu^\beta} x_l^\alpha x_k^\beta h_i^k h^{lj} - \tilde{f}_{\nu^\beta} x_k^\beta h_{i;l}^k g^{lj} + \sigma \tilde{f}_{\nu^\alpha} \nu^\alpha h_i^k h_k^j \\
& + \dot{\Phi} F^{kl,rs} h_{kl;i} h_{rs;}{}^j + 2 \dot{\Phi} F^{kl} \bar{R}_{\alpha\beta\gamma\delta} x_m^\alpha x_i^\beta x_k^\gamma x_r^\delta h_l^m g^{rj} \\
& - \dot{\Phi} F^{kl} \bar{R}_{\alpha\beta\gamma\delta} x_m^\alpha x_k^\beta x_r^\gamma x_l^\delta h_i^m g^{rj} - \dot{\Phi} F^{kl} \bar{R}_{\alpha\beta\gamma\delta} x_m^\alpha x_k^\beta x_i^\gamma x_l^\delta h^{mj} \\
& + \sigma \dot{\Phi} F^{kl} \bar{R}_{\alpha\beta\gamma\delta} \nu^\alpha x_k^\beta \nu^\gamma x_l^\delta h_i^j - \sigma \dot{\Phi} F \bar{R}_{\alpha\beta\gamma\delta} \nu^\alpha x_i^\beta \nu^\gamma x_m^\delta g^{mj} \\
& + \sigma(\Phi - \tilde{f}) \bar{R}_{\alpha\beta\gamma\delta} \nu^\alpha x_i^\beta \nu^\gamma x_m^\delta g^{mj} + \ddot{\Phi} F_i F^j \\
& + \dot{\Phi} F^{kl} \bar{R}_{\alpha\beta\gamma\delta;\epsilon} \{ \nu^\alpha x_k^\beta x_l^\gamma x_i^\delta x_m^\epsilon g^{mj} + \nu^\alpha x_i^\beta x_k^\gamma x_m^\delta x_l^\epsilon g^{mj} \}.
\end{aligned}
\tag{4.36}
$$

The proof is identical to that of Lemma 4.1; we only have to keep in mind that f now also depends on the normal.

If we had assumed F to be homogeneous of degree d_0 instead of 1, then, we would have to replace the explicit term F—occurring twice in the preceding lemma—by $d_0 F$.

LEMMA 4.9 (Evolution of $\tilde{v}$). *Consider the flow* (3.7) *in a Lorentzian space N such that the spacelike flow hypersurfaces can be written as graphs over $\mathcal{S}_0$. Then, $\tilde{v}$ satisfies the evolution equation*

$$
\begin{aligned}
\dot{\tilde{v}} - \dot{\Phi} F^{ij} \tilde{v}_{ij} = {} & - \dot{\Phi} F^{ij} h_{ik} h_j^k \tilde{v} + [(\Phi - \tilde{f}) - \dot{\Phi} F] \eta_{\alpha\beta} \nu^\alpha \nu^\beta \\
& - 2 \dot{\Phi} F^{ij} h_j^k x_i^\alpha x_k^\beta \eta_{\alpha\beta} - \dot{\Phi} F^{ij} \eta_{\alpha\beta\gamma} x_i^\beta x_j^\gamma \nu^\alpha \\
& - \dot{\Phi} F^{ij} \bar{R}_{\alpha\beta\gamma\delta} \nu^\alpha x_i^\beta x_k^\gamma x_j^\delta \eta_\epsilon x_l^\epsilon g^{kl} \\
& - \tilde{f}_\beta x_i^\beta x_k^\alpha \eta_\alpha g^{ik} - \tilde{f}_{\nu^\beta} x_k^\beta h^{ik} x_i^\alpha \eta_\alpha,
\end{aligned}
\tag{4.37}
$$

where η is the covariant vector field $(\eta_\alpha) = e^\psi(-1, 0, \dots, 0)$.

The proof is identical to the proof of Lemma 4.4.

In the Riemannian case we have:

LEMMA 4.10 (Evolution of v). *Consider the flow* (3.7) *in a normal Gaussian coordinate system (x^α), where the $M(t)$ can be written as graphs of a function $u(t)$ over some compact Riemannian manifold $\mathcal{S}_0$. Then the quantity*

$$
v = \sqrt{1 + |Du|^2} = (r_\alpha \nu^\alpha)^{-1} \tag{4.38}
$$

satisfies the evolution equation

$$\begin{aligned}\dot{\upsilon} - \dot{\Phi}F^{ij}\upsilon_{ij} = &- \dot{\Phi}F^{ij}h_{ik}h_j^k\upsilon - 2\upsilon^{-1}\dot{\Phi}F^{ij}\upsilon_i\upsilon_j \\ &+ [(\Phi - f) - \dot{\Phi}F]r_{\alpha\beta}\nu^\alpha\nu^\beta\upsilon^2 \\ &+ 2\dot{\Phi}F^{ij}h_j^k x_i^\alpha x_k^\beta r_{\alpha\beta}\upsilon^2 + \dot{\Phi}F^{ij}r_{\alpha\beta\gamma}x_i^\beta x_j^\gamma\nu^\alpha\upsilon^2 \\ &+ \dot{\Phi}F^{ij}\bar{R}_{\alpha\beta\gamma\delta}\nu^\alpha x_i^\beta x_k^\gamma x_j^\delta r_\epsilon x_l^\epsilon g^{kl}\upsilon^2 \\ &+ \tilde{f}_\beta x_i^\beta x_k^\alpha r_\alpha g^{ik}\upsilon^2 + \tilde{f}_{\nu^\beta}x_k^\beta h^{ik}x_i^\alpha r_\alpha\upsilon^2,\end{aligned} \tag{4.39}$$

where $r = x^0$ *and* $(r_\alpha) = (1, 0, \dots, 0)$.

5. Stability of the limit hypersurfaces

DEFINITION 5.1. Let N be semi-Riemannian, F a curvature operator, and $M \subset N$ a compact, spacelike hypersurface, such that M is admissible and F^{ij}, evaluated at (h_{ij}, g_{ij}), the second fundamental form and metric of M, is divergence free, then M is said to be a *stable* solution of the equation

$$F_{|M} = f, \tag{5.1}$$

where $f = f(x)$ is defined in a neighbourhood of M, if the first eigenvalue λ_1 of the linearization, which is the operator in (3.33) on page 122, is non-negative, or equivalently, if the quadratic form

$$\int_M F^{ij}u_iu_j - \sigma\int_M \{F^{ij}h_i^k h_{kj} + F^{ij}\bar{R}_{\alpha\beta\gamma\delta}\nu^\alpha x_i^\beta\nu^\gamma x_j^\delta + f_\alpha\nu^\alpha\}u^2 \tag{5.2}$$

is non-negative for all $u \in C^2(M)$.

It is well-known that the corresponding eigenspace is then onedimensional and spanned by a strictly positive eigenfunction η

$$-F^{ij}\eta_{ij} - \sigma\{F^{ij}h_i^k h_{kj} + F^{ij}\bar{R}_{\alpha\beta\gamma\delta}\nu^\alpha x_i^\beta\nu^\gamma x_j^\delta + f_\alpha\nu^\alpha\}\eta = \lambda_1\eta. \tag{5.3}$$

Notice that F^{ij} is supposed to be divergence free, which will be the case, if $F = H_k$, $1 \le k \le n$, and the ambient space has constant curvature, as we shall prove at the end of this section. If $k = 1$, then $F^{ij} = g^{ij}$ and N can be arbitrary, while in case $k = 2$, we have

$$F^{ij} = Hg^{ij} - h^{ij}, \tag{5.4}$$

hence N Einstein will suffice.

To simplify the formulation of the assumptions let us define:

DEFINITION 5.2. A curvature function F is said to be of class (D), if for every admissible hypersurface M the tensor F^{ij}, evaluated at M, is divergence free.

We shall prove in this section that the limit hypersurface of a converging curvature flow will be a stable stationary solution, if the initial flow velocity has a weak sign.

THEOREM 5.3. *Suppose that the curvature flow* (3.7) *on page* 118 *exists for all time, and that the leaves $M(t)$ converge in C^4 to a hypersurface M, where the curvature function F is supposed to be of class (D). Then M is a stable solution of the equation*

$$F_{|M} = f \tag{5.5}$$

provided the velocity of the flow has a weak sign

$$\Phi - \tilde{f} \ge 0 \quad \vee \quad \Phi - \tilde{f} \le 0 \tag{5.6}$$

at $t = 0$ and $M(0)$ is not already a solution of (5.5).

PROOF. Convergence of a subsequence of the $M(t)$ would actually suffice for the proof, however, the assumption (5.6) immediately implies that the flow converges, if a subsequence converges and a priori estimates in $C^{4,\alpha}$ are valid.

The starting point is the evolution equation (3.20) on page 120 from which we deduce in view of the parabolic maximum principle that $\Phi - \tilde{f}$ has a weak sign during the evolution, cf. [**18**, Proposition 2.7.1], i.e., if we assume without loss of generality that at $t = 0$

$$\Phi - \tilde{f} \ge 0, \tag{5.7}$$

then this inequality will be valid for all t. Moreover, there holds

$$\int_{M(t)} (\Phi - \tilde{f}) > 0 \qquad \forall\, 0 \le t < \infty \tag{5.8}$$

if this relation is valid for $t = 0$, as we shall prove in the lemma below.

On the other hand, the assumption

$$\int_{M(0)} (\Phi - \tilde{f}) > 0 \tag{5.9}$$

is a natural assumption, for otherwise the initial hypersurface would already be a stationary solution which of course may not be stable.

Notice also that apart from the factor $\dot{\Phi}$ the equation (3.20) looks like the parabolic version of the linearization of $(F - f)$. If the technical function $\Phi = \Phi(r)$ is not the trivial one $\Phi(r) = r$, then we always assume that $f > 0$ and that this is also valid for the limit hypersurface M. Only in case $\Phi(r) = r$ and $F = H$, we allow f to be arbitrary.

Thus, our assumptions imply that in any case

$$\dot{\Phi} > \epsilon_0 > 0 \qquad \forall t \in \mathbb{R}_+. \tag{5.10}$$

Furthermore, we derive from (3.20) that not only the elliptic part converges to 0 but also

$$(\Phi - \tilde{f})' = \dot{\Phi}\dot{F} + \sigma\dot{\Phi}(f) f_\alpha \nu^\alpha (\Phi - \tilde{f}), \tag{5.11}$$

i.e.,

$$\lim \dot{F} = 0. \tag{5.12}$$

Suppose now that M is not stable, then the first eigenvalue λ_1 is negative and there exists a strictly positive eigenfunction η solving the equation (5.3) evaluated at M. Let $\mathcal{U}$ be a tubular neighbourhood of M with a corresponding future directed normal Gaussian coordinate system (x^α) and extend η to $\mathcal{U}$ by setting

$$\eta(x^0, x) = \eta(x), \tag{5.13}$$

where, by a slight abuse of notation, we also denote (x^i) by x. Thus there holds

$$\eta_\alpha \nu^\alpha = 0 \tag{5.14}$$

in M, and choosing $\mathcal{U}$ sufficiently small, we may assume

$$|\eta_\alpha \nu^\alpha| < \eta \tag{5.15}$$

for all hypersurfaces $M(t) \subset \mathcal{U}$.

Now consider the term

$$\int_{M(t)} \dot{\Phi}^{-1}(\Phi - \tilde{f})\eta \tag{5.16}$$

for large t, which converges to 0. Since it is positive, in view of (5.8), there must exist a sequence of t, not explicitly labelled, tending to infinity such that

$$\begin{aligned} 0 &\geq \frac{d}{dt}\int_{M(t)} \dot{\Phi}^{-1}(\Phi - \tilde{f})\eta \\ &= \int_{M(t)} -\dot{\Phi}^{-2}\ddot{\Phi}\dot{F}(\Phi - \tilde{f})\eta + \int_{M(t)} \dot{\Phi}^{-1}(\Phi - \tilde{f})'\eta \\ &\quad - \sigma\int_{M(t)} \dot{\Phi}^{-1}\eta_\alpha\nu^\alpha(\Phi - \tilde{f})^2 - \sigma\int_{M(t)} \dot{\Phi}^{-1}(\Phi - \tilde{f})^2 H\eta, \end{aligned} \tag{5.17}$$

where we used the relation (3.8) on page 118 to derive the last integral.

The rest of the proof is straight-forward. Multiply the equation (3.20) by $\dot{\Phi}^{-1}\eta$ and integrate over $M(t)$ for those values of t satisfying the preceding inequality to deduce

$$\tag{5.18} \begin{aligned} -\int_{M(t)}\dot{\Phi}^{-1}(\Phi-\tilde{f})' &= \int_{M(t)} -F^{ij}\eta_{ij}(\Phi-\tilde{f}) \\ -\sigma\int_{M(t)}\{F^{ij}h_i^kh_{kj} &+ F^{ij}\bar{R}_{\alpha\beta\gamma\delta}\nu^\alpha x_i^\beta\nu^\gamma x_j^\delta+\dot{\Phi}^{-1}\dot{\Phi}(f)f_\alpha\nu^\alpha\}\eta(\Phi-\tilde{f}), \end{aligned}$$

and conclude further that the right-hand side can be estimated from above by

$$\tag{5.19} \frac{\lambda_1}{2}\int_{M(t)}\eta(\Phi-\tilde{f})$$

for large t, while the left-hand side can be estimated from below by

$$\tag{5.20} -\epsilon(t)\int_{M(t)}(\Phi-\tilde{f})\eta$$

such that

$$\tag{5.21} \epsilon(t)>0 \quad\wedge\quad \lim\epsilon(t)=0$$

in view of (5.17), where we used (5.12), (5.15) as well as

$$\tag{5.22} \lim(\Phi-\tilde{f})=0;$$

a contradiction because of (5.8). □

Lemma 5.4. *Let $M(t)$ be a solution of the curvature flow (3.7) on page 118 defined on a maximal time interval $[0,T^*)$, $0<T^*\le\infty$, and suppose that $\Phi-\tilde{f}$ has a weak sign at $t=0$, e.g.,*

$$\tag{5.23} (\Phi-\tilde{f})\ge 0$$

and suppose furthermore that

$$\tag{5.24} \int_{M(0)}(\Phi-\tilde{f})>0,$$

then

$$\tag{5.25} \int_{M(t)}(\Phi-\tilde{f})>0 \qquad \forall\, 0\le t<T^*.$$

Proof. Let M_0 be an abstract compact Riemannian manifold that is being isometrically embedded in N with image $M(0)$. Let (ξ^i) be a generic coordinate system for M_0 and abbreviate $(\Phi-\tilde{f})$ by u. The evolution equation (3.20) can then be looked at as a linear parabolic equation for

$u = u(t,\xi)$ on M_0 with time dependent coefficients and time dependent Riemannian metric $g_{ij}(t,\xi)$.

By assumption u doesn't vanish identically at $t = 0$, i.e., there exists a ball $B_\rho = B_\rho(\xi_0)$ such that

$$u(0,\xi) > 0 \qquad \forall \xi \in \bar{B}_\rho(\xi_0). \tag{5.26}$$

Let $\mathcal{C}$ be the cylinder

$$\mathcal{C} = [0, T^*) \times \bar{B}_\rho \tag{5.27}$$

and assume that there exists a first $t_0 > 0$ such that

$$\inf_{\bar{B}_\rho} u(t_0, \cdot) = 0 = u(t_0, \xi_1). \tag{5.28}$$

We shall show that this is not possible: If $\xi_1 \in B_\rho$, then this contradicts the strong parabolic maximum principle, cf. [**18**, Lemma 2.7.1], and if $\xi_1 \in \partial B_\rho$, then we deduce from [**18**, Lemma 2.7.4] (a parabolic version of the Hopf Lemma)

$$\frac{\partial u}{\partial \nu}(t_0, \xi_1) < 0, \tag{5.29}$$

where ν is the exterior normal of the ball B_ρ in ξ_1, contradicting the fact that the gradient of $u(t_0, \cdot)$ vanishes in ξ_1 because it is a minimum point; notice that we already know $u \geq 0$ in $[0, T^*) \times M_0$. $\square$

For some curvature operators one can prove a priori estimates for the second fundamental form only for stationary solutions and not for the leaves of a corresponding curvature flow. In order to use a curvature flow to obtain a stationary solution one uses „ϵ-regularization", i.e., instead of the curvature function F one considers

$$\tilde{F}(h_{ij}) = F(h_{ij} + \epsilon H g_{ij}) \tag{5.30}$$

for $\epsilon > 0$, and starts a curvature flow with $\tilde{F}$ and fixed $\epsilon > 0$.

A priori estimates for the regularized flow are usually fairly easily derived, since

$$\tilde{F}^{ij} = F^{ij} + \epsilon F^{kl} g_{kl} g^{ij}, \tag{5.31}$$

but of course the estimates depend on ϵ. Having uniform estimates one can deduce that the flow—or at least a subsequence—converges to a limit hypersurfaces M_ϵ satisfying

$$\tilde{F}_{|M_\epsilon} = f. \tag{5.32}$$

Then, if uniform $C^{4,\alpha}$-estimates for the M_ϵ can be derived, a subsequence will converge to a solution M of

$$F_{|M} = f, \tag{5.33}$$

cf. [**13**], where this method has been used to find hypersurfaces of prescribed scalar curvature in Lorentzian manifolds, see also Theorem 6.9 on page 152.

We shall now show that the solutions M obtained by this approach are all stable, if F is of class (D) and the initial velocities of the regularized flows have a weak sign. Notice that the curvature functions $\tilde{F}$ are in general not of class (D).

THEOREM 5.5. *Let F be of class (D), then any solution M of*

$$F_{|M} = f \tag{5.34}$$

obtained by a regularized curvature flow as described above is stable, provided the initial velocity of the regularized flow has a weak sign, i.e., it satisfies

$$\Phi - \tilde{f} \geq 0 \quad \vee \quad \Phi - \tilde{f} \leq 0 \tag{5.35}$$

at $t = 0$ and the flow hypersurfaces converge to the stationary solution in C^4.

PROOF. Let M_ϵ be the limit hypersurfaces of the regularized flow for $\epsilon > 0$, and assume that the M_ϵ satisfy uniform $C^{4,\alpha}$-estimates such that a subsequence, not relabelled, converges in C^4 to a compact spacelike hypersurface M solving the equation

$$F_{|M} = f. \tag{5.36}$$

Assume that M is not stable so that the first eigenvalue of the linearization is negative and there exists a strictly positive eigenfunction η satisfying (5.3). Extend η in a small tubular neighbourhood $\mathcal{U}$ of M such that (5.15) is valid for all M_ϵ, if ϵ is small, $\epsilon < \epsilon_0$.

For those ϵ we then deduce

$$\begin{aligned} &-\tilde{F}^{ij}\eta_{ij} - \tilde{F}^{ij}_{;ij}\eta - 2\tilde{F}^{ij}_{;j}\eta_i \\ &\qquad - \sigma\{\tilde{F}^{ij}h^k_i h_{kj} + \tilde{F}^{ij}\bar{R}_{\alpha\beta\gamma\delta}\nu^\alpha x^\beta_i \nu^\gamma x^\delta_j + f_\alpha \nu^\alpha\}\eta < \frac{\lambda_1}{2}\eta, \end{aligned} \tag{5.37}$$

where the inequality is evaluated at M_ϵ and where we used the convergence in C^4.

Now, fix ϵ, $\epsilon < \epsilon_0$, then the preceding inequality is also valid for the flow hypersurfaces $M(t)$ converging to M_ϵ, if t is large, and the same arguments as at the end of the proof of Theorem 5.3 lead to a contradiction. Hence, M has to be a stable solution. □

Knowing that a solution is stable often allows to deduce further geometric properties of the underlying hypersurface like that it is either strictly stable or totally geodesic especially if the curvature function is the mean curvature, cf. e.g., [**29**], where the stability property has been extensively used to deduce geometric properties.

We want to prove that a neighbourhood of stable solutions can be foliated by a family of hypersurfaces satisfying the equation modulo a constant.

THEOREM 5.6. *Let $M \subset N$ be compact, spacelike, orientable and a stable solution of*

$$G_{|_M} \equiv (F - f)_{|_M} = 0, \tag{5.38}$$

where F is of class (D) and M as well as F, f are of class $C^{m,\alpha}$, $2 \le m \le \infty$, $0 < \alpha < 1$, then a neighbourhood of M can be foliated by a family

$$\Lambda = \{\, M_\epsilon \colon |\epsilon| < \epsilon_0 \,\} \tag{5.39}$$

of spacelike $C^{m,\alpha}$-hypersurfaces satisfying

$$G_{|_{M_\epsilon}} = \tau(\epsilon), \tag{5.40}$$

where τ is a real function of class $C^{m,\alpha}$. The M_ϵ can be written as graphs over M in a tubular neighbourhood of M

$$M_\epsilon = \{\, (u(\epsilon, x), x) \colon x \in M \,\} \tag{5.41}$$

such that u is of class $C^{m,\alpha}$ in both variables and there holds

$$\dot{u} > 0. \tag{5.42}$$

PROOF. (i) Let us assume that M is strictly stable. Consider a tubular neighbourhood of M with corresponding normal Gaussian coordinates (x^α) such that $M = \{x^0 = 0\}$. The nonlinear operator G can then be viewed as an elliptic operator

$$G : B_\rho(0) \subset C^{m,\alpha}(M) \to C^{m-2,\alpha}(M) \tag{5.43}$$

where ρ is so small that all corresponding graphs are admissible.

In a smaller ball DG is a topological isomorphism, since M is strictly stable, and hence G is a diffeomorphism in a neighbourhood of the origin, and there exist smooth unique solutions

$$M_\epsilon = \{\, u(\epsilon, x) \colon x \in M \,\} \qquad |\epsilon| < \epsilon_0 \tag{5.44}$$

of the equations

$$G_{|_{M_\epsilon}} = \epsilon \tag{5.45}$$

such that $u \in C^{m,\alpha}((-\epsilon_0, \epsilon_0) \times M)$.

Differentiating with respect to ϵ yields

$$DG\dot{u} = 1. \tag{5.46}$$

Let us consider this equation for $\epsilon = 0$, i.e., on M, and define

$$\eta = \min(\dot{u}, 0). \tag{5.47}$$

Then we deduce

$$0 \le \int_M \langle DG\eta, \eta\rangle = \int_M \eta \le 0, \tag{5.48}$$

and hence there holds

$$\dot{u} \ge 0, \tag{5.49}$$

because of the strict stability of M.

Applying then the maximum principle to (5.46), we deduce further

$$\inf_M \dot{u} > 0, \tag{5.50}$$

hence the hypersurfaces form a foliation if ϵ_0 is chosen small enough such that

$$\inf_{M_\epsilon} \dot{u}(\epsilon, \cdot) > 0 \qquad \forall\, |\epsilon| < \epsilon_0. \tag{5.51}$$

(ii) Assume now that M is not strictly stable. After introducing coordinates corresponding to a tubular neighbourhood $\mathcal{U}$ of M as in part (i) any function $u \in C^{m,\alpha}(M)$ with $|u|_{m,\alpha}$ small enough defines an admissible hypersurface

$$M(u) = \operatorname{graph} u \subset \mathcal{U} \tag{5.52}$$

such that $G_{|_{M(u)}}$ can be expressed as

$$G_{|_{M(u)}} = G(u). \tag{5.53}$$

Let

$$A = DG(0), \tag{5.54}$$

then A is self-adjoint, monotone

$$\langle Au, u\rangle \ge 0 \qquad \forall\, u \in H^{1,2}(M) \tag{5.55}$$

and the smallest eigenvalue of A is equal to zero, the corresponding eigenspace spanned by a strictly positive eigenfunction η.

Similarly as in [**2**, p. 621] we consider the operator

$$\Psi(u, \tau) = (G(u) - \tau, \varphi(u)) \tag{5.56}$$

defined in $B_\rho(0) \times \mathbb{R}$, $B_\rho(0) \subset C^{m,\alpha}(M)$ for small $\rho > 0$, where φ is a linear functional

$$\varphi(u) = \int_M \eta u \tag{5.57}$$

Ψ is of class $C^{m,\alpha}$ and maps

$$\Psi : B_\rho(0) \times \mathbb{R} \to C^{m-2,\alpha}(M) \times \mathbb{R}, \tag{5.58}$$

such that

$$D\Psi = \begin{pmatrix} DG & -1 \\ \varphi & 0 \end{pmatrix} \tag{5.59}$$

evaluated at $(0,0)$ is bijective as one easily checks. Indeed let (u,ϵ) satisfy

$$D\Psi(u,\epsilon) = (0,0), \tag{5.60}$$

then

$$Au = DGu = \epsilon \quad \wedge \quad \int_M \eta u = 0, \tag{5.61}$$

hence

$$\epsilon \int_M \eta = \langle Au, \eta \rangle = \langle u, A\eta \rangle = 0 \tag{5.62}$$

and we conclude $\epsilon = 0$ as well as $u = 0$.

To prove the surjectivity, let $(w,\delta) \in C^{m-2,\alpha}(M) \times \mathbb{R}$ be arbitrary. Choosing

$$\epsilon = -\frac{\int_M w\eta}{\int_M \eta} \tag{5.63}$$

we deduce

$$\int_M (\epsilon + w)\eta = 0, \tag{5.64}$$

hence there exists $\bar{u} \in C^{m,\alpha}(M)$ solving

$$A\bar{u} = \epsilon + w \tag{5.65}$$

and

$$u = \bar{u} + \lambda\eta \tag{5.66}$$

with

$$\lambda = \delta - \int_M \eta\bar{u} \tag{5.67}$$

then satisfies

$$\int_M \eta u = \epsilon, \tag{5.68}$$

i.e.,

$$D\Psi(u,\epsilon)=(w,\delta). \tag{5.69}$$

Applying the inverse function theorem we conclude that there exists $\epsilon_0>0$ and functions $(u(\epsilon,x),\tau(\epsilon))$ of class $C^{m,\alpha}$ in both variables such that

$$G(u(\epsilon))=\tau(\epsilon)\quad\wedge\quad\int_M\eta u(\epsilon)=\epsilon\quad\forall\,|\epsilon|<\epsilon_0; \tag{5.70}$$

$\tau(\epsilon)$ is constant for fixed ϵ.

The hypersurfaces

$$\varLambda=\{\,M_\epsilon=M(u(\epsilon))\colon|\epsilon|<\epsilon_0\,\} \tag{5.71}$$

will form a foliation, if we can show that

$$\dot u\neq 0. \tag{5.72}$$

Differentiating the equations in (5.70) with respect to ϵ and evaluating the result at $\epsilon=0$ yields

$$A\dot u(0)=\dot\tau(0)\quad\wedge\quad\int_M\eta\dot u(0)=1 \tag{5.73}$$

and we deduce further

$$\dot\tau(0)\int_M\eta=\langle A\dot u(0),\eta\rangle=\langle\dot u(0),A\eta\rangle=0 \tag{5.74}$$

and thus

$$\dot\tau(0)=0\quad\wedge\quad\dot u(0)=\eta>0, \tag{5.75}$$

if η is normalized such that $\langle\eta,\eta\rangle=1$, i.e., we have $\dot u(\epsilon)>0$, if ϵ_0 is chosen small enough. □

Remark 5.7. Let M be a stable solution of

$$G_{|M}=0 \tag{5.76}$$

as in the preceding theorem, but not strictly stable and let M_ϵ be a foliation of a neighbourhood of M such that

$$G_{|M_\epsilon}=\tau(\epsilon)\qquad\forall\,|\epsilon|<\epsilon_0. \tag{5.77}$$

If M is the limit hypersurface of a curvature flow as in Theorem 5.3, then

$$\tau(\epsilon)>0\quad\forall\,0<\epsilon<\epsilon_0, \tag{5.78}$$

if the flow hypersurfaces $M(t)$ converge to M from above, which is tantamount to

$$\Phi(F) - \tilde{f} \geq 0, \tag{5.79}$$

or we have

$$\tau(\epsilon) < 0 \qquad \forall -\epsilon_0 < \epsilon < 0, \tag{5.80}$$

if

$$\Phi(F) - \tilde{f} \leq 0, \tag{5.81}$$

in which case the flow hypersurfaces converge to M from below.

The direction "above" is defined by the region the normal $\sigma\nu$ of M points to.

PROOF. Let us assume that the flow hypersurfaces satisfy (5.79) and fix $0 < \epsilon < \epsilon_0$. We may also suppose that the initial hypersurface $M(0)$ doesn't intersect the tubular neighbourhood of M which is being foliated by M_ϵ. Now, fix $0 < \epsilon < \epsilon_0$, then there must be a first $t > 0$ such that $M(t)$ touches M_ϵ from above which yields, in view of the maximum principle,

$$G_{|M_\epsilon} = \tau(\epsilon) > 0, \tag{5.82}$$

since $\tau(\epsilon) \leq 0$ would imply $\tau(\epsilon) = 0$ and $M_\epsilon = M(t)$, cf. [**18**, Theorem 2.7.9], i.e., $M(t)$ would be a stationary solution, which is impossible as we have proved in Lemma 5.4. □

Finally, let us show that the symmetric polynomials H_k, $1 \leq k \leq n$, are of class (D), if the ambient space has constant curvature.

LEMMA 5.8. *Let N be a semi-Riemannian space of constant curvature, then the symmetric polynomials $F = H_k$, $1 \leq k \leq n$, are of class (D). In case $k = 2$ it suffices to assume N Einstein.*

PROOF. We shall prove the result by induction on k. First we note that the cones of definition $\Gamma_k \subset \mathbb{R}^n$ of the H_k form an ordered chain

$$\Gamma_k \subset \Gamma_{k-1} \qquad \forall 1 < k \leq n, \tag{5.83}$$

cf. [**7**], so that a hypersurface admissible for H_k is also admissible for H_{k-1}.

For $k = 1$ we have

$$F^{ij} = g^{ij} \tag{5.84}$$

and the result is obviously valid for arbitrary N.

Thus let us assume that the result is already proved for $1 \leq k < n$. Set $F = H_{k+1}$, $\hat{F} = H_k$ and let M be an admissible hypersurface for F with principal curvatures κ_i.

From the definition of the H_k's we immediately deduce

$$\hat{F} = \frac{\partial F}{\partial \kappa_i} + \kappa_i \frac{\partial \hat{F}}{\partial \kappa_i} \tag{5.85}$$

for fixed i, no summation over i, or equivalently,

$$\hat{F} g^{ij} = F^{ij} + \hat{F}^{jm} h^i_m, \tag{5.86}$$

notice that the last term is a symmetric tensor, since for any symmetric curvature function F F^{ij} and h_{ij} commute, cf. [**18**, Lemma 2.1.9]. Thus there holds

$$F^{ij} = \hat{F} g^{ij} - \hat{F}^{jm} h^i_m \tag{5.87}$$

and we deduce, using the induction hypothesis,

$$\begin{aligned} F^{ij}_{;j} &= \hat{F}^i - \hat{F}^{jm} h^i_{m;j} = \hat{F}^i - \hat{F}^{jm} h_{mj;}{}^i \\ &= \hat{F}^i - \hat{F}^i = 0, \end{aligned} \tag{5.88}$$

where we applied the Codazzi equations at one point.

If $F = H_2$, then

$$F^{ij} = H g^{ij} - h^{ij} \tag{5.89}$$

and the assumption N Einstein suffices to conclude that F^{ij} is divergence free. □

6. Existence results

From now on we shall assume that ambient manifold N is Lorentzian, or more precisely, that it is smooth, globally hyperbolic with a compact, connected Cauchy hypersurface. Then there exists a smooth future oriented time function x^0 such that the metric in N can be expressed in Gaussian coordinates (x^α) as

$$d\bar{s}^2 = e^{2\psi}\{-(dx^0)^2 + \sigma_{ij} dx^i dx^j\}, \tag{6.1}$$

where x^0 is the time function and the (x^i) are local coordinates for

$$\mathcal{S}_0 = \{x^0 = 0\}. \tag{6.2}$$

$\mathcal{S}_0$ is then also a compact, connected Cauchy hypersurface. For a proof of the splitting result see [**4**, Theorem 1.1], and for the fact that all Cauchy hypersurfaces are diffeomorphic and hence $\mathcal{S}_0$ is also compact and connected, see [**3**, Lemma 2.2].

One advantage of working in globally hyperbolic spacetimes with a compact Cauchy hypersurface is that all compact, connected spacelike C^m-hypersurfaces M can be written as graphs over $\mathcal{S}_0$.

LEMMA 6.1. *Let N be as above and $M \subset N$ a connected, spacelike hypersurface of class C^m, $1 \le m$, then M can be written as a graph over $\mathcal{S}_0$*

$$M = \operatorname{graph} u_{|\mathcal{S}_0} \tag{6.3}$$

with $u \in C^m(\mathcal{S}_0)$.

We proved this lemma under the additional hypothesis that M is achronal, [**10**, Proposition 2.5], however, this assumption is unnecessary as has been shown in [**25**, Theorem 1.1].

We are looking at the curvature flow (3.7) on page 118 and want to prove that it converges to a stationary solution hypersurface, if certain assumptions are satisfied.

The existence proof consists of four steps:

(i) Existence on a maximal time interval $[0, T^*)$.
(ii) Proof that the flow stays in a compact subset.
(iii) Uniform a priori estimates in an appropriate function space, e.g., $C^{4,\alpha}(\mathcal{S}_0)$ or $C^\infty(\mathcal{S}_0)$, which, together with (ii), would imply $T^* = \infty$.
(iv) Conclusion that the flow—or at least a subsequence of the flow hypersurfaces—converges if t tends to infinity.

The existence on a maximal time interval is always guaranteed, if the data are sufficiently regular, since the problem is parabolic. If the flow hypersurfaces can be written as graphs in Gaussian coordinate system, as will always be the case in a globally hyperbolic spacetime with a compact Cauchy hypersurface in view of Lemma 6.1, the conditions are better than in the general case:

THEOREM 6.2. *Let $4 \le m \in \mathbb{N}$ and $0 < \alpha < 1$, and assume the semi-Riemannian space N to be of class $C^{m+2,\alpha}$. Let the strictly monotone curvature function F, the functions f and Φ be of class $C^{m,\alpha}$ and let $M_0 \in C^{m+2,\alpha}$ be an admissible compact, spacelike, connected, orientable*[3] *hypersurface. Then the curvature flow* (3.7) *on page* 118 *with initial hypersurface M_0 exists in a maximal time interval $[0, T^*)$, $0 < T^* \le \infty$, where in case that the flow hypersurfaces cannot be expressed as graphs they are supposed to be smooth, i.e, the conditions should be valid for arbitrary $4 \le m \in \mathbb{N}$ in this case.*

A proof can be found in [**18**, Theorem 2.5.19, Lemma 2.6.1].

The second step, that the flow stays in a compact set, can only be achieved by barrier assumptions, cf. Definition 2.1. Thus, let $\Omega \subset N$ be open and precompact such that $\partial\Omega$ has exactly two components

$$\partial\Omega = M_1 \,\dot\cup\, M_2 \tag{6.4}$$

[3]Recall that oriented simply means there exists a continuous normal, which will always be the case in a globally hyperbolic spacetime.

where M_1 is a lower barrier for the pair (F, f) and M_2 an upper barrier. Moreover, M_1 has to lie in the past of M_2

$$M_1 \subset I^-(M_2), \tag{6.5}$$

cf. [**18**, Remark 2.7.8].

Then the flow hypersurfaces will always stay inside $\bar{\Omega}$, if the initial hypersurface M_0 satisfies $M_0 \subset \Omega$, [**18**, Theorem 2.7.9]. This result is also valid if M_0 coincides with one the barriers, since then the velocity $(\Phi - \tilde{f})$ has a weak sign and the flow moves into Ω for small t, if it moves at all, and the arguments of the proof are applicable.

In Lorentzian manifolds the existence of barriers is associated with the presence of past and future singularities. In globally hyperbolic spacetimes, when N is topologically a product

$$N = I \times \mathcal{S}_0, \tag{6.6}$$

where $I = (a, b)$, singularities can only occur, when the endpoints of the interval are approached. A singularity, if one exists, is called a *crushing singularity*, if the sectional curvatures become unbounded, i.e.,

$$\bar{R}_{\alpha\beta\gamma\delta}\bar{R}^{\alpha\beta\gamma\delta} \to \infty \tag{6.7}$$

and such a singularity should provide a future resp. past barrier for the mean curvature function H.

DEFINITION 6.3. Let N be a globally hyperbolic spacetime with compact Cauchy hypersurface $\mathcal{S}_0$ so that N can be written as a topological product $N = I \times \mathcal{S}_0$ and its metric expressed as

$$d\bar{s}^2 = e^{2\psi}(-(dx^0)^2 + \sigma_{ij}(x^0, x)dx^i dx^j). \tag{6.8}$$

Here, x^0 is a globally defined future directed time function and (x^i) are local coordinates for $\mathcal{S}_0$. N is said to have a *future* resp. *past mean curvature barrier*, if there are sequences M_k^+ resp. M_k^- of closed, spacelike, admissible hypersurfaces such that

$$\lim_{k\to\infty} H_{|_{M_k^+}} = \infty \quad \text{resp.} \quad \lim_{k\to\infty} H_{|_{M_k^-}} = -\infty \tag{6.9}$$

and

$$\limsup \inf_{M_k^+} x^0 > x^0(p) \qquad \forall p \in N \tag{6.10}$$

resp.

$$\liminf \sup_{M_k^-} x^0 < x^0(p) \qquad \forall p \in N, \tag{6.11}$$

If one stipulates that the principal curvatures of the M_k^+ resp. M_k^- tend to plus resp. minus infinity, then these hypersurfaces could also serve as barriers for other curvature functions. The past barriers would most certainly be non-admissible for any curvature function except H.

REMARK 6.4. Notice that the assumptions (6.9) alone already implies (6.10) resp. (6.11), if either

$$\limsup \inf_{M_k^+} x^0 > a \tag{6.12}$$

resp.

$$\liminf \sup_{M_k^-} x^0 < b \tag{6.13}$$

where $(a, b) = x^0(N)$, or, if

$$\bar{R}_{\alpha\beta}\nu^\alpha\nu^\beta \geq -\Lambda \qquad \forall \langle \nu, \nu \rangle = -1. \tag{6.14}$$

where $\Lambda \geq 0$.

PROOF. It suffices to prove that the relation (6.10) is automatically satisfied under the assumptions (6.12) or (6.14) by switching the light cone and replacing x^0 by $-x^0$ in case of the past barrier.

Fix k, and let

$$\tau_k = \inf_{M_k} x^0, \tag{6.15}$$

then the coordinate slice

$$M_{\tau_k} = \{x^0 = \tau_k\} \tag{6.16}$$

touches M_k from below in a point $p_k \in M_k$ where $\tau_k = x^0(p_k)$ and the maximum principle yields that in that point

$$H_{|M_{\tau_k}} \geq H_{|M_k}, \tag{6.17}$$

hence, if k tends to infinity the points (p_k) cannot stay in a compact subset, i.e.,

$$\limsup x^0(p_k) \to b \tag{6.18}$$

or

$$\limsup x^0(p_k) \to a. \tag{6.19}$$

We shall show that only (6.18) can be valid. The relation (6.19) evidently contradicts (6.12).

In case the assumption (6.14) is valid, we consider a fixed coordinate slice $M_0 = \{x^0 = \text{const}\}$, then all hypersurfaces M_k satisfying

$$H_{|M_0} < \inf_{M_k} H \quad \wedge \quad \sqrt{n\Lambda} < \inf_{M_k} H \tag{6.20}$$

have to lie in the future of M_0, cf. [**18**, Lemma 4.7.1], hence the result. □

A future mean curvature barrier certainly represents a singularity, at least if N satisfies the condition

$$\bar{R}_{\alpha\beta}\nu^\alpha\nu^\beta \ge -\Lambda \qquad \forall \langle \nu, \nu\rangle = -1 \tag{6.21}$$

where $\Lambda \ge 0$, because of the future timelike incompleteness, which is proved in [**1**], and is a generalization of Hawking's earlier result for $\Lambda = 0$, [**24**]. But these singularities need not be crushing, cf. [**15**, Section 2] for a counterexample.

The uniform a priori estimates for the flow hypersurfaces are the hardest part in any existence proof. When the flow hypersurfaces can be written as graphs it suffices to prove C^1 and C^2 estimates, namely, the induced metric

$$g_{ij}(t,\xi) = \langle x_i, x_j\rangle \tag{6.22}$$

where $x = x(t,\xi)$ is a local embedding of the flow, should stay uniformly positive definite, i.e., there should exist positive constants c_i, $1 \le i \le 2$, such that

$$c_1 g_{ij}(0,\xi) \le g_{ij}(t,\xi) \le c_2 g_{ij}(0,\xi), \tag{6.23}$$

or equivalently, that the quantity

$$\tilde{v} = \langle \eta, \nu\rangle, \tag{6.24}$$

where ν is the past directed normal of $M(t)$ and η the vector field

$$\eta = (\eta_\alpha) = e^\psi(-1, 0, \dots, 0), \tag{6.25}$$

is uniformly bounded, which is achieved with the help of the parabolic equation (4.37) on page 128, if it is possible at all.

However, in some special situations C^1-estimates are automatically satisfied, cf. Theorem 6.11 at the end of this section.

For the C^2-estimates the principle curvatures κ_i of the flow hypersurfaces have to stay in a compact set in the cone of definition Γ of F, e.g., if F is the Gaussian curvature, then $\Gamma = \Gamma_+$ and one has to prove that there are positive constants k_i, $i = 1, 2$ such that

$$k_1 \le \kappa_i \le k_2 \qquad \forall\, 1 \le i \le n \tag{6.26}$$

uniformly in the cylinder $[0, T^*) \times M_0$, where M_0 is any manifold that can serve as a base manifold for the embedding $x = x(t,\xi)$.

The parabolic equations that are used for these curvature estimates are (4.36) on page 128, usually for an upper estimate, and (4.33) on page 127 for the lower estimate. Indeed, suppose that the flow starts at the upper barrier, then

$$F \geq f \tag{6.27}$$

at $t = 0$ and this estimate remains valid throughout the evolution because of the parabolic maximum principle, use (4.36). Then, if upper estimates for the κ_i have been derived and if $f > 0$ uniformly, then we conclude from (6.27) that the κ_i stay in a compact set inside the open cone Γ, since

$$F_{|\partial\Gamma} = 0. \tag{6.28}$$

To obtain higher order estimates we are going to exploit the fact that the flow hypersurfaces are graphs over $\mathcal{S}_0$ in an essential way, namely, we look at the associated scalar flow equation (4.21) on page 125 satisfied by u. This equation is a nonlinear uniformly parabolic equation, where the operator $\Phi(F)$ is also concave in h_{ij}, or equivalently, convex in u_{ij}, i.e., the $C^{2,\alpha}$-estimates of Krylov and Safonov, [**26**, Chapter 5.5] or see [**28**, Chapter 10.6] for a very clear and readable presentation, are applicable, yielding uniform estimates for the standard parabolic Hölder semi-norm

$$[D^2u]_{\beta,\bar{Q}_T} \tag{6.29}$$

for some $0 < \beta \leq \alpha$ in the cylinder

$$Q = [0, T) \times \mathcal{S}_0, \tag{6.30}$$

independent of $0 < T < T^*$, which in turn will lead to $H^{m+2+\alpha,\frac{m+2+\alpha}{2}}(\bar{Q}_T)$ estimates, cf. [**18**, Theorem 2.5.9, Remark 2.6.2].

$H^{m+2+\alpha,\frac{m+2+\alpha}{2}}(\bar{Q}_T)$ is a parabolic Hölder space, cf. [**27**, p. 7] for the original definition and [**18**, Note 2.5.4] in the present context.

The estimate (6.29) combined with the uniform C^2-norm leads to uniform $C^{2,\beta}(\mathcal{S}_0)$-estimates independent of T.

These estimates imply that $T^* = \infty$.

Thus, it remains to prove that $u(t,\cdot)$ converges in $C^{m+2}(\mathcal{S}_0)$ to a stationary solution $\tilde{u}$, which is then also of class $C^{m+2,\alpha}(\mathcal{S}_0)$ in view of the Schauder theory.

Because of the preceding a priori estimates $u(t,\cdot)$ is precompact in $C^2(\mathcal{S}_0)$. Moreover, we deduce from the scalar flow equation (4.21) on page 125 that $\dot{u}$ has a sign, i.e., the $u(t,\cdot)$ converge monotonely in $C^0(\mathcal{S}_0)$ to $\tilde{u}$ and therefore also in $C^2(\mathcal{S}_0)$.

To prove that graph $\tilde{u}$ is a solution, we again look at (4.21) and integrate it with respect to t to obtain for fixed $x \in \mathcal{S}_0$

$$|\tilde{u}(x) - u(t,x)| = \int_t^\infty e^{-\psi}v|\Phi - \tilde{f}|, \tag{6.31}$$

where we used that $(\Phi-\tilde f)$ has a sign, hence $(\Phi-\tilde f)(t,x)$ has to vanish when t tends to infinity, at least for a subsequence, but this suffices to conclude that $\operatorname{graph}\tilde u$ is a stationary solution and

$$\lim_{t\to\infty}(\Phi-\tilde f)=0. \tag{6.32}$$

Using the convergence of u to $\tilde u$ in C^2, we can then prove:

THEOREM 6.5. *The functions $u(t,\cdot)$ converge in $C^{m+2}(\mathcal{S}_0)$ to $\tilde u$, if the data satisfy the assumptions in Theorem 6.2, since we have*

$$u\in H^{m+2+\beta,\frac{m+2+\beta}{2}}(\bar Q), \tag{6.33}$$

where $Q=Q_\infty$.

PROOF. Out of convention let us write α instead of β knowing that α is the Hölder exponent in (6.29).

We shall reduce the Schauder estimates to the standard Schauder estimates in $\mathbb{R}^n$ for the heat equation with a right-hand side by using the already established results (6.29) and

$$u(t,\cdot)\underset{C^2(\mathcal{S}_0)}{\rightarrow}\tilde u\in C^{m+2,\alpha}(\mathcal{S}_0). \tag{6.34}$$

Let (U_k) be a finite open covering of $\mathcal{S}_0$ such that each U_k is contained in a coordinate chart and

$$\operatorname{diam}U_k<\rho, \tag{6.35}$$

ρ small, ρ will be specified in the proof, and let (η_k) be a subordinate finite partition of unity of class $C^{m+2,\alpha}$.

Since

$$u\in H^{m+2+\alpha,\frac{m+2+\alpha}{2}}(\bar Q_T) \tag{6.36}$$

for any finite T, cf. [**18**, Lemma 2.6.1], and hence

$$u(t,\cdot)\in C^{m+2,\alpha}(\mathcal{S}_0)\qquad\forall\, 0\le t<\infty \tag{6.37}$$

we shall choose $u_0=u(t_0,\cdot)$ as initial value for some large t_0 such that

$$|a^{ij}(t,\cdot)-\tilde a^{ij}|_{0,\mathcal{S}_0}<\epsilon/2\qquad\forall\, t\ge t_0, \tag{6.38}$$

where

$$a^{ij}=v^2\dot\Phi F^{ij} \tag{6.39}$$

and $\tilde a^{ij}$ is defined correspondingly for $\tilde M=\operatorname{graph}\tilde u$.

However, making a variable transformation we shall always assume that $t_0=0$ and $u_0=u(0,\cdot)$.

We shall prove (6.33) successively.

(i) Let us first show that

$$D_x u \in H^{2+\alpha,\frac{2+\alpha}{2}}(\bar{Q}). \tag{6.40}$$

This will be achieved, if we show that for an arbitrary $\xi \in C^{m+1,\alpha}(T^{1,0}(\mathcal{S}_0))$

$$\varphi = D_\xi u \in H^{2+\alpha,\frac{2+\alpha}{2}}(\bar{Q}), \tag{6.41}$$

cf. [**18**, Remark 2.5.11].

Differentiating the scalar flow equation (4.21) on page 125 with respect to ξ we obtain

$$\dot{\varphi} - a^{ij}\varphi_{ij} + b^i\varphi_i + c\varphi = f, \tag{6.42}$$

where of course the symbol f has a different meaning then in (4.21).

Later we want to apply the Schauder estimates for solutions of the heat flow equation with right-hand side. In order to use elementary potential estimates we have to cut off φ near the origin $t = 0$ by considering

$$\tilde{\varphi} = \varphi\theta, \tag{6.43}$$

where $\theta = \theta(t)$ is smooth satisfying

$$\theta(t) = \begin{cases} 1, & t > 1, \\ 0, & t \le \frac{1}{2}. \end{cases} \tag{6.44}$$

This modification doesn't cause any problems, since we already have a priori estimates for finite t, and we are only concerned about the range $1 \le t < \infty$. $\tilde{\varphi}$ satisfies the same equation as φ only the right-hand side has the additional summand $w\dot{\theta}$.

Let $\eta = \eta_k$ be one of the members of the partition of unity and set

$$w = \tilde{\varphi}\eta, \tag{6.45}$$

then w satisfies a similar equation with slightly different right-hand side

$$\dot{w} - a^{ij}w_{ij} + b^i w_i + cw = \tilde{f} \tag{6.46}$$

but we shall have this in mind when applying the estimates.

The $w(t,\cdot)$ have compact support in one of the U_k's, hence we can replace the covariant derivatives of w by ordinary partial derivatives without changing the structure of the equation and the properties of the right-hand side, which still only depends linearly on φ and $D\varphi$.

We want to apply the well-known estimates for the ordinary heat flow equation

$$\dot{w} - \Delta w = \hat{f} \tag{6.47}$$

where w is defined in $\mathbb{R} \times \mathbb{R}^n$.

To reduce the problem to this special form, we pick an arbitrary $x_0 \in U_k$, set $z_0 = (0, x_0)$, $z = (t, x)$ and consider instead of (6.46)

$$\begin{aligned} \dot{w} - a^{ij}(z_0)w_{ij} &= \hat{f} \\ &= [a^{ij}(z) - a^{ij}(z_0)]w_{ij} - b^i w_i - cw + \tilde{f}, \end{aligned} \tag{6.48}$$

where we emphasize that the difference

$$|a^{ij}(z) - a^{ij}(z_0)| \tag{6.49}$$

can be made smaller than any given $\epsilon > 0$ by choosing $\rho = \rho(\epsilon)$ in (6.35) and $t_0 = t_0(\epsilon)$ in (6.38) accordingly. Notice also that this equation can be extended into $\mathbb{R} \times \mathbb{R}^n$, since all functions have support in $\{t \ge \frac{1}{2}\}$.

Let $0 < T < \infty$ be arbitrary, then all terms belong to the required function spaces in $\bar{Q}_T$ and there holds

$$[w]_{2+\alpha,Q_T} \le c[\hat{f}]_{\alpha,Q_T}, \tag{6.50}$$

where $c = c(n, \alpha)$. The brackets indicate the standard unweighted parabolic semi-norms, cf. [**18**, Definition 2.5.2], which are identical to those defined in [**27**, p. 7], but there the brackets are replaced by kets.

Thus, we conclude

$$\begin{aligned} [w]_{2+\alpha,Q_T} \le c \sup_{U_k \times (0,T)} &|a^{ij}(z) - a^{ij}(z_0)|[D^2 w]_{\alpha,Q_T} + c[f]_{\alpha,Q_T} \\ + c_1\{[D^2 u]_{\alpha,Q_+} &+ [Du]_{\alpha,Q_T} + [u]_{\alpha,Q_T} + |w|_{0,Q_T} + |D^2 w|_{0,Q_T}\}, \end{aligned} \tag{6.51}$$

where c_1 is independent of T, but dependent on η_k. Here we also used the fact that the lower order coefficients and $\varphi, D\varphi$ are uniformly bounded.

Choosing now $\epsilon > 0$ so small that

$$c\epsilon < \tfrac{1}{2} \tag{6.52}$$

and ρ, t_0 accordingly such the difference in (6.49) is smaller than ϵ, we deduce

$$\begin{aligned} &[w]_{2+\alpha,Q_T} \le 2c[f]_{\alpha,Q_T} \\ &+ 2c_1\{[D^2 u]_{\alpha,Q_+} + [Du]_{\alpha,Q_T} + [u]_{\alpha,Q_T} + |w|_{0,Q_T} + |D^2 w|_{0,Q_T}\}. \end{aligned} \tag{6.53}$$

Summing over the partition of unity and noting that ξ is arbitrary we see that in the preceding inequality we can replace w by Du everywhere resulting in the estimate

$$\begin{aligned} [Du]_{2+\alpha,Q_T} &\le c_1[f]_{\alpha,Q_T} \\ &+ c_1\{[D^2 u]_{\alpha,Q_T} + [Du]_{\alpha,Q_T} + [u]_{\alpha,Q_T} + |Du|_{0,Q_T} + |D^3 u|_{0,Q_T}\} \end{aligned} \tag{6.54}$$

where c_1 is a new constant still independent of T.

Now the only critical terms on the right-hand side are $|D^3u|_{0,Q_T}$, which can be estimated by (6.57), and the Hölder semi-norms with respect to t

$$[Du]_{\frac{\alpha}{2},t,Q_T} + [u]_{\frac{\alpha}{2},t,Q_T}. \tag{6.55}$$

The second one is taken care of by the boundedness of $\dot{u}$, see (4.21) on page 125, while the first one is estimated with the help of equation (6.42) revealing

$$|D\dot{u}| \le c\{\sup_{[0,T]}|u|_{3,\mathcal{S}_0} + |f|_{0,Q_T}\}, \tag{6.56}$$

since for fixed but arbitrary t we have

$$|u|_{3,\mathcal{S}_0} \le \epsilon[D^3u]_{\alpha,\mathcal{S}_0} + c_\epsilon|u|_{0,\mathcal{S}_0}, \tag{6.57}$$

where c_ϵ is independent of t.

Hence we conclude

$$|Du|_{2+\alpha,Q_T} \le \text{const} \tag{6.58}$$

uniformly in T.

(ii) Repeating these estimates successively for $2 \le l \le m$ we obtain uniform estimates for

$$\sum_{l=2}^{m}[D_x^l u]_{2+\alpha,Q_T}, \tag{6.59}$$

which, when combined with the uniform C^2-estimates, yields

$$|u(t,\cdot)|_{m+2,\alpha,\mathcal{S}_0} \le \text{const} \tag{6.60}$$

uniformly in $0 \le t < \infty$.

Looking at the equation (4.21) we then deduce

$$|\dot{u}(t,\cdot)|_{m,\alpha,\mathcal{S}_0} \le \text{const} \tag{6.61}$$

uniformly in t.

(iii) To obtain the estimates for $D_t^r u$ up to the order

$$[\tfrac{m+2+\alpha}{2}] \tag{6.62}$$

we differentiate the scalar curvature equation with respect to t as often as necessary and also with respect to the mixed derivatives $D_t^r D_x^s$ to estimate

$$\sum_{1\le 2r+s<m+2+\alpha} D_t^r D_x^s u \tag{6.63}$$

using (6.60), (6.61) and the results from the prior differentiations.

Combined with the estimates for the heat equation in $\mathbb{R} \times \mathbb{R}^n$ these estimates will also yield the necessary a priori estimates for the Hölder semi-norms in $\bar{Q}$, where again the smallness of (6.49) has to be used repeatedly. □

REMARK 6.6. The preceding regularity result is also valid in Riemannian manifolds, if the flow hypersurfaces can be written as graphs in a Gaussian coordinate system. In fact the proof is unaware of the nature of the ambient space.

With the method described above the following existence results have been proved in globally hyperbolic spacetimes with a compact Cauchy hypersurface. $\Omega \subset N$ is always a precompact domain the boundary of which is decomposed as in (6.4) and (6.5) into an upper and lower barrier for the pair (F, f). We also apply the stability results from Section 5 and the just proved regularity of the convergence and formulate the theorems accordingly.

By convergence of the flow in C^{m+2} we mean convergence of the leaves $M(t) = \operatorname{graph} u(t, \cdot)$ in this norm.

THEOREM 6.7. *Let M_1, M_2 be lower resp. upper barriers for the pair (H, f), where $f \in C^{m,\alpha}(\bar{\Omega})$ and the M_i are of class $C^{m+2,\alpha}$, $4 \le m$, $0 < \alpha < 1$, then the curvature flow*

$$\begin{aligned} \dot{x} &= (H - f)\nu \\ x(0) &= x_0, \end{aligned} \tag{6.64}$$

where x_0 is an embedding of the initial hypersurface $M_0 = M_2$ exists for all time and converges in C^{m+2} to a stable solution M of class $C^{m+2,\alpha}$ of the equation

$$H_{|M} = f, \tag{6.65}$$

provided the initial hypersurface is not already a solution.

The existence result was proved in [**11**, Theorem 2.2], see also [**18**, Theorem 4.2.1] and the remarks following the theorem. Notice that f isn't supposed to satisfy any sign condition.

For spacetimes that satisfy the timelike convergence condition and for functions f with special structural conditions existence results via a mean curvature flow were first proved in [**6**].

The Gaussian curvature or the curvature functions F belonging to the larger class (K^*), see [**10**] for a definition, require that the admissible hypersurfaces are strictly convex.

Moreover, proving a priori estimates for the second fundamental form of a hypersurface M in general semi-Riemannian manifolds, when the curvature function is not the mean curvature, or does not behave similar to it, requires that a strictly convex function χ is defined in a neighbourhood of the hypersurface, see Lemma 2.2 on page 116 where sufficient assumptions are stated which imply the existence of strictly convex functions.

Furthermore, when we consider curvature functions of class (K^*), notice that the Gaussian curvature belongs to that class, then the right-hand side f can be defined in $T(\bar{\Omega})$ instead of $\bar{\Omega}$, i.e., in a local trivialization of the tangent bundle f can be expressed as

$$f = f(x, \nu) \quad \wedge \quad \nu \in T_x(N). \tag{6.66}$$

We shall formulate the existence results with this more general assumption, though of course any stability claim only makes sense for $f = f(x)$.

THEOREM 6.8. *Let $F \in C^{m,\alpha}(\Gamma_+)$, $4 \le m$, $0 < \alpha < 1$, be a curvature function of class (K^*), let $0 < f \in C^{m,\alpha}(T(\bar{\Omega}))$, and let M_1, M_2 be lower resp. upper barriers for (F, f) of class $C^{m+2,\alpha}$. Then the curvature flow*

$$\begin{aligned} \dot{x} &= (\Phi - \tilde{f})\nu \\ x(0) &= x_0 \end{aligned} \tag{6.67}$$

where $\Phi(r) = \log r$ and x_0 is an embedding of $M_0 = M_2$, exists for all time and converges in C^{m+2} to a stationary solution $M \in C^{m+2,\alpha}$ of the equation

$$F_{|M} = f \tag{6.68}$$

provided the initial hypersurface M_2 is not already a stationary solution and there exists a strictly convex function $\chi \in C^2(\bar{\Omega})$.

When $f = f(x)$ and F is of class (D), then M is stable.

The theorem was proved in [**10**] when f is only defined in $\bar{\Omega}$ and in the general case in [**18**, Theorem 4.1.1].

When $F = H_2$ is the scalar curvature operator, then the requirement that f is defined in the tangent bundle and not merely in N is a necessity, if the scalar curvature is to be prescribed. To prove existence results in this case, f has to satisfy some natural structural conditions, namely,

$$0 < c_1 \le f(x, \nu) \qquad \text{if} \quad \langle \nu, \nu \rangle = -1, \tag{6.69}$$

$$|||f_\beta(x, \nu)||| \le c_2(1 + |||\nu|||^2), \tag{6.70}$$

and

$$|||f_{\nu^\beta}(x, \nu)||| \le c_3(1 + |||\nu|||), \tag{6.71}$$

for all $x \in \bar{\Omega}$ and all past directed timelike vectors $\nu \in T_x(\Omega)$, where $|||\cdot|||$ is a Riemannian reference metric.

Applying a curvature flow to obtain stationary solutions requires to approximate F and f by functions F_ϵ and f_k and to use these functions for the flow. The F_ϵ are the ϵ-regularizations of F, which we already discussed before, cf. (5.30) on page 133. Let us also write $\tilde{F}$ instead of F_ϵ as before.

The functions f_k have the property that $|||f_{k\beta}|||$ only grows linearly in $|||\nu|||$ and $|||f_{k\nu^\beta}(x,\nu)|||$ is bounded. To simplify the presentation we shall therefore assume that f satisfies

$$|||f_\beta(x,\nu)||| \le c_2(1+|||\nu|||), \tag{6.72}$$

$$|||f_{\nu^\beta}(x,\nu)||| \le c_3, \tag{6.73}$$

and also

$$0 < c_1 \le f(x,\nu) \qquad \forall\, \nu \in T_x(N),\ \langle\nu,\nu\rangle < 0, \tag{6.74}$$

although the last assumption is only a minor point that can easily be dealt with, see [**13**, Remark 2.6], and [**13**, Section 7 and 8] for the other approximations of f.

The barriers M_i, $i = 1,2$, for (F,f) satisfy the barrier condition of course only weakly, i.e., no strict inequalities; however, because of the ϵ-regularization we need strict inequalities, so that the M_i's are also barriers for $(\tilde{F}, f)$, if ϵ is small. In [**13**, Remark 2.4 and Lemma 2.5] it is shown that strict inequalities for the barriers may be assumed without loss of generality.

Now, we can formulate the existence result for the scalar curvature operator $F = H_2$ under these provisions.

THEOREM 6.9. *Let $f \in C^{m,\alpha}(T(\bar{\Omega}))$, $4 \le m$, $0 < \alpha < 1$, satisfy the conditions* (6.72), (6.73) *and* (6.74), *and let M_1, M_2 be strict lower resp. upper barriers of class $C^{m+2,\alpha}$ for (F,f). Let $\tilde{F}$ be the ϵ-regularization of the scalar curvature operator F, then the curvature flow for $\tilde{F}$*

$$\begin{aligned} \dot{x} &= (\Phi - \tilde{f}) \\ x(0) &= x_0 \end{aligned} \tag{6.75}$$

where $\Phi(r) = r^{\frac{1}{2}}$ and x_0 is an embedding of $M_0 = M_2$, exists for all time and converges in C^{m+2} to a stationary solution $M_\epsilon \in C^{m+2,\alpha}$ of

$$\tilde{F}_{|M_\epsilon} = f \tag{6.76}$$

provided there exists a strictly convex function $\chi \in C^2(\bar{\Omega})$ and $0 < \epsilon$ is small.

The M_ϵ then converge in C^{m+2} to a solution $M \in C^{m+2,\alpha}$ of

$$F_{|M} = f. \tag{6.77}$$

If $f = f(x)$ and N Einstein, then M is stable.

These statements, except for the stability and the convergence in C^{m+2}, are proved in [**13**].

REMARK 6.10. Let us now discuss the pure mean curvature flow

$$\dot{x} = H\nu \tag{6.78}$$

with initial spacelike hypersurface M_0 of class $C^{m+2,\alpha}$, $m \geq 4$ and $0 < \alpha < 1$. From the corresponding scalar curvature flow (4.21) on page 125 we immediately infer that the flow moves into the past of M_0, if

$$H_{|M_0} \geq 0 \tag{6.79}$$

and into its future, if

$$H_{|M_0} \leq 0. \tag{6.80}$$

Let us only consider the case (6.79) and also assume that M_0 is not maximal. From the a priori estimates in [**11**, Section 3 and Section 4] we then deduce that the flow remains smooth as long as it stays in a compact set of N, and if a compact, spacelike hypersurface M_1 of class C^2 satisfying

$$H_{|M_1} \leq 0 \tag{6.81}$$

lies in the past of M_0, then the flow will exist for all time and converge in C^{m+2} to a stable maximal hypersurface M, hence a neighbourhood of M can be foliated by CMC hypersurfaces, where those in the future of M have positive mean curvature, in view of Remark 5.7 on page 138.

Thus, the flow will converge if and only if such a hypersurface M_1 lies in the past of M_0.

Conversely, if there exists a compact, spacelike hypersurface M_1 in N satisfying (6.81), and there is no stable maximal hypersurface in its future, then this is a strong indication that N has no future singularity, assuming that such a singularity would produce spacelike hypersurfaces with positive mean curvature.

An example of such a spacetime is the $(n+1)$-dimensional de Sitter space which is geodesically complete and has exactly one maximal hypersurface M which is also totally geodesic but not stable, and the future resp. past of M are foliated by coordinate slices with negative resp. positive mean curvature.

To conclude this section let us show which spacelike hypersurfaces satisfy C^1-estimates automatically.

THEOREM 6.11. *Let $M = \operatorname{graph} u_{|\mathcal{S}_0}$ be a compact, spacelike hypersurface represented in a Gaussian coordinate system with unilateral bounded principal curvatures, e.g.,*

$$\kappa_i \geq \kappa_0 \quad \forall\, i. \tag{6.82}$$

Then, the quantity $\tilde{v} = \frac{1}{\sqrt{1-|Du|^2}}$ can be estimated by

$$\tilde{v} \leq c(|u|, \mathcal{S}_0, \sigma_{ij}, \psi, \kappa_0), \tag{6.83}$$

where we assumed that in the Gaussian coordinate system the ambient metric has the form as in (6.1).

PROOF. We suppose as usual that the Gaussian coordinate system is future oriented, and that the second fundamental form is evaluated with respect to the past directed normal. We observe that

$$\|Du\|^2 = g^{ij}u_iu_j = e^{-2\psi}\frac{|Du|^2}{v^2}, \tag{6.84}$$

hence, it is equivalent to find an a priori estimate for $\|Du\|$.

Let λ be a real parameter to be specified later, and set

$$w = \tfrac{1}{2}\log\|Du\|^2 + \lambda u. \tag{6.85}$$

We may regard w as being defined on $\mathcal{S}_0$; thus, there is $x_0 \in \mathcal{S}_0$ such that

$$w(x_0) = \sup_{\mathcal{S}_0} w, \tag{6.86}$$

and we conclude

$$0 = w_i = \frac{1}{\|Du\|^2}u_{ij}u^j + \lambda u_i \tag{6.87}$$

in x_0, where the covariant derivatives are taken with respect to the induced metric g_{ij}, and the indices are also raised with respect to that metric.

Expressing the second fundamental form of a graph with the help of the Hessian of the function

$$e^{-\psi}v^{-1}h_{ij} = -u_{ij} - \bar{\Gamma}^0_{00}u_iu_j - \bar{\Gamma}^0_{0i}u_j - \bar{\Gamma}^0_{0j}u_i - \bar{\Gamma}^0_{ij}. \tag{6.88}$$

we deduce further

$$\begin{aligned} \lambda\|Du\|^4 &= -u_{ij}u^iu^j \\ &= e^{-\psi}\tilde{v}h_{ij}u^iu^j + \bar{\Gamma}^0_{00}\|Du\|^4 \\ &\quad + 2\bar{\Gamma}^0_{0j}u^j\|Du\|^2 + \bar{\Gamma}^0_{ij}u^iu^j. \end{aligned} \tag{6.89}$$

Now, there holds

$$u^i = g^{ij}u_j = e^{-2\psi}\sigma^{ij}u_jv^{-2}, \tag{6.90}$$

and by assumption,

$$h_{ij}u^iu^j \ge \kappa_0\|Du\|^2, \tag{6.91}$$

i.e., the critical terms on the right-hand side of (6.89) are of fourth order in $\|Du\|$ with bounded coefficients, and we conclude that $\|Du\|$ can't be too large in x_0 if we choose λ such that

$$\lambda \le -c|||\bar{\Gamma}^0_{\alpha\beta}||| - 1 \tag{6.92}$$

with a suitable constant c; w, or equivalently, $\|Du\|$ is therefore uniformly bounded from above. □

Especially for convex graphs over $\mathcal{S}_0$ the term $\tilde{v}$ is uniformly bounded as long as they stay in a compact set.

7. The inverse mean curvature flow

Let us now consider the inverse mean curvature flow (IMCF)

$$\dot{x} = -H^{-1}\nu \tag{7.1}$$

with initial hypersurface M_0 in a globally hyperbolic spacetime N with compact Cauchy hypersurface $\mathcal{S}_0$.

N is supposed to satisfy the timelike convergence condition

$$\bar{R}_{\alpha\beta}\nu^\alpha\nu^\beta \geq 0 \qquad \forall\, \langle\nu,\nu\rangle = -1. \tag{7.2}$$

Spacetimes with compact Cauchy hypersurface that satisfy the timelike convergence condition are also called *cosmological spacetimes*, a terminology due to Bartnik.

In such spacetimes the inverse mean curvature flow will be smooth as long as it stays in a compact set, and, if $H_{|M_0} > 0$ and if the flow exists for all time, it will necessarily run into the future singularity, since the mean curvature of the flow hypersurfaces will become unbounded and the flow will run into the future of M_0. Hence the claim follows from Remark 6.4 on page 143.

However, it might be that the flow will run into the singularity in finite time. To exclude this behaviour we introduced in [**15**] the so-called *strong volume decay condition*, cf. Definition 7.2. A strong volume decay condition is both necessary and sufficient in order that the IMCF exists for all time.

THEOREM 7.1. *Let N be a cosmological spacetime with compact Cauchy hypersurface $\mathcal{S}_0$ and with a future mean curvature barrier. Let M_0 be a closed, connected, spacelike hypersurface with positive mean curvature and assume furthermore that N satisfies a future volume decay condition. Then the IMCF (7.1) with initial hypersurface M_0 exists for all time and provides a foliation of the future $D^+(M_0)$ of M_0.*

The evolution parameter t can be chosen as a new time function. The flow hypersurfaces $M(t)$ are the slices $\{t = \text{const}\}$ and their volume satisfies

$$|M(t)| = |M_0|e^{-t}. \tag{7.3}$$

Defining a new time function τ by choosing

$$\tau = 1 - e^{-\frac{1}{n}t} \tag{7.4}$$

we obtain $0 \le \tau < 1$,

$$|M(\tau)| = |M_0|(1-\tau)^n, \tag{7.5}$$

and the future singularity corresponds to $\tau = 1$.

Moreover, the length $L(\gamma)$ *of any future directed curve* γ *starting from* $M(\tau)$ *is bounded from above by*

$$L(\gamma) \le c(1-\tau), \tag{7.6}$$

where $c = c(n, M_0)$. *Thus, the expression* $1-\tau$ *can be looked at as the radius of the slices* $\{\tau = \text{const}\}$ *as well as a measure of the remaining life span of the spacetime.*

Next we shall define the strong volume decay condition.

Definition 7.2. Suppose there exists a time function x^0 such that the future end of N is determined by $\{\tau_0 \le x^0 < b\}$ and the coordinate slices $M_\tau = \{x^0 = \tau\}$ have positive mean curvature with respect to the past directed normal for $\tau_0 \le \tau < b$. In addition the volume $|M_\tau|$ should satisfy

$$\lim_{\tau \to b} |M_\tau| = 0. \tag{7.7}$$

A decay like that is normally associated with a future singularity and we simply call it *volume decay.* If (g_{ij}) is the induced metric of M_τ and $g = \det(g_{ij})$, then we have

$$\log g(\tau_0, x) - \log g(\tau, x) = \int_{\tau_0}^{\tau} 2e^{\psi} \bar{H}(s, x) \quad \forall\, x \in \mathcal{S}_0, \tag{7.8}$$

where $\bar{H}(\tau, x)$ is the mean curvature of M_τ in (τ, x). This relation can be easily derived from the relation (3.8) on page 118 and Remark 3.6 on page 120. A detailed proof is given in [**12**].

In view of (7.7) the left-hand side of this equation tends to infinity if τ approaches b for a.e. $x \in \mathcal{S}_0$, i.e.,

$$\lim_{\tau \to b} \int_{\tau_0}^{\tau} e^{\psi} \bar{H}(s, x) = \infty \quad \text{for a.e. } x \in \mathcal{S}_0. \tag{7.9}$$

Assume now, there exists a continuous, positive function $\varphi = \varphi(\tau)$ such that

$$e^{\psi} \bar{H}(\tau, x) \ge \varphi(\tau) \qquad \forall\, (\tau, x) \in (\tau_0, b) \times \mathcal{S}_0, \tag{7.10}$$

where

$$\int_{\tau_0}^{b} \varphi(\tau) = \infty, \tag{7.11}$$

then we say that the future of N satisfies a *strong volume decay condition.*

REMARK 7.3. (i) By approximation we may assume that the function φ above is smooth.

(ii) A similar definition holds for the past of N by simply reversing the time direction. Notice that in this case the mean curvature of the coordinate slices has to be negative.

LEMMA 7.4. *Suppose that the future of N satisfies a strong volume decay condition, then there exist a time function $\tilde{x}^0 = \tilde{x}^0(x^0)$, where x^0 is the time function in the strong volume decay condition, such that the mean curvature $\bar{H}$ of the slices $\tilde{x}^0 = \text{const}$ satisfies the estimate*

$$e^{\tilde{\psi}}\bar{H} \geq 1. \tag{7.12}$$

The factor $e^{\tilde{\psi}}$ is now the conformal factor in the representation

$$d\bar{s}^2 = e^{2\tilde{\psi}}(-(d\tilde{x}^0)^2 + \sigma_{ij}dx^i dx^j). \tag{7.13}$$

The range of $\tilde{x}^0$ is equal to the interval $[0, \infty)$, i.e., the singularity corresponds to $\tilde{x}^0 = \infty$.

A proof is given in [**15**, Lemma 1.4].

REMARK 7.5. Theorem 7.1 can be generalized to spacetimes satisfying

$$\bar{R}_{\alpha\beta}\nu^\alpha\nu^\beta \geq -\Lambda \qquad \forall \langle \nu, \nu \rangle = -1 \tag{7.14}$$

with a constant $\Lambda \geq 0$, if the mean curvature of the initial hypersurface M_0 is sufficiently large

$$H_{|M_0} > \sqrt{n\Lambda}, \tag{7.15}$$

cf. [**25**]. In that thesis it is also shown that the future mean curvature barrier assumption can be dropped, i.e., the strong volume decay condition is sufficient to prove that the IMCF exists for all time and provides a foliation of the future of M_0. Hence, the strong volume decay condition already implies the existence of a future mean curvature barrier, since the leaves of the IMCF define such a barrier.

8. The IMCF in ARW spaces

In the present section we consider spacetimes N satisfying some structural conditions, which are still fairly general, and prove convergence results for the leaves of the IMCF.

Moreover, we define a new spacetime $\hat{N}$ by switching the light cone and using reflection to define a new time function, such that the two spacetimes N and $\hat{N}$ can be pasted together to yield a smooth manifold having a metric singularity, which, when viewed from the region N is a big crunch, and when viewed from $\hat{N}$ is a big bang.

The inverse mean curvature flows in N resp. $\hat{N}$ correspond to each other via reflection. Furthermore, the properly rescaled flow in N has a natural smooth extension of class C^3 across the singularity into $\hat{N}$. With respect to this natural diffeomorphism we speak of a transition from big crunch to big bang.

DEFINITION 8.1. A globally hyperbolic spacetime N, $\dim N = n+1$, is said to be *asymptotically Robertson-Walker* (ARW) with respect to the future, if a future end of N, N_+, can be written as a product $N_+ = [a,b)\times\mathcal{S}_0$, where $\mathcal{S}_0$ is a Riemannian space, and there exists a future directed time function $\tau = x^0$ such that the metric in N_+ can be written as

$$d\breve{s}^2 = e^{2\tilde{\psi}}\{-(dx^0)^2 + \sigma_{ij}(x^0,x)dx^idx^j\}, \tag{8.1}$$

where $\mathcal{S}_0$ corresponds to $x^0 = a$, $\tilde{\psi}$ is of the form

$$\tilde{\psi}(x^0,x) = f(x^0) + \psi(x^0,x), \tag{8.2}$$

and we assume that there exists a positive constant c_0 and a smooth Riemannian metric $\bar{\sigma}_{ij}$ on $\mathcal{S}_0$ such that

$$\lim_{\tau\to b} e^{\psi} = c_0 \quad \wedge \quad \lim_{\tau\to b}\sigma_{ij}(\tau,x) = \bar{\sigma}_{ij}(x), \tag{8.3}$$

and

$$\lim_{\tau\to b} f(\tau) = -\infty. \tag{8.4}$$

Without loss of generality we shall assume $c_0 = 1$. Then N is ARW with respect to the future, if the metric is close to the Robertson-Walker metric

$$d\bar{s}^2 = e^{2f}\{-dx^{0^2} + \bar{\sigma}_{ij}(x)dx^idx^j\} \tag{8.5}$$

near the singularity $\tau = b$. By *close* we mean that the derivatives of arbitrary order with respect to space and time of the conformal metric $e^{-2f}\breve{g}_{\alpha\beta}$ in (8.1) should converge to the corresponding derivatives of the conformal limit metric in (8.5) when x^0 tends to b. We emphasize that in our terminology Robertson-Walker metric does not imply that $(\bar{\sigma}_{ij})$ is a metric of constant curvature, it is only the spatial metric of a warped product.

We assume, furthermore, that f satisfies the following five conditions

$$-f' > 0, \tag{8.6}$$

there exists $\omega\in\mathbb{R}$ such that

$$n+\omega-2 > 0 \quad \wedge \quad \lim_{\tau\to b}|f'|^2e^{(n+\omega-2)f} = m > 0. \tag{8.7}$$

Set $\tilde{\gamma} = \frac{1}{2}(n + \omega - 2)$, then there exists the limit

$$\lim_{\tau \to b} (f'' + \tilde{\gamma}|f'|^2) \tag{8.8}$$

and

$$|D_\tau^m (f'' + \tilde{\gamma}|f'|^2)| \le c_m |f'|^m \qquad \forall m \ge 1, \tag{8.9}$$

as well as

$$|D_\tau^m f| \le c_m |f'|^m \qquad \forall m \ge 1. \tag{8.10}$$

If $\mathcal{S}_0$ is compact, then we call N a *normalized* ARW spacetime, if

$$\int_{\mathcal{S}_0} \sqrt{\det \bar{\sigma}_{ij}} = |S^n|. \tag{8.11}$$

REMARK 8.2. (i) If these assumptions are satisfied, then the range of τ is finite, hence, we may—and shall—assume w.l.o.g. that $b = 0$, i.e.,

$$a < \tau < 0. \tag{8.12}$$

(ii) Any ARW spacetime with compact $\mathcal{S}_0$ can be normalized as one easily checks. For normalized ARW spaces the constant m in (8.7) is defined uniquely and can be identified with the mass of N, cf. [**20**].

(iii) In view of the assumptions on f the mean curvature of the coordinate slices $M_\tau = \{x^0 = \tau\}$ tends to ∞, if τ goes to zero.

(iv) ARW spaces with compact $\mathcal{S}_0$ satisfy a strong volume decay condition, cf. Definition 7.2 on page 156.

(v) Similarly one can define N to be ARW with respect to the past. In this case the singularity would lie in the past, correspond to $\tau = 0$, and the mean curvature of the coordinate slices would tend to $-\infty$.

We assume that N satisfies the timelike convergence condition and that $\mathcal{S}_0$ is compact. Consider the future end N_+ of N and let $M_0 \subset N_+$ be a spacelike hypersurface with positive mean curvature $\breve{H}_{|M_0} > 0$ with respect to the past directed normal vector $\breve{\nu}$—it will become apparent in a moment why we use the symbols $\breve{H}$ and $\breve{\nu}$ and not the usual ones H and ν. Then, as we have proved in the preceding section, the inverse mean curvature flow

$$\dot{x} = -\breve{H}^{-1}\breve{\nu} \tag{8.13}$$

with initial hypersurface M_0 exists for all time, is smooth, and runs straight into the future singularity.

If we express the flow hypersurfaces $M(t)$ as graphs over $\mathcal{S}_0$

$$M(t) = \operatorname{graph} u(t, \cdot), \tag{8.14}$$

then we have proved in [**14**].

THEOREM 8.3. (i) *Let N satisfy the above assumptions, then the range of the time function x^0 is finite, i.e., we may assume that $b = 0$. Set*

$$\tilde{u} = ue^{\gamma t}, \tag{8.15}$$

where $\gamma = \frac{1}{n}\tilde{\gamma}$, then there are positive constants c_1, c_2 such that

$$-c_2 \le \tilde{u} \le -c_1 < 0, \tag{8.16}$$

and $\tilde{u}$ converges in $C^\infty(\mathcal{S}_0)$ to a smooth function, if t goes to infinity. We shall also denote the limit function by $\tilde{u}$.

(ii) *Let $\breve{g}_{ij}$ be the induced metric of the leaves $M(t)$, then the rescaled metric*

$$e^{\frac{2}{n}t}\breve{g}_{ij} \tag{8.17}$$

converges in $C^\infty(\mathcal{S}_0)$ to

$$(\tilde{\gamma}m)^{\frac{1}{\tilde{\gamma}}}(-\tilde{u})^{\frac{2}{\tilde{\gamma}}}\bar{\sigma}_{ij}. \tag{8.18}$$

(iii) *The leaves $M(t)$ get more umbilical, if t tends to infinity, namely, there holds*

$$\breve{H}^{-1}|\breve{h}_i^j - \tfrac{1}{n}\breve{H}\delta_i^j| \le ce^{-2\gamma t}. \tag{8.19}$$

In case $n + \omega - 4 > 0$, we even get a better estimate

$$|\breve{h}_i^j - \tfrac{1}{n}\breve{H}\delta_i^j| \le ce^{-\frac{1}{2n}(n+\omega-4)t}. \tag{8.20}$$

To prove the convergence results for the inverse mean curvature flow, we consider the flow hypersurfaces to be embedded in N equipped with the conformal metric

$$d\bar{s}^2 = -(dx^0)^2 + \sigma_{ij}(x^0, x)dx^i dx^j. \tag{8.21}$$

Though, formally, we have a different ambient space we still denote it by the same symbol N and distinguish only the metrics $\breve{g}_{\alpha\beta}$ and $\bar{g}_{\alpha\beta}$

$$\breve{g}_{\alpha\beta} = e^{2\tilde{\psi}}\bar{g}_{\alpha\beta} \tag{8.22}$$

and the corresponding geometric quantities of the hypersurfaces $\breve{h}_{ij}, \breve{g}_{ij}, \breve{\nu}$ resp. h_{ij}, g_{ij}, ν, etc., i.e., the standard notations now apply to the case when N is equipped with the metric in (8.21).

The second fundamental forms $\breve{h}_i^j$ and h_i^j are related by

$$e^{\tilde{\psi}}\breve{h}_i^j = h_i^j + \tilde{\psi}_\alpha\nu^\alpha\delta_i^j \tag{8.23}$$

and, if we define F by

$$F = e^{\tilde{\psi}} \breve{H}, \tag{8.24}$$

then

$$F = H - n\tilde{v}f' + n\psi_\alpha \nu^\alpha, \tag{8.25}$$

where

$$\tilde{v} = v^{-1}, \tag{8.26}$$

and the evolution equation can be written as

$$\dot{x} = -F^{-1}\nu, \tag{8.27}$$

since

$$\breve{\nu} = e^{-\tilde{\psi}}\nu. \tag{8.28}$$

The flow exists for all time and is smooth, due to the results in the preceding section.

Next, we want to show how the metric, the second fundamental form, and the normal vector of the hypersurfaces $M(t)$ evolve by adapting the general evolution equations in Section 3 on page 117 to the present situation.

LEMMA 8.4. *The metric, the normal vector, and the second fundamental form of $M(t)$ satisfy the evolution equations*

$$\dot{g}_{ij} = -2F^{-1}h_{ij}, \tag{8.29}$$

$$\dot{\nu} = \nabla_M(-F^{-1}) = g^{ij}(-F^{-1})_i x_j, \tag{8.30}$$

and

$$\dot{h}_i^j = (-F^{-1})_i^j + F^{-1}h_i^k h_k^j + F^{-1}\bar{R}_{\alpha\beta\gamma\delta}\nu^\alpha x_i^\beta \nu^\gamma x_k^\delta g^{kj} \tag{8.31}$$

$$\dot{h}_{ij} = (-F^{-1})_{ij} - F^{-1}h_i^k h_{kj} + F^{-1}\bar{R}_{\alpha\beta\gamma\delta}\nu^\alpha x_i^\beta \nu^\gamma x_j^\delta. \tag{8.32}$$

Since the initial hypersurface is a graph over $\mathcal{S}_0$, we can write

$$M(t) = \operatorname{graph} u(t)_{|\mathcal{S}_0} \quad \forall\, t \in I, \tag{8.33}$$

where u is defined in the cylinder $\mathbb{R}_+ \times \mathcal{S}_0$. We then deduce from (8.27), looking at the component $\alpha = 0$, that u satisfies a parabolic equation of the form

$$\dot{u} = \frac{\tilde{v}}{F}, \tag{8.34}$$

where we emphasize that the time derivative is a total derivative, i.e.

$$\dot{u} = \frac{\partial u}{\partial t} + u_i \dot{x}^i. \tag{8.35}$$

Since the past directed normal can be expressed as

$$(\nu^\alpha) = -e^{-\psi} v^{-1}(1, u^i), \tag{8.36}$$

we conclude from (8.34)

$$\frac{\partial u}{\partial t} = \frac{v}{F}. \tag{8.37}$$

For this new curvature flow the necessary decay estimates and convergence results can be proved, which in turn can be immediately translated to corresponding convergence results for the original IMCF.

Transition from big crunch to big bang. With the help of the convergence results in Theorem 8.3, we can rescale the IMCF such that it can be extended past the singularity in a natural way.

We define a new spacetime $\hat{N}$ by reflection and time reversal such that the IMCF in the old spacetime transforms to an IMCF in the new one.

By switching the light cone we obtain a new spacetime $\hat{N}$. The flow equation in N is independent of the time orientation, and we can write it as

$$\dot{x} = -\breve{H}^{-1}\breve{\nu} = -(-\breve{H})^{-1}(-\breve{\nu}) \equiv -\hat{H}^{-1}\hat{\nu}, \tag{8.38}$$

where the normal vector $\hat{\nu} = -\breve{\nu}$ is past directed in $\hat{N}$ and the mean curvature $\hat{H} = -\breve{H}$ negative.

Introducing a new time function $\hat{x}^0 = -x^0$ and formally new coordinates $(\hat{x}^\alpha)$ by setting

$$\hat{x}^0 = -x^0, \quad \hat{x}^i = x^i, \tag{8.39}$$

we define a spacetime $\hat{N}$ having the same metric as N—only expressed in the new coordinate system—such that the flow equation has the form

$$\dot{\hat{x}} = -\hat{H}^{-1}\hat{\nu}, \tag{8.40}$$

where $M(t) = \operatorname{graph} \hat{u}(t)$, $\hat{u} = -u$, and

$$(\hat{\nu}^\alpha) = -\tilde{v} e^{-\tilde{\psi}}(1, \hat{u}^i) \tag{8.41}$$

in the new coordinates, since

$$\hat{\nu}^0 = -\breve{\nu}^0 \frac{\partial \hat{x}^0}{\partial x^0} = \nu^0 \tag{8.42}$$

and

$$\hat{\nu}^i = -\breve{\nu}^i. \tag{8.43}$$

The singularity in $\hat{x}^0 = 0$ is now a past singularity, and can be referred to as a big bang singularity.

The union $N \cup \hat{N}$ is a smooth manifold, topologically a product $(-a, a) \times \mathcal{S}_0$—we are well aware that formally the singularity $\{0\} \times \mathcal{S}_0$ is not part of the union; equipped with the respective metrics and time orientation it is a spacetime which has a (metric) singularity in $x^0 = 0$. The time function

$$\hat{x}^0 = \begin{cases} x^0, & \text{in } N, \\ -x^0, & \text{in } \hat{N}, \end{cases} \tag{8.44}$$

is smooth across the singularity and future directed.

$N \cup \hat{N}$ can be regarded as a *cyclic universe* with a contracting part $N = \{\hat{x}^0 < 0\}$ and an expanding part $\hat{N} = \{\hat{x}^0 > 0\}$ which are joined at the singularity $\{\hat{x}^0 = 0\}$.

It turns out that the inverse mean curvature flow, properly rescaled, defines a natural C^3- diffeomorphism across the singularity and with respect to this diffeomorphism we speak of a transition from big crunch to big bang.

Using the time function in (8.44) the inverse mean curvature flows in N and $\hat{N}$ can be uniformly expressed in the form

$$\dot{\hat{x}} = -\hat{H}^{-1}\hat{\nu}, \tag{8.45}$$

where (8.45) represents the original flow in N, if $\hat{x}^0 < 0$, and the flow in (8.40), if $\hat{x}^0 > 0$.

Let us now introduce a new flow parameter

$$s = \begin{cases} -\gamma^{-1}e^{-\gamma t}, & \text{for the flow in } N, \\ \gamma^{-1}e^{-\gamma t}, & \text{for the flow in } \hat{N}, \end{cases} \tag{8.46}$$

and define the flow $y = y(s)$ by $y(s) = \hat{x}(t)$. $y = y(s, \xi)$ is then defined in $[-\gamma^{-1}, \gamma^{-1}] \times \mathcal{S}_0$, smooth in $\{s \neq 0\}$, and satisfies the evolution equation

$$y' \equiv \tfrac{d}{ds}y = \begin{cases} -\hat{H}^{-1}\hat{\nu}e^{\gamma t}, & s < 0, \\ \hat{H}^{-1}\hat{\nu}e^{\gamma t}, & s > 0. \end{cases} \tag{8.47}$$

In [**14**] we proved:

THEOREM 8.5. *The flow* $y = y(s, \xi)$ *is of class* C^3 *in* $(-\gamma^{-1}, \gamma^{-1}) \times \mathcal{S}_0$ *and defines a natural diffeomorphism across the singularity. The flow parameter* s *can be used as a new time function.*

REMARK 8.6. The regularity result for the transition flow is optimal, i.e., given any $0 < \alpha < 1$, then there is an ARW space such that the transition flow is not of class $C^{3,\alpha}$, cf. [**19**].

REMARK 8.7. Since ARW spaces have a future mean curvature barrier, a future end can be foliated by CMC hypersurfaces the mean curvature of which can be used as a new time function., see [**9**] and [**22**]. In [**8**] we study this foliation a bit more closely and prove that, when writing the CMC hypersurfaces as graphs $M_\tau = \operatorname{graph} \varphi(\tau, \cdot)$ in the special coordinate system of the ARW space, where τ is the mean curvature, of M_τ then

$$\tau(-\varphi)^{1+\tilde{\gamma}^{-1}} \to \text{const} > 0, \tag{8.48}$$

notice that $\varphi < 0$, and hence

$$\lim_{\tau \to \infty} \frac{\varphi(\tau, x)}{\varphi(\tau, y)} = 1 \qquad \forall x, y \in \mathcal{S}_0. \tag{8.49}$$

Moreover, the new time function

$$s = -\tau^{-q}, \quad q = \frac{\tilde{\gamma}}{1 + \tilde{\gamma}} \tag{8.50}$$

can be extended to the mirror universe $\hat{N}$ by odd reflection as a function of class C^3 across the singularity with non-vanishing gradient.

References

[1] Lars Andersson and Gregory Galloway, *Ds/cft and spacetime topology*, Adv. Theor. Math. Phys. **68** (2003), 307–327, hep-th/0202161, 17 pages.

[2] Robert Bartnik, *Remarks on cosmological spacetimes and constant mean curvature surfaces*, Comm. Math. Phys. **117** (1988), no. 4, 615–624.

[3] Antonio N. Bernal and Miguel Sanchez, *On smooth Cauchy hypersurfaces and Geroch's splitting theorem*, Commun. Math. Phys. **243** (2003), 461–470, arXiv:gr-qc/0306108.

[4] ———, *Smoothness of time functions and the metric splitting of globally hyperbolic spacetimes*, Comm. Math. Physics **257** (2005), 43–50, arXiv:gr-qc/0401112.

[5] Shiu Yuen Cheng and Shing Tung Yau, *On the regularity of the solution of the n-dimensional Minkowski problem*, Comm. Pure Appl. Math. **29** (1976), no. 5, 495–516.

[6] Klaus Ecker and Gerhard Huisken, *Parabolic methods for the construction of spacelike slices of prescribed mean curvature in cosmological spacetimes.*, Commun. Math. Phys. **135** (1991), no. 3, 595–613.

[7] Lars Gårding, *An inequality for hyperbolic polynomials*, J. Math. Mech. **8** (1959), 957–965.

[8] Claus Gerhardt, *Properties of the CMC foliation of ARW spaces*, in preparation.
[9] ———, *H-surfaces in Lorentzian manifolds*, Commun. Math. Phys. **89** (1983), 523–553.
[10] ———, *Hypersurfaces of prescribed curvature in Lorentzian manifolds*, Indiana Univ. Math. J. **49** (2000), 1125–1153, arXiv:math.DG/0409457.
[11] ———, *Hypersurfaces of prescribed mean curvature in Lorentzian manifolds*, Math. Z. **235** (2000), 83–97, arXiv:math.DG/0409465.
[12] ———, *Estimates for the volume of a Lorentzian manifold*, Gen. Relativity Gravitation **35** (2003), 201–207, math.DG/0207049.
[13] ———, *Hypersurfaces of prescribed scalar curvature in Lorentzian manifolds*, J. reine angew. Math. **554** (2003), 157–199, math.DG/0207054.
[14] ———, *The inverse mean curvature flow in ARW spaces - transition from big crunch to big bang*, 2004, arXiv:math.DG/0403485, 39 pages.
[15] ———, *The inverse mean curvature flow in cosmological spacetimes*, 2004, arXiv:math.DG/0403097, 24 pages.
[16] ———, *Minkowski type problems for convex hypersurfaces in the sphere*, 2005, arXiv:math.DG/0509217, 30 pages.
[17] ———, *Analysis II*, International Series in Analysis, International Press, Somerville, MA, 2006, 395 pp.
[18] ———, *Curvature Problems*, Series in Geometry and Topology, vol. 39, International Press, Somerville, MA, 2006, 323 pp.
[19] ———, *The inverse mean curvature flow in Robertson-Walker spaces and its application to cosmology*, Methods Appl. Analysis **13** (2006), no. 1, 19–28, gr-qc/0404112.
[20] ———, *The mass of a Lorentzian manifold*, Adv. Theor. Math. Phys. **10** (2006), 33–48, math.DG/0403002.
[21] ———, *Minkowski type problems for convex hypersurfaces in hyperbolic space*, 2006, arXiv:math.DG/0602597, 32 pages.
[22] ———, *On the CMC foliation of future ends of a spacetime*, Pacific J. Math. **226** (2006), no. 2, 297–308, math.DG/0408197.
[23] Bo Guan and Pengfei Guan, *Convex hypersurfaces of prescribed curvatures.*, Ann. Math. **156** (2002), 655–673.
[24] S. W. Hawking and G. F. R. Ellis, *The large scale structure of space-time*, Cambridge University Press, London, 1973.
[25] Heiko Kröner, *Der inverse mittlere Krümmungsfluß in Lorentz Mannigfaltigkeiten*, 2006, Diplomthesis, Heidelberg University.
[26] N. V. Krylov, *Nonlinear elliptic and parabolic equations of the second order*, Mathematics and its Applications (Soviet Series), vol. 7, D. Reidel Publishing Co., Dordrecht, 1987.
[27] O. A. Ladyzhenskaya, V. A. Solonnikov, and N. N. Ural'tseva, *Linear and quasi-linear equations of parabolic type. Translated from the Russian by S. Smith*, Translations of Mathematical Monographs. 23. Providence, RI: American Mathematical Society (AMS). XI, 648 p., 1968 (English).
[28] Oliver C. Schnürer, *Partielle Differentialgleichungen 2*, 2006, pdf file, Lecture notes.
[29] Shing-Tung Yau, *Geometry of three manifolds and existence of Black Hole due to boundary effect*, arXiv:math.DG/0109053.

Ruprecht-Karls-Universität, Institut für Angewandte Mathematik, Im Neuenheimer Feld 294, 69120 Heidelberg, Germany
E-mail address: gerhardt@math.uni-heidelberg.de
URL: http://www.math.uni-heidelberg.de/studinfo/gerhardt/

Global Regularity and Singularity Development for Wave Maps

J. Krieger

1. The wave maps problem

Let (M,g) be a Riemannian manifold, and denote the standard Minkowski space by $\mathbb{R}^{n+1}$, $n \geq 1$, equipped with the Minkowski metric $m_{\alpha\beta} = m^{\alpha\beta} = \text{diag}(-1,1,\ldots,1)$. A map $u : \mathbb{R}^{n+1} \to (M,g)$ is called a *wave map* provided it is formally critical with respect to the Lagrangian action functional[1]

$$\mathcal{L}(u) := \int_{\mathbb{R}^{n+1}} \langle \partial_\alpha u, \partial^\alpha u \rangle_g d\sigma, \; \partial^\alpha = m^{\alpha\beta}\partial_\beta$$

Here $d\sigma = \Pi_{\alpha=0}^{n} dx_\alpha$ denotes the standard volume element on $\mathbb{R}^{n+1}$. In local coordinates, one finds the equation

$$\Box u^i + \Gamma^i_{jk}\partial_\alpha u^j \partial^\alpha u^k = 0, \; \Box = \partial_\alpha \partial^\alpha \tag{1.1}$$

We shall refer to this as the local coordinate formulation of the wave maps condition. Another possibility is to isometrically embed the target into a Euclidean space $M \hookrightarrow \mathbb{R}^k$, and then work in terms of the ambient coordinates. This is seen to lead to the condition

$$\Box u \perp TM_u \tag{1.2}$$

which leads to

$$\Box u^i + S^i_{jk}(u)\partial_\alpha u^j \partial^\alpha u^k = 0 \tag{1.3}$$

in terms of the ambient coordinates, where S^i_{jk} is the 2nd fundamental form of the embedding $M \hookrightarrow \mathbb{R}^k$.

[1]Throughout, the Einstein summation convention is in force.

Finally, in case the target is parallelizable[2], and assuming $\{e_i\}_{i=1}^k$ is a global orthonormal frame, we can introduce the family of functions ϕ_α^i via the relations

$$u_*(\partial_\alpha) = \phi_\alpha^i e_i$$

Then one deduces a divergence-curl system for the quantities ϕ_α^i of the form

$$\partial_\alpha \phi_\beta^i - \partial_\beta \phi_\alpha^i = C_{jk}^i \phi_\alpha^i \phi_\beta^k \tag{1.4}$$

$$\partial_\alpha \phi^{i\alpha} = -\Gamma_{jk}^i \phi_\alpha^j \phi^{k\alpha} \tag{1.5}$$

Here we have the relations $C_{jk}^i = \Gamma_{jk}^i - \Gamma_{kj}^i$, while the Γ_{jk}^i are the Christoffel symbols with respect to the frame $\{e_i\}$, i.e. $\nabla_{e_j} e_k = \Gamma_{jk}^i e_i$. While the former relation (1.4) arises simply from the fact that the ϕ_α^i represent the push-forward of the ∂_α under u, the latter relation (1.5) encodes the wave maps condition. We shall refer to this system as the *intrinsic formulation* of wave maps.

In some sense, the intrinsic formulation is the most natural one, as it exemplifies the *gauge invariance* inherent in the problem: choosing a gauge means fixing an orthonormal frame $\{e_i\}_{i=1}^k$ for TM. Choosing this frame judiciously may improve the features of the equations. Indeed, this gauge freedom was exploited first for the elliptic analogue of wave maps, harmonic maps (for example [**17**]), and then in [**11**], and implicitly in [**61, 62**], which then led to explicit applications in [**27, 33, 42, 50**].

EXAMPLES.

(i) let $M = \mathbb{R}$. A wave map $u : \mathbb{R}^{n+1} \to \mathbb{R}$ is a free wave $\Box u = 0$.

(ii) let $M = S^{k-1} \subset \mathbb{R}^k$ equipped with the standard metric. Then the wave maps condition can be cast in terms of ambient coordinates as follows:

$$\Box u = -u u_{,\alpha}^t u^{,\alpha}, \quad u \in \mathbb{R}^k$$

Here $u \in \mathbb{R}^k$ is a column vector, and u^t its transpose.

(iii) Now let $M = \mathbf{H}^2$, the hyperbolic plane with metric $dh = \frac{dx^2+dy^2}{y^2}$, $y > 0$. Choose the frame $e_1 = -y\frac{d}{dx}$, $e_2 = -y\frac{d}{dy}$. Then the wave maps condition, cast in terms of the associated family ϕ_α^i, $\alpha = 0, 1, 2$, $i = 1, 2$, becomes

$$\partial_\beta \phi_\alpha^1 - \partial_\alpha \phi_\beta^1 = \phi_\alpha^1 \phi_\beta^2 - \phi_\alpha^2 \phi_\beta^1, \; \partial_\beta \phi_\alpha^2 - \partial_\alpha \phi_\beta^2 = 0 \tag{1.6}$$

$$\partial_\alpha \phi^{1\alpha} = -\phi_\alpha^1 \phi^{2\alpha}, \; \partial_\alpha \phi^{2\alpha} = \phi_\alpha^1 \phi^{1\alpha} \tag{1.7}$$

[2] In a lot of cases, one may reduce to this situation, for example by finding an isometric embedding of M into an $\mathbb{R}^k$ equipped with a metric which renders the embedding totally geodesic; see [**8**].

The basic question one would like to answer is the *Cauchy problem* for wave maps: given initial data $(u, \partial_t u) : \mathbb{R}^n \longrightarrow M \times TM$, is there a global-in-time wave map extending them? If not, explain the breakdown.

As posed the question is too vague. Indeed, a first issue to answer is what are the minimal requirements on regularity of the data to even obtain unique local solutions. It is by now well understood that there is a satisfactory local well-posedness theory provided the data are of regularity $(u, \partial_t u) \in H^s \times H^{s-1}$, $s > \frac{n}{2}$, $n \geq 2$. See [**24, 28**]. Moreover, there are examples demonstrating that uniqueness of (distributional) solutions fails provided the data are of class H^s, $s < \frac{n}{2}$, see [**1**]. On the positive side, global distributional solutions of energy class (H^1) have been constructed in the case $n = 2$ in [**41**].

Solutions of the optimal C^∞ smoothness are also called *classical wave maps*. We also observe here in passing that the Cauchy problem for the case $n = 1$ is well understood (see [**16, 60**]) and globally well-posed, and so we omit this case from now on. Thus a more precise formulation of our main problem is as follows: *given initial data* $(u, \partial_t u) \in H^s \times H^{s-1}$, *for some* $s > \frac{n}{2}$, $n \geq 2$, *decide if they can be globally extended to a wave map.*

In spite of a lot of recent progress, this question is largely open at this time.

Motivation for studying wave maps

(a) Particle physics: The wave maps problem appears in [**12**] with target $SU(2) = S^3$ and domain $\mathbb{R}^{3+1}$, also referred to as nonlinear sigma model.

(b) Einstein's equations: following Choquest-Bruhat and Moncrief [**7**], consider a space-time 4 manifold $M = M_1 \times \mathbb{R}$, where M_1 is a principal $U(1)$-bundle over a surface Σ, with projection $\pi : M \to \Sigma \times \mathbb{R}$. Then consider metrics on M of the Kaluza-Klein form

$$g = \pi^*(\tilde{g}) + \pi^*(e^{2\gamma})\Theta^2$$

for some $U(1)$-connection 1-form $\Theta = A_\alpha dx^\alpha$, $A_\alpha \in u(1) = i\mathbb{R}$. Choosing suitable local coordinates we may write $\Theta = dx^3 + \sum_{\alpha=0}^{2} A_\alpha dx^\alpha$, where $x_{1,2}$ are coordinates on Σ. Then the curvature is given by $F := d\Theta = F_{\alpha\beta} dx^\alpha dx^\beta$, and Einstein's equations may be seen to imply that defining[3] $e^{3\gamma}(*F) = E$, we have $dE = 0$. If for example $\Sigma = \mathbb{R}^2$ topologically, we have $E = d\omega$ for some function ω. Then it may be seen that the map

$$(\gamma, \omega) : (\Sigma \times \mathbb{R}, \tilde{g}) \longrightarrow (\mathbf{H}^2, h)$$

is a wave map. The metric $\tilde{g}$ in turn is driven by the wave map. This suggests as a model problem wave maps $u : \mathbb{R}^{2+1} \longrightarrow \mathbf{H}^2$. The case $n = 2$ is also referred to as the *energy critical case*, as explained below.

[3]Here $*F = \frac{1}{2}\epsilon_{\alpha\beta\gamma}F^{\beta\gamma}dx^\alpha$ for the antisymmetric symbol $\epsilon_{\alpha\beta\gamma}$.

(c) Wave maps are the natural hyperbolic analogues of the much studied *harmonic map heat flow*, which in local coordinates is described by

$$\partial_t u^i = \triangle u^i + \sum_{\alpha=1}^{n} \Gamma^i_{jk}(u)\partial_\alpha u^j \partial^\alpha u^k$$

2. Well-posedness typology for nonlinear wave equations

Consider a general problem of the form

$$\Box u = N(u, \nabla u),\ (u, \partial_t u)|_{t=0} = (u_0, u_1) \tag{2.1}$$

for some smooth $N(.,.)$. Of course, wave maps in local coordinates fall into this category. Assume that the set of solutions $u(t,x)$ is invariant under the scaling transformation $u(t,x) \to \lambda^\alpha u(\lambda t, \lambda x)$. For wave maps, for example, we have $\alpha = 0$. Then one introduces the *critical Sobolev index* $s_c = \frac{n}{2} - \alpha$. Observe that the norm

$$||u_0||_{\dot{H}^{s_c}} + ||u_1||_{\dot{H}^{s_c-1}}$$

is left invariant under the re-scaling. We are interested in *(i) local well-posedness*, and, having established (i), *(ii) global well-posedness*. In particular, we shall use the following definitions:

DEFINITION 2.1. *(strong local well-posedness) We call the problem* (2.1) *strongly locally well-posed in the Sobolev space $H^s(\mathbb{R}^n)$, which we recall is the space of all tempered distributions f satisfying $\int_{\mathbb{R}^n} |\hat{f}(\xi)|^2(1+|\xi|^2)^s d\xi < \infty$, provided the following holds: given data $(u_0, u_1) \in H^s \times H^{s-1}$, there is a time interval*[4] *$[0,T]$, $T = T(||u_0||_{H^s} + ||u_1||_{H^{s-1}}) > 0$ and a unique distributional solution $u(t,x) \in C([0,T], H^s) \cap C^1([0,T], H^{s-1})$ of* (2.1) *on $[0,T] \times \mathbf{R}^n$, which depends continuously on the data, and preserves higher regularity of the data.*

DEFINITION 2.2. *(weak global well-posedness for small data) We call* (2.1) *weakly globally well-posed for small data (u_0, u_1) in $\dot{H}^{s_c} \times \dot{H}^{s_c-1}$, provided there exists some $\epsilon > 0$ such that for all $(u_0, u_1) \in H^s \times H^{s-1}$, $s > s_c$ with $||u_0||_{\dot{H}^{s_c}} + ||u_1||_{\dot{H}^{s_c-1}} < \epsilon$, there exists a unique global solution $u(t,x) \in C((-\infty,\infty), H^s) \cap C^1((-\infty,\infty), H^{s-1})$.*

The general method for establishing strong local well-posedness is by means of iteration in a suitable Banach space, while the global (weak) well-posedness at the critical Sobolev index follows usually by proving a priori bounds.

[4]It is understood that $T(.)$ is a decreasing function of its argument.

The significance of the critical Sobolev index becomes clear on account of the following

Expected Optimal Well-posedness:

Scaling constraint

Supercritical: If $s < s_c$, *one expects ill-posedness in* H^s. Assume for example $s_c > 0$, and there exist data $(u_0, u_1) \in H^s \times H^{s-1}$, $s < s_c$, such that the corresponding solution exists on a maximal interval $[0, T)$, $T < \infty$. By finite propagation speed we may assume that they are compactly supported. The data $(u_{0,i}, u_{1,i})$, where $(v_{0,i}, v_{1,i})(x) = (2^{\alpha i} v_0(2^i x), 2^{(\alpha+1)i} v_1(2^i x))$, $i \geq 1$, result in solutions with maximal interval of existence $[0, 2^{-i}T)$. Then letting $(a_0, a_1) := \sum_{i \geq 1} T_i(u_{0,i}, u_{1,i})$ for a suitable translation operator T_i, $T_i(v(x)) := v(x + b_i)$, chosen such that the corresponding solutions do not overlap on their interval of existence, we have

$$\left\| \sum_{i \geq 1} T_i(u_{0,i}, u_{1,i}) \right\|_{\dot{H}^s \times \dot{H}^{s-1}} \leq \sum_{i \geq 1} 2^{i(s-s_c)} < \infty$$

for any $s < s_c$. Hence we have $|| \sum_{i \geq 1} T_i(u_{0,i}, u_{1,i})||_{H^s} < \infty$, $s < s_c$, but the solution corresponding to $\sum_{i \geq 1} T_i(u_{0,i}, u_{1,i})$ cannot exist on any interval $[0, T_0)$, $T_0 > 0$.

Critical: If $s = s_c$, *one expects weak global well-posedness for data small in the critical space* $\dot{H}^{s_c} \times \dot{H}^{s_c - 1}$. Note that re-scaling does not change $||u_0||_{\dot{H}^{s_c} \times \dot{H}^{s_c-1}}$, so a local well-posedness result translates into a global well-posedness result. Strong local well-posedness at the critical level is not expected due to negative results in special cases, see e.g. [**60**], where failure of uniformly continuous dependence on the data is shown. Thus the interval of existence $[0, T)$ is expected to depend on the profile of the data, and not just $||(u_0, u_1)||_{\dot{H}^{s_c} \times \dot{H}^{s_c-1}}$.

Subcritical: If $s > s_c$, *one expects strong local well-posedness in* $H^s \times H^{s-1}$, *for 'well behaved equations'*. However, for generic equations, this intuition fails, see [**39**]; in particular, one may have to impose $s > s_1 > s_c$. It is correct for geometric wave equations with null-structure such as WM [**24**], YM [**31**], MKG [**40**].

Note that $s_c = \frac{n}{2}$ for wave maps in the local coordinate formulation.

Energy constraint

Now assume that our problem of type (2.1) also admits an energy type conserved functional, i.e., a quantity $E[u] \gtrsim ||u||_{\dot{H}^{s_0}} + ||u_t||_{\dot{H}^{s_0 - 1}}$ which is preserved under the flow.

EXAMPLES. For the equation $\Box u = |u|^{p-1}u$, $p \geq 1$, on $\mathbb{R}^{n+1}$, we may put

$$E[u] := \int_{\mathbb{R}^n} \left(\frac{1}{2}[u_t^2 + |\nabla_x u|^2] + \frac{1}{p+1}|u|^{p+1} \right) dx,$$

whence we have $E[u] \gtrsim ||u||_{\dot{H}^1}$.

For the wave maps problem, if we embed isometrically $M \hookrightarrow \mathbb{R}^N$, we may define

$$E[u] := \int_{\mathbb{R}^n} \left(|u_t|^2 + |\nabla_x u|^2\right) dx$$

Alternatively, using assuming TM parallelizable and ϕ^i_α as in the gauge invariant formulation of wave maps, we may define

$$E[u] = \sum_{i,\alpha} \int_{\mathbb{R}^n} (\phi^i_\alpha)^2 dx$$

Then one distinguishes between the following cases:

Energy subcritical $s_c < s_0$: one expects global well-posedness, provided strong local well-posedness in the full subcritical range, or also just for some $s_c < s < s_0$. Assume for example $s_0 \geq 1$. By finite propagation speed, to control the evolution on any finite time interval we may assume the data to be compactly supported. Then a priori control over $||u||_{\dot{H}^{s_0}} + ||u_t||_{\dot{H}^{s_0-1}}$ in conjunction with Sobolev's embedding and Hoelder's inequality results in an a priori bound on $||u||_{L^2} + ||u_t||_{L^2}$ on fixed time slices. Hence we get a priori bounds on $||u||_{H^s} + ||u_t||_{H^{s-1}}$, $s \leq s_0$, which allows us to extend the local solutions to global ones.

Energy critical $s_c = s_0$. Global well-posedness hinges on fine structure of equation. Assume that weak global well-posedness at the critical Sobolev regularity obtains. Then finite propagation speed reduces establishing global well-posedness (in the sense of regularity preservation above the critical regularity) to excluding an energy concentration scenario.

EXAMPLES. The semilinear energy critical equation $\Box u = u^5$ on $\mathbb{R}^{3+1}$ is globally well-posed, see e.g., [**14, 49, 56**]. Proof of this fact relies on establishing a Morawetz type inequality, i.e., a priori estimate for a suitable space-time average of $|u|$.

The wave maps problem becomes energy critical in dimension $n = 2$. Then the global behavior hinges on the geometric structure of the target. We have that suitable wave maps $u : \mathbb{R}^{2+1} \longrightarrow S^2$ break down in finite time, see [**37, 45**] and below, while it is conjectured that we have global well-posedness when the target is $\mathbf{H}^2$, see e.g., [**22**].

Energy supercritical $s_c > s_0$. No global well-posedness for generic large data expected. This is under the assumption that there is no 'better energy type functional' which provides a priori control for $||u||_{\dot{H}^s}$, some $s > s_c$ in the local well-posedness range. Indeed, it appears that for all examples falling under this heading no positive well-posedness results for large data are known.

Example: The wave maps equation with $n \geq 3$ is energy supercritical. Explicit blow up examples with a rather eclectic list of targets can be found in [**6, 47**].

3. The local existence theory for wave maps

Recall the equation (1.1), which will suffice for the purposes of constructing a local solution, provided we further assume the data to map into a single coordinate chart. Schematically, the equation is thus of the form $\Box u = \Gamma(u)(\nabla u)^2$, $(u, u_t)|_{t=0} = (u_0, u_1)$. This crude formulation suffices for strong local well-posedness in H^s for the range $s > \frac{n}{2} + 1$; recall that scaling is $s_c = \frac{n}{2}$. To see this, recall the Sobolev embedding $H^{\frac{n}{2}+}(\mathbb{R}^n) \subset L^\infty(\mathbb{R}^n)$, whence

$$||\Gamma(u)(\nabla u)^2||_{H^{\frac{n}{2}+\epsilon}(\mathbb{R}^n)} \lesssim ||\nabla u||^2_{H^{\frac{n}{2}+\epsilon}(\mathbb{R}^n)}, \ \epsilon > 0$$

by using the fractional Leibnitz' rule. Then using that

$$u(t,x) = S(t)u_0 + U(t)u_1 + \int_0^t U(t-s)[\Gamma(u)(\nabla u)^2](s)ds,$$

where $S(t)u_0$, $U(t)u_1$ are the free propagators for initial data $(u_0, 0)$, $(0, u_1)$, respectively, as well as the energy inequality

$$||S(t)u_0 + U(t)u_1||_{\dot{H}^s(\mathbb{R}^n)} \lesssim ||u_0||_{\dot{H}^s(\mathbb{R}^n)} + ||u_1||_{\dot{H}^{s-1}(\mathbb{R}^n)}, \ \epsilon \in \mathbb{R},$$

we can run a Banach iteration on sufficiently small time intervals to get a unique fixed point: $u(t,x) \in C([0,T], H^{\frac{n}{2}+1+\epsilon}(\mathbb{R}^n)) \cap C^1([0,T], H^{\frac{n}{2}+\epsilon}(\mathbb{R}^n))$, $\epsilon > 0$, $T > 0$ sufficiently small.

The main issue for the local theory then is how to improve this argument to require as little smoothness as possible. Recall from the preceding section that we expect the optimal local well-posedness threshold to be $s_c = \frac{n}{2}$. Indeed, a well-known example of Nirenberg, see e.g. [**51, 53**], shows that this cannot be improved in general: consider the scalar problem

$$\Box u = -(u_t)^2 + \sum_{i=1}^n (\partial_{x_i} u)^2 \tag{3.1}$$

Note that using the coordinates $(\log r, \theta)$ on $\mathbb{R}^2 \backslash \{0\}$ with polar coordinates (r, θ), the map $(u, 0) : \mathbb{R}^{n+1} \longrightarrow \mathbb{R}^2 \backslash \{0\}$ is then a wave map. This is due to the fact that (3.1) implies $\Box(e^u) = 0$. Now choose a free wave v on $\mathbb{R}^{n+1}$, $n \geq 2$, such that $||v(0,.)||_{L^\infty_x} < < 1$ but such that $||v(t,.)||_{L^\infty_x} > 1$, and $\sup_{t \in \mathbb{R}} ||v(t,.)||_{H^{\frac{n}{2}}} < \epsilon$, for arbitrarily small $\epsilon > 0$. Then $u = -\log(1-v)$ is defined as long as $v < 1$, but becomes singular in finite time. Indeed, by rescaling $u(t,x) \to u(\lambda t, \lambda x)$, $\lambda > 1$ (which will not increase the $||.||_{H^{\frac{n}{2}}}$-norm), we can achieve breakdown on arbitrarily small time intervals, even when restricting to small $H^{\frac{n}{2}}$-norm. This shows that we cannot expect (1.1) to be strongly locally well-posed at the critical level (or below)[5].

[5]One may object here that the issue comes from the geodesic incompleteness of the target $\mathbb{R}^2 \backslash \{0\}$. However, there are examples of failure of local well-posedness below the critical regularity for target S^2, see [**1**].

It turns out that the counterexample provided above is sharp, due to the following theorem of Klainerman-Machedon [**24**] and Klainerman-Selberg [**28**]: we formulate it as follows (the original work only referred to wave maps confined to single coordinate charts).

THEOREM 3.1. *Let (M, g) be uniformly isometrically embeddable*[6] *into a Euclidean space. Then the wave maps problem is strongly locally well-posed in $H^{\frac{n}{2}+\epsilon}$, any $\epsilon > 0$, $n \geq 2$. Here the norm $||u||_{H^{\frac{n}{2}+\epsilon}}$ is defined by using ambient coordinates.*

We outline here roughly the procedure for obtaining this theorem. Observe that the preceding argument established *fixed time slice estimates* on ∇u, even though the logic of the preceding argument only presupposes $L^1([0,T], H^{\frac{n}{2}+\epsilon})$ control of the nonlinearity $\Gamma(u)(\nabla u)^2$. In effect, it is more natural to work with L^2 based spaces, on account of Plancherel's theorem. We then seek spaces X, Y (the latter to hold the nonlinearity) of functions defined (possibly locally) on space-time and satisfying the following requirements:

Abstract function space requirements:

(i) $X \subset C([0,T], H^{\frac{n}{2}+\epsilon}) \cap C^1([0,T], H^{\frac{n}{2}-1+\epsilon})$, $\epsilon > 0$
(ii) $X \times X \subset X$
(iii) $\nabla_{x,t} X \times \nabla_{x,t} X \subset Y$, $Y \times X \subset Y$
(iv) $\Box^{-1} Y \subset X$
(v) $S(t)[H^{\frac{n}{2}+\epsilon}] \subset X$, $U(t)[H^{\frac{n}{2}-1+\epsilon}] \subset X$.

It turns out that these requirements cannot quite work. First, due to explicit examples in [**39**], the equation $\Box u = u_t^2$ on $\mathbb{R}^{3+1}$ is ill-posed in H^2. Wave maps have the benefit of exhibiting a special algebraic cancellation structure, following Klainerman [**20**], called a *null-structure*. Hence (iii) needs to replaced by the requirement

(iii') $Q_0(X, X) \subset Y$, $Y \times X \subset Y$, where $Q_0(u,v) = \sum_{\nu=0}^{n} \partial_\nu u \partial^\nu v$.

Further, the fact that free waves correspond to measures supported on the light cone on the Fourier side lead to difficulties when working with $L^2(\mathbb{R}^{n+1})$ based functions. Hence using a trick of Bourgain, working on a finite time interval $[0,T]$, $T > 0$, one introduces a smooth cutoff $\chi_T(t)$ supported on a dilate of $[0,T]$ and satisfying $\chi_T|_{[0,T]} = 1$. Then we require [7]

(iv') $\chi_T \Box^{-1} Y \subset X$
(v') $\chi_T S(t)[H^{\frac{n}{2}+\epsilon}] \subset X$, $\chi_T U(t)[H^{\frac{n}{2}-1+\epsilon}] \subset X$.

[6]This means we assume that there exists an isometric embedding such that the 2nd fundamental form has bounded derivatives with respect to geodesic coordinates on M. Thus one has (1.3), which has the same structure as (1.1).

[7]The implied constants depend on T.

Function spaces X, Y satisfying the requirements (i), (ii), (iii'), (iv'), (v'), were introduced in [**24**], in fashion similar to the $X^{s,b}$ or Bourgain spaces introduced in the context of the nonlinear Schrodinger equation. The idea is to introduce a family of spaces analogous to Sobolev spaces but with weights on the Fourier side reflecting the geometry of the light cone.

Definition: Define $X^{s,b}$ to be the completion of $\mathcal{S}(\mathbb{R}^{n+1})$ with respect to the norm

$$||u||^2_{X^{s,b}} := \int_{\mathbb{R}^{n+1}} \left(1 + [|\xi| + |\tau|]^2\right)^s \left(1 + ||\tau| - |\xi||^2\right)^b |\tilde{u}|^2(\tau, \xi) d\tau d\xi$$

Here we denote the space-time Fourier transform by[8]

$$\tilde{u}(\tau, \xi) := \int_{\mathbb{R}^{n+1}} u(t, x) e^{-i(t\tau + x\cdot\xi)} dt dx$$

Then we have the theorem of Klainerman-Machedon and Klainerman-Selberg [**24, 28, 29**]:

THEOREM 3.2. *Let $b > \frac{1}{2}$, $s = \frac{n}{2} + \epsilon$, $n \geq 2$. Then the embeddings (i), (ii), (iii'), (iv'), (v') are true.*

One version of the proof proceeds by first showing it for truncated free waves as inputs[9] and then using the following **superposition principle for $\boldsymbol{X^{s,b}}$:**

LEMMA 3.3. *Let $u \in X^{s,b}$, $b > \frac{1}{2}$. Then we can write*

$$u = \int_{\mathbb{R}} u_a da \text{ with } \int_{\mathbb{R}} ||u_a||_{L^\infty_t H^s_x} da \lesssim ||u||_{X^{s,b}},$$

where each function $u_a = e^{iat}\phi_a$, with $\Box \phi_a = 0$.

PROOF. This is easily seen by writing

$$\tilde{u}(\tau, \xi) = \int_{\mathbb{R}} \delta(\tau - |\xi| - a)\tilde{u}(|\xi| + a, \xi) da, \ \widetilde{u_a} := \delta(\tau - |\xi| - a)\tilde{u}(|\xi| + a, \xi)$$

and then, using $b > \frac{1}{2}$ and the Cauchy-Schwarz inequality,

$$\begin{aligned}
&\int_{\mathbb{R}} \left(\int_{\mathbb{R}^n} (1 + |\xi|)^{2s} |\tilde{u}|^2(|\xi| + a, \xi) d\xi \right)^{\frac{1}{2}} da \\
&= \int_{\mathbb{R}} (1 + |a|)^{-b} \left(\int_{\mathbb{R}^n} (1 + |\xi|)^{2s} (1 + |a|)^{2b} |\tilde{u}|^2(|\xi| + a, \xi) d\xi \right)^{\frac{1}{2}} da \\
&\leq \left(\int_{\mathbb{R}} (1 + |a|)^{-2b} da \right)^{\frac{1}{2}} \left(\int_{\mathbb{R}^n} (1 + |\xi|)^{2s} (1 + |a|)^{2b} |\tilde{u}|^2(|\xi| + a, \xi) d\xi da \right)^{\frac{1}{2}},
\end{aligned}$$

whence $\int_{\mathbb{R}} ||\tilde{u}(|\xi| + a, \xi)||_{H^s} da \lesssim ||u||_{X^{s,b}}$. □

[8] By contrast, we denote the spatial Fourier transform $\int_{\mathbb{R}^n} e^{-ix\cdot\xi} f(x) dx$ by $\hat{f}(\xi)$.

[9] Thus one considers for example $Q_0(u, v)$ where u, v are free waves.

The role of the null-structure is easily seen by noting the identity $2\partial_\nu u\partial^\nu v = \Box(uv) - (\Box u)v - u(\Box v)$. Hence one essentially reduces the proof of theorem 3.2 to establishing (i), (ii), and the 2nd part of (iii'). We refer to [**29**] for the details. The local solution is then constructed by a simple iteration procedure in the Banach space X. Another version of the proof proceeds via (space-time) frequency localization as well as Strichartz estimates, as detailed in the next section, which deals with well-posedness at the critical level. See e.g., [**29**].

4. Small data global existence theory

One strategy toward global existence is using Klainerman's method of commuting vector fields, which is well adapted to equations exhibiting a null-form structure, such as wave maps. See, e.g., [**21, 53**]. The difficulty with this approach is that the smallness requirement for the data is too stringent to use it as stepping stone toward the large data global existence problem in the pivotal energy critical case $n = 2$. Indeed, for energy critical problems(like the $2+1$-dimensional wave maps), the most natural approach, see [**8**], is to first establish weak global well-posedness for small data at the energy level and then show, using a Morawetz type monotonicity statement, that the energy cannot concentrate inside a backward light cone. This suffices due to the following

LEMMA 4.1. *Assume that the Cauchy problem for wave maps* $u : \mathbb{R}^{2+1} \longrightarrow M$ *is weakly globally well-posed for data of small energy. Also, assume that the energy cannot concentrate in light cones. This means that for every light cone*[10] $C^{\pm}_{t_0,x_0} = \{(t,x) | t - t_0 = \pm\, |x - x_0|\}$, *we have* $\lim_{t\to t_0} \int_{t\times\mathbb{R}^n\cap C^{\pm}_{t_0,x_0}} [|u_t|^2 + |\nabla u|^2]dx = 0$, *see comments in the proof below.Then the problem is globally well-posed in the sense of global regularity preservation for data in* $H^s \times H^{s-1}$, $s > 1$.

PROOF. Assume that the wave map breaks down at time $0 < t = T < \infty$, and exists for $t < T$. We may truncate the data at time $t = 0$ to compact support. We then claim that if $\epsilon > 0$ is such that $||u_0||_{\dot{H}^1} + ||u_1||_{L^2} < \epsilon$ ensures global regularity preservation, there exists a time $T_0 < T$ such that

$$\int_{|x-x_0|\leq 2(T-T_0)} [|u_t|^2 + |\nabla u|^2](T_0, x)dx < \epsilon,\ \forall x_0 \in \mathbb{R}^n$$

Here we assume $M \hookrightarrow \mathbb{R}^k$ via isometric embedding. Indeed, this follows easily from a compactness argument and finite propagation speed, as well as the local energy inequality. By global weak well-posedness for small data[11],

[10] Here the choice of the sign depends on what side of the plane $t = t_0$ the wave map exists, at least locally.

[11] Strictly speaking, one has to truncate the solution on the slice $t = T_0$ to define it on all of $\mathbb{R}^n$ with small energy. This may be done upon choosing the embedding $M \hookrightarrow \mathbb{R}^k$ to have bounded image, for exampe, see [**43**].

we may extend u of the same regularity as the initial data to the cone $|x-x_0|=2(T-T_0)-(t-T_0)$, $T_0\le t\le 2T-T_0$. The solutions constructed on the cones as x_0 varies agree on the overlap by finite propagation speed, and hence we can extend the solution past the purported blow up time, contradicting the assumption. □

4.1. Strong global wellposedness in the critical Besov space. A first step in this direction was taken by Tataru, who proved the following result, [**66, 67**]:

THEOREM 4.2. *(Tataru) The problem* (1.1) *is globally strongly well-posed for small data* (u_0,u_1) *in the Besov space* $\dot{B}^{\frac{n}{2},1}\times\dot{B}^{\frac{n}{2}-1,1}$, $n\ge 2$.

The definition of the spaces $\dot{B}^{\frac{n}{2},1}$ involves Littlewood-Paley frequency localizers as follows: choose a nonnegative $\chi\in C_0^\infty$ with support in $[\frac{1}{4},4]$ and such that $\sum_{k\in\mathbf{Z}}\chi(\frac{x}{2^k})=1$ for all $x\in\mathbb{R}_{>0}$. Then we define

$$P_k f(x)=(2\pi)^{-n}\int_{\mathbb{R}^n}\chi\left(\frac{\xi}{2^k}\right)\hat{f}(\xi)e^{ix\cdot x}d\xi,$$

where $\hat{f}(\xi)=\int_{\mathbb{R}^n}f(x)e^{ix\cdot\xi}dx$. Note that provided $f\in C_0^\infty(\mathbb{R}^n)$, we have $f=\sum_{k\in\mathbf{Z}}P_kf$. One then defines $||\phi||_{\dot{B}^{\frac{n}{2},1}(\mathbb{R}^n)}:=\sum_{k\in\mathbf{Z}}2^{\frac{nk}{2}}||P_k\phi||_{L^2(\mathbb{R}^n)}$ for smooth compactly supported ϕ, say, and denotes by $\dot{B}^{\frac{n}{2},1}$ the closure. We immediately observe the important embedding $\dot{B}^{\frac{n}{2},1}(\mathbb{R}^n)\subset L^\infty(\mathbb{R}^n)$. In particular, enforcing a smallness condition on the initial data in the Besov space ensures that the wave map will remain in a single coordinate patch, whence one may work globally in time with the formulation (1.1). The crux toward establishing Tataru's result in high dimensions, i.e., $n\ge 4$, consists in *frequency localization* and exploitation of *Strichartz estimates*, which are quite useful in a high dimensional setting. We recall here, see for example [**51, 53**]

THEOREM 4.3. (*Strichartz estimates in high dimensions*) *Assume* $\Box u=0$, $(u,u_t)|_{t=0}=(u_0,u_1)$, $u\in C^\infty(\mathbb{R}^{n+1})$ *and compactly supported on fixed time slices. Then we have for* $n\ge 4$

$$\sup_{\frac{1}{p}+\frac{n-1}{2q}\le\frac{n-1}{4},\,p\ge 2}||P_0u||_{L_t^pL_x^q(\mathbb{R}^{n+1})}\lesssim||u_0||_{L^2}+||u_1||_{L^2}$$

Now consider the nonlinearity in (1.1), which we simplify to $2Q_0(u,u)=\Box(u^2)-2(\Box u)u$. As the free wave propagator respects frequency localization, one first tries to control the frequency localized constituents of the nonlinearity, i.e., the expressions $P_k[\Box(u^2)-2(\Box u)u]$. For this one needs to find a suitable Banach space to iterate in. The idea here is to place each frequency localized component of u in a Banach space which is compatible with the scaling underlying the equation. A natural candidate for such a space is provided by the space

$$X=\left\{\phi|\sum_{k\in\mathbf{Z}}\sup_{\frac{1}{p}+\frac{n-1}{2q}\leq\frac{n-1}{4},p\geq 2}2^{\left[\frac{1}{p}+\frac{n}{q}\right]k}||P_k\phi||_{L_t^pL_x^q}<\infty\right\}$$

The problem here is that this space is not well-behaved with respect to applying the operator $\Box$. The solution here is to shrink the space by imposing a weak type of $X^{s,b}$ control in addition: introduce the dyadic space-time frequency localizers Q_j via $\widetilde{Q_j\phi}=\chi\left(\frac{||\tau|-|\xi||}{2^j}\right)\tilde{\phi}(\tau,\xi)$, $j\in\mathbf{Z}$. Hence again $\sum_{j\in\mathbf{Z}}Q_j\phi=\phi$ provided $\phi\in C_0^\infty(\mathbb{R}^{n+1})$, say. Then our new space X is defined to be

$$X=\left\{\phi|\sum_{k\in\mathbf{Z}}\sup_{\frac{1}{p}+\frac{n-1}{2q}\leq\frac{n-1}{4},p\geq 2}2^{\left[\frac{1}{p}+\frac{n}{q}\right]k}||P_k\phi||_{L_t^pL_x^q}<\infty\right\}$$
$$\cap\left\{\phi|\sum_{k\in\mathbf{Z}}\sup_{j\in\mathbf{Z}}2^{\frac{j}{2}+\frac{nk}{2}}||P_kQ_j\phi||_{L_t^2L_x^2}<\infty\right\}$$

The correct space to place the nonlinearity in is then, for example, the following (see [**66**]) :

$$Y=\left\{\phi|\sum_{k\in\mathbf{Z}}[2^{\left(\frac{n}{2}-1\right)k}||P_k\phi||_{L_t^1L_x^2}<\infty\right\}$$

Then one can show the embeddings[12,13,14]

$$\Box IX\times X\subset Y,\ \Box^{-1}Y\subset X,\ I=\sum_{k\in\mathbf{Z}}P_kQ_{<k}$$

These in conjunction with the algebra estimates $X\times X\subset X$, $X\times Y\subset Y$ and the structure of Q_0 suffice to set up an iteration scheme for small data. To see how the embedding $\Box IX\times X\subset Y$ can hold, let us consider a simplified situation in which both factors are of unit frequency. Thus $Iu=P_0Q_{\leq 0}u$, $v=P_0v$. Then

$$\sum_{k\in\mathbf{Z}}2^{\left(\frac{n}{2}-1\right)k}||P_k[\Box Iuv]||_{L_t^1L_x^2}\lesssim||\Box Iu||_{L_t^2L_x^2}||v||_{L_t^2L_x^\infty}\lesssim||u||_X||v||_X$$

---

[12]The operator $\Box^{-1}$ refers to the solution of the inhomogeneous problem given by the Duhamel parametrix.

[13]One uses the fact that $P_k(L_t^1L_x^2)\subset\dot{X}_k^{0,-\frac{1}{2}}=\{\phi|\sup_{j\in\mathbf{Z}}2^{-\frac{j}{2}}||P_kQ_j\phi||_{L_t^2L_x^2}<\infty\}$.

[14]We define $Q_{<k}:=\sum_{j<k}Q_j$. The point of the operator I is to ensure restriction to the hyperbolic region where the distance to the light cone is at most comparable to the spatial frequency. In the opposite case, expanding the null-form $\partial_\nu u\partial^\nu u$ as above may be harmful.

Here one uses that the summation over k is effectively only over $k \leq O(1)$, due to our frequency support assumptions. The fact that $\Box^{-1}Y \subset X$ is partly due to the Strichartz estimates. We refer for a detailed proof of Theorem 4.2 in case $n \geq 4$ to [**66**].

The cases $n = 2, 3$, treated in [**67**], are significantly harder, and involve a different kind of decomposition of u into travelling waves. We shall discuss this in more detail below.

4.2. Weak global well-posedness in the critical Sobolev space. The first result establishing weak global well-posedness for small energy data in the case $n = 2$ (as well as $n = 3, 4$) was obtained by Tao in [**62**], in the case when the target $M = S^k$, $k \geq 1$. This was preceded by [**61**] which dealt with weak global well-posedness for data small in $\dot{H}^{\frac{n}{2}}$, $n \geq 5$. The challenge in improving Tataru's result to data in $\dot{H}^{\frac{n}{2}}$ lies in dealing with what one may refer to as the *summation problem*: recall that the strategy for proving Theorem 4.2 was to control the frequency localized components of the nonlinearity and then invoke the Besov structure to sum over all frequencies[15] involved. However, when working with Sobolev spaces, one can only assume *square summability* of the various frequency components, and hence summing over all possible frequency interactions becomes quite challenging. Tao's method takes advantage of the intrinsic Gauge freedom of the problem, albeit in somewhat complicated form. This was clarified in [**27, 42**] and especially [**50**], where Tao's result was extended to a much more general class of target manifolds and the case $n \geq 4$. As hinted at in the preceding section, the case $n \geq 4$ is significantly simpler than the case $n = 3$ and especially the case $n = 2$, as the dispersive effects of free waves become weaker in lower dimensions and hence estimates from the linear theory become increasingly scarce, forcing one to exploit more and more of the fine structure of the problem. Indeed, in [**50**] it became clear that in dimensions $n \geq 4$, one does not have to invoke the somewhat difficult $X^{s,b}$-space framework, but Strichartz estimates alone suffice. We cite

THEOREM 4.4. [**27**] *($n \geq 5$, boundedly parallelizable)* [**42**] *($n \geq 4$, compact symmetric space target),* [**50**] *($n \geq 4$, uniformly isometrically embeddable targets) Let $n \geq 4$ and let $(M, g) \hookrightarrow \mathbb{R}^k$ be uniformly isometrically embeddable into a Euclidean space*[16]*. Then wave maps $u : \mathbb{R}^{n+1} \longrightarrow M$ are weakly globally well-posed for data small in $\dot{H}^{\frac{n}{2}}$. More precisely, this statement applies to the equation for u obtained in terms of the ambient coordinates on $\mathbb{R}^k$,* (1.3).

[15]The only delicate issue then is to deal with destructive resonance phenomena, i.e., two high frequency waves resulting in a cascade of low frequencies. This turns out to be fairly simple.

[16]This means we assume that there exists an isometric embedding such that the 2nd fundamental form has bounded derivatives with respect to geodeic coordinates on M of all orders.

The method of Klainerman-Rodnianski [**27**], Nahmod-Stefanov-Uhlenbeck [**42**] and Shatah and Struwe in [**50**] exploits the Gauge freedom as in Tao's breakthrough work [**61, 62**], but from the point of view of the intrinsic formulation (1.4), (1.5). Observe that from there one may infer a system of wave equations for the ϕ^i_α, which may be seen to be of the form

$$\Box\phi^i_\alpha = -2\Gamma^i_{jk}\phi^k_\nu\partial^\nu\phi^j_\alpha + "\phi^3"$$

Using Strichartz estimates alone and without introduction of a null-from structure, the quadratic term on the right cannot be estimated. The trick then consists in realizing that the product $A^i_{j\nu} := -2\Gamma^i_{jk}\phi^k_\nu$ is effectively a skew-symmetric (in i, j) connection form. This is of course a reflection of the fact that $\{e_i\}$ is an orthonormal frame. Now one changes the Gauge, i.e., replaces $\{e_i\}$ by $\{\tilde{e}_i\} := \{ge_i\}$ for a suitable $SO(k)$-valued function g (with k the dimension of M), and $(A^i_{j\nu})_{ij} =: A_\nu$ by $\tilde{A}_\nu := g^{-1}\partial_i g + g^{-1}A_\nu g$. Then the corresponding equation for $\tilde{\phi}^i_\alpha$ becomes

$$\Box\tilde{\phi}^i_\alpha = \tilde{A}^i_{j\nu}\partial^\nu\tilde{\phi}^j_\alpha + "\tilde{\phi}^3"$$

Imposing the Coulomb condition $\sum_{i=1}^n \partial_i\tilde{A}_i = 0$, and using the crucial fact that *the curvature of the original connection* $A_\nu = (A^i_{j\nu})_{ij}$, $F_{\alpha\beta} = \partial_\alpha A_\beta - \partial_\beta A_\alpha + [A_\alpha, A_\beta]$, *is quadratic in* ϕ *thanks to* (1.4), one can solve an elliptic system of equations for the $\tilde{A}_\alpha$, resulting schematically in $\tilde{A} = \nabla^{-1}(\phi^2)$, where ∇^{-1} stands for operators of the form $\nabla\triangle^{-1}$. This Gauge condition is of course quite classical, and was treated in the seminal paper [**69**] by Karen Uhlenbeck. Thus the new leading term in the nonlinearity is now of the form

$$\nabla\phi\nabla^{-1}(\phi^2),$$

which is trilinear, and assuming the factors $\nabla\phi$, $\nabla^{-1}(\phi^2)$ to be of the same frequency, of essentially cubic form ϕ^3 (we are careless here about the distinction of $\tilde{\phi}$, ϕ). One can now easily build a Banach space X based on Strichartz estimates alone and of essentially the kind displayed in the preceding subsection, with

$$||\nabla\phi\nabla^{-1}(\phi^2)||_{L^1_t\dot{H}^{\frac{n}{2}-1}} \lesssim ||\phi||^3_X,$$

as long as $n \geq 4$, and Strichartz estimates allow one to close a bootstrapping argument. We refer to [**50**] for details, but note that the use of Lorentz spaces there is easily avoidable due to the $L^2_tL^6_x$-endpoint for Strichartz estimates in $n = 4$ dimensions.

The extension of theorem 4.4 to the significantly more difficult cases $n = 2, 3$ and target different than S^k was achieved in [**33, 35, 68**]. We state

THEOREM 4.5. [**33, 35, 68**]. *Let the target* M *be uniformly isometrically embeddable into a Euclidean space or let* M *be the hyperbolic plane. Then* *the Cauchy problem for wave maps originating on* $\mathbb{R}^{n+1}$, $n=2,3$, *is weakly globally well-posed for data of small critical Sobolev norm.*

We explain here the method of [**33**], where the Gauge freedom becomes particularly transparent[17] when the target is $\mathbf{H}^2$. We state the following

THEOREM 4.6. [**33**] *The Cauchy problem for wave maps* $u : \mathbb{R}^{2+1} \longrightarrow \mathbf{H}^2$ *is globally well-posed for data of small energy. More precisely, for given smooth data* $(u_0, u_1) : \mathbb{R}^2 \longrightarrow \mathbf{H}^2 \times T\mathbf{H}^2$, *which, upon writing* $u_0(x) = (\mathbf{x}, \mathbf{y})(x)$, $\mathbf{y} > 0$, $u_1(x) = (\tilde{\mathbf{x}}, \tilde{\mathbf{y}})(x)$ *(where we use the identification* $T_{(\mathbf{x},\mathbf{y})}\mathbf{H}^2 = \mathbb{R}^2$ *via* $(\tilde{\mathbf{x}}, \tilde{\mathbf{y}}) \to \tilde{\mathbf{x}}\frac{\partial}{\partial \mathbf{x}} + \tilde{\mathbf{y}}\frac{\partial}{\partial \mathbf{y}}$*) satisfy the inequality*

$$\int_{\mathbb{R}^2} \left(\sum_{i=1,2} \left[\frac{\partial_i \mathbf{x}}{\mathbf{y}}\right]^2 (x) \right) + \left(\sum_{i=1,2} \left[\frac{\partial_i \mathbf{y}}{\mathbf{y}}\right]^2 (x) + \left[\frac{\tilde{\mathbf{x}}}{\mathbf{y}}\right]^2 (x) + \left[\frac{\tilde{\mathbf{y}}}{\mathbf{y}}\right]^2 (x) \right) dx < \epsilon$$

for $\epsilon > 0$ *sufficiently small, there exists a unique smooth wave map* $u : \mathbb{R}^{2+1} \longrightarrow \mathbf{H}^2$ *extending the data globally in time; thus* $(u, u_t)|_{t=0} = (u_0, u_1)$.

The proof departs from the intrinsic formulation in terms of the global orthonormal frame $\{e_1, e_2\} := \{-\mathbf{y}\frac{\partial}{\partial \mathbf{x}}, -\mathbf{y}\frac{\partial}{\partial \mathbf{y}}\}$, where we use the identification $\mathbf{H}^2 = \{(\mathbf{x}, \mathbf{y}) | \mathbf{y} > 0\}$. Then we have (1.6), (1.7), which can be used to deduce a system of wave equations for the quantities ϕ^i_α, of the following schematic form:

$$\Box \phi^1_\alpha = -2\phi^1_\nu \partial^\nu \phi^2_\alpha + "\phi^3", \ \Box \phi^2_\alpha = +2\phi^1_\nu \partial^\nu \phi^1_\alpha + "\phi^3",$$

where the fine structure of the terms "ϕ^3" has been suppressed. Given the scarcity of Strichartz estimates in $2+1$ dimensions (recall that one can only control norms $||u||_{L^p_t L^q_x}$ where $\frac{1}{p} + \frac{n-1}{2q} \le \frac{n-1}{4}$, which for $n=2$ translates into $p \ge 4$), there is no hope to control the quadratic terms on the right hand side using Strichartz estimates alone. Indeed, there is no hope to place these in the energy space $L^1_t L^2_x$ even at the fixed frequency $k=0$.

One then passes into the Coulomb Gauge following the procedure detailed above in the case $n \ge 4$. Specifically, we introduce the complex valued functions

$$\psi_\alpha = \psi^1_\alpha + i\psi^2_\alpha := (\phi^1_\alpha + i\phi^2_\alpha) e^{-i\Delta^{-1} \sum_{j=1,2} \partial_j \phi^1_j}$$

These quantities then satisfy a remarkable divergence curl system of the following form:

$$\begin{aligned} \partial_\alpha \psi_\beta - \partial_\beta \psi_\alpha = \, & i\psi_\beta \Delta^{-1} \sum_{j=1,2} \partial_j (\psi^1_\alpha \psi^2_j - \psi^2_\alpha \psi^1_j) \\ & - i\psi_\alpha \Delta^{-1} \sum_{j=1,2} \partial_j (\psi^1_\beta \psi^2_j - \psi^2_\beta \psi^1_j) \end{aligned} \tag{4.1}$$

[17]One is aided by the fact that the Gauge group $U(1)$ is abelian in this case.

$$\partial_\nu \psi^\nu = i\psi^\nu \triangle^{-1} \sum_{j=1,2} \partial_j(\psi^1_\nu \psi^2_j - \psi^2_\nu \psi^1_j). \tag{4.2}$$

From here one can deduce a system of wave equations as follows:

$$\begin{aligned} \Box\psi_\alpha = {} & i\partial^\beta \left[\psi_\alpha \triangle^{-1} \sum_{j=1}^{2} \partial_j \left[\psi^1_\beta \psi^2_j - \psi^2_\beta \psi^1_j\right]\right] \\ & - i\partial^\beta \left[\psi_\beta \triangle^{-1} \sum_{j=1}^{2} \partial_j \left[\psi^1_\alpha \psi^2_j - \psi^2_\alpha \psi^1_j\right]\right] \\ & + i\partial_\alpha \left[\psi_\nu \triangle^{-1} \sum_{j=1}^{2} \partial_j \left[\psi^{1\nu} \psi^2_j - \psi^{2\nu} \psi^1_j\right]\right]. \end{aligned} \tag{4.3}$$

The virtue of these equations is that the right hand side is roughly speaking of the form $\nabla[\psi\nabla^{-1}[\psi^2]]$, and hence of cubic form, which in some sense provides more room for estimates. Nonetheless, finding a space X holding the functions ψ_α as well as a space Y holding the nonlinearity still appears extraordinarily challenging, and the strategy in [**33**] uses a refined version of the spaces in [**62**], which in turn are based on Tataru's travelling wave decompositions in [**67**]. The rough idea behind these spaces is as follows: using the schematic form $\nabla[\psi\nabla^{-1}[\psi^2]]$ for the nonlinearity for now, localize it to frequency $k=0$, i.e. replace it by $P_0\nabla[\psi\nabla^{-1}[\psi^2]]$. Ideally, assuming control over a Strichartz type norm $||P_k\psi||_{L^2_t L^\infty_x}$, we could at least in principle try to place this expression into $L^1_t L^2_x$ by using an estimate of the form

$$||P_0\nabla[\psi\nabla^{-1}[\psi^2]]||_{L^1_t L^2_x} \lesssim ||\psi||_{L^2_t L^\infty_x} ||\nabla^{-1}[\psi^2]||_{L^2_t L^2_x}$$

Here the 2nd term could be estimated, at least when restricted to frequencies ≥ 0, by

$$||P_{\geq 0}\nabla^{-1}[\psi^2]||_{L^2_t L^2_x} \lesssim ||\psi||_{L^2_t L^\infty_x} ||\psi||_{L^\infty_t L^2_x}$$

This procedure unfortunately fails for several reasons. Most prominently, in $n=2$ spatial dimensions, there is no hope to recover control over $||\psi||_{L^2_t L^\infty_x}$ using Strichartz estimates, as the best such estimate available is for $||\psi||_{L^4_t L^\infty_x}$, see theorem 4.3.

In spite of this, it turns out and is well-known (e.g., [**70**]) that in certain situations, *it is possible to obtain control over* $||\psi_1\psi_2||_{L^2_t L^2_x}$, *provided* ψ_1, ψ_2 *are free waves satisfying a certain angular separation condition for their space-time frequency supports.* Indeed, letting $\psi_{1,2}$ be free waves, say at frequency 0, write $\psi_1 = \widetilde{\mu_1}$, $\psi_2 = \widetilde{\mu_2}$, where $\tilde{}$ denotes space-time Fourier transform, and $\mu_{1,2} = \delta(\tau - |\xi|)f_1(\xi)$, $\mu_2 = \delta(\tau - |\xi|)f_2(\xi)$, are measures supported on the light cone. Then, assuming the angular supports of $f_{1,2}$ to be separated, we have $\widetilde{\psi_1\psi_2} = \mu_1 * \mu_2$, whence $||\widetilde{\psi_1\psi_2}||_{L^1_\tau L^1_\xi} \leq ||f_1||_{L^1}||f_2||_{L^1}$,

$||\widetilde{\psi_1\psi_2}||_{L^\infty_\tau L^\infty_\xi} \lesssim ||f_1||_{L^\infty}||f_2||_{L^\infty}$, and from here one gets $||\psi_1\psi_2||_{L^2_t L^2_x} \lesssim ||\psi_1||_{L^\infty_t L^2_x}||\psi_2||_{L^\infty_t L^2_x}$ by interpolation and Plancherel's theorem.

The issue then becomes how to deal with bilinear interactions $\psi_1\psi_2$ *where no angular separation condition for the Fourier support of the factors is fulfilled.* The standard way to deal with such a situation is to exploit an additional cancellation stemming from a null-form structure in the equation at hand, for example one of type $Q_{ij}(\psi_1, \psi_2) = \partial_i\psi_1\partial_j\psi_2 - \partial_j\psi_1\partial_i\psi_2$. However, such a structure is not apparent in (4.3).

To remedy this situation, one resorts to a Hodge type decomposition, which works due to (4.1), see [**33**]. This trick is inspired due to a similar procedure in [**26**]. Specifically, one writes $\psi_\nu = R_\nu\psi + \chi_\nu$ where $R_\nu = \partial_\nu\sqrt{-\triangle}^{-1}$, $\nu = 0, 1, 2$, and we impose the vanishing divergence condition $\sum_{i=1,2}\partial_i\chi_i = 0$. From this one easily deduces relations of the schematic form $\chi = \nabla^{-1}[\psi\nabla^{-1}[\psi^2]]$. Now one substitutes $\psi_\nu = R_\nu\psi + \chi_\nu$ into (4.3). The null-structure emerges if one only substitutes gradient type terms $R_\nu\psi$. On the other hand, the strategy implemented in [**33**] is that substituting at least one term χ_ν for ψ_ν, one has 'more room' for estimates due to the inherently multilinear structure of χ.

On a more technical level, the above intuition needs to be translated to a context in which the factors $\psi_{1,2}$ etc are no longer free waves. A reasonable procedure is to use function spaces whose elements are built up from pieces which in some way behave like free waves. The breakthrough paper [**67**], which also provided the basis for much of the harmonic analysis in [**62**], achieves this by decompositions into functions replicating the behavior of travelling waves. Specifically, the intuition again comes from free waves: let $\Box u = 0$, $(u, u_t)|_{t=0} = (0, g)$, in the context of $\mathbb{R}^{2+1}$. Then we have

$$
\begin{aligned}
u(t,x) &= c\int_0^{2\pi}\int_0^\infty \frac{e^{it|\xi|} - e^{-it|\xi|}}{2|\xi|} e^{ix\cdot\xi}\hat{g}(|\xi|,\omega)|\xi|d|\xi|d\omega \\
&= \int_0^{2\pi} a_\omega(t + x\cdot\omega)d\omega - \int_0^{2\pi} b_\omega(-t + x\cdot\omega)d\omega
\end{aligned}
$$

where we define $a_\omega(s) = c\int_0^\infty e^{is|\xi|}\hat{g}(|\xi|,\omega)d|\xi|$, and similarly for $b_\omega(s)$. Note that each of the functions $(t,x) \to a_\omega(t + x\cdot\omega)$, $(t,x) \to b_\omega(-t + x\cdot\omega)$, is a travelling wave. Furthermore, from Plancherel's theorem and Holder's inequality, we can conclude that if the support of $\hat{g}(\xi)$ is contained in the region $|\xi| \sim 1$, we have

$$
\int_0^{2\pi} ||a_\omega(.)||_{L^2_s}d\omega \lesssim ||g||_{L^2_x}, \int_0^{2\pi} ||b_\omega(.)||_{L^2_s}d\omega \lesssim ||g||_{L^2_x}
$$

Introduce the 'null-frame coordinates'

$$
t_\omega := \frac{1}{\sqrt{2}}(1,\omega)\cdot(t,x),\ x_\omega = (t,x) - \frac{t_\omega}{\sqrt{2}}(1,\omega),
$$

Then if we abuse notation and write $a_\omega(t,x) := a_\omega(t + x \cdot \omega)$ etc, we have just observed that $\int_0^{2\pi} ||a_\omega(t,x)||_{L^2_{t_\omega} L^\infty_{x_\omega}} d\omega \lesssim ||g||_{L^2}$, and similarly for b_ω. In other words, free waves can be written as superpositions of travelling waves which satisfy the preceding relations. To move beyond the context of free waves and yet retain the property of decomposability into functions behaving in some sense like travelling waves, one introduces an atomic Banach space. More specifically, for reasons which will become apparent soon, one introduces a Banach space for each angular sector $\kappa \subset S^1$, as follows: define $PW[\kappa]$ to be the atomic Banach space whose atoms are functions ψ satisfying $\inf_{\omega\in\kappa} ||\psi||_{L^2_{t_\omega} L^\infty_{x_\omega}} \leq 1$. This means that we define the norm $||.||_{PW[\kappa]}$ as follows: assuming $\psi \in C_0^\infty(\mathbb{R}^n)$

$$||\psi||_{PW[\kappa]} := \inf_{\int_\kappa \phi_\omega d\omega = \psi} \int_\kappa ||\phi_\omega||_{L^2_{t_\omega} L^\infty_{x_\omega}},$$

where the infimum is over all representations $\int_\kappa \phi_\omega d\omega = \psi$ with $\omega \to C_0^\infty(\mathbb{R}^{n+1})$ continuous, say. The norms $||.||_{PW[\kappa]}$ are effectively our substitute for the missing $||.||_{L^2_t L^\infty_x}$.

Now consider a product $\phi_1\phi_2$ where the factors have angularly separated Fourier support, say along sectors $\kappa_1, \kappa_2 \subset S^1$. Also, suppose that the distance is much larger than $|\kappa_2|$. Assuming control over $||\phi_1||_{PW[\kappa_1]}$, we see that it suffices to control a norm such as $\sup_{\omega\notin 2\kappa_2} ||\phi_2||_{L^\infty_{t_\omega} L^2_{x_\omega}}$ in order to be able to bound $||\phi_1\phi_2||_{L^2_t L^2_x}$. That a slight modification of this indeed works is suggested by the following lemma, which tests this ansatz by looking at essentially free waves again:

LEMMA 4.7. *Let ψ be a temporally truncated free wave with compact spatial support on fixed time slices in the context of $\mathbb{R}^{2+1}$. Thus we have $\psi(t,x) = \chi(t)\tilde{\psi}(t,x)$ where $\Box\tilde{\psi} = 0$, and $\chi(.)$ smoothly localizes to a compact time interval $[-T,T]$, say. Letting $P_{0,\kappa}$ denote a Fourier multiplier which localizes smoothly to (logarithmic) frequency ~ 0 and angular sector κ, and further $Q^+_{<-2l-10}$ a space-time frequency multiplier smoothly restricting to distance $\leq 2^{-2l-10}$ to the light cone, $l \in \mathbf{N}$, as well as the upper half space $\tau > 0$, we have the inequality*

$$\sup_{\omega\notin 2\kappa} dist(\omega,\kappa)||P_{0,\kappa}Q^+_{<-2l-10}\psi||_{L^\infty_{t_\omega} L^2_{x_\omega}} \lesssim ||\psi||_{\dot{X}_0^{1,\frac{1}{2},1}} = \sum_{j\in\mathbf{Z}} 2^{\frac{j}{2}} ||Q_j\psi||_{L^2_t L^2_x}$$

The latter quantity is bounded by the energy of $\tilde{\psi}$.

PROOF. Note that for $j < -2l-10$, using Hoelder's and Young's inequality

$$||P_{0,\kappa}Q^+_j\psi||_{L^\infty_{t_\omega} L^2_{x_\omega}} \leq ||P_{0,\kappa}Q^+_j\psi||_{L^2_{\xi_\omega} L^1_{\tau_\omega}} \lesssim [\text{dist}(\kappa,\omega)]^{-1} 2^{\frac{j}{2}} ||Q_j\psi||_{L^2_t L^2_x}$$

To see this, one needs to obtain a restriction on the interval for $\tau_\omega = \frac{1}{\sqrt{2}}(1,\omega)\cdot(\tau,\xi)$ on the space time Fourier support of $P_{0,\kappa}Q_j^+\psi$: note that for fixed $\xi_\omega = \left(\tau - \frac{\tau_\omega}{\sqrt{2}}, \xi - \frac{\tau_\omega\omega}{\sqrt{2}}\right) := (\xi_\omega^1, \xi_\omega^2)$, one has

$$2^j \sim \left|\left(\xi_\omega^1 + \frac{\tau_\omega}{\sqrt{2}}\right)^2 - \left|\xi_\omega^2 + \frac{\tau_\omega\omega}{\sqrt{2}}\right|^2\right| = \left|(\xi_\omega^1)^2 - \left|\xi_\omega^2\right|^2 + \frac{\tau_\omega}{\sqrt{2}}(\tau - \xi\cdot\omega)\right|$$

Further, we can bound from below

$$|\tau - \xi\cdot\omega| = ||\tau - |\xi| + |\xi| - \xi\cdot\omega|| \gtrsim [\mathrm{dis}(\omega,\kappa)]^2$$

on account of $||\tau - |\xi|| \ll 2^{-2l} \leq [\mathrm{dis}(\omega,\kappa)]^2$. This implies that τ_ω is restricted to an interval of length $\sim (\mathrm{dist}(\omega,\kappa))^{-2}2^j$. □

The preceding observations render the functional framework below plausible. We need to define it carefully to state the precise null-form estimate, theorem 4.8, which is at the heart of the paper [**33**]. First, we define (using notation from [**62**])

$$||\psi||_{NFA[\kappa]^*} := \sup_{\omega\notin 2\kappa} \mathrm{dist}(\omega,\kappa)||\psi||_{L^\infty_{t_\omega}L^2_{x_\omega}}$$

Then we define

$$||\psi||_{S[k,\kappa]} := ||\psi||_{L^\infty_t L^2_x} + 2^{-\frac{k}{2}}|\kappa|^{-\frac{1}{2}}||\psi||_{PW[\kappa]} + ||\psi||_{NFA[\kappa]^*}$$

The spaces: For every integer $l < -10$, subdivide S^1 into a uniformly finitely overlapping collection K_l of caps κ of diameter 2^l. Also, for every integer λ with $-10 \geq \lambda \geq l$, we subdivide the angular sector $\{\xi \in \mathbf{R}^2 | \frac{\xi}{|\xi|} \in \kappa,\ |\xi| \sim 2^k\}$ into a uniformly finitely overlapping collection $C_{k,\kappa,\lambda}$ of slabs R of width $2^{k+\lambda}$. We introduce various localization operators associated with these regions: for each $\kappa \in K_l$, choose a smooth cutoff $a_\kappa : S^1 \to \mathbf{R}_{\geq 0}$ supported on a dilate of κ. These are to be chosen such that $\sum_{\kappa\in K_l} a_\kappa = 1$. We also introduce cutoffs $m_R(.) : \mathbf{R}_{>0} \to \mathbf{R}_{\geq 0}$ such that the cutoff $m_R(|\xi|)a_\kappa\left(\frac{\xi}{|\xi|}\right)$ localizes to a dilate of the slab R. Also, we require that $\sum_{R\in C_{k,\kappa,\lambda}} m_R(|\xi|) = \chi_0\left(\frac{|\xi|}{2^k}\right)$. We have the associated pseudo differential operator $\tilde{P}_R\psi$:

$$\widehat{\tilde{P}_R\psi}(t,\xi) = m_R(|\xi|)a_\kappa\left(\frac{\xi}{|\xi|}\right)\hat{\psi}(\xi)$$

We also have the ΨDO's $P_{k,\kappa}$ associated with multiplier $a_\kappa\big(\frac{\xi}{|\xi|}\big)\chi_0\big(\frac{|\xi|}{2^k}\big)$. Then, define

(4.4)

$$||\psi||_{S[k]} := ||\psi||_{L_t^\infty L^2} + ||\psi||_{\dot{X}_k^{0,\frac{1}{2},\infty}} + ||\psi||_{\dot{X}_k^{-\frac{1}{2},1,2}}$$

$$+ \sup_{\pm} \sup_{l<-10} \sup_{-10\geq\lambda\geq l} |\lambda|^{-1} \left(\sum_{\kappa\in K_l} \sum_{R\in C_{k,\kappa,\lambda}} ||\tilde{P}_R Q^{\pm}_{<k+2l}\psi||^2_{S[k,\pm\kappa]} \right)^{\frac{1}{2}}.$$

Here we use the notation $||\phi||_{\dot{X}_k^{a,b,c}} := 2^{ak}[\sum_{j\in\mathbf{Z}}(2^{jb}||Q_j\phi||_{L_t^2L_x^2})^c]^{\frac{1}{c}}$. This is the norm used to control the frequency-k piece of ψ.

Next, let $N[k]$ be the atomic Banach space whose atoms are Schwartz functions $F\in\mathcal{S}(\mathbf{R}^{2+1})$ with spatial Fourier support contained in the region $|\xi|\sim 2^k$ and

(1) $||F||_{L_t^1\dot{H}^{-1}}\leq 1$ and F has modulation $<2^{k+100}$.

(2) F is at modulation[18] $\sim 2^j$ and satisfies $||F||_{L_t^2L_x^2}\leq 2^{\frac{j}{2}}2^k$.

(3) F satisfies $||F||_{\dot{X}_k^{-\frac{1}{2},-1,2}}\leq 1$.

(4) There exists an integer $l<-10$, and Schwartz functions F_κ with Fourier support in the region

$$\left\{(\tau,\xi)\,|\,\pm\tau>0,\ ||\tau|-|\xi||\leq 2^{k-2l-100},\ \frac{\xi}{|\xi|}\in\pm\kappa\right\}$$

with the properties

$$F=\sum_{\kappa\in K_l} F_\kappa,\quad \left(\sum_{\kappa\in K_l} ||F_\kappa||^2_{NFA[\kappa]}\right)^{\frac{1}{2}} \leq 2^k$$

In the last inequality, $NFA[\kappa]$ denotes the dual of $NFA[\kappa]^*$ (the completion of $\mathcal{S}(\mathbf{R}^{2+1})$ with respect to $||.||_{NFA[\kappa]^*}$) used in the definition of $S[k,\kappa]$: Thus $NFA[\kappa]$ is the atomic Banach space whose atoms F satisfy

$$\frac{1}{\mathrm{dist}(\omega,\,\kappa)}||F||_{L^1_{t_\omega}L^2_{x_\omega}}\leq 1$$

for some $\omega\notin 2\kappa$.

We can now state the core null-form estimate of [**33**], which is the crux for estimating the right hand side of (4.3). Note that the expressions estimated below, up to the frequency localizations, are obtained by substituting the gradients $R_\nu\psi$ for ψ_ν on the right of (4.3):

[18]By this it is meant that $||\tau|-|\xi||\sim 2^j$ on the space-time Fourier support of F. The term 'modulation' is adopted from [**62**].

THEOREM 4.8. **[33]** *Let* $I = \sum_{k\in\mathbf{Z}} P_k Q_{<k+100}$. *Then there exist numbers* $\delta_{1,2} > 0$ *such that*

(4.5)
$$\Bigg\| \partial^\beta P_k \left[R_\alpha P_{k_1}\psi_1 \triangle^{-1} \sum_{j=1}^{2} \partial_j I [R_\beta P_{k_2}\psi_2 R_j P_{k_3}\psi_3 - R_\beta P_{k_3}\psi_3 R_j P_{k_2}\psi_2] \right] + \partial_\alpha P_k \times \left[R_\beta P_{k_1}\psi_1 \triangle^{-1} \sum_{j=1}^{2} \partial_j I \left[R^\beta P_{k_2}\psi_2 R_j P_{k_3}\psi_3 - R_j P_{k_2}\psi_2 R^\beta P_{k_3}\psi_3 \right] \right] \Bigg\|_{N[k]}$$
$$\leq C 2^{\delta_1 \min\{-\min\{k_1-k,k_2-k,k_3-k\},0\}} \times \prod_i 2^{\delta_2 \min\{\max_{j\neq i}\{k_i-k,k_i-k_j\},0\}} \prod_l \|P_{k_l}\psi_l\|_{S[k_l]},$$

(4.6)
$$\Bigg\| P_k \partial^\beta \left[R_\beta P_{k_1}\psi_1 \triangle^{-1} \sum_j \partial_j I \left[R_\alpha P_{k_2}\psi_2 R_j P_{k_3}\psi_3 - R_j P_{k_2}\psi_2 R_\alpha P_{k_3}\psi_3 \right] \right] \Bigg\|_{N[k]}$$
$$\leq C 2^{\delta_1 \min\{-\min\{k_1-k,k_2-k,k_3-k\},0\}} \times \prod_i 2^{\delta_2 \min\{\max_{j\neq i}\{k_i-k,k_i-k_j\},0\}} \prod_l \|P_{k_l}\psi_l\|_{S[k_l]}.$$

We observe that this theorem effectively deals with the *summation problem*, as summing over the indices k_i for fixed k is reduced to adding the squares of $\|P_{k_i}\psi\|_{S[k_i]}$, as is easily verified. Full details are to be found in **[33]**.

5. Approaching the large data problem in the critical dimension $n = 2$ and hyperbolic target

With the result of the preceding section as well as lemma 4.1 in hand, global regularity for wave maps $u : \mathbb{R}^{2+1} \longrightarrow \mathbf{H}^2$ with large data will follow if we establish a non-concentration result for the energy inside light cones. Unfortunately, there does not seem to be a straightforward procedure for this at the present time, and the most promising strategy may be a variant of *Bourgain's induction on energy method*, **[4]**: the idea here is to not only show global regularity preservation, but furthermore obtain global control over a suitable space-time norm of the wave map, in particular ensuring a type of asymptotic growth control. To implement Bourgain's method, one first needs to establish a good perturbation theory for wave maps, of the following kind. We let $u : [-T, T] \times \mathbb{R}^2 \longrightarrow \mathbf{H}^2$, $T \in (0, \infty]$, a smooth wave map, with derivative components ϕ with respect to the standard frame and ψ in the Coulomb Gauge:

Perturbative setup:

Find a family of norms $||\psi||_{X([-T,T]\times\mathbb{R}^2}$, scaling like $||.||_{L_t^\infty L_x^2}$, and increasing with respect to T, with the properties that

(i) $||\psi||_{X([-\infty,\infty]\times\mathbb{R}^2)} < \infty$ *whenever* ψ *smooth and the energy is small enough.*

(ii) $\sup_{T>0}||\psi||_{X([-T,T]\times\mathbb{R}^2} < \infty$ *and smooth data implies* ψ *globally smooth.*

(iii) *assuming* $||\psi||_{X([-\infty,\infty]\times\mathbb{R}^2} < \infty$, *there exists an open neighborhood* U *of the data* (u_0, u_1) *in the energy topology such that for a wave map* $(\tilde{u}, \tilde{\phi}, \tilde{\psi})$ *with data* $(\tilde{u}_0, \tilde{u}_1) \in U$, *we have* $||\tilde{\psi}||_{X([-\infty,\infty]\times\mathbb{R}^2} < \infty$.

Goal: show that $\forall E_0 \geq 0$, $\sup_{E[u]\leq E_0}||\psi||_{X([-\infty,\infty]\times\mathbb{R}^2)} = C(E_0) < \infty$. Here the supremum is taken over all smooth wave maps $u : \mathbb{R}^{2+1} \longrightarrow \mathbf{H}^2$ of energy $\leq E_0$.

Note that the perturbative setup ensures that the set of energies for which $C(E_0) < \infty$ is nonempty and open. Hence if the Goal is false, there exists a least energy for which $C(E_0) = \infty$. Ideally, one wants to show that there exists a wave map of energy E_0 for which $\sup_{T>0}||\psi||_{X([-T,T]\times\mathbb{R}^2} = \infty$. This wave map being a least energy blow up solution has to be extremely special, and one hopes to be able to rule out such solutions. More specifically, the next section will discuss the 'bubbling off of a harmonic map' scenario in a symmetric situation, but a modification of which is believed to also apply in the general case, see the discussion below. In particular, the non-existence of finite energy harmonic maps from $\mathbb{R}^2$ to $\mathbf{H}^2$ should then rule out blow up for target $\mathbf{H}^2$.

While there is no result yet outlining a general perturbation theory as in the above setup, the functional framework detailed before should suffice for this task. Moreover, due to the abelian nature of the Gauge group, the Coulomb Gauge can be canonically constructed for target $\mathbf{H}^2$ even for large energy wave maps. The more general case of targets M of dimension $k \geq 3$ and hence nonabelian Gauge group offers the additional challenge of constructing canonical Gauges substitutable for the Coulomb Gauge, which is no longer canonically constructible for large data then. In the case of negative curvature, a promising candidate for such a Gauge has been announced in [**64**].

Finally, we mention a partial large energy perturbative result, [**36**], which is based on the fact, to be discussed in the next section, that the Cauchy problem for radial data is globally well-posed, [**8, 9**]: we have the

THEOREM 5.1. [**36**] *Let* $(u_0, u_1) : \mathbb{R}^2 \longrightarrow \mathbf{H}^2 \times T\mathbf{H}^2$ *be smooth, compactly supported spherically symmetric Cauchy data. Then for any* $\sigma > 0$, *there exists* $\epsilon > 0$ *such that for all no longer necessarily spherically symmetric* *smooth initial data* $(\tilde{u}_0, \tilde{u}_1) : \mathbb{R}^2 \longrightarrow \mathbf{H}^2 \times T\mathbf{H}^2$ *which are* ϵ*-close to* (u_0, u_1) *in the* $H^{1+\sigma}$*-topology, one has global existence.*

The proof of this theorem exploits additional information about the pointwise decay of radial wave maps into geodesically convex targets, which was derived in [**9**]. Part of the challenge consists in proving that the large energy radial wave maps can be bounded with respect to the norms discussed above. A further crucial ingredient in the proof is control over a certain range of *subcritical Sobolev norms* for the radial wave map, which is a priori only globally bounded in energy. For details, we refer to [**36**].

6. Imposing symmetry: radial and equivariant wave maps in the case $n = 2$

6.1. Radial wave maps. A wave map $u : \mathbb{R}^{2+1} \longrightarrow M$ is called radial, provided that it only depends on r if we equip $\mathbb{R}^2$ with polar coordinates (r, θ). In this case, we have the following theorem due to Christodoulou-Tahvildar-Zadeh and Struwe:

THEOREM 6.1. [**8, 58, 59**] *Let* $(u_0, u_1) : \mathbb{R}^2 \longrightarrow M \times TM$ *be smooth radial data,* (M, g) *a smooth Riemannian manifold which is either compact or satisfies a suitable geodesic convexity type condition. Then there exists a unique smooth and global-in-time wave map* $u : \mathbb{R}^{2+1} \longrightarrow M$ *extending* (u_0, u_1).

This theorem was first proved in [**8**] for targets satisfying a type of geodesic convexity condition, and then later relaxed to general targets in [**59**], by means of a careful blow up analysis. We shortly explain the outline of the argument in [**8**], which of course parallels the strategy for the general case explained before:

(i) Establish a small data global well-posedness result.
(ii) Show that an energy concentration scenario is impossible.

An advantage of the radiality assumption is that if a singularity forms (whence the energy concentrates in a light cone), this can only possibly occur at the spatial origin $r = 0$. This is on account of the conservation of energy. Furthermore, assuming an energy concentration scenario in a backward light cone centered at the space-time origin ($(t, r) = (0, 0)$, which we may always arrange), pointwise estimates on certain components the energy momentum tensor become possible. The latter is a family of functions $T_{\mu\nu}$, $0 \leq \mu, \nu \leq 2$, associated with the wave map u as follows:

$$T_{\mu\nu} = g_{ij} \left(\partial_\mu u^i \partial_\nu u^j - \frac{1}{2} m_{\mu\nu} \partial_\alpha u^i \partial^\alpha u^j \right), \; m_{\mu\nu} = \text{Minkowski metric}$$

Note that the wave maps condition implies the vanishing divergence equation

$$\nabla^\nu T_{\mu\nu} = 0, \tag{6.1}$$

where we stress the use of covariant differentiation as this is to be valid for arbitrary systems of coordinates on $\mathbb{R}^{2+1}$, in particular the polar ones (t, r, θ). In the radial context, the equations (6.1) can be integrated along characteristics to yield pointwise bounds inside the light cone and away from the singularity, which then imply the vanishing of the kinetic part of the energy in an averaged sense, see [**8**]:

$$\lim_{T \to 0} \frac{1}{|T|} \int_{0<|t|<|T|} \int_{|x| \leq |t|} |u_t|^2 dx = 0 \tag{6.2}$$

Furthermore, one can show that the the energy has to concentrate in arbitrarily narrow neighborhoods of the axis $r=0$:

$$\lim_{|t| \to 0} \int_{|t|>|x|>\lambda|t|} [|u_t|^2 + |u_r|^2] dx = 0 \, \forall \lambda > 0 \tag{6.3}$$

Assuming a kind of geodesic convexity assumption, one can bound the spatial part of the local energy $\int_{|x|<|t|} |u_r|^2 dx$ in terms of the temporal part $\int_{|x|<|t|} |u_t|^2$, which together with (6.2) and the monotonicity of the local energy implies the impossibility of an energy concentration scenario in the radial case.

This argument is extended in [**58, 59**] to the context of general compact smooth targets. The argument in [**58**] is especially intuitive, and parallels developments detailed below in the equivariant case: one uses (6.2), as well as (6.3), to infer, using suitable rescalings of the wave map, existence of a nontrivial finite energy harmonic map $u_0 : \mathbb{R}^2 \longrightarrow S^k$, which also needs to be radial. This however is impossible, ruling out blow up.

We also mention here the paper [**9**], which provides detailed pointwise asymptotics of radial wave maps, under a geodesic convexity type assumption on the target. This work provides the basis for theorem 6.1. The interesting issue remains as to whether the solutions constructed in [**58, 59**], satisfy similar asymptotics.

6.2. Equivariant wave maps. Assume that the target M admits a smooth S^1 action, $\rho : S^1 \to \text{Isom}(M)$. A wave map $u : \mathbb{R}^{2+1} \longrightarrow M$ is called equivariant with respect to this action, provided we have

$$u(t, \omega x) = \rho(\omega) u(t, x), \; \forall \omega \in S^1 \tag{6.4}$$

Here S^1 acts on $\mathbb{R}^2$ in the canonical fashion as rotations. Paralleling the developments for radial wave maps in [**8**], we have the following result by Shatah-Tahvildar-Zadeh.

THEOREM 6.2. [**48**] *Let the target (M, g) be a warped product manifold satisfying a suitable geodesic convexity condition. Then equivariant wave maps $u : \mathbb{R}^{2+1} \longrightarrow M$ with smooth data stay globally regular.*

We refer to [**48**] for a detailed statement, as well as further results, including optimal local well-posedness results which are now subsumed by the more general theory expounded in section 4.

Of particular interest is the case of equivariant wave maps $u : \mathbb{R}^{2+1} \longrightarrow S^2$, which does not satisfy the hypotheses of the preceding theorem. We let S^1 act on S^2 by means of rotations around the z-axis via $\rho(\omega) = k\omega$, $k \in \mathbf{Z}\backslash\{0\}$, $\omega \in S^1$. Fixing a k, the wave map is then determined in terms of the polar angle, and becomes a scalar equation on $\mathbb{R}^{1+1}$ as follows:

$$-u_{tt} + u_{rr} + \frac{1}{r}u_r = k^2\frac{\sin(2u)}{2r^2}$$

The case $k=1$ in particular is called *co-rotational*, and has aroused a lot of interest, due to numerical experiments [**3, 18**], which suggested development of singularities within finite time. A rigorous result establishing the development of singularities [**37**] will be discussed below.

Recall that every homotopy class of maps $u : S^2 \longrightarrow S^2$ has a harmonic representative. Indeed, in terms of polar coordinates on $S^2 = \mathbb{R}^2 \cup \{\infty\}$ as well as spherical coordinates on S^2 these are given by

$$Q_k : (r, \theta) \longrightarrow (2\arctan(r^k), k\theta),$$

where $k \in \mathbf{Z} \cong \pi_2(S^2)$. Alternatively, one may view Q_k as the composition of $z^k : \mathbb{C} \cup \{\infty\} \rightarrow \mathbb{C} \cup \{\infty\}$ with the (essentially) stereographic projection Q_1. Each of these harmonic maps is of course a *static wave map*.

We observe that due to weak global well-posedness for small data as well as energy conservation, a singularity in this context can only form at $r=0$. A general 'bubbling off' theorem is due to Struwe in the co-rotational case $k=1$:

THEOREM 6.3. [**57**] *Let the wave map $u : \mathbb{R}^{2+1} \longrightarrow S^2$ be co-rotational and become singular at $(t, r) = (0, 0)$. Then there exists a sequence of times $t_i \rightarrow 0$, as well as scaling parameters λ_i with $|\lambda_i| \rightarrow \infty$, and further $|\lambda_i t_i| \rightarrow \infty$, such that we have*

$$u(t_i, r) = Q_1(\lambda_i r) + \epsilon(t_i, r) \tag{6.5}$$

where the local energy of $\epsilon(t_i, .)$, $E_{loc}(\epsilon(t_i, .))$, converges to 0. *Explicitly, this means*

$$\lim_{i\to\infty} \int_{r<|t_i|} \left(\frac{1}{2}[\epsilon_t^2(t_i, r) + \epsilon_r^2(t_i, r)] + \frac{\sin^2(\epsilon(t_i, r))}{r^2}\right) rdr = 0$$

In other words, if a co-rotational wave map $u : \mathbb{R}^{2+1} \longrightarrow S^2$ develops a singularity, there exists a sequence of times t_i approaching the singularity such that, upon restricting the wave map to the light cone centered at the singularity, it can be decoupled into a re-scaled harmonic map $Q_1(\lambda_i r)$, as well as a radiation part, which converges to zero in the energy topology. Furthermore, the re-scaling is strictly faster than self-similar, whence the energy concentrates in a cuspidal region $r<\frac{1}{\lambda_i}$ (on each time slice $t=t_i$). Note that the fact that $|\lambda_i t_i|\rightarrow\infty$ is a reflection of (6.3), which can also be proven for equivariant wave maps, see [**48**], by the same technique as in the radial case.

Note that Struwe's theorem does not make an assertion about the actual occurence of singularity formation, nor about the rate at which $|\lambda_i|\rightarrow\infty$. For example, it is an open issue whether the above decomposition (6.5) holds on each time sufficiently close to the blow up.

6.3. Consequences for non-symmetric wave maps in the critical dimension. Theorem 6.3 indicates that singularity formation for wave maps may be tied to an inherent lack of compactness, due to the scaling invariance. Removing the equivariance enlarges this lack of compactness, by adding translations, as well as Lorentz boosts, for example of the form

$$(t,x_1,x_2)\longrightarrow\left(\frac{ct-x_1}{\sqrt{c^2-1}},\frac{t-cx_1}{\sqrt{c^2-2}},x_2\right),\quad c>1$$

Fortunately, unlike for example in the context of the critical defocusing Schrodinger equation on $\mathbb{R}^{3+1}$, where the issue of translation invariance poses significant difficulties, see [**10**], the finite propagation speed of wave maps, as well as energy conservation, imply that for general wave maps without symmetry assumptions, singularity formation can only happen via a finite number of energy concentration scenarios within a forward light cone, say (posing data at time $t=0$). One may then concentrate on what happens within a single such light cone. Note that each Lorentz boost centered at the tip of the light cone leaves the latter invariant, but moves its time axis. Then the following general blow up scenario may be plausible:

Conjectured general bubbling off scenario: Let $u:\mathbb{R}^{2+1}\longrightarrow S^2$ be a wave map, no longer necessarily equivariant, which becomes singular at $(t,x)=(0,0)$. Then there exists N and a sequence of times $t_i\rightarrow 0$, and for each i Lorentz boosts T_i^k, $k=1,\dots,N$, as well as parameters λ_i^k with $|t_i\lambda_i^k|\rightarrow\infty$ and harmonic maps $Q_k(x)$, such that the wave map, restricted to the half of the light cone on which it is defined, may be decomposed into

$$u(t_i,x)=\sum_{k=1}^{N}T_i^k[Q_k(\lambda_i^k x)]+\epsilon(t_i,x),$$

where we have $E_{loc}(\epsilon(t_i,x))\rightarrow 0$. Here the wave map is represented in terms of the ambient coordinates $S^2\hookrightarrow\mathbb{R}^3$.

7. Singularity formation in the critical dimension

Recall from section 2 that due to the supercritical nature of wave maps $\mathbb{R}^{n+1} \longrightarrow M$, $n \geq 3$, generic targets M, one expects singularity formation for large data. This has been achieved for a rather eclectic list of targets in [**6, 47**], via a self-similar ansatz, i.e., writing $u(t,x) = v(\frac{x}{t})$, and then solving for v, which leads to an elliptic problem for v that can be solved by variational techniques in certain cases. In particular, it is known that self-similar blow up wave maps with smooth data exist with domain $\mathbb{R}^{3+1}$ and target S^3. In the latter case, the explicit self-similar solutions are believed to represent the generic blow up phenomenon, see e.g. [**2**], although this appears out of reach of present techniques. No other types of blow up solutions are known in the supercritical regime. As seen below, recent blow up results in the critical dimension are of a strikingly different character. In particular, the rates are never self-similar.

7.1. The co-rotational case. In the critical dimension $n = 2$, existence of singular wave maps with target S^2 has been conjectured for a while, on account of numerical experiments, see e.g. [**3**]. These have been for the most part restricted to co-rotational wave maps $u : \mathbb{R}^{2+1} \longrightarrow S^2$, see preceding section, and thus deal with the scalar equation (with u the polar angle)

$$-u_{tt} + u_{rr} + \frac{1}{r}u_r = \frac{\sin(2u)}{2r^2} \tag{7.1}$$

Recall that these admit the static solution $u(t,r) = Q_1(r) = 2\arctan(r)$. It was observed in [**3**] that sufficiently large perturbations of these appear to lead to singularities in finite time, while small enough perturbations lead to global existence and scattering. Rigorous existence of blow up solutions with precise blow up dynamics is provided by the following

THEOREM 7.1. [**37**] *Let $\nu > \frac{1}{2}$, and let $t_0 > 0$ sufficiently small. Define $\lambda(t) = t^{-1-\nu}$, and fix a large integer N. Then there exists a function u^e satisfying*

$$u^e \in C^{\nu + \frac{1}{2}+}(t_0 > t > 0, |x| \leq t), E_{loc}(u^e)(t) \lesssim [\lambda(t)t]^{-2} |\log t|^2$$

and a blow up solution $u(t,r)$ to (7.1) on $[0,t_0]$ which has the form

$$u(r,t) = Q_1(\lambda(t)r) + u^e(r,t) + \epsilon(r,t),\ 0 \leq r \leq t,$$

with $\epsilon \in t^N H^{1+\nu-}_{loc}(\mathbb{R}^2)$, $\epsilon_t \in t^{N-1} H^{\nu-}_{loc}(\mathbb{R}^2)$, $E_{loc}(\epsilon) \lesssim t^N$. The corresponding solution $u(t,x)$ can be extended of class $H^{1+\nu-}$ to all of $[0,t_0] \times \mathbb{R}^2$.

A surprising feature is that there is a continuum of blow up rates. Indeed, the fact that a nonlinear equation may admit such a family of blow up rates seems to have been observed here for the first time. Furthermore, the fact that the energy of the static solutions $Q_k(.)$ is $4k\pi$, together with the conjectured general bubbling off scenario from above, appear to suggest that theorem 7.1 is optimal in some sense: any initial data of energy strictly below that of $Q_1(.)$, which is 4π, should lead to globally regular solutions, while the theorem produces blow up solutions of energy arbitrarily close to 4π.

We now outline the strategy for proving this theorem: one may be tempted to try a naive perturbative ansatz, namely

$$u(t,r) = Q(\lambda(t)r) + \epsilon(t,r), \tag{7.2}$$

where $\epsilon(t,.)$ is small in a suitable sense, and where $\lambda(t)$ is chosen to blow up at $t=0$. In particular, one may try to construct ϵ by means of an iterative procedure. This ansatz however appears to fail: note that substituting $Q(\lambda(t)r)$ into (7.1), one obtains an error of the form

$$\ddot{\lambda}(t)rQ'(\lambda(t)r) + (\dot{\lambda})^2 r^2 Q''(\lambda(t)r) \tag{7.3}$$

However, note that this function is not in $L^2(rdr)$, unless $\lambda(t) = t^{-1}$, which we know is excluded from Struwe's theorem 6.3. Furthermore, this function is not going to be small, and hence we cannot expect that adding on a small function ϵ, which is obtained by means of a direct iteration, will counteract this error. The fact that (7.3) is not L^2-integrable is related to an interesting feature of the co-rotational case $k = \pm 1$, namely the fact that *the spectrum of the linearization around the static solution $Q_1(.)$ has a resonance at its endpoint $\xi = 0$*. On the other hand, for equivariance indices $|k| \geq 2$, the linearization around $Q_k(.)$ has an eigenvalue at $\xi = 0$. Indeed, the resonance and eigenfunctions are given by $\frac{\partial}{\partial\lambda}Q_k(\lambda r)|_{\lambda=1} = rQ_k'(r)$.

Note that one may be tempted to deal with the non-square-integrability of (7.3) by truncation at infinity. Indeed, one is really only interested in constructing solutions on the backward light cone $r \leq t$, as one may extend these arbitrarily outside the light cone. Hence one may try to replace $Q(\lambda(t)r)$ by $\chi(\frac{r}{t})Q(\lambda(t)r)$ for some smooth cutoff $\chi(\frac{r}{t})$, where we write $Q_1 =: Q$ for the remainder of this subsection. Indeed, rather than imposing a fixed form of χ, one may try to substitute an ansatz $\chi(\frac{r}{t})Q(\lambda(t)r) + \epsilon(t,r)$ into (7.1) and thereby obtain the correct form of χ. Indeed, it turns out that one obtains an equation of the form

$$(a^2-1)\chi_{aa} + \left[2a\left(1 - \frac{\lambda_t t}{\lambda}\right) + a^{-1}\right]\chi_a = (\text{independent of } \chi) + O(\chi^2), \tag{7.4}$$

where we have introduced the self-similar variable $a = \frac{r}{t}$. The operator $(a^2-1)\partial_a^2 + [2a(1 - \frac{\lambda_t t}{\lambda}) + a^{-1}]\partial_a$ is a singular linear differential operator, which for $\lambda(t) = t^{-1-\nu}$ admits a fundamental system of solutions of regularity $C^{\nu+\frac{1}{2}+}$ across $a = 1$, which corresponds to the light cone. Unfortunately, equation (7.4) is effectively inconsistent, as the right hand side depends on t in addition to a, and hence we really would have to choose χ of the form $\chi(\frac{r}{t}, t)$, introducing additional error terms due to time differentiation. Nevertheless, these observations motivate one to look for an approximate solution (which is the $Q(\lambda(t)r) + u^e(t,r)$ in the theorem), which is in effect a large profile modification of $Q(\lambda(t)r)$ for fixed times t away from $t = 0$, and obtained by solving certain elliptic problems on fixed time slices which in some sense approximate the hyperbolic problem.

The precise procedure for proving theorem 7.1 is then as follows: to find u^e, one solves a sequence of elliptic problems on fixed time slices(thus in some sense neglecting time derivatives) which improve the accuracy of the problem near the symmetry axis $r = 0$, as well as the light cone $r = t$. The latter is achieved by working with the coordinates (a, t) instead of (r, t), as in the preceding paragraph. Specifically, if u is an approximate solution of (7.1), an exact solution $u + \epsilon$ is given with ϵ satisfying

$$\left(-\partial_t^2 + \partial_r^2 + \frac{1}{r}\partial_r\right)\epsilon - \frac{\cos(2u)}{2r^2}\sin(2\epsilon) + \frac{\sin(2u)}{2r^2}(1 - \cos(2\epsilon)) = e,$$

where the error e generated by u is given by

$$e = \left(-\partial_t^2 + \partial_r^2 + \frac{1}{r}\partial_r\right)u - \frac{\sin(2u)}{2r^2}$$

We approximate $\sin(2\epsilon)$ by 2ϵ, and moving nonlinear (in ϵ) terms to the right, we encounter the problem

$$\left(-\partial_t^2 + \partial_r^2 + \frac{1}{r}\partial_r - \frac{\cos(2u)}{r^2}\right)\epsilon = \text{small}$$

Near the origin $r = 0$, one expects the time derivatives to play less of a role. Moreover, the smaller t becomes the less u should differ from $Q(\lambda(t)r)$, which suggests replacing u by the latter. Hence one obtains the problem

$$\left(\partial_r^2 + \frac{1}{r}\partial_r - \frac{\cos[Q(\lambda(t)r)]}{r^2}\right)\epsilon = \text{small}$$

On the other hand, near $r = t$, one expects the time derivatives to play a role, and replaces the above by

$$\left(-\partial_t^2 + \partial_r^2 + \frac{1}{r}\partial_r - \frac{1}{r^2}\right)\epsilon = \text{small}$$

In short, one constructs $u^e = \sum_{k=1}^{M} v_k$, where we put

$$\left(\partial_r^2 + \frac{1}{r}\partial_r - \frac{\cos[Q(\lambda(t)r)]}{r^2}\right)v_{2k+1} = e_{2k} \tag{7.5}$$

$$\left(-\partial_t^2 + \partial_r^2 + \frac{1}{r}\partial_r - \frac{1}{r^2}\right)v_{2k} = e_{2k+1} \tag{7.6}$$

where the errors e_k generated become increasingly smaller. It turns out that (7.6), when translated into the coordinates $a = \frac{r}{t}$, t, leads to a problem of the form (7.4) for a suitable auxiliary function. For details we refer to section 2 in [**37**]. Choosing M large enough, one may achieve $e_M = O\Big(\frac{1}{[\lambda(t)t]^N}\Big)$, for arbitrary N.

Having constructed an approximate solution $Q(\lambda(t)r) + u^e(t,r)$ with

$$\left(-\partial_t^2 + \partial_r^2 + \frac{1}{r}\partial_r\right)[Q(\lambda(t)r) + u^e(t,r)] - \frac{\sin[2(Q(\lambda(t)r) + u^e(t,r))]}{2r^2} = O\left(\frac{1}{[\lambda(t)t]^N}\right)$$

one needs to correct it to a precise solution $u = Q(\lambda(t)r) + u^e(t,r) + \epsilon(t,r)$, where $\epsilon(t,r)$ is to be solved for by means of iteration. We now consider the equation for ϵ. In order to avoid a time-dependent elliptic operator, one replaces the coordinates (t,r) by the new ones $\tau = -\int_t^1 \lambda(s)ds + \frac{1}{\nu}$, and $R = \lambda(t)r$. Also, we replace ϵ by $\tilde{\epsilon}(\tau,R) := R^{\frac{1}{2}}\epsilon(t,r)$. Then one obtains the problem

$$\left[-\left(\partial_\tau + \frac{\lambda_\tau}{\lambda}R\partial_R\right)^2 + \frac{1}{4}\left(\frac{\lambda_\tau}{\lambda}\right)^2 + \frac{1}{2}\partial_\tau\left(\frac{\lambda_\tau}{\lambda}\right)\right]\tilde{\epsilon} - \mathcal{L}\tilde{\epsilon} = \text{small}, \tag{7.7}$$

where

$$\mathcal{L} := -\partial_R^2 + \frac{3}{4R^2} - \frac{8}{(1+R^2)^2}$$

The latter operator has the property that in spite of being defined on the half-line, it is self-adjoint even without the extra imposition of a boundary condition at $R=0$. This is due to the fact that the singularity at $R=0$ already forces a vanishing condition at $R=0$ for functions in its domain. The theory of operators of the type of $\mathcal{L}$, so called *strongly singular operators*, is developed in [**13**]. We need the fact that the spectrum of $\mathcal{L}$ consists of $[0,\infty)$, and 0 is a resonance for $\mathcal{L}$, which is now considered as an operator on $L^2([0,\infty))$. Indeed, we have that

$$\mathcal{L}\left(\frac{R^{\frac{3}{2}}}{1+R^2}\right) = 0$$

It turns out that $\mathcal{L}$ admits for each $z \in \mathbb{C}$ a fundamental system of solutions $\phi(R,z)$, $\theta(R,z)$ for the eigenvalue problem $(\mathcal{L} - z)f = 0$, which obey the asymptotics

$$\phi(R,z) \sim R^{\frac{3}{2}},\ \theta(R,z) \sim R^{-\frac{1}{2}}$$

One also says that $\mathcal{L} - z$ is in the limit point case at $R=0$, see e.g. [**38**], and $\phi(R,z)$ is the Weyl-Titchmarsh solution of $\mathcal{L} - z$ at $R=0$.

Further, assuming now $\operatorname{Im} z > 0$, the Weyl-Titchmarsh solution at $R = \infty$ is given asymptotically by

$$\psi^+(R,z) \sim z^{-\frac{1}{4}} e^{iz^{\frac{1}{2}}R}, \quad \operatorname{Im} z^{\frac{1}{2}} > 0,$$

and letting $\operatorname{Im} z \to 0$, one obtains an oscillatory (at $R = \infty$) fundamental system of solutions $\psi^+(R,\xi), \psi^-(R,\xi) = \overline{\psi_+(R,\xi)}$, $\xi > 0$. We then also have the condition

$$W(\theta,\phi) = 1, \; W(f,g) = \left(\frac{d}{dR}f\right)g - \left(\frac{d}{dR}g\right)f$$

With these tools, one can then introduce a distorted Fourier transform, as follows:

$$f(R) \longrightarrow \hat{f}(\xi) := \int_0^\infty \phi(R,\xi) f(R) dR$$

which is to be interpreted in a weighted L^2-sense, similarly to the ordinary Fourier transform. Further, one has the inversion relation

$$f(R) = \int_0^\infty \phi(R,\xi)\hat{f}(\xi)\rho(\xi)d\xi, \; \rho(\xi) = \frac{1}{\pi}\operatorname{Im}\left(m(\xi + i0)\right),$$

where the spectral density ρ is defined in terms of

$$m(\xi) := \frac{W(\theta(.,\xi), \psi^+(.,\xi))}{W(\psi^+(.,\xi), \phi(.,\xi))}$$

The construction of $\tilde{\epsilon}(\tau, R)$ as above now proceeds via representing

$$\tilde{\epsilon}(\tau, R) = \int_0^\infty x(\tau,\xi)\phi(R,\xi)\rho(\xi)d\xi,$$

and working with the Fourier coefficients $x(\tau,\xi)$ instead. The difficulty one encounters here is that it is not immediate (as in the case of the free d'Alembertian $\Box$) to deduce a transport equation for $x(\tau,\xi)$ from (7.7). Nevertheless, neglecting the terms $\frac{\lambda_\tau}{\lambda} R\partial_R$ in $(\partial_\tau + \frac{\lambda_\tau}{\lambda} R\partial_R)$ allows one to obtain an approximate transport equation for the $x(\tau,\xi)$, which is enough thanks to the rapid decay of the error (namely $(\lambda(t)t)^{-N}$). For details, we again refer to [**37**].

We conclude this subsection by noting that interesting open questions remain:

(i) What are the stability properties of the solutions constructed? One may conjecture that there is a high co-dimensional manifold of data resulting in the same blow up, which would be somewhat analogous to [**5**].

(ii) Can the same construction be carried out for the higher homotopy indices $k \geq 2$? It appears that the resonance for the linearization around the ground state plays a fundamental role.

(iii) Are there C^∞-smooth data resulting in the kind of blow up rates constructed in the theorem?
(iv) What other targets admit such blow up solutions? Can one say that the existence of non-trivial finite-energy harmonic maps originating on $\mathbb{R}^2$ to the target suffices?

7.2. Higher equivariance classes. Now consider equivariant wave maps $u:\mathbb{R}^{2+1}\longrightarrow S^2$ but of index $k\geq 4$. Then we have the following result due to Rodnianski and Sterbenz:

THEOREM 7.2. [**45**] *For $k\geq 4$, there exists an open(with respect to a suitable Sobolev topology) set of k-equivariant initial data (in particular C^∞-data) arbitrarily close to $Q_k(.)$, resulting in blow up solutions of the form*

$$u(t,r)=Q_k(\lambda(t)r)+\epsilon(t,r),$$

where we have $\lambda(t)\sim\frac{\sqrt{|\log(T-t)|}}{T-t}$, and T is the blow up time.

Observe that this theorem guarantees an open (within the k-equivariant category) set of data resulting in a kind of *stable blow up*. Not surprisingly, the method of proof here is quite different than the one of theorem 7.1, and deduces the blow up rate from monotonicity type arguments and an orthogonality relation, rather than imposing it. Moreover, rather than constructing $\epsilon(t,r)$ via iteration, Rodnianski-Sterbenz control it by means of a Morawetz type estimate (for a time dependent wave operator!), hence via a priori type estimates. The restriction $k\geq 4$ should be relaxable to $k\geq 2$, while the case $k=1$ appears unreachable, as the method heavily relies on the L^2-integrability of the zero mode $\frac{\partial}{\partial_\lambda}[Q_k(\lambda r)]|_{\lambda=1}$. We refer to [**45**] for details. It remains an interesting open issue to see whether the methods of theorem 7.1, theorem 7.2 can be combined to deduce a stable blow up regime in the co-rotational case $k=1$. Further, the issue of *whether any of these blow up solutions remain stable in the full category of (non-equivariant) wave maps $u:\mathbb{R}^{2+1}\longrightarrow S^2$ appears a quite difficult open problem.*

Acknowledgment

The author thanks the referee for pointing out numerous improvements.

References

[1] P. D'Ancona, V. Georgiev, *On the continuity of the solution operator of the wave maps system*, preprint.
[2] P. Bizon, Comm. Math. Phys. 215 (2000), 45.
[3] P. Bizon, T. Chmaj, T. Zbislaw, *Formation of singularities for equivariant* $(2+1)$-*dimensional wave maps into the 2-sphere.*, Nonlinearity **14** (2001), no. 5, 1041–1053.
[4] J. Bourgain, *Global wellposedness of defocusing critical nonlinear Schršdinger equation in the radial case.*, J. Amer. Math. Soc. **12** (1999), no. 1, 145–171.

[5] J. Bourgain, W. Wang, *Construction of blowup solutions for the nonlinear Schroedinger equation with critical nonlinearity*, Annali Scuola Ecole Normale Cl.(4) **25** (1997), no.1–2, 197–215.

[6] Cazenave, Thierry; Shatah, Jalal; Tahvildar-Zadeh, A. Shadi, *Harmonic Maps and the development of singularities in Wave Maps and Yang-Mills fields.*, Ann. Inst. H. Poincare Phys. Theor. **68** (1998), no. 3, 315–349.

[7] Y. Choquet-Bruhat, V. Moncrief, *Existence theorem for solutions of Einstein's equations with* 1 *parameter spacelike isometry groups*, Quantization, nonlinear partial differential equations, and operator algebra(Cambridge, MA, 1994), 67–80, Proc. Sympos. Pure Math., 59, Amer. Math. Soc., Providence, RI, 1996.

[8] D. Christodoulou, A. Tahvildar-Zadeh, *On the regularity of spherically symmetric wave maps*, C.P.A.M., **46** (1993), 1041–1091.

[9] D. Christodoulou, A. Tahvildar-Zadeh, *On the asymptotic behavior of spherically symmetric wave maps*, Duke Math. J. **71** (1993), no. 1, 31–69.

[10] J. Colliander, M. Keel, G. Staffilani, H. Takaoka, T. Tao, *Global well-posedness and scattering in the energy space for the critical nonlinear Schrodinger equation in* R^3, preprint.

[11] A. Freire, S. Muller, M. Struwe, *Weak convergence of wave maps from (1+2)-dimensional Minkowski space to Riemannian manifolds*, Inventiones Math. **130** (1997), 589–617.

[12] M. Gell-Mann, M. Levy, *The axial vector current in beta decay*, Nuovo Cim. **16** (1960), 705.

[13] F. Gesztesy, M. Zinchenko, *On spectral theory for Schrödinger operators with strongly singular potentials.* preprint 2005.

[14] M. Grillakis, *Regularity and asymptotic behaviour of the wave equation with a critical nonlinearity.*, Ann. of Math. (2) **132** (1990), no. 3, 485–509.

[15] M. Guenther, *Isometric embeddings of Riemannian manifolds*, Proceedings Interntl. Congress of Mathematicians, Vol. 1–2 (Kyoto 1990), p. 1137–1143, Tokyo, 1991, Math. Soc. Japan.

[16] C.-H. Gu, *On the Cauchy problem for harmonic maps defined on two-dimensional Minkowski space*, Comm. Pure Appl. Math. **33** 727–737, 1980.

[17] F. Helein, *Regularite des applications faiblement harmoniques entre une surface et une varietee Riemanienne*, C. R. Acad. Sci. Paris Ser. 1 Math. **312** (1991), 591–596.

[18] J. Isenberg, S. Liebling, *Singularity formation in* $2+1$ *wave maps*, J. Math. Phys. **43** (2002) no. 1, 678–683.

[19] M. Keel, T. Tao, *Endpoint Strichartz estimates*, Amer. J. Math. **120** (1998), no. 5, 955–980.

[20] S. Klainerman, *The null condition and global existence to nonlinear wave equations*, Nonlinear systems of partial differential equations in applied mathematics, Part 1 (Santa Fe, N. M., 1984), 293–326, Lectures in Appl. Math., 23, Amer. Math. Soc., Providence, RI, 1986.

[21] S. Klainerman, *Uniform decay estimates and the Lorentz invariance of the classical wave equation.*, Comm. Pure Appl. Math. **38** (1985), no. 3, 321–332.

[22] S. Klainerman, *UCLA lectures on nonlin. wave eqns.*, preprint (2001).

[23] S. Klainerman, D. Foschi, *Bilinear Space-Time Estimates for Homogeneous Wave Equations*, Ann. Scient. Ec. Norm. Sup., 4e serie, t. **33** (2000), 211–274.

[24] S. Klainerman, M. Machedon, *Smoothing estimates for null forms and applications*, Duke Math. J. **81** (1995), 99–133.

[25] S. Klainerman, M. Machedon, *On the algebraic properties of the* $H^{\frac{n}{2},\frac{1}{2}}$ *spaces*, I.M.R.N. **15** (1998), 765–774.

[26] S. Klainerman, M. Machedon, *On the regularity properties of a model problem related to wave maps*, Duke Math. J., **87** (1997), 553–589.

[27] S. Klainerman, I. Rodnianski, *On the global regularity of wave maps in the critical Sobolev norm*, I.M.R.N. **13** (2001), 655–677.

[28] S. Klainerman, S. Selberg, *Remark on the optimal regularity for equations of wave maps type*, C.P.D.E. **22** (1997), 901–918.

[29] S. Klainerman, S. Selberg, *The spaces $H^{s,\theta}$ and applications to nonlinear wave equations*, preprint.

[30] S. Klainerman, S. Selberg, *Bilinear estimates and applications to nonlinear wave equations*, preprint.

[31] S. Klainerman, D. Tataru, *On the optimal regularity for the Yang-Mills equations in* $\mathbf{R}^{4+1}$, Journal of the American Math. Soc., **12** (1999), 93–116.

[32] J. Krieger, *Global Regularity of Wave Maps in 2 and 3 spatial dimensions*, Ph. D. Thesis, Princeton University (2003).

[33] J. Krieger, *Global regularity of Wave Maps from* $\mathbf{R}^{3+1}$ *to surfaces*, CMP 238/1–2 (2003), 333–366.

[34] J. Krieger, *Null-Form estimates and nonlinear waves*, Adv. Differemtial Equations **8** (2003), no. 10, 1193–1236.

[35] J. Krieger, *Global regularity of Wave Maps from* $\mathbf{R}^{2+1}$ *to* $\mathbf{H}^2$, CMP **250** (2004), 507–580.

[36] J. Krieger, *Stability of Spherically Symmetric Wave Maps*, Memoirs of the AMS Vol. **181** (2006), no. 853.

[37] J. Krieger, W. Schlag, D. Tataru, *Renormalization and blow up for charge one equivariant critical wave maps*, preprint.

[38] B. Levitan, I. Sargsjan, *Sturm-Liouville and Dirac operators.* Translated from the Russian. Mathematics and its Applications (Soviet Series), 59. Kluwer Academic Publishers Group, Dordrecht, 1991.

[39] H. Lindblad, *A sharp counterexample to the local existence of low-regularity solutions to nonlinear wave equations*, Duke Math. J. **72** (1993), 503–539.

[40] M. Machedon, J. Sterbenz, *Almost optimal local well-posedness for the* $(3+1)$*-dimensional Maxwell-Klein-Gordon equations.*, J. Amer. Math. Soc. **17** (2004), no. 2, 297–359.

[41] S. Muller, M. Struwe, *Global existence of wave maps in 1+2 dimensions with finite energy data*, Topological methods in nonlinear analysis **7** (1996), 245–259.

[42] A. Nahmod, A. Stefanov, K. Uhlenbeck, *On the well-posedness of the wave maps problem in high dimensions*, Comm. Anal. Geom. **11** (2003) no. 1, 49–83.

[43] J. Nash, *John The imbedding problem for Riemannian manifolds.*, Ann. of Math. (2) **63** (1956), 20–63.

[44] I. Rodnianski, *The wave map problem small data critical regularity*, Seminaire Bourbaki, 58eme annee, 2005-2006, no. 965.

[45] I. Rodnianski, J. Sterbenz, *On the formation of singularities in the critical* $O(3)$ σ*-model*, preprint.

[46] S. Selberg, *Multilinear space-time estimates and applications to local existence theory for nonlinear wave equations*, Ph.D. thesis, Princeton University, 1999.

[47] J. Shatah, *Weak solutions and development of singularities of the* SU(2) σ*-model.*, Comm. Pure Appl. Math. **41** (1988), no. 4, 459–469.

[48] J. Shatah, A. Tahvildar-Zadeh, *On the Cauchy Problem for Equivariant Wave Maps*, Comm. Pure Appl. Math. **47** (1994), 719–754.

[49] J. Shatah, M. Struwe, *Regularity results for nonlinear wave equations.*, Ann. of Math. (2) **138** (1993), no. 3, 503–518.

[50] J. Shatah, M.Struwe, *The Cauchy problem for wave maps*, I.M.R.N. **11** (2002), 555–571.

[51] J. Shatah, M. Struwe, *Geometric Wave Equations*, AMS Courant Lecture Notes 2.

[52] T. Sideris, *Global existence of harmonic maps in Minkowski space*, Comm. Pure Appl. Math. **42** (1989), no. 1, 1–13.

[53] C. D. Sogge, *Lectures on nonlinear wave equations*, Monographs in Analysis II, Interntl. Press, Boston, MA, 1995.
[54] E. Stein, *Harmonic Analysis: Real-Variable Methods, Orthogonality and Oscillatory Integrals*, Princeton University Press, Princeton, NJ, 1993.
[55] J. Sterbenz, *Angular regularity and Strichartz estimates for the wave equation*, preprint.
[56] M. Struwe, *Globally regular solutions to the* u^5 *Klein-Gordon equation.* Ann. Scuola Norm. Sup. Pisa Cl. Sci. (4) **15** (1988), no. 3, 495–513 (1989).
[57] M. Struwe, *Equivariant Wave Maps in 2 space dimensions*, preprint.
[58] M. Struwe, *Radially Symmetric Wave Maps from 1+2 dimensional Minkowski space to the sphere*, Math. Z. **242** (2002).
[59] M. Struwe, *Radially symmetric wave maps from* $(1+2)$*-dimensional Minkowski space to general targets.*, Calc. Var. Partial Differential Equations **16** (2003), no. 4, 431–437.
[60] T. Tao, *Ill-posedness for one-dimensional Wave Maps at the critical regularity*, Am. Journal of Math. **122** (2000), no. 3, 451–463.
[61] T. Tao, *Global regularity of wave maps I*, I.M.R.N. **6** (2001), 299–328.
[62] T. Tao, *Global regularity of wave maps II*, Comm. Math. Phys. **224** (2001), 443–544.
[63] T. Tao, *Counterexamples to the* $n = 3$ *endpoint Strichartz estimate for the wave equation*, preprint.
[64] T. Tao, *Geometric renormalization of large energy wave maps*, Journees Equations aux derives partielles, Forges les Eaux, 7–11 June 2004, XI 1–32.
[65] T. Tao, *Nonlinear dispersive equations: local and global analysis*, CBMS Regional Conference Series in Mathematics no. **106** (2006), pp. 373.
[66] D. Tataru, *Local and global results for wave maps I*, Comm. PDE **23** (1998), 1781–1793.
[67] D. Tataru, *On global existence and scattering for the wave maps equation*, Amer. Journal. Math. **123** (2001), no. 1, 37–77.
[68] D. Tataru, *Rough solutions for the Wave Maps equation*, preprint.
[69] K. Uhlenbeck, *Connections with* L^p*-bounds on curvature*, CMP **83** (1982), no. 1, 31–42.
[70] T. Wolff, *A sharp bilinear cone restriction estimate.*, Ann. of Math. (2) **153** (2001), no. 3, 661–698.

HARVARD UNIVERSITY, DEPT. OF MATHEMATICS, SCIENCE CENTER, 1 OXFORD STREET, CAMBRIDGE, MA 02138, U.S.A.
E-mail address: jkrieger@math.harvard.edu

Relativistic Teichmüller Theory – A Hamilton-Jacobi Approach to $2+1$-Dimensional Einstein Gravity

Vincent Moncrief

ABSTRACT. We consider vacuum spacetimes in $2+1$ dimensions defined on manifolds of the form $M=\Sigma\times\mathbf{R}$ where Σ is a compact, orientable surface of genus >1. By exploiting the harmonicity properties of the Gauss map for an arbitrary constant mean curvature (CMC) slice in such a spacetime we relate the Hamiltonian dynamics of the corresponding reduced Einstein equations to some fundamental results in the Teichmüller theory of harmonic maps. In particular we show, expanding upon an argument sketched by Puzio, that a global complete solution to the Hamilton-Jacobi equation for the reduced Einstein equations can be expressed in terms of the Dirichlet energy for harmonic maps defined over the surface Σ. While in principle this complete solution to the Hamilton-Jacobi equation determines all the solution curves to the reduced Einstein equations, one can derive a more explicit characterization of these curves through the solution of an associated (parametrized) Monge-Ampêre equation. Using the latter we define a corresponding family of "ray structures" on the Teichmüller space of the chosen 2-manifold Σ. These ray structures are similar but complementary to a different family of such ray structures defined by M. Wolf and we herein derive a "relativistic interpretation" of both sets. We also use our Hamilton-Jacobi results to define complementary families of Lagrangian foliations of the cotangent bundle of the Teichmüller space of Σ and to provide the corresponding "relativistic interpretation" of the leaves of these foliations.

1. Introduction

If Einstein's vacuum field equations are formulated for 3-dimensional Lorentzian metrics on manifolds of the form $M=\Sigma\times\mathbf{R}$, where Σ is a compact surface, then it is hardly surprising to find the Teichmüller space of Σ

playing an important role in the analysis. Indeed, it can be shown from several independent points of view [**1, 2, 3**] that this Teichmüller space, $\mathcal{T}(\Sigma)$, serves as the natural reduced configuration space and its cotangent bundle $T^*\mathcal{T}(\Sigma)$ the natural reduced phase space for Einstein's equations treated as a Hamiltonian dynamical system. That this system is only finite dimensional, in contrast to the situation for higher dimensional spacetimes, is an immediate consequence of the fact that a vanishing Einstein tensor (i.e., the vacuum condition) implies a vanishing Riemann tensor in 3 dimensions and hence the absence of any local degrees of freedom for the gravitational field. Only certain global degrees of freedom remain and these can be identified with the Teichmüller parameters describing the conformal geometry induced on Σ by the spacetime metric at any given "time". The evolution of these Teichmüller parameters as one sweeps through the leaves of a foliation of the spacetime by suitably chosen (hyper-) surfaces is the finite dimensional dynamical system that we are interested in.

In higher dimensions, the vanishing of the Einstein tensor leaves open the possibility of non-vanishing curvature and in fact, one can show that this latter tensor satisfies a hyperbolic equation whose (non-stationary) solutions can be thought of as describing gravitational waves. Here, too, it is possible to show that the conformal geometry induced on the leaves of a foliation by Cauchy hypersurfaces provides the natural reduced configuration degrees of freedom but, in contrast to the 3-dimensional case, the associated Teichmüller-like space of conformal structures is always infinite dimensional (to accommodate the gravitational waves) and the corresponding reduced field equations take the form of a hyperbolic/elliptic system of partial differential equations [**4, 5**]. Only in 3 dimensions (the lowest, nontrivial possibility) do the Einstein field equations reduce (after solution of the elliptic constraints and imposition of suitable coordinate gauge conditions) to ordinary differential equations unless some additional restriction, such as spatial homogeneity, is imposed upon the higher dimensional metrics under study.

In this article, we are primarily interested in studying the reduced Einstein equations in so-called CMCSH (constant-mean-curvature-spatially-harmonic) gauge which is defined by the requirements that the level surfaces of a suitably chosen time function, which are each Cauchy (hyper-) surfaces diffeomorphic to Σ, satisfy the CMC (constant-mean-curvature) condition and that the induced Riemannian metric g on each such Cauchy surface is such that the identity map from (Σ, g) to $(\Sigma, \tilde{g})$, for some conveniently chosen target metric $\tilde{g}$ is harmonic. This latter condition is well-known to depend only upon the conformal class of the domain metric g and hence only upon the Teichmüller parameters of this slice dependent variable. The reduced Einstein equations, which can be expressed in Hamiltonian form, give the evolution of these Teichmüller parameters together with their canonically conjugate momenta (which, taken together, provide coordinates for $T^*\mathcal{T}(\Sigma)$).

One of our main results is to show that the foliation of such vacuum metrics on $\Sigma \times \mathbf{R}$ by CMC slices is globally determined through the solution of an associated (fully non-linear) elliptic equation of Monge-Ampêre type which depends parametrically upon the mean curvature variable τ (which plays the role of "time") and upon the choice of an arbitrary point of $T^*\mathcal{T}(\Sigma)$ which can be thought of as an asymptotic data point in the reduced phase space. By exploiting the method of continuity, we prove that every solution of this Monge-Ampêre equation (each of which is fixed by prescribing asymptotic data at $\tau = 0$) extends globally to a solution for all τ in the interval $(-\infty, 0]$. The limit $\tau \searrow -\infty$ corresponds to a big bang singularity at which the geometric area of Σ tends to zero and for which, generically, the corresponding solution curve runs off-the-edge of Teichmüller space. The opposite limit $\tau \nearrow 0$ corresponds to that of infinite cosmological expansion for which the geometric area of Σ blows up but for which the induced conformal geometry always asymptotes to an interior point of Teichmüller space (which together with an associated asymptotic "velocity" is determined by the chosen point of $T^*\mathcal{T}(\Sigma)$). It is known from earlier work that the range $(-\infty, 0)$ always exhausts the maximal Cauchy development for each vacuum solution [**6**].

A closely related result that we shall derive shows how the Dirichlet energy for a suitably defined harmonic map (the Gauss Map for a CMC slice of the associated, flat spacetime) can be exploited to yield a global, complete solution to the Hamilton-Jacobi equation for the Hamiltonian system defined by the reduced field equations. A sketch of how to relate the Dirichlet energy for the Gauss map to a complete solution for the Hamilton-Jacobi equation was given earlier by Puzio [**7**]. To make his insight more precise, we fill in some of the details that were not provided in Puzio's argument and, in particular, show how the partial derivatives of the Dirichlet energy (with respect to the Teichmüller parameters) are related to the momenta of the reduced Hamiltonian formalism.

The complete solution to the reduced Einstein-Hamilton-Jacobi equation that we obtain, allows us to define a set of "ray-structures" on Teichmüller space that are similar but complementary to the well-known ray structures defined by Michael Wolf [**8**]. In our formulation each ray corresponds to (the projection of) a solution curve of the reduced Einstein equations and the collection of such curves yielding a particular ray structure corresponds to the collection of all those curves having the same asymptotic conformal geometry as $\tau \nearrow 0$. By varying this target point over the interior of Teichmüller space one obtains all of the ray structures defined by the complete solution to the Hamilton-Jacobi equation. By contrast to this, one can also give a relativistic interpretation to Wolf's rays but each such ray corresponds to the locus of endpoints defined by a one-parameter family of solutions to Einstein's equations defined by fixing (at say $\tau = -1$), the conformal geometry, but scaling up the traceless part of the second fundamental form (a holomorphic quadratic differential in Wolf's terminology) by a (spatially constant) multiplicative factor.

Wolf was able to prove that (holding the domain conformal geometry fixed) he could define a global chart for the (topologically trivial) Teichmüller space of Σ by varying the holomorphic quadratic differential over the full vector space of such objects. In a complementary way here, we are able to exploit the Monge-Ampêre analysis mentioned above to define a family of global charts for Teichmüller space in each of which the target conformal geometry is held fixed and one varies a holomorphic quadratic differential defined relative to this fixed target. Wolf gets a family of such charts by varying the domain conformal geometry whereas we get a family by varying the target.

Our approach to this latter problem is quite different from that of B. Tabak [**9**]. She also holds the target (of a family of harmonic maps) fixed but exploits the properties of so-called subsonic ϱ-holomorphic quadratic differentials to develop a family of global charts for Teichmüller space. The concept of subsonic ϱ-holomorphic quadratic differentials was introduced by L.M. Sibner and R.J. Sibner in connection with a certain hydrodynamics problem wherein they showed that these objects were expressible in terms of a certain non-linear generalization of harmonic one-forms on a compact manifold [**10**]. Since the dimension of the space of such generalized harmonic one-forms coincides with that given by Hodge theory (the first Betti number of the manifold) when these objects are non-singular it is necessary to allow forms with certain well-defined singularities in order to match the correct dimension of Teichmüller space in the higher genus cases for surfaces. Tabak gives a careful study of such singularities that need be allowed. By contrast though, our approach only requires globally regular holomorphic quadratic differentials of the conventional type, in close parallel to Wolf's treatment.

Hamilton-Jacobi theory is closely connected to the construction of Lagrangian foliations of the associated phase space, in our case $T^*\mathcal{T}(\Sigma)$ the cotangent bundle of Teichmüller space. Since we are in the ideal situation of having a global, complete solution to the Hamilton-Jacobi equation, we can exploit this close connection to define two (one-parameter families of) global Lagrangian foliations of $T^*\mathcal{T}(\Sigma)$ and to give the leaves of these foliations a "natural" interpretation in terms of corresponding families of solutions to Einstein's equations.

2. Preliminary Computations

Let Σ be a compact connected, orientable surface of genus >1 and set $M = \Sigma \times \mathbf{R}$. Relative to a "time" function t defined on M whose level surfaces are diffeomorphic to Σ we can express Lorentzian metrics on M in the Arnowitt, Deser and Misner (ADM) form

$$\begin{aligned} ds^2 &= {}^{(3)}g_{\mu\nu}dx^\mu dx^\nu \\ &= -N^2dt^2 + g_{ab}(dx^a + X^a dt)(dx^b + X^b dt). \end{aligned} \tag{2.1}$$

Here $\mu, \nu \ldots$ range over $\{0,1,2\}$, where $x^0 = t$ is the time, and $a, b, \ldots$ range over $\{1,2\}$ where $\{x^1, x^2\}$ are the spatial coordinates. Induced upon each level surface of t is a Riemannian metric g_t (the first fundamental form) given by

$$g_t = g_{ab} dx^a \otimes dx^b \tag{2.2}$$

where here and below we suppress the spacetime coordinate dependence of component expressions such as g_{ab} to simplify the notation. N is a positive function on M (the "lapse") and $X^a \frac{\partial}{\partial x^a}$ is an (in general t-dependent) vector field tangent to the level surfaces of t (the "shift").

The covariant derivative of the unit normal field ("future" directed towards increasing t) to the surfaces of constant t determines, in the usual way, another symmetric two-tensor k_t (the second fundamental form) on each such surface which we shall write in component form as

$$k_t = k_{ab} dx^a \otimes dx^b. \tag{2.3}$$

Writing $\mu_g = \sqrt{\det g_{ab}}$ for the area element of g_t, we define the gravitational momentum π_t (a symmetric, contravariant, two-tensor density with components π^{ab}) by

$$\pi^{ab} = -\mu_g (k^{ab} - g^{ab} tr_g k) \tag{2.4}$$

where $g^{ab} = (g_t^{-1})^{ab}$ are the components of the inverse metric to g_t, two-dimensional indices are raised and lowered using g_t and g_t^{-1} and where $tr_g k = g^{ab} k_{ab}$, the trace of k_t. This latter quantity, the mean curvature of the $t =$ constant hypersurfaces, will play an important role in what follows and we shall often designate it by the symbol τ. Thus, from the formulas above

$$\tau := tr_g k = g^{ab} k_{ab} = \frac{tr_g \pi}{\mu_g} = \frac{g_{ab} \pi^{ab}}{\mu_g}. \tag{2.5}$$

The ADM action for Einstein's equations is given by

$$I_{ADM} = \int_{\mathcal{I} \times \Sigma} d^3x \{\pi^{ab} g_{ab,t} - N\mathcal{H} - X^a J_a\} \tag{2.6}$$

where $\mathcal{I} = [t_0, t_1] \subset \mathbf{R}$ is an arbitrary closed interval and where

$$\mathcal{H} = \mathcal{H}(g, \pi) = \frac{1}{\mu_g}(\pi^{ab}\pi_{ab} - (tr_g \pi)^2) - \mu_g^{(2)} R(g) \tag{2.7a}$$

$$J_a = J_a(g, \pi) = -2\pi_a^b|_b = -2^{(2)}\nabla_b \pi_a^b. \tag{2.7b}$$

Here ${}^{(2)}R(g)$ is the scalar curvature of the Riemannian metric g and $|$ or ${}^{(2)}\nabla$ designates covariant differentiation with respect to this metric.

Variation of I_{ADM} with respect to the lapse N and shift X^a yields the Einstein constraint equations

$$\mathcal{H}(g,\pi) = 0, \quad J_a(g,\pi) = 0 \tag{2.8}$$

whereas variation with respect to g and π yield evolution equations (in Hamiltonian form) for these "canonical" variables. There are no equations determining N and X^a and these quantities can be specified freely to determine a coordinate system on the developing spacetime. Two choices which we shall make use of are the gaussian normal choice $N=1, X^a=0$ (in which the spatial coordinates are held constant along the normal geodesics from an initial slice and t is the metrically determined proper time along those geodesics) and another in which the hypersurfaces of constant t are required to each have constant mean curvature and the spatial coordinates are required to satisfy a harmonic condition defined below. Imposing these conditions upon the spacetime coordinates leads to a system of linear elliptic equations for N and X^a which determines these quantities uniquely in terms of the remaining (canonical) data $\{g_t, \pi_t\}$. The Bianchi identities ensure that the constraints (2.8) are conserved by the evolution equations in essentially an arbitrary coordinate gauge.

The Einstein evolution equations for vacuum $2+1$ gravity simplify greatly when expressed in gaussian normal coordinates (gnc) and in fact reduce to a decoupled system of ordinary differential equations for $\{g_{ab}, \pi^{ab}\}$ along each normal geodesic in the evolving flat spacetime. Insofar as these geodesics are initially diverging the absence of spacetime curvature (which results from the fact that vanishing Einstein tensor implies vanishing Riemann tensor in 3 dimensions) ensures that these "straight lines" will never cross to the future of the initial surface and thus that the gnc coordinate system will never break down in this temporal direction (though in general it does break down in the opposite direction). As we shall show later by analyzing the constraint equations in detail, the future directed normals to a constant mean curvature hypersurface having $\tau =$ constant <0 (and where "future" designates the direction of increasing τ) are always diverging and thus the gnc coordinate systems developed from such an initial surface cover the entire spacetime to the future of this surface. Indeed, by exploiting the simplified form of the evolution equations, one can construct the spacetime metric essentially explicitly in gaussian normal coordinates.

To see this, set $N=1$ and $X^a=0$ in the evolution equations and derive easily that

$$\begin{aligned} g_{ab,t} &= \frac{2}{\mu_g}(\pi_{ab} - g_{ab}tr_g\pi) \\ \partial_t\left(\pi^b_a - \frac{1}{2}\delta^b_a tr_g\pi\right) &= 0 \\ \partial_t(tr_g\pi) &= \frac{\pi^d_c\pi^c_d - (tr_g\pi)^2}{\mu_g} := \mathcal{K} \end{aligned} \tag{2.9}$$

from which follow

$$\partial_t \mu_g = -tr_g \pi, \tag{2.10}$$
$$\partial_t \mathcal{K} = \frac{\partial}{\partial t}\left(\frac{\pi^d_c \pi^c_d - (tr_g \pi)^2}{\mu_g}\right) = 0,$$

and, upon using the momentum constraint,

$$\partial_t ({\mu_g}^{(2)} R(g)) = 0. \tag{2.11}$$

Taking, with no real loss of generality $t = 0$ on the initial (CMC) surface, one sees immediately that $\mathcal{K} = \overset{o}{\mathcal{K}} = \mathcal{K}|_{t=0}$ and finds by straightforward integration that

$$\mu_g(t) = \overset{o}{\mu}_g - \left\{ \overset{o}{tr}_g \pi t + \frac{1}{2}\overset{o}{\mathcal{K}} t^2 \right\} \tag{2.12}$$
$$tr_g \pi(t) = \overset{o}{tr}_g \pi + \overset{o}{\mathcal{K}} t$$

where $\overset{o}{tr}_g \pi := tr_g \pi|_{t=0}, \overset{o}{\mu}_g = \mu_g|_{t=0}$. The same evolution equations give $\frac{\partial}{\partial t} J_a(g, \pi) = 0$, and, if $J_a = 0$, that $\partial_t \mathcal{H}(g, \pi) = 0$ as well. In fact, as noted above, one has separately that $\partial_t ({\mu_g}^{(2)} R(g)) = 0$ and $\partial_t \mathcal{K} = 0$ where $\mathcal{H} = \mathcal{K} - {\mu_g}^{(2)} R(g)$. Clearly, Eq. (2.9b) also gives conservation of the traceless part of π^b_a (i.e., that $\pi^b_a - \frac{1}{2}\delta^b_a tr_g \pi = (\pi^b_a - \frac{1}{2}\delta^b_a tr_g \pi)|_{t=0}$).

Knowing the solution for μ_g, $tr_g \pi$ and $\pi^b_a - \frac{1}{2}\delta^b_a tr_g \pi$ we need only solve for the (conformally invariant) density $\mu_g g^{ab}$ which satisfies the differential equation

$$\partial_t (\mu_g g^{ab}) = -2(\mu_g g^{ac}) \left(\frac{\pi^b_c - \frac{1}{2}\delta^b_c tr_g \pi}{\mu_g} \right). \tag{2.13}$$

The details of this solution are straightforward but uninteresting so we here give only the result that

(2.14)
$$\mu_g g^{ab}(t) = \cosh(R(t))(\mu_g g^{ab})\bigg|_{t=0} + \sinh(R(t)) \left(\frac{\pi^{ab} - \frac{1}{2} g^{ab} tr_g \pi}{\sqrt{\frac{1}{2}\frac{\lambda^d_c \lambda^c_d}{(\mu_g)^2}}} \right)\Bigg|_{t=0}$$

where

$$\lambda^b_a = \pi^b_a - \frac{1}{2}\delta^b_a tr_g \pi \tag{2.15}$$

and

$$e^{R(t)} = \left(\frac{2ct + b + \sqrt{-q}}{2ct + b - \sqrt{-q}} \right) \left(\frac{b - \sqrt{-q}}{b + \sqrt{-q}} \right) \tag{2.16}$$

with

$$b = -\left(\frac{tr_g\pi}{\mu_g}\right)\Big|_{t=0} = -\tau\Big|_{t=0} \tag{2.17}$$
$$c = -\frac{1}{2}\left(\frac{\mathcal{K}}{\mu_g}\right)\Big|_{t=0}$$
$$q = -2\left(\frac{\lambda_c^d\lambda_d^c}{(\mu_g)^2}\right)\Big|_{t=0}$$

and where, if we also assume that $\tau|_{t=0} = \text{constant}$ on the initial surface λ_a^b has zero divergence with respect to $\overset{o}{g}_{ab} = g_{ab}|_{t=0}$ as well as zero trace.

It is straightforward to verify using the explicit formulas above, that $g_{ab}(t)$ is a smooth Riemannian metric $\forall\ t \geq 0$ provided that the initial data satisfies the conditions that

(i) $g_{ab}\big|_{t=0}$ is smooth and Riemannian on Σ,
(ii) $\tau\big|_{t=0}$ is smooth with $\tau\big|_{t=0} < 0$ on Σ,
(iii) $\mathcal{K}\big|_{t=0}$ is smooth with $\mathcal{K}|_{t=0} < 0$ on Σ.

We shall later impose the restriction that $\tau|_{t=0}$ be a negative constant on Σ and find that the constraint equations force condition (iii) to hold provided the remaining Cauchy data λ_a^b is also smooth. Note that the zeros of λ_a^b do not disturb the smoothness of $\mu_g g^{ab}(t)$ since the factor $\sinh(R(t))$ vanishes whenever λ_a^b does.

The gaussian normal slicing is in general not a CMC slicing (except when $\tau|_{t=0} = \text{constant}$ and $\lambda_a^b = 0$) but one nevertheless has

$$t\tau(t) = -2\frac{\left\{\frac{\mathcal{K}}{\mu_g}\Big|_{t=0} + \frac{1}{t}\tau\Big|_{t=0}\right\}}{\left\{\frac{\mathcal{K}}{\mu_g}\Big|_{t=0} + \frac{2}{t}\tau\Big|_{t=0} - \frac{2}{t^2}\right\}} = -2 + O\left(\frac{1}{t}\right). \tag{2.18}$$

Of even more interest to us is the limiting behavior of the corresponding rescaled metric, $\frac{2}{t^2}g_{ab}(t)$. This metric also has a limit (as a Riemannian metric on Σ) as $t \to \infty$. Explicit computation gives

$$\rho_{ab} := \lim_{t\to\infty}\left(\frac{2}{t^2}g_{ab}\right) = \left\{(k_c^d k_d^c)g_{ab} + 2(tr_g k)\left(k_{ab} - \frac{1}{2}g_{ab}(tr_g k)\right)\right\}\Big|_{t=0} \tag{2.19}$$

which is clearly smooth and symmetric; positive definiteness depends upon the condition that $\mathcal{K}\Big|_{t=0} < 0$ on Σ. As we have mentioned, this will be shown to follow from the constraint equations at least when the initial data surface is CMC.

In fact, when the constraint equations (2.8) hold for the initial data $\{g_{ab}, k_{ab}\}\big|_{t=0}$ the metric ρ_{ab} actually has constant (negative) curvature, with ${}^{(2)}R(\rho) = -1$. This can be shown by an explicit calculation or, less directly, by the following argument. Conservation of the constraints, taken together with the gnc result that $\partial_t \mathcal{K} = 0$, yields

$$(2.20) \qquad {}^{(2)}R\left(\frac{2}{t^2}g(t)\right) = \frac{1}{2}t^{2\,(2)}R(g(t)) = \frac{\frac{1}{2}t^2\mathcal{K}\big|_{t=0}}{\left(\mu_g\big|_{t=0} - t(tr_g\pi)\big|_{t=0} - \frac{1}{2}t^2\mathcal{K}\big|_{t=0}\right)} \underset{t\to\infty}{\longrightarrow} -1$$

since, once again, $\overset{o}{\mathcal{K}} = \mathcal{K}\big|_{t=0} < 0$ on Σ. Thus ${}^{(2)}R(\rho) = \lim_{t\to\infty} {}^{(2)}R(\frac{2}{t^2}g(t)) = -1$.

Another remarkable property obtains if we restrict the initial surface to have constant mean curvature. The identity map from $(\Sigma, \overset{o}{g}_{ab}) = (\Sigma, g_{ab}\big|_{t=0})$ to (Σ, ρ_{ab}) is in fact a harmonic map. This can be shown by explicit computation of the quantity

$$(2.21) \qquad V^c := \overset{o}{g}{}^{ab}(\Gamma^c_{ab}(\overset{o}{g}) - \tilde{\Gamma}^c_{ab}(\rho))$$

where $\Gamma^c_{ab}(\overset{o}{g})$ and $\tilde{\Gamma}^c_{ab}(\rho)$ are the Christoffel connection components of $\overset{o}{g}_{ab}$ and ρ_{ab} respectively. Harmonicity of the identity mapping corresponds precisely to the vanishing of the vector field V^c and this in turn follows from imposition of the constraints and the additional condition that $\overset{o}{\tau} = \tau\big|_{t=0} =$ constant < 0 on Σ.

Again there is a less direct argument which shows why this should be true and which traces back to a well-known result of Ruh and Vilms [**11**] in the purely Riemannian case. One can show that the Gauss map from a CMC hypersurface in a flat spacetime is harmonic [**7**]. Our spacetime metric ${}^{(3)}g_{\mu\nu}$ is flat since the (vacuum) Einstein equations in three dimensions imply that ${}^{(3)}g_{\mu\nu}$ has vanishing curvature and we are now imposing the restriction that the initial slice be CMC. That the identity map from $(\Sigma, \overset{o}{g}_{ab})$ to (Σ, ρ_{ab}) realizes the Gauss map in this case follows from the construction using (appropriately enough) gaussian normal coordinates as we have done. To each (future directed) normal to the initial surface one assigns a point in the hyperbolic space (Σ, ρ) by following the normal geodesic in that direction to its ideal endpoint. We use the starting point and ideal ending point of each such normal geodesic to identify the two copies of Σ and appeal to the results given above to recognize that (Σ, ρ) is indeed hyperbolic.

The data $\{g_{ab}, k_{ab}\}\big|_{t=0}$ are assumed to satisfy the constraints and to have $\tau\big|_{t=0} = tr_g k\big|_{t=0} =$ constant on Σ. We can remove this implicit restriction by appealing to the standard (conformal) method for solving the constraints as follows. First, it follows from the classical uniformization theorem that any metric g_{ab} on a higher genus surface Σ is uniquely and smoothly globally conformal to another metric γ_{ab} which has constant curvature, ${}^{(2)}R(\gamma) = -1$. Writing $g_{ab} = e^{2\lambda}\gamma_{ab}$, for some smooth function λ defined on Σ and imposing the CMC condition that $\frac{tr_g \pi}{\mu_g} = tr_g k = \tau = \text{constant}$ on Σ one finds that the momentum constraint, $J_a(g,\pi) = -2^{(2)}\nabla_b \pi_a^b = 0$, can now be expressed

$$(2.22) \qquad {}^{(2)}\tilde{\nabla}_b(\gamma)\left(\pi_a^b - \frac{1}{2}\delta_a^b tr_g \pi\right) = {}^{(2)}\tilde{\nabla}_b(\gamma)\lambda_a^b = 0$$

where ${}^{(2)}\tilde{\nabla}_b(\gamma)$ signifies covariant differentiation with respect to the conformal metric γ_{ab}. In other words, the traceless tensor density λ_a^b should also be "transverse" (i.e. divergence free) with respect to the conformal metric γ_{ab}.

The Hamiltonian constraint, $\mathcal{H}(g,\pi) = 0$, can now be expressed as a non-linear elliptic equation (the "Lichnerowicz equation" in relativity literature) for the conformal factor λ. In the notation above, this equation takes the form

$$(2.23) \qquad \begin{aligned} {}^{(2)}\Delta_\gamma \lambda &= \gamma^{ab(2)}\tilde{\nabla}_a(\gamma)^{(2)}\tilde{\nabla}_b(\gamma)\lambda \\ &= \frac{1}{4}\tau^2 e^{2\lambda} - \frac{1}{2}\frac{\lambda_a^b\lambda_b^a}{(\mu_\gamma)^2}e^{-2\lambda} - \frac{1}{2} \end{aligned}$$

where μ_γ is the area element of the conformal metric γ_{ab}, λ_a^b is transverse-traceless with respect to this metric (with $\gamma_{bc}\lambda_a^b$ symmetric) and ${}^{(2)}R(\gamma) = -1$.

Upon integration over Σ it is easy to see that Eq. (2.23) has no solutions if $\tau = 0$ on Σ. For any non-zero constant τ however, one can show (using for example the method of sub and super solutions [**12, 13**]) that Eq. (2.23) always has a unique, smooth, globally defined solution λ on Σ. To summarize, the general solution to the constraint equations for a CMC slice (with $\tau = \text{constant} \neq 0$) can be expressed in terms of the free data

$$\{(\tau, \gamma_{ab}, \lambda_a^b)\,|\,\tau = \text{const} \neq 0, \gamma_{ab} \text{ a hyperbolic metric on } \Sigma \text{ with } {}^{(2)}R(\gamma) = -1, \lambda_a^b \text{ a } TT \text{ symmetric tensor density w.r.t. } \gamma\}$$

by setting

$$(2.24) \qquad g_{ab} = e^{2\lambda}\gamma_{ab}, \; \pi_a^b = \lambda_a^b + \frac{1}{2}\tau\mu_g\delta_a^b$$

where λ is the solution of Lichnerowicz's equation (2.23).

Since the constraint equations are naturally covariant with respect to diffeomorphisms of Σ (which automatically conserve the constancy of τ) one can, without any essential loss of generality, pass to the quotient,

"Teichmüller," space

$$\begin{aligned} \mathcal{T}(\Sigma) &\approx \mathcal{M}_{-1}(\Sigma)/\mathcal{D}_0(\Sigma) \\ &\approx \mathbf{R}^{6\ \mathrm{genus}(\Sigma)-6} \end{aligned} \tag{2.25}$$

where $\mathcal{M}_{-1}(\Sigma)$ designates the space of Riemannian metrics γ_{ab} on Σ which have ${}^{(2)}R(\gamma) = -1$ and where $\mathcal{D}_0(\Sigma)$ signifies the group of diffeomorphisms of Σ isotopic to the identity. As is known from the work of Eells, Earle and Sampson [**14**], which we shall recall in more detail below, one can represent Teichmüller spaces as a global cross section of the (trivial) $\mathcal{D}_0(\Sigma)$ bundle

$$\mathcal{M}_{-1}(\Sigma) \longrightarrow \mathcal{M}_{-1}(\Sigma)/\mathcal{D}_0(\Sigma). \tag{2.26}$$

Such cross sections can be constructed through the use of harmonic maps. Restricting the metric γ_{ab} to lie in such a global cross section and recalling that λ_a^b is TT (transverse-traceless) with respect to γ_{ab} one can regard the space of pairs $\{\gamma_{ab}, \lambda_a^b\}$ as a representation of the cotangent bundle, $T^*\mathcal{T}(\Sigma)$, of the Teichmüller space of Σ [**2**].

Noting that

$$\begin{aligned} \frac{g_{bc}}{\mu_g}\lambda_a^c = \frac{g_{bc}}{\mu_g}\left(\pi_a^c - \frac{1}{2}\delta_a^c tr_g\pi\right) \\ = -\left(k_{ab} - \frac{1}{2}g_{ab}tr_g k\right) \\ = -\left(k_{ab} - \frac{1}{2}g_{ab}\tau\right) \end{aligned} \tag{2.27}$$

and that $\frac{g_{ab}}{\mu_g} = \frac{\gamma_{ab}}{\mu_\gamma}$ one sees that $(k_{ab} - \frac{1}{2}g_{ab}\tau)$ is transverse-traceless with respect to γ_{ab} (or, in fact, to any metric conformal to γ_{ab} such as g_{ab}). Writing k_{ab}^{TT} for $(k_{ab} - \frac{1}{2}g_{ab}\tau)$ we can re-express Eq. (2.19) in the form

$$\rho_{ab} = \left\{\left(e^{-4\lambda}\gamma^{de}\gamma^{cf}k_{ce}^{TT}k_{df}^{TT} + \frac{1}{2}\tau^2\right)e^{2\lambda}\gamma_{ab} + 2\tau k_{ab}^{TT}\right\} \tag{2.28}$$

with now both ρ_{ab} and γ_{ab} hyperbolic (with unit negative scalar curvature) and with (since harmonicity depends only upon the conformal structure of the domain metric) the identity map from (Σ, γ) to (Σ, ρ) harmonic. Note that one can absorb τ into the remaining variables by defining

$$\tau^2 e^{2\lambda} = e^{2\tilde{\lambda}}, \tau k_{ab}^{TT} = \tilde{k}_{ab}^{TT} \tag{2.29}$$

so that

$$\rho_{ab} = \left\{\left(e^{-4\tilde{\lambda}}\gamma^{de}\gamma^{cf}\tilde{k}_{ce}^{TT}\tilde{k}_{df}^{TT} + \frac{1}{2}\right)e^{2\tilde{\lambda}}\gamma_{ab} + 2\tilde{k}_{ab}^{TT}\right\} \tag{2.30}$$

with $\tilde{\lambda}$ satisfying the (τ-autonomous) equation

$$ {}^{(2)}\Delta_\gamma \tilde{\lambda} = \frac{1}{4} e^{2\tilde{\lambda}} - \frac{1}{2} \frac{\tilde{\lambda}_a^b \tilde{\lambda}_b^a}{(\mu_\gamma)^2} e^{-2\tilde{\lambda}} - \frac{1}{2} \tag{2.31} $$

where $\tilde{\lambda}_a^b := \tau \lambda_a^b$ (so that $\frac{g_{bc}}{\mu_g} \tilde{\lambda}_a^c = \frac{\gamma_{bc}}{\mu_\gamma} \tilde{\lambda}_a^c = -\tilde{k}_{ab}^{TT}$).

Equation (2.30) is equivalent to a remarkable, well-known formula in the theory of harmonic maps which allows one, if γ_{ab} is held fixed, to parameterize target metrics ρ_{ab} in terms of transverse-traceless symmetric tensors (holomorphic quadratic differentials in the mathematics literature) in such a way that the identity map from (Σ, γ) to (Σ, ρ) is automatically harmonic. Indeed this gives a particular means of constructing a global cross section of the bundle

$$ \mathcal{M}_{-1}(\Sigma) \longrightarrow \mathcal{M}_{-1}(\Sigma)/\mathcal{D}_0(\Sigma) \tag{2.32} $$

and thus a concrete model for the Teichmüller space $\mathcal{T}(\Sigma)$. We shall expand upon this connection to the conventional harmonic maps approach to Teichmüller theory in a subsequent section, but for now, return to the main thread of our discussion.

We conclude this section with the proof, promised above, that $\mathcal{K}$ defined by

$$ \begin{aligned} \mathcal{K} &:= \left(\frac{\pi_b^a \pi_a^b - (tr_g \pi)^2}{\mu_g} \right) \\ &= \mu_g (k_{ab} k^{ab} - (tr_g k)^2) \end{aligned} \tag{2.33} $$

satisfies $\mathcal{K} < 0$ on a CMC slice satisfying the initial value constraints. The momentum constraint is equivalent to

$$ k_{a|b}^b - (tr_g k)_{|a} = 0 \tag{2.34} $$

which may be re-expressed as the Codazzi condition

$$ k_{b|c}^a - k_{c|b}^a = 0. \tag{2.35} $$

Taking the divergence of this equation,

$$ k_{b|c}^a{}^{|c} - k_{c|b}^a{}^{|c} = 0, \tag{2.36} $$

commuting covariant derivatives in the second term and reexpressing the curvature of g through

$$ {}^{(2)}R_{dabc} = \frac{1}{2} {}^{(2)}R(g)(g_{db} g_{ac} - g_{cd} g_{ab}) \tag{2.37} $$

one derives

$$ k_{ab|c}{}^{|c} - \left(k_{ac}{}^{|c}\right)_{|b} = {}^{(2)}R(g) \left(k_{ab} - \frac{1}{2} g_{ab} tr_g k\right). \tag{2.38} $$

Imposing the CMC condition $tr_g k = \tau =$ constant and the momentum constraint and decomposing k_{ab} via

$$k_{ab} = k_{ab}^{TT} + \frac{1}{2} g_{ab} \tau \tag{2.39}$$

one arrives at

$$k_{ab|c}^{TT\ |c} = {}^{(2)}R(g) k_{ab}^{TT}. \tag{2.40}$$

It follows easily from Eq. (2.40), upon using now the Hamiltonian constraint

$$\mu_g \left(k_{ab}^{TT} k^{TT ab} - \frac{1}{2}\tau^2 \right) = \mu_g\, {}^{(2)}R(g), \tag{2.41}$$

that

$$\begin{aligned} &{}^{(2)}\Delta_g \left(\left|k^{TT}\right|_g^2 - \frac{1}{2}\tau^2 \right) - 2 \left|k^{TT}\right|_g^2 \left(\left|k^{TT}\right|_g^2 - \frac{1}{2}\tau^2 \right) \\ &= 2 k_{ab|c}^{TT} k^{TT ab|c} \geq 0 \end{aligned} \tag{2.42}$$

where

$$\left|k^{TT}\right|_g^2 := k_{ab}^{TT} k^{TT ab}. \tag{2.43}$$

The strong maximum principle applies to this equation and implies that

$$\left|k^{TT}\right|_g^2 - \frac{1}{2}\tau^2 < 0 \tag{2.44}$$

on the surface Σ [**15**]. This strict inequality gives $\mathcal{K} < 0$ on Σ and hence implies the global regularity of the gnc solutions presented above to the future of an initial CMC slice.

3. The Dirichlet energy of the Gauss map

Let any two Riemannian metrics g and ρ, defined on Σ, be expressed in local coordinates $\{x^a\}$ and $\{\psi^A\}$ as

$$\begin{aligned} g &= g_{ab}(x) dx^a \otimes dx^b \\ \rho &= \rho_{AB}(\psi) d\psi^A \otimes d\psi^B \end{aligned} \tag{3.1}$$

and suppose that a mapping $\psi : (\Sigma, g) \to (\Sigma, \rho)$, expressible locally by giving $\psi^A(x^b)$, is to be harmonic. Then ψ must satisfy the Euler-Lagrange equations which result from varying the "action" functional

$$\mathcal{A}(g, \rho, \psi) := \frac{1}{2} \int_\Sigma d\mu_g g^{ab} \psi^A_{,a} \psi^B_{,b} \rho_{AB} \tag{3.2}$$

with respect to ψ. These "harmonic map" equations take the form

$$ {}^{(2)}\Delta_g \psi^C + g^{ab}\psi^A_{,a}\psi^B_{,b}\tilde{\Gamma}^C_{AB}(\psi) = 0 \tag{3.3} $$

where the $\tilde{\Gamma}^A_{BC}$ are the Christoffel symbols of ρ_{AB} and ${}^{(2)}\Delta_g$ is the Laplacian of g. The action and its Euler-Lagrange equations are invariant with respect to conformal transformations of the metric g, where $g_{ab} \to e^{2\omega} g_{ab}$, and thus depend only upon the conformal class of g.

In the previous section we found that if g and ρ are related by Eq. (2.28) then the identity map from (Σ, g) to (Σ, ρ), expressible locally as $\psi^A(x) = x^A$, is harmonic. Evaluating the action $\mathcal{A}$ on this mapping yields

$$ \begin{aligned} \mathcal{A}(g, \rho, Id) &= \frac{1}{2}\int_\Sigma d\mu_g g^{ab}\rho_{ab} \\ &= \int_\Sigma d\mu_g g^{ab}\left\{\frac{1}{2}k^d_c k^c_d g_{ab} + (tr_g k)\left(k_{ab} - \frac{1}{2}g_{ab}(tr_g k)\right)\right\} \\ &= \int_\Sigma d\mu_g k^d_c k^c_d. \end{aligned} \tag{3.4} $$

Recalling that the Hamiltonian constraint satisfied by the data $\{g, k\}$ takes the form

$$ \begin{aligned} \mathcal{H} &= \mu_g\left[k^d_c k^c_d - (tr_g k)^2\right] - \mu_g {}^{(2)}R(g) \\ &= 0 \end{aligned} \tag{3.5} $$

one sees that Eq. (3.4) can also be written as

$$ \begin{aligned} \mathcal{A}(g, \rho, Id) &= \int_\Sigma \tau^2 d\mu_g + \int_\Sigma d\mu_g {}^{(2)}R(g) \\ &= \int_\Sigma d\mu_g k^d_c k^c_d = \int_\Sigma d\mu_g \left[k^{TTd}_c k^{TTc}_d + \frac{1}{2}\tau^2\right] \end{aligned} \tag{3.6} $$

where $k^{TTd}_c = g^{de}k^{TT}_{ce}$ and $\tau = tr_g k$ as before. Thus one also has

$$ \frac{1}{2}\int_\Sigma \tau^2 d\mu_g = \int_\Sigma d\mu_g k^{TTd}_c k^{TTc}_d - \int_\Sigma d\mu_g {}^{(2)}R(g) \tag{3.7} $$

where the second term on the right hand side is constant by the Gauss-Bonnet theorem. When we turn to the study of the (reduced) Einstein equations in CMC (as opposed to gnc) gauge the quantity $\int_\Sigma \tau^2 d\mu_g$, when re-expressed in terms of the variables $\{\tau, \gamma_{ab}, \lambda^b_a\}$ via $g_{ab} = e^{2\lambda}\gamma_{ab}$ (with λ determined by the Lichnerowicz equation) will play the role of a (reduced) Hamiltonian for the Einstein "flow" on $T^*\mathcal{T}(\Sigma) \times \mathbf{R}$. Note that, from Eq. (3.7), the infimum of this quantity, which results from setting $k^{TTd}_c = 0$, is always given by the Gauss-Bonnet invariant (i.e., the Euler characteristic of Σ).

4. The reduced Hamiltonian

A local existence theorem for the vacuum Einstein equations in CMCSH (constant-mean-curvature-spatially-harmonic) gauge was proven in Ref. [**4**] for spatially compact spacetimes of dimension $n+1$ for arbitrary $n \geq 2$. The proof involved various higher order energy estimates to control the (Sobolev space) norms of solutions to the gauge-fixed field equations. The case $n=2$ is very special however, and can be treated by much simpler ODE methods once it is realized that the gauge-fixed field equations describe dynamics in a finite dimensional phase space - the cotangent bundle of the Teichmüller space $\mathcal{T}(\Sigma)$. The local existence result for this problem can be treated by the methods developed in Ref. [**2**] after only a slight modification to impose the CMCSH gauge conditions under consideration here.

The local result derived in [**2**] was subsequently extended to a global one in Ref. [**6**] wherein a non-zero cosmological constant was also allowed for. The main technique for this argument involved the use of the Dirichlet energy on Teichmüller space, exploiting its known properties as a proper function, to bound the motion to the interior of Teichmüller space for all values of mean curvature τ in the range $(-\infty, 0)$ and then to show that this motion captures the maximal Cauchy development of every solution. Except for a lower dimensional subset of "trivial" solutions which are known explicitly, all solutions run "off-the-edge" of Teichmüller space in the limit as $\tau \searrow -\infty$ which corresponds to the big-bang singularities of these (vacuum) $2+1$ dimensional cosmological models. The opposite limit, $\tau \nearrow 0$, corresponds to the limit of infinite cosmological expansion wherein, however, the motion remains confined to the interior of Teichmüller space. We shall show below that in fact every solution tends to a limit in $\mathcal{T}(\Sigma)$ as $\tau \nearrow 0$.

The arguments of Ref. [**6**] only exploited the Dirichlet energy in a rather crude way. In the present paper, we shall significantly refine the application of this energy by showing how it yields a complete solution to the Hamilton-Jacobi equation for the reduced Einstein equations in CMCSH gauge. As a first step in this direction, let us briefly recall some of the key results of Ref. [**2**] with the notation of that paper modified to conform to that used here and with the gauge conditions adjusted to agree with the CMCSH choice made here.

As we have already shown in Sec. 2 above, the general solution to the constraints, $\mathcal{H}(g,\pi) = J_a(g,\pi) = 0$, in CMC gauge can be expressed (c.f. Eq. (2.24)) as:

$$g_{ab} = e^{2\lambda}\gamma_{ab}, \ \pi^b_a = \lambda^b_a + \frac{1}{2}\tau\mu_g\delta^b_a \tag{4.1}$$

where ${}^{(2)}R(\gamma) = -1$, λ^b_a is a symmetric TT tensor density with respect to γ_{ab}, $\tau < 0$ is constant on Σ and λ is the corresponding unique solution to the Lichnerowicz equation (2.23). Defining, as in Ref. [**2**],

$$p^{TTab} = \gamma^{ac}\lambda^b_c \tag{4.2}$$

we thus get

$$g_{ab} = e^{2\lambda}\gamma_{ab}, \ \pi^{ab} = e^{-2\lambda}p^{TTab} + \frac{1}{2}\tau\mu_\gamma\gamma^{ab} \tag{4.3}$$

for this general solution.

When the constraints are satisfied along a differentiable curve (that need not be a solution of the remaining field equations), the ADM action (2.6) reduces to

$$I^*_{\text{ADM}} = \int_{\mathcal{I}\times\Sigma} d^3x\{\pi^{ab}g_{ab,t}\}. \tag{4.4}$$

We substitute the foregoing expressions for π^{ab} and g_{ab} into this formula but restrict the conformal metric to lie in global cross-section of the $\mathcal{D}_0(\Sigma)$ bundle $\mathcal{M}_{-1}(\Sigma) \to \mathcal{M}_{-1}(\Sigma)/\mathcal{D}_0(\Sigma) \approx \mathcal{T}(\Sigma) \approx \mathbf{R}^{6\text{genus}(\Sigma)-6}$ determined by the requirement that the identity map from (Σ,γ) to (Σ,ρ) be harmonic. Here ρ is an arbitrary metric satisfying ${}^{(2)}R(\rho) = -1$. Later we shall allow ρ itself to vary over another suitably chosen cross-section of this same bundle and thus consider a parametrized family of such reduced actions but, for now, simply regard ρ as fixed. That the cross-sections defined by the requirements that Id : $(\Sigma,\gamma) \to (\Sigma,\rho)$ be harmonic are indeed global, was proven in Ref. [**14**].

By choosing global coordinates $\{q^\alpha \mid \alpha = 1,\ldots, 6\text{genus}(\Sigma) - 6\}$ on the topologically trivial space $\mathcal{T}(\Sigma) \approx R^{6\text{genus}(\Sigma)-6}$ and lifting these up to the chosen cross section, one can express the metrics realizing this cross section as smooth functions of the q^α's and hence as $\gamma_{ab}(x^c, q^\alpha)$ relative to local coordinates $\{x^a\}$ on Σ. Along a differential curve of such metrics, one thus has

$$\frac{\partial\gamma_{ab}}{\partial t} = \frac{\partial\gamma_{ab}}{\partial q^\alpha}\dot{q}^\alpha \tag{4.5}$$

where, by construction, the tensor fields $\left\{\frac{\partial\gamma_{ab}}{\partial q^\alpha} \mid \alpha = 1,\ldots, 6\text{genus}(\Sigma) - 6\right\}$ provide a basis to the tangent space to the cross section at any point thereof. As discussed in Ref. [**2**] (c.f. Eq. (2.22) and associated references) each such tangent vector has a unique L^2-orthogonal (relative to $\gamma_{ab}(x^c, q^a)$) decomposition of the form

$$\frac{\partial\gamma_{ab}}{\partial q^\alpha} = k^{TT}_{(\alpha)ab} + \left(\mathcal{L}_{{}^{(2)}X_{(\alpha)}}\gamma\right)_{ab} \tag{4.6}$$

where $k^{TT}_{(\alpha)}$ is a TT symmetric tensor with respect to γ and ${}^{(2)}X_{(\alpha)}$ a vector field on Σ (with $\mathcal{L}_{{}^{(2)}X_{(\alpha)}}$ signifying its Lie derivative).

Exploiting this decomposition, it is straightforward to show that there exists a smoothly varying dual basis $\{m^{TTab(\beta)}(x^c, q^\gamma)\}$ of symmetric TT tensor density fields defined on Σ such that

$$\int_\Sigma m^{TTab(\beta)}k^{TT}_{(\alpha)ab}d^2x = \int_\Sigma m^{TTab(\beta)}\frac{\partial\gamma_{ab}}{\partial q^\alpha}d^2x = \delta^\beta_\alpha. \tag{4.7}$$

In terms of this natural dual basis for the cotanent space to $\mathcal{T}(\Sigma)$, one can express an arbitrary TT tensor density p^{TT} as

$$p^{TTab}(x^c, q^\alpha, p_\alpha) = p_\alpha m^{TTab(\alpha)}(x^c, q^\alpha) \tag{4.8}$$

for suitable coefficients $\{p^\alpha\}$. The coordinates $\{q^\alpha, p_\alpha\}$ may be regarded, as we shall see below, as a (global) canonical chart for $T^*\mathcal{T}(\Sigma)$.

Substituting the foregoing expressions into the reduced action and carrying out the steps displayed in Eq. (2.21) of Ref. [**2**], one arrives at

$$I^*_{ADM} = \int_{\mathcal{I}} dt \left\{ p_\alpha \frac{dq^\alpha}{dt} - \frac{d\tau}{dt} \int_\Sigma \mu_g d^2x \right\} + \int_\Sigma d^2x [\tau \mu_g] \,|_{t_0}^{t_1} \tag{4.9}$$

wherein we recognize the canonical character of the coordinates $\{q^\alpha, p^\alpha\}$. The boundary term on the right hand side of Eq. (4.9) makes no contribution to the equations of motion. We therefore drop it and define

$$I^*_{ADM} \,|_{\text{reduced}} = \int_{\mathcal{I}} dt \left\{ p_\alpha \dot{q}^\alpha - \frac{d\tau}{dt} \int_\Sigma \mu_g d^2x \right\}. \tag{4.10}$$

For our purposes the most natural choice of time function (which differs from that made in Ref. [**2**]) corresponds to setting $t = -\frac{1}{\tau}$, so that $\frac{d\tau}{dt} = \tau^2$, which thus yields a reduced Hamiltonian

$$H^*_{ADM|\text{reduced}} := H_{\text{reduced}} = \tau^2 \int_\Sigma \mu_g d^2x = \tau^2 \int_\Sigma e^{2\lambda} \mu_\gamma d^2x. \tag{4.11}$$

Here H_{reduced} is regarded as a (globally defined) function on $T^*\mathcal{T}(\Sigma) \times \mathbf{R}^+$, $H(q^\alpha, p_\alpha, t)$, determined by expressing

$$\begin{aligned} \gamma_{ab} = \gamma_{ab}(x^c, q^\alpha),\ p^{TTab} &= p^{TTab}(x^c, q^\alpha, p_\alpha) \\ &= p_\alpha m^{TTab(\alpha)}(x^c, q^\alpha), \end{aligned} \tag{4.12}$$

solving the Lichnerowicz equation (2.23) for $\lambda = \lambda(x^c, q^\alpha, p_\alpha, t)$ and carrying out the integral over Σ.

As discussed in Ref. [**2**] the resulting Hamiltonian is independent of the choice of representative cross section used in its construction (e.g., independent of the metric ρ) and describes dynamics on the natural reduced phase space $T^*\mathcal{T}(\Sigma)$ in terms of canonical coordinates $\{q^\alpha, p_\alpha\}$ on that space.

On the other hand, since our ultimate aim is to reconstruct (from solution curves $\{q^\alpha(t), p_\alpha(t)\}$) vacuum metrics on $\Sigma \times \mathbf{R}$ we are more directly interested in the lifts of curves back up to the chosen cross section where they yield evolving sets of ADM data

$$\begin{aligned} g_{ab}(x^c, t) &= e^{2\lambda} \gamma_{ab}(x^c, q^\alpha(t)) \\ \pi^{ab}(x^c, t) &= e^{-2\lambda} p^{TTab}(x^c, q^\alpha(t), p_\alpha(t)) \\ &\quad + \frac{1}{2} \tau(t) (\mu_\gamma \gamma^{ab})(x^c, q^\alpha(t)) \end{aligned} \tag{4.13}$$

expressible in terms of the solution $\lambda = \lambda(x^c, t, q^\alpha(t), p_\alpha(t))$ to Lichnerowicz's equation.

To complete the formula for the spacetime metric we also need the lapse function N and shift field X^a. As discussed in Ref. [**2**] and shown more explicitly in Ref. [**4**], these are uniquely fixed by the elliptic equations defined by the requirements that the gauge conditions be preserved in time. These latter correspond to

$$\tau = \frac{tr_g \pi}{\mu_g} = -\frac{1}{t} \tag{4.14}$$
$$V^c = g^{ab}(\Gamma^c_{ab}(g) - \tilde{\Gamma}^c_{ab}(\rho)) = 0$$

where $\Gamma^c_{ab}(g)$ and $\tilde{\Gamma}^c_{ab}(\rho)$ are the Christoffel connection components of the metrics g and ρ respectively. Computing the t-derivatives of τ and V^c and setting these equal to τ^2 and 0 respectively leads to

$$\tau^2 = \frac{\partial \tau}{\partial t} = -\Delta_g N + \frac{N}{(\mu_g)^2} \pi^b_a \pi^a_b \tag{4.15}$$

which can also be written as

$$e^{2\lambda}\tau^2 = -\Delta_\gamma N + N \left\{ \frac{e^{-2\lambda}}{(\mu_\gamma)^2} \left[\gamma_{ab}\gamma_{cd} p^{TTac} p^{TTbd} + \frac{1}{2} e^{4\lambda} (\mu_\gamma)^2 \tau^2 \right] \right\} \tag{4.16}$$

and

$$0 = -g^{ad} g^{be} h_{de} \left(\Gamma^c_{ab}(g) - \tilde{\Gamma}^c_{ab}(\rho) \right) \tag{4.17}$$
$$+ \frac{1}{2} g^{ab} g^{ce} (h_{ae|b} + h_{be|a} - h_{ab|e})$$

where

$$h_{ab} = \frac{2N}{\mu_g}(\pi_{ab} - g_{ab} tr_g \pi) + X_{a|b} + X_{b|a}. \tag{4.18}$$

The existence and uniqueness of solutions to these equations was established in Ref. [**4**] together with the fact that imposing this choice for $\{N, X^a\}$ suffices to preserve the gauge conditions (4.14).

A remarkable feature of the reduced Hamiltonian, established in Ref. [**6**], is that it monotonically decays for all solutions except the trivial ones corresponding to $p^{TTab} = 0$ for which it stays constant. We shall show later that every solution tends, as $t \to \infty$ (or $\tau \nearrow 0$) to one such that H_{reduced} always achieves its infinum (identified below Eq. (3.7)) in the limit of infinite cosmological expansion.

Given initial data $(\overset{o}{\gamma}_{ab}, \overset{o}{k}{}^{TT}_{ab}, \overset{o}{\tau} = \text{constant } < 0$ with ${}^{(2)}R(\overset{o}{\gamma}) = -1)$ we can set $\lambda^c_a = -\mu_{\overset{o}{\gamma}} \overset{o}{\gamma}{}^{cb} \overset{o}{k}{}^{TT}_{ab}$ (c.f. Eq. (2.27)) and solve the Lichnerowicz equation (2.23) for the function $\overset{o}{\lambda}$ and thereby compute the corresponding target hyperbolic metric $\overset{o}{\rho}_{ab}$ using equation (2.28). Since we know that the identity map from $(\Sigma, \overset{o}{\gamma})$ to $(\Sigma, \overset{o}{\rho})$ is harmonic, it seems natural to now fix the CMCSH gauge precisely by requiring that the evolving conformal metric γ_τ continue to lie in the (Eells, Earle) global cross section of the bundle

$$\mathcal{M}_{-1}(\Sigma) \to \mathcal{M}_{-1}(\Sigma)/\mathcal{D}_0(\Sigma) \approx \mathcal{T}(\Sigma) \tag{4.19}$$

defined by

$$\{\gamma \mid {}^{(2)}R(\gamma) = -1, \mathrm{Id} : (\Sigma, \gamma) \to (\Sigma, \overset{o}{\rho}) \text{ is harmonic}\} \tag{4.20}$$

during the subsequent evolution. Taking into account the conformal invariance of the harmonicity condition with respect to the domain metric, the results of Ref. [**4**] show that this gauge condition (combined with the CMC condition which fixes the lapse) uniquely determines the shift vector field and conversely, that when the shift is fixed by the associated elliptic equation, the CMCSH gauge conditions will continue to hold during the evolution.

However, we could now compute a (potentially) different, target hyperbolic metric ρ_τ on any CMC slice of the subsequent evolution by simply evaluating the right hand side of Eq. (2.19) for the geometric data (g_τ, k_τ, τ) induced on that slice. By construction γ_τ (which can always be recovered from g_τ by uniformization) will satisfy harmonicity of the mapping

$$\mathrm{Id} : (\Sigma, \gamma_\tau) \to (\Sigma, \rho_\tau) \tag{4.21}$$

as well as the CMCSH gauge condition which ensures harmonicity of

$$\mathrm{Id} : (\Sigma, \gamma_\tau) \to (\Sigma, \overset{o}{\rho}). \tag{4.22}$$

But what is the relationship between ρ_τ and $\overset{o}{\rho} := \rho_{\overset{o}{\tau}}$? We shall conclude this section by showing that $\rho_\tau = \overset{o}{\rho}$ and thus that the metric ρ_τ, defined on each slice by Eq. (2.19), is in fact a constant of the motion for the particular CMCSH gauge under consideration.

First recall that, in a flat spacetime, the traces of the holonomics defined via the parallel propagation of vectors around incontractible loops are invariant with respect to arbitrary, continuous deformations of these loops within the spacetime. Thus if one evaluates any collection of such traces on say a given CMC slice and then deforms the chosen loops continuously to corresponding loops in another such slice, one will necessarily obtain the same values for the traces under study. We shall see however, that these same values for the traces may also be computed by a corresponding holonomy calculation carried out in a certain "reference spacetime" $(\Sigma \times \mathbb{R}^+, {}^{(3)}\eta_\tau)$ defined, for any fixed hyperbolic metric ρ_τ of the family described above, by the flat metric

$${}^{(3)}\eta_\tau = -dt \otimes dt + \frac{t^2}{2}(\rho_\tau(x^c))_{ab} dx^a \otimes dx^b. \tag{4.23}$$

But since these traces are independent of the value of τ used to compute $\rho_{\mathcal{T}}$ and since a complete, independent set of (6 genus(Σ) $-$ 6) such traces determines the $\rho_{\mathcal{T}}$ appearing in Eq. (4.23) up to isometry it will follow that $\rho_{\mathcal{T}}$ can only have the form $\rho_{\mathcal{T}} = \varphi_{\mathcal{T}}^{*}\overset{o}{\rho} = \varphi_{\mathcal{T}}^{*}\rho_{\overset{o}{\mathcal{T}}}$ for some (possibly non-trivial) diffeomorphism $\varphi_{\mathcal{T}} : \Sigma \to \Sigma$ defined for each value of τ achieved during the (non-singular) CMCHS evolution. Finally, however, it will follow from the particular choice of CMCHS gauge that we have made to specify that evolution, that $\varphi_{\mathcal{T}} = \mathrm{Id}$ is the only possibility and thus that $\rho_{\mathcal{T}} = \overset{o}{\rho}$ for every allowed value of τ.

For any given CMC slice of the spacetime under study, we can compute the future evolution from that slice in gnc coordinates via Eqs. (2.12)–(2.18). Parallel propagation of a vector v around an arbitrary loop chosen (for convenience) to lie in a surface of constant gaussian time t is determined by solving

$$\frac{dv^{\mu}}{d\lambda} + {}^{(3)}\Gamma^{\mu}_{\alpha\beta}\, v^{\alpha}\frac{dx^{\beta}}{d\lambda} = 0, \tag{4.24}$$

where ${}^{(3)}\Gamma^{\mu}_{\alpha\beta}$ are the Christoffel symbols of the spacetime metric ${}^{(3)}g_{\alpha\beta}$ expressed in the chosen gnc coordinates. Taking $t(\lambda) = t =$ constant and writing out this equation for the rescaled vector $\tilde{v}$ defined by

$$\tilde{v}^{0} = v^{0}, \tilde{v}^{a} = tv^{a} \tag{4.25}$$

one gets

$$\begin{aligned} &\frac{d\tilde{v}^{0}}{d\lambda} + \frac{1}{2}\left(\frac{g_{ab,t}}{t}\right)\tilde{v}^{a}\frac{dx^{b}}{d\lambda} = 0, \\ &\frac{d\tilde{v}^{a}}{d\lambda} + {}^{(2)}\Gamma^{a}_{bc}(g)\tilde{v}^{b}\frac{dx^{c}}{d\lambda} + \frac{1}{2}(t^{2}g^{ad})\left(\frac{g_{cd,t}}{t}\right)\tilde{v}^{0}\frac{dx^{c}}{d\lambda} = 0 \end{aligned} \tag{4.26}$$

where ${}^{(2)}\Gamma^{a}_{bc}(g)$ are the Christoffel symbols of the metric g defined (from the chosen CMC initial data) by Eqs. (2.12)–(2.18).

On an arbitrary t = constant slice these equations are difficult to analyze but since we know that traces of resultant holonomy calculations will automatically be independent of the gnc slice chosen we can evaluate them in the limit as $t \to \infty$ by exploiting the facts (easily derived from Eqs. (2.12)–(2.18)) that

$$\begin{aligned} &\lim_{t\to\infty}\left(\frac{2}{t^{2}}g_{ab}\right) = (\rho_{\mathcal{T}})_{ab} \\ &\lim_{t\to\infty}\left(\frac{t^{2}}{2}g^{ab}\right) = (\rho_{\mathcal{T}})^{ab} \end{aligned} \tag{4.27}$$

$$\lim_{t\to\infty}\left(\frac{g_{ab,t}}{t}\right) = (\rho_{\mathcal{T}})_{ab}$$
$$\lim_{t\to\infty}\ {}^{(2)}\Gamma^a_{bc}(g) = \lim_{t\to\infty}\ {}^{(2)}\Gamma^a_{bc}\left(\frac{2}{t^2}g\right)$$
$$= {}^{(2)}\Gamma^a_{bc}(\rho_{\mathcal{T}})$$

and that the solutions of the linear equations (4.26) vary continuously with the coefficients. But the limiting equations so obtained,

$$\text{(4.28)}\qquad \begin{aligned} &\frac{d\tilde{v}^0}{d\lambda} + \frac{1}{2}(\rho_{\mathcal{T}})_{ab}\tilde{v}^a\frac{dx^b}{d\lambda} = 0,\\ &\frac{d\tilde{v}^a}{d\lambda} +^{(2)}\Gamma^a_{bc}(\rho_{\mathcal{T}})\tilde{v}^b\frac{dx^c}{d\lambda} + \tilde{v}^0\frac{dx^a}{d\lambda} = 0, \end{aligned}$$

are equivalent to the parallel propagation equations one would obtain (at an arbitrary instant of gnc time $t > 0$) for the "reference metric" ${}^{(3)}\eta_{\mathcal{T}}$ given above by Eq. (4.23).

But a flat metric of the form (4.23) defined on $\Sigma \times \mathbb{R}^+$ is isometric to a quotient of the interior of the future light cone of a point in 3-dimensional Minkowski space by a discrete subgroup of the (proper orthochronos) Lorentz group that fixes that point. The subgroup in question must be homomorphic to a representation of the fundamental group $\pi_1(\Sigma)$, of the higher genus surface Σ and can be recovered from $(\Sigma \times \mathbb{R}^+, {}^{(3)}\eta_{\mathcal{T}})$ by the computation of a complete, independent set of $(6\,\text{genus}\,(\Sigma) - 6)$ traces of holonomies. Conversely, a specification of these traces determines the $\rho_{\mathcal{T}}$ needed in the metric form (4.23) up to isometry (i.e., up to the pull back action by a diffeomorphism of Σ).

But this implies, as stated above, that the target metric $\rho_{\mathcal{T}}$ computed from an arbitrary CMC slice in the evolving spacetime $(\Sigma \times \mathbb{R}^+, {}^{(3)}g)$ must satisfy $\rho_{\mathcal{T}} = \varphi^*_{\mathcal{T}}\rho_{\overset{\circ}{\mathcal{T}}} = \varphi^*_{\mathcal{T}}\rho$ for some diffeomorphism $\varphi_{\mathcal{T}} : \Sigma \to \Sigma$ of Σ which, by continuity of the evolution, is necessarily isotopic to the identity. The foregoing implies that the corresponding curve of (uniformized) metric $\gamma_{\mathcal{T}}$ satisfies both (4.21) and (4.22) with $\rho_{\mathcal{T}} = \varphi^*_{\mathcal{T}}\overset{\circ}{\rho}$. But, if $\varphi_{\mathcal{T}} \neq \text{Id}$, the Eells, Earle global cross section of the bundle (4.19) is disjoint from that obtained upon replacing $\overset{\circ}{\rho}$ with $\rho_{\mathcal{T}} = \varphi^*_{\mathcal{T}}\overset{\circ}{\rho}$ since, in view of the covariance of the construction, the latter cross section is obtained from the former by pulling back each metric by the same diffeomorphism. It follows that since $\gamma_{\mathcal{T}}$ satisfies both (4.21) and (4.22), we must have $\varphi_{\mathcal{T}} = \text{Id}$ throughout the evolution.

It is worth mentioning here that the metric ${}^{(3)}g$ takes the special form ${}^{(3)}\eta_{\mathcal{T}}$ (for which the gaussian time slices are also CMC) only if the initial data for ${}^{(3)}g$ satisfies $\overset{\circ}{k}{}^{TT}_{ab} = 0$ which, of course, is generically not the case. Roughly speaking, the "reference metric" ${}^{(3)}\eta_{\mathcal{T}}$ is constructed using only half of the data needed for the specification of the actual spacetime metric ${}^{(3)}g$. There are different ways of invariantly prescribing the "missing" data needed to fully characterize ${}^{(3)}g$. For example, in the Witten approach (cf. Ref [**1**]), one supplements the (linear) holonomies discussed above with certain translational holonomies defined in regarding $(\Sigma \times \mathbb{R}^+, {}^{(3)}g)$ as a quotient of Minkowski space by a suitably chosen discrete subgroup of the full inhomogeneous Lorentz (or Poincaré) group. We shall not pursue that approach here, but instead, following the proposal of Puzio [**7**], will characterize the remaining data in terms of the conserved quantities defined by a complete solution of the associated Hamilton-Jacobi equation.

5. Hamilton-Jacobi theory and the Dirichlet energy

In this section we shall establish a remarkable relationship between the Dirichlet energy for the Gauss map discussed in Sec. 3 and solutions to the Hamilton-Jacobi equation for the reduced Einstein equations discussed in the previous section. In fact, we shall derive a dynamically complete solution to the Hamilton-Jacobi equation by exploiting this relationship and thereby arrive at an implicit formula for the general solution to the reduced Einstein equations in CMCSH gauge.

In Sec. 3 we found that whenever any two metrics γ and ρ (satisfying ${}^{(2)}R(\gamma) = {}^{(2)}R(\rho) = -1$) are related by Eq. (2.28), for some choice of $\tau = \text{constant} < 0, k^{TT}$ a TT tensor with respect to γ and λ determined uniquely by Eq. (2.23), then the identity map from (Σ, γ) to (Σ, ρ) is harmonic and its Dirichlet energy (which depends only upon the conformal class of γ) is expressible as

$$\mathcal{A}(\gamma, \rho, Id) = \frac{1}{2} \int_\Sigma d\mu_\gamma \gamma^{ab} \rho_{ab}. \tag{5.1}$$

Recalling the change of notation defined by Eq. (2.29), one can re-express the relationship between γ and ρ through equations (2.30) and (2.31) which are autonomous relative to τ. In Ref. [**8**], Michael Wolf used these latter equations (with no apparent relativistic or associated Gauss map interpretation) to prove that, for any such fixed metric γ (with ${}^{(2)}R(\gamma) = -1$), one could smoothly parametrize the space of metrics ρ satisfying

(i) ${}^{(2)}R(\rho) = -1$, and
(ii) $\text{Id} : (\Sigma, \gamma) \to (\Sigma, \rho)$ is harmonic

by the space of TT tensors relative to γ (i.e., by the tensors $\tilde{k}^{TT}_{ab}$ appearing in these formulas, normally referred to as holomorphic quadratic differentials in the mathematics literature). More precisely, Wolf showed that the foregoing formulas define a global diffeomorphism between the space of TT tensors defined relative to a fixed γ and the space of uniformized metrics ρ satisfying (i) and (ii) above. Thus for any fixed pair of metrics γ and ρ (each having scalar curvature $= -1$ and satisfying (ii) above) and any choice of $\tau = \text{constant} < 0$ there exists a unique TT tensor k^{TT} such that equations (2.28) and (2.23) hold.

We want to compute the variation of the energy defined by Eq. (5.1) above, holding ρ fixed and allowing γ to vary over that global cross section of $\mathcal{M}_{-1}(\Sigma)$ defined by the requirement that $\mathrm{Id} : (\Sigma, \gamma) \to (\Sigma, \rho)$ be harmonic. Writing $\gamma_{ab} = \gamma_{ab}(x^c, q^\alpha)$ as in the previous section we evaluate the energy $\mathcal{A}(\gamma(q), \rho, \mathrm{Id})$ and compute its partial derivatives with respect to the $\{q^\alpha\}$. The result is

$$\frac{\partial \mathcal{A}(\gamma(q), \rho, \mathrm{Id})}{\partial q^\alpha} = -\frac{1}{2}\int_\Sigma d\mu_\gamma \left(\gamma^{ac}\gamma^{bd} - \frac{1}{2}\gamma^{cd}\gamma^{ab}\right)\rho_{ab}\frac{\partial \gamma_{cd}}{\partial q^\alpha}. \tag{5.2}$$

But substituting the expression (2.28) for ρ (as justified by Wolf's result) one arrives at

$$\begin{aligned} \frac{\partial \mathcal{A}}{\partial q^\alpha}(\gamma(q), \rho, \mathrm{Id}) &= -\int_\Sigma d\mu_\gamma \tau k^{TT}_{ab}\gamma^{ac}\gamma^{bd}\frac{\partial \gamma_{cd}}{\partial q^\alpha} \\ &= -\tau\int_\Sigma p^{TT cd}\frac{\partial \gamma_{cd}}{\partial q^\alpha} = \tau\int_\Sigma p_\beta m^{TT cd(\beta)}\frac{\partial \gamma_{cd}}{\partial q^\alpha} \\ &= \tau p_\alpha \end{aligned} \tag{5.3}$$

where we have used equations (2.4), (4.3), and (4.8) to simplify the result.

It thus follows that the gradient of the rescaled Dirichlet energy function, $\frac{1}{\tau}\mathcal{A}(\gamma(q), \rho, \mathrm{Id})$, with respect to the coordinates $\{q^\alpha\}$ (a global chart for $\mathcal{T}(\Sigma)$) yields precisely the canonical momentum components $\{p_\alpha\}$ such that the vacuum Einstein spacetime determined by the data $\{q^\alpha, p_\alpha, t = -\frac{1}{\tau}\}$ has asymptotic rescaled metric (c.f., Eq. (2.19)) given by ρ. As we saw in Sec. 3, this metric ρ corresponds to the asymptotic conformal geometry invariantly defined by the linear holonomies of the chosen vacuum spacetime.

For later convenience, let us now write T (instead of the more generic t), for the preferred time coordinate $-\frac{1}{\tau}$ and define

$$\mathcal{S}(q^\alpha, \rho, T) = -T\left[\mathcal{A}(\gamma(q), \rho, \mathrm{Id}) - \int_\Sigma d\mu_\gamma\, {}^{(2)}R(\gamma)\right] \tag{5.4}$$

wherein $\int_\Sigma d\mu_\gamma\, {}^{(2)}R(\gamma)$ of course is just the Gauss-Bonnet invariant of Σ. One now has

$$\begin{aligned} p_\alpha &= \frac{\partial \mathcal{S}}{\partial q^\alpha}(q^\alpha, \rho, T), \\ -\frac{\partial \mathcal{S}}{\partial T} &= \mathcal{A}(\gamma(q), \rho, \mathrm{Id}) - \int_\Sigma d\mu_\gamma^{(2)} R(\gamma) \end{aligned} \tag{5.5}$$

but it follows from Eqs. (3.6) and (5.3) that the right hand side of this last equation is equal to $H_{\mathrm{reduced}}(q^\alpha, p_\alpha, T)\Big|_{p_\alpha = \frac{\partial \mathcal{S}}{\partial q^\alpha}}$. In other words, that $\mathcal{S}(q^\alpha, \rho, T)$, for fixed ρ satisfies the Hamilton-Jacobi equation for the reduced Einstein equations in CMCSH gauge

$$-\frac{\partial \mathcal{S}}{\partial T} = H_{\mathrm{reduced}}\left(q^\alpha, \frac{\partial \mathcal{S}}{\partial q^\alpha}, T\right). \tag{5.6}$$

This $\mathcal{S}$ is just a particular solution determined by the chosen target metric ρ but the choice of ρ was arbitrary. We are free to let ρ range over $\mathcal{M}_{-1}(\Sigma)$ but, in view of the $\mathcal{D}_0(\Sigma)$ invariance of the Dirichlet energy functional, there is no essential loss of generality in restricting ρ to lie in a global cross section for the bundle

$$\mathcal{M}_{-1}(\Sigma) \to \mathcal{M}_{-1}(\Sigma)/\mathcal{D}_0(\Sigma) \approx \mathcal{T}(\Sigma) \tag{5.7}$$

and hence in a model for Teichmüller space. Choosing global coordinates $\{Q^\alpha\}$ for this model one can now write, with a slight abuse of notation

$$\mathcal{S}(q^\alpha, Q^\alpha, T) = \mathcal{S}(q^\alpha, \rho(Q), T) \tag{5.8}$$

and regard this function $\mathcal{S}$ as globally defined on $\mathcal{T}(\Sigma) \times \mathcal{T}(\Sigma) \times \mathbf{R}^+$.

To see that this $\mathcal{S}$ is dynamically complete, we note that the coordinates $\{q^\alpha\}$ label the arbitrary (at time $T \in \mathbf{R}^+$) initial conformal geometry which, upon making an arbitrary choice of global cross section, gets represented by the metric $\gamma_{ab}(x^c, q^\alpha)$ lying in that cross section. For fixed γ the freedom to allow the target metric $\rho(Q)$ to vary over an independent global cross section of $\mathcal{M}_{-1}(\Sigma)$ (in particular over all those metrics for which $\mathrm{Id} : (\Sigma, \gamma) \to (\Sigma, \rho)$ is harmonic) provides precisely (via Wolf's diffeomorphism result) the freedom to complement γ_{ab} with an arbitrary TT tensor k^{TT}_{ab}. Thus we get fully general Cauchy data sets $\{\gamma_{ab}, k^{TT}_{ab}, T = -\frac{1}{\tau}\}$ by varying $\{q^\alpha, Q^\alpha\}$ freely.

A well known result in Hamilton-Jacobi theory is that one can derive a complementary set of constants of the motion to the $\{Q^\alpha\}$'s by differentiating the complete solution $\mathcal{S}(q^\alpha, Q^\alpha, T)$ with respect to these Q^α's. More precisely the quantities P_α defined by

$$\begin{aligned} P_\alpha &= -\frac{\partial \mathcal{S}}{\partial Q^\alpha}(q^\alpha, Q^\alpha, T) \\ &= +T\frac{\partial}{\partial Q^\alpha}(\mathcal{A}(\gamma(q), \rho(Q), \mathrm{Id})) \end{aligned} \tag{5.9}$$

are constants of the motion for the solutions of the reduced Hamilton equations and together with the $\{Q^\alpha\}$ form a complete set of canonically conjugate variables that are all constants of the notion. Since the solution curves are now implicitly determined by setting $\{Q^\alpha, P_\alpha\}$ equal to suitable values and solving equations (5.9) for $\{q^\alpha(T, Q, P)\}$ with the p_α's then given by $p_\alpha(T, Q, P) = \frac{\partial \mathcal{S}}{\partial q^\alpha}(q(T, Q, P), Q)$ it is of interest to express these equations more explicitly.

Writing $\rho_{ab}(x^c, Q^\alpha)$ for the ρ_{ab} in Eq. (5.1) and differentiating this formula with respect to the Q^α's yields

$$\frac{P_\alpha}{T} = \frac{1}{2}\int_\Sigma d\mu_\gamma \gamma^{ab}\frac{\partial \rho_{ab}}{\partial Q^\alpha}(x^c, Q^\alpha) \tag{5.10}$$

with $\gamma_{ab}=\gamma_{ab}(x^c,q^\alpha)$ in the above. Since ρ_{ab} is only varying over metrics satisfying ${}^{(2)}R(\rho)= -1$ and in fact only over a global cross section representing Teichmüller space the partial derivatives are always expressible in the form

$$\frac{\partial \rho_{ab}}{\partial Q^\alpha}(x^c,Q^\alpha)=\tilde{\ell}^{TT(\alpha)}_{ab}(x^c,Q^\alpha)+(\mathcal{L}_{{}^{(2)}X(x^c,Q^\alpha)}\rho(x^c,Q^\alpha))_{ab} \tag{5.11}$$

(c.f. Eq. (2.22) of Ref. (2) and associated footnotes) where here the $\tilde{\ell}^{TT(\alpha)}_{ab}$ (x^c,Q^α) provide, at each fixed $\{Q^\alpha\}$ a basis for the TT symmetric tensors with respect to $\rho_{ab}(x^c,Q^\alpha)$ and where the vector fields ${}^{(2)}X(x^c,Q^\alpha)$ depend upon the chosen cross-section but are determined by that choice uniquely. We shall now show that the contributions from these Lie-derivative terms drop out of the formula for P_α and hence are ignorable in the following.

For any ${}^{(2)}X$ we can integrate by parts to get

$$\begin{aligned}\int_\Sigma d\mu_\gamma\gamma^{ab}\,(\mathcal{L}_{{}^{(2)}X}\rho)_{ab} &= -\int_\Sigma(\mathcal{L}_{{}^{(2)}X}\mu_\gamma\gamma^{-1})^{ab}\rho_{ab}\\ &= -\int_\Sigma d\mu_\gamma\left\{\left(\frac{1}{2}\gamma^{cd}\gamma^{ab}-\gamma^{ac}\gamma^{bd}\right)(X_{c|d}+X_{d|c})\rho_{ab}\right\}\\ &= +\int_\Sigma d\mu_\gamma\{[X^{a|b}+X^{b|a}-\gamma^{ab}X^c_{|c}]\rho_{ab}\}\\ &= \int_\Sigma d\mu_\gamma\{[X^{a|b}+X^{b|a}-\gamma^{ab}X^c_{|c}]2\tau k^{TT}_{ab}\}\\ &= 0\end{aligned} \tag{5.12}$$

where we have used Eq. (2.28) for ρ_{ab} in the last step, exploited the tracelessness of $(\mathcal{L}_{{}^{(2)}X}\mu_\gamma\gamma^{-1})^{ab}$ and the transverse-tracelessness of k^{TT}_{ab} with respect to γ_{ab} to complete the reduction. Thus we get

$$\frac{P_\alpha}{T}=\frac{1}{2}\int_\Sigma d\mu_\gamma\gamma^{ab}\tilde{\ell}^{TT(\alpha)}_{ab}(x^c,Q^\alpha)=-\tau P_\alpha \tag{5.13}$$

where $\gamma_{ab}=\gamma_{ab}(x^c,q^\alpha)$ and the $\{\tilde{\ell}^{TT(\alpha)}_{ab}(x^c,Q^\alpha)\}$ yield a basis for the TT tensors with respect to $\rho_{ab}(x^c,Q^\alpha)$.

We shall see below that every solution has the property that

$$\gamma_{ab}(x^c,q^\alpha(T,Q,P))\longrightarrow\rho_{ab}(x^c,Q^\alpha)\text{ as }T\longrightarrow\infty \tag{5.14}$$

which is clearly compatible with Eq. (5.13) above. Indeed this equation contains in principle complete information about the solutions to the reduced Einstein equations but, lacking a more explicit representation for the family of metrics $\gamma_{ab}(x^c,q^\alpha)$ which fill out a cross section defined by the property that $\mathrm{Id}:(\Sigma,\gamma)\to(\Sigma,\rho)$ be harmonic, we cannot convert this result into a very explicit formula for the solution curves. Wolf's diffeomorphism result provides a coordinate system for metrics satisfying this condition but under the restriction that γ is held fixed and ρ varies. We want the opposite. In the following sections we shall show how to derive a complementary result to that of Wolf in which the (Teichmüller) space of metrics γ satisfying harmonicity of $\mathrm{Id} : (\Sigma, \gamma) \to (\Sigma, \rho)$ for fixed ρ is globally parametrized by the space of TT tensors with respect to the fixed metric ρ. This will lead us to a more explicit representation of the solution curves but one which necessitates the solution of an associated Monge-Ampêre equation.

6. Analysis of the Gauss Map equation

As discussed in Sec. 2 the Gauss map equation takes the form

$$\rho_{ab} = (k_c^d k_d^c) g_{ab} + 2(tr_g k)\left(k_{ab} - \frac{1}{2} g_{ab} tr_g k\right) \tag{6.1}$$

where g_{ab} and $k_{ab} = g_{bc}k_a^c$ are respectively the first and second fundamental forms induced on a CMC slice in a vacuum Einstein spacetime with Cauchy surfaces diffeomorphic to a higher genus surface Σ. The momentum constraint on the Cauchy data (g_{ab}, k_{ab}) is equivalent (when the mean curvature $tr_g k := \tau$ is constant as we have assumed) to harmonicity of the identity mapping from (Σ, g) to (Σ, ρ) and the Hamiltonian constraint is equivalent to the equation ${}^{(2)}R(\rho) = -1$ which we also assume imposed.

By the uniformization theorem we can always set $g_{ab} = e^{2\lambda}\gamma_{ab}$ for some uniquely determined metric γ which satisfies $R(\gamma) = -1$ and uniquely determined smooth function λ. When $\tau = tr_g k = g^{ab}k_{ab}$ is constant the traceless part of k_{ab} defined by

$$k_{ab}^{TT} = k_{ab} - \frac{1}{2} g_{ab} g^{cd} k_{cd} \tag{6.2}$$

is in fact transverse-traceless (by the momentum constraint) with respect to g or, by the conformal invariance of this condition, with respect to any metric conformal to g such as γ. Thus we can rewrite the Gauss map equation in the form

$$\rho_{ab} = \left(e^{-2\lambda}\gamma^{ce}\gamma^{df} k_{cd}^{TT} k_{ef}^{TT} + \frac{1}{2}\tau^2 e^{2\lambda}\right)\gamma_{ab} + 2\tau k_{ab}^{TT} \tag{6.3}$$

where

$$\begin{gathered} {}^{(2)}R(\rho) = -1, {}^{(2)}R(\gamma) = -1 \\ \tau = tr_g k = g^{ab} k_{ab} = \text{constant} \end{gathered} \tag{6.4}$$

and where

$$k_{ab}^{TT} = k_{ab} - \frac{1}{2} g_{ab} tr_g k \tag{6.5}$$

is transverse-traceless with respect to γ_{ab}. The Hamiltonian constraint corresponds to the satisfaction of Lichnerowicz's equation by the conformal factor λ.

In the gauge we have chosen, the metric ρ, which is determined up to isometry by the linear holonomies of the vacuum spacetime under study, remains fixed while the metric γ evolves within a (global) cross-section of the space $\mathcal{M}_{-1}(\Sigma)$ which represents the Teichmüller space $\mathcal{T}(\Sigma) \approx \mathcal{M}_{-1}(\Sigma)/\mathcal{D}_0(\Sigma)$. The cross-section in question is the smooth submanifold of $\mathcal{M}_{-1}(\Sigma)$ consisting of those metrics γ' such that the identity map from (Σ, γ') to (Σ, ρ) is harmonic. In this same (CMCSH) gauge τ plays the role of time and labels the CMC slices of a global foliation of the spacetime. The lapse function N and shift field X are uniquely determined throughout the evolution by the elliptic equations (4.15) and (4.17) discussed in Sec. 4.

Since ρ_{ab} is fixed during evolution in the chosen gauge and since λ is uniquely determined in terms of $(\gamma_{ab}, k_{ab}^{TT}, \tau)$ from Lichnerowicz's equation it appears that the curve of TT tensors k_{ab}^{TT} should be determined, via Eq. (6.3), from the curve of metrics γ_{ab} evolving within the chosen cross-section of $\mathcal{M}_{-1}(\Sigma)$. To see this more explicitly, compute

$$\begin{aligned} tr_{\gamma}\rho &:= \gamma^{ab}\rho_{ab} \\ &= 2\left(e^{-2\lambda}\left|k^{TT}\right|_{\gamma}^{2} + \frac{1}{2}\tau^{2}e^{2\lambda}\right) \end{aligned} \tag{6.6}$$

where

$$\left|k^{TT}\right|_{\gamma}^{2} = \gamma^{cd}\gamma^{ef}k_{ce}^{TT}k_{df}^{TT} \tag{6.7}$$

and rewrite Eq. (6.3) in the form

$$\rho_{ab}^{tr} := \rho_{ab} - \frac{1}{2}\gamma_{ab}\gamma^{ef}\rho_{ef} = 2\tau k_{ab}^{TT}. \tag{6.8}$$

Thus, up to a factor of 2τ, k_{ab}^{TT} is simply the traceless part, ρ_{ab}^{tr}, of the fixed metric ρ_{ab} computed with respect to the moving metric γ_{ab}. Rewriting Eq. (6.6) in this notation yields

$$tr_{\gamma}\rho := \gamma^{ab}\rho_{ab} = \tau^{2}e^{2\lambda} + \frac{\left|\rho^{tr}\right|_{\gamma}^{2}e^{-2\lambda}}{2\tau^{2}} \tag{6.9}$$

where

$$\left|\rho^{tr}\right|_{\gamma}^{2} = \gamma^{df}\gamma^{ce}\rho_{cf}^{tr}\rho_{de}^{tr}. \tag{6.10}$$

Solving the associated quadratic equation for $e^{2\lambda}$ then gives

$$\tau^{2}e^{2\lambda} = \frac{tr_{\gamma}\rho + \sqrt{(tr_{\gamma}\rho)^{2} - 2|\rho^{tr}|_{\gamma}^{2}}}{2} \tag{6.11}$$

which shows how λ is determined by ρ and the moving metric γ. Recalling Eq. (3.5) we also have

$$\mu_\rho = \sqrt{\det \rho_{ab}} = -\mu_g {}^{(2)}R(g) \tag{6.12}$$

and hence

$$ {}^{(2)}R(g) = -\frac{\mu_\rho}{\mu_g} = -\frac{\mu_\rho e^{-2\lambda}}{\mu_\gamma}. \tag{6.13}$$

It was argued in Sec. 5 that the motion of γ within the chosen cross section of $\mathcal{M}_{-1}(\Sigma)$ is determined implicitly by the equation

$$-\tau P_\alpha = \frac{1}{2}\int_\Sigma d\mu_\gamma\, \gamma^{ab}\, \tilde{\ell}^{TT}_{ab(\alpha)} \tag{6.14}$$

where $\{\tilde{\ell}^{TT}_{(\alpha)} \mid \alpha = 1, \ldots, 6\,\text{genus}(\Sigma) - 6\}$ is a fixed basis of TT tensors with respect to ρ_{ab} and $\{P_\alpha\}$ is a set of 6 genus $(\Sigma) - 6$ arbitrary constants which, together with the complimentary 6 genus $(\Sigma) - 6$ independent holonomies that determine ρ_{ab}, form a complete set of constants of the motion determining a vacuum spacetime. To convert this formula into a more explicit characterization of the motion of γ it is essential to develop a more explicit characterization of the cross-section of metrics in which γ is moving.

We begin by setting:

$$\zeta^b_a := \mu_g k^{TTb}_a = \mu_g g^{bc} k^{TT}_{ac} \tag{6.15}$$

and solving Eq. (6.3) algebraically for the (inverse-) metric g^{ab}. This solution can be expressed as

$$\mu_g g^{a\ell} = 2\tau\rho^{ab}\zeta^\ell_b + \sqrt{1 + \frac{2\tau^2|\zeta|^2}{(\mu_\rho)^2}}\,\mu_\rho\rho^{a\ell} \tag{6.16}$$

$$\tau^2\mu_g = \mu_\rho\left(1 + \sqrt{1 + \frac{2\tau^2|\zeta|^2}{(\mu_\rho)^2}}\right)$$

where

$$|\zeta|^2 = \zeta^f_e \zeta^e_f. \tag{6.17}$$

Now, ζ^b_a is required to be transverse traceless (and symmetric) with respect to g_{ab} (or any metric γ_{ab} conformal to g_{ab}) but we should like to re-express these conditions on ζ^b_a relative to the fixed metric ρ_{ab}. First of all note that

$$\mu_g g^{a\ell}\zeta^m_a = 2\tau\rho^{ab}\zeta^\ell_b\zeta^m_a + \sqrt{1 + \frac{2\tau^2|\zeta|^2}{(\mu_\rho)^2}}\,\mu_\rho\rho^{a\ell}\zeta^m_a \tag{6.18}$$

so that, since the middle term is automatically symmetric, it follows that the contravariant form of ζ_a^m relative to the g metric is symmetric if and only if its contravariant form relative to the ρ metric is symmetric. The condition that ζ_a^m be transverse is equivalent (when $\tau =$ constant) to the momentum constraint and this in turn, as we saw above, is equivalent to the requirement that the identity map from (Σ, g) to (Σ, ρ) be harmonic. One can express this latter formulation as the requirement that

$$\tilde{\nabla}_b(\rho)(\mu_g g^{ab}) = 0 \tag{6.19}$$

where $\tilde{\nabla}_b(\rho)$ signifies covariant differentiation with respect to ρ_{ab}. Computing the ρ-divergence of Eq. (6.16) and imposing the condition (6.19) leads immediately to

$$2\tau\tilde{\nabla}_\ell(\rho)\left(\frac{\zeta_b^\ell}{\mu_\rho}\right) + \partial_b\sqrt{1 + \frac{2\tau^2|\zeta|^2}{(\mu_\rho)^2}} = 0 \tag{6.20}$$

as a differential condition on ζ_a^m which only involves the metric ρ_{ab}. Thus ζ_a^m is required to be a traceless tensor density which is both symmetric with respect to ρ and satisfies Eq. (6.20).

To solve this equation we first decompose ζ_b^ℓ into a transverse-traceless summand (relative to ρ) and an L^2-orthogonal conformal Killing form

$$\zeta_b^\ell = \zeta_b^{TT\ell} + \mu_\rho\rho_{ab}(\tilde{\nabla}^a(\rho)Y^\ell + \tilde{\nabla}^\ell(\rho)Y^a - \rho^{a\ell}\tilde{\nabla}_m(\rho)Y^m) \tag{6.21}$$

where $\tilde{\nabla}^m(\rho) = \rho^{mr}\tilde{\nabla}_r(\rho)$.

Recalling that ρ has constant curvature (since ${}^{(2)}R(\rho) = -1$) one finds readily that, in terms of the foregoing expansion of ζ_b^ℓ

$$\tilde{\nabla}_\ell(\rho)\left(\frac{\zeta_b^\ell}{\mu_\rho}\right) = \tilde{\nabla}_\ell(\rho)\tilde{\nabla}^\ell(\rho)(\rho_{bc}Y^c) - \frac{1}{2}\rho_{bc}Y^c. \tag{6.22}$$

Now the vector field Y^c has an L^2-orthogonal expansion of the form

$$Y^c = Y^{tr_c} + \rho^{cd}\Lambda_{,d} \tag{6.23}$$

where $\tilde{\nabla}_c(\rho)Y^{tr_c} = 0$ but it is straightforward to show that the second order elliptic operator on the right hand side of Eq. (6.22) maps divergence free vectors to the space orthogonal to gradients and furthermore, that this operator has trivial kernel within the space of divergence free vector fields.

Thus the only possible solutions of Eq. (6.20) must be expressible in the form of Eq. (6.21) where however, $Y^c = \rho^{cd}\Lambda_{,d}$ for some function Λ.

Substituting this result into Eq. (6.20) one gets the following equation for Λ

$$\tilde{\nabla}_\ell(\rho)\left(\frac{\zeta_b^\ell}{\mu_\rho}\right) = \tilde{\nabla}_\ell(\rho)\tilde{\nabla}^\ell(\rho)(\Lambda_{,b}) - \frac{1}{2}\Lambda_{,b} \tag{6.24}$$
$$= -\frac{1}{2\tau}\partial_b\sqrt{1+\frac{2\tau^2|\zeta|^2}{(\mu_\rho)^2}}$$

where

$$\frac{\zeta_b^\ell}{\mu_\rho} = \left(\frac{\zeta_b^{TT\ell}}{\mu_\rho}\right) + 2\tilde{\nabla}_b(\rho)\tilde{\nabla}^\ell(\rho)\Lambda - \delta_b^\ell(\tilde{\nabla}_r(\rho)\tilde{\nabla}^r(\rho)\Lambda). \tag{6.25}$$

Fortunately, one easily proves again using the constancy of curvature of ρ, that

$$\tilde{\nabla}_\ell(\rho)\tilde{\nabla}^\ell(\rho)\Lambda_{,b} = \partial_b\left[\tilde{\nabla}_\ell(\rho)\tilde{\nabla}^\ell(\rho)\Lambda - \frac{1}{2}\Lambda\right] \tag{6.26}$$

so that Eq. (6.24) becomes

$$\partial_b\left\{\tilde{\nabla}_\ell(\rho)\tilde{\nabla}^\ell(\rho)\Lambda - \Lambda + \frac{1}{2\tau}\sqrt{1+\frac{2\tau^2|\zeta|^2}{(\mu_\rho)^2}}\right\} = 0. \tag{6.27}$$

Thus we must have

$$2\tau(\tilde{\nabla}_\ell(\rho)\tilde{\nabla}^\ell(\rho)\Lambda - \Lambda) + \sqrt{1+\frac{2\tau^2|\zeta|^2}{(\mu_\rho)^2}} = C = \text{constant} \tag{6.28}$$

where C could only depend upon τ at most. Note however that, if we define

$$\Lambda^\dagger = 2\tau\Lambda + C \tag{6.29}$$

then $\Lambda^\dagger$ satisfies

$$\tilde{\nabla}_\ell(\rho)\tilde{\nabla}^\ell(\rho)\Lambda^\dagger - \Lambda^\dagger + \sqrt{1+\frac{1}{2}\left|\frac{2\tau\zeta}{\mu_\rho}\right|^2} = 0 \tag{6.30}$$

where

$$2\tau\frac{\zeta_b^\ell}{\mu_\rho} = \left(\frac{2\tau\zeta_b^{TT\ell}}{\mu_\rho}\right) + 2\tilde{\nabla}_b(\rho)\tilde{\nabla}^\ell(\rho)\Lambda^\dagger - \delta_b^\ell\left(\tilde{\nabla}_m(\rho)\tilde{\nabla}^m(\rho)\Lambda^\dagger\right) \tag{6.31}$$

and wherein the undetermined C plays no role.

Equation (6.30) is, in view of the Hessian of $\Lambda^\dagger$ which appears in the expression (6.31) for ζ^ℓ_b, a fully non-linear equation for $\Lambda^\dagger$. We shall show however that it always has a unique, smooth solution for an arbitrary choice of the TT tensor $\left(\frac{\zeta_b^{TT\ell}}{\mu_\rho}\right)$ which therefore parametrizes the space of solutions $\Lambda^\dagger$ and hence, through Eq. (6.31), $\left(\frac{\zeta_b^\ell}{\mu_\rho}\right)$.

Let us first, however, show that the density $\zeta_b^{TT\,\ell}$ is necessarily a constant of the motion (i.e., satisfies $\partial_\tau\zeta_b^{TT\,\ell}=0$) for any solution of the reduced field equations. This follows from substituting the expression for $\mu_\gamma\gamma^{a\ell}=\mu_g g^{a\ell}$ given by Eq. (6.16) into Eq. (6.14), expanding out ζ^ℓ_b through the use of Eq. (6.21) and then exploiting the transverse-traceless character of $\tilde{\ell}^{TT}_{a\ell(\alpha)}$ (with respect to the fixed metric ρ_{ab}) to finally derive that

$$-P_\alpha = \int_\Sigma \tilde{\ell}^{TT}_{a\ell(\alpha)}\rho^{ab}\zeta_b^{TT\,\ell}. \tag{6.32}$$

Thus the components of $\zeta_b^{TT\,\ell}$ in the basis defined by the time-independent family of (transverse-traceless) tensors $\{\tilde{\ell}^{TT}_{(\alpha)} \mid \alpha = 1, \dots 6\,\mathrm{genus}\,(\Sigma) - 6\}$ are simply the constants of the motion $\{-P_\alpha\}$ identified previously.

7. Some basic properties of the $\Lambda^\dagger$ equation

At a point $x_{\min}\in\Sigma$ where a solution $\Lambda^\dagger$ to equation (6.30) achieves its minimum value $\Lambda^\dagger_{\min}$ one evidently has (since $\tilde{\nabla}_\ell(\rho)\tilde{\nabla}^\ell(\rho)\Lambda^\dagger(x_{\min})\geq 0$)

$$\Lambda^\dagger_{\min} = \Lambda^\dagger(x_{\min}) \geq \sqrt{1+\frac{1}{2}\left|\frac{2\tau\zeta}{\mu_\rho}\right|^2}(x_{\min}) \geq 1. \tag{7.1}$$

Thus $\Lambda^\dagger_{\min}\geq 1$ for any solution. On the other hand, if we square equation (6.30) and integrate over Σ, we obtain the integral formula

$$\int_\Sigma \left(\tilde{\nabla}_\ell(\rho)\tilde{\nabla}^\ell(\rho)\Lambda^\dagger - \Lambda^\dagger\right)^2 d\mu_\rho = \int_\Sigma \left(1+\frac{1}{2}\left|\frac{2\tau\zeta}{\mu_\rho}\right|^2\right) d\mu_\rho \tag{7.2}$$

which is also satisfied by an arbitrary solution. Upon substituting the expression (6.31) for $\frac{2\tau\zeta}{\mu_\rho}$ and simplifying the result through the use of ${}^{(2)}R(\rho)=-1$ and the transverse-traceless character of $\zeta_b^{TT_\ell}$ one easily derives

$$\int_\Sigma \left\{[(\Lambda^\dagger)^2-1] + (\tilde{\nabla}_\ell(\rho)\Lambda^\dagger)(\tilde{\nabla}^\ell(\rho)\Lambda^\dagger)\right\} d\mu_\rho = 2\tau^2\int_\Sigma \left(\frac{\zeta_b^{TT_\ell}}{\mu_\rho}\frac{\zeta_\ell^{TT_b}}{\mu_\rho}\right) d\mu_\rho \tag{7.3}$$

from which, rather surprisingly, all second derivatives have canceled.

An immediate consequence of Eq. (7.3) is that if either $\tau = 0$ or $\tau \neq 0$ but $\zeta_b^{TT\ell} = 0$, so that the right hand side vanishes, then since $\Lambda^\dagger \geq 1$ a priori, we get that $\Lambda^\dagger = 1$ identically on Σ. We are interested in curves of solutions to Eq. (6.30) determined by data $\{\rho_{ab}, \zeta_b^{TT\ell}\}$ specified on Σ and parametrized by $\tau \in [0, -\infty)$. The above result has shown that any such curve will necessarily satisfy $\Lambda^\dagger|_{\tau=0} = 1$ on Σ.

To establish the uniqueness of solutions to the $\Lambda^\dagger$ equation (for arbitrary but fixed ρ_{ab}, $\zeta_a^{TT\ell}$, and τ) assume that $\Lambda^\dagger = \theta$ and $\Lambda^\dagger = \psi$ are any two such solutions (corresponding to the same given data) and consider the curve of functions $\chi_t = t\theta + (1-t)\psi$ for $0 \leq t \leq 1$. Clearly $\chi_0 = \psi, \chi_1 = \theta$ and $\frac{d\chi}{dt} = \theta - \psi$. Noting that

$$\int_0^1 dt \frac{d}{dt}\left\{\tilde{\nabla}_\ell(\rho)\tilde{\nabla}^\ell(\rho)\chi_t - \chi_t + \sqrt{1 + \frac{1}{2}\left|\frac{2\tau\zeta_t}{\mu_\rho}\right|^2}\right\} \tag{7.4}$$
$$= \left[\tilde{\nabla}_\ell(\rho)\tilde{\nabla}^\ell(\rho)\chi_t - \chi_t + \sqrt{1 + \frac{1}{2}\left|\frac{2\tau\zeta_t}{\mu_\rho}\right|^2}\right]\Bigg|_{t=0}^{t=1} = 0$$

(where ζ_t is the expression for ζ with $\Lambda^\dagger$ replaced by χ_t) we carry out the t-differentiation explicitly on the left hand side and express the result as

$$\int_0^1 dt \left\{\rho^{\ell m} + \rho^{mb} \frac{\left(\frac{2\tau}{\mu_\rho}\zeta_t\right)_b^\ell}{\sqrt{1 + \frac{1}{2}\left(\frac{2\tau}{\mu_\rho}\zeta_t\right)^2}}\right\} \tilde{\nabla}_\ell(\rho)\tilde{\nabla}_m(\rho)(\theta - \psi) - (\theta - \psi) = 0. \tag{7.5}$$

Our aim is to show that this is an elliptic equation for $(\theta - \psi)$ of the form

$$\bar{g}^{\ell m}\tilde{\nabla}_\ell(\rho)\tilde{\nabla}_m(\rho)(\theta - \psi) - (\theta - \psi) = 0 \tag{7.6}$$

where $\bar{g}^{\ell m}$ is a positive definite (inverse) metric on Σ. It will then follow from an elementary maximum principle argument that both $\theta - \psi \geq 0$ on Σ and $\psi - \theta \geq 0$ on Σ from which one thus gets that $\theta = \psi$ on Σ and hence that solutions are unique.

The set of (inverse) Riemannian metrics on Σ is an open cone in the set of symmetric tensor fields so that to show that

$$\bar{g}^{\ell m} := \int_0^1 dt \left\{\rho^{\ell m} + \rho^{mb} \frac{\left(\frac{2\tau}{\mu_\rho}\zeta_t\right)_b^\ell}{\sqrt{1 + \frac{1}{2}\left(\frac{2\tau}{\mu_\rho}\zeta_t\right)^2}}\right\} \tag{7.7}$$

is Riemannian it suffices to show that

$$\bar{g}_t^{\ell m} := \rho^{\ell m} + \frac{\frac{2\tau\zeta_t^{\ell m}}{\mu_\rho}}{\sqrt{1 + \frac{1}{2}\left|\frac{2\tau\zeta_t}{\mu_\rho}\right|^2}} \tag{7.8}$$

is positive definite for each $t \in [0,1]$ (where here we write $\zeta_t^{\ell m}$ for $\rho^{mb}(\zeta_t)^{\ell}_{b}$). At any point of Σ let $v^a = \rho^{ab} v_b$ be an arbitrary unit vector with respect to ρ (i.e., $\rho_{ab} v^a v^b = 1$) and compute $\bar{g}_t^{\ell m} v_\ell v_m$ using the formula above. For convenience choose coordinates so that $\rho^{\ell n} = \delta^{\ell n}$ at the chosen point and, if needed, make a further, orthogonal transformation to diagonalize the (real-symmetric-traceless) matrix $\frac{2\tau\zeta_t^{\ell m}}{\mu_\rho}$ at the chosen point, writing the result as

$$\left(\frac{2\tau\zeta_t^{\ell n}}{\mu_\rho}\right) = \begin{pmatrix} u & 0 \\ 0 & -u \end{pmatrix} \tag{7.9}$$

with $\rho^{ab} v_a v_b = v_1^2 + v_2^2 = 1$ at the chosen point. Evaluating $\bar{g}_t^{\ell n} v_\ell v_n$ at this point we get

$$\bar{g}_t^{\ell m} v_\ell v_m = 1 + \frac{u(v_1^2 - v_2^2)}{\sqrt{1+u^2}}. \tag{7.10}$$

Since $\left|v_1^2 - v_2^2\right| \leq 1$ and $|\frac{u}{\sqrt{1+u^2}}| < 1$ it follows that $\bar{g}_t^{\ell n}$ is positive definite at the chosen point which was an arbitrary point of Σ.

The foregoing calculation is also directly relevant to establishing the ellipticity of the equation for $\Lambda^\dagger$, a result we shall need for proving the existence of solutions via the method for continuity. To see this define, for fixed $\{\rho_{ab}, \tau, \zeta_b^{TT\ell}\}$, the non-linear operator

$$\mathcal{F}(\cdot, \Lambda^\dagger) = \tilde{\nabla}_\ell(\rho)\tilde{\nabla}^\ell(\rho)\Lambda^\dagger - \Lambda^\dagger + \sqrt{1 + \frac{1}{2}\left|\frac{2\tau\zeta}{\mu_\rho}\right|^2} \tag{7.11}$$

where, as before,

$$\frac{2\tau\zeta_b^\ell}{\mu_\rho} = \frac{2\tau\zeta_b^{TT\ell}}{\mu_\rho} + 2\tilde{\nabla}_b(\rho)\tilde{\nabla}^\ell(\rho)\Lambda^\dagger - \delta_b^\ell\tilde{\nabla}_m(\rho)\tilde{\nabla}^m(\rho)\Lambda^\dagger \tag{7.12}$$

and compute the first variation about an arbitrary C^2 configuration $\Lambda^\dagger$. Designating the variation of $\Lambda^\dagger$ by $\delta\Lambda^\dagger$ one gets

$$\begin{aligned} D\mathcal{F}(\cdot, \Lambda^\dagger) \cdot \delta\Lambda^\dagger &= \left\{\rho^{\ell m} + \frac{\frac{2\tau\zeta^{\ell m}}{\mu_\rho}}{\sqrt{1 + \frac{1}{2}\left|\frac{2\tau\zeta}{\mu_\rho}\right|^2}}\right\}\tilde{\nabla}_\ell(\rho)\tilde{\nabla}_m(\rho)\delta\Lambda^\dagger - \delta\Lambda^\dagger \\ &= \tilde{g}^{\ell m}\tilde{\nabla}_\ell(\rho)\tilde{\nabla}_m(\rho)\delta\Lambda^\dagger - \delta\Lambda^\dagger \end{aligned} \tag{7.13}$$

from which the ellipticity (i.e., injectivity of the principle symbol) follows from the calculation done just above which showed that

$$\tilde{g}^{\ell m} = \rho^{\ell m} + \frac{\frac{2\tau\zeta^{\ell m}}{\mu_\rho}}{\sqrt{1 + \frac{1}{2}\left|\frac{2\tau\zeta}{\mu_\rho}\right|^2}} \tag{7.14}$$

is Riemannian.

Since the operator $\tilde{g}^{\ell m}\tilde{\nabla}_\ell(\rho)\tilde{\nabla}_m(\rho)$ clearly plays an important role in our analysis it is of some interest to explore its relationship to other natural elliptic operators arising in this context. Note that, for any arbitrary C^2-function λ we can write, using Eq. (6.16)

$$\mu_g g^{a\ell}\lambda_{,a} = 2\tau\rho^{ab}\lambda_{,a}\zeta^\ell_b + \sqrt{1 + \frac{1}{2}\left|\frac{2\tau\zeta}{\mu_\rho}\right|^2}\,\mu_\rho\rho^{a\ell}\lambda_{,a}. \tag{7.15}$$

Since $\mu_g g^{a\ell}\lambda_{,a}$ is a vector density its ordinary divergence $\partial_\ell(\mu_g g^{a\ell}\lambda_{,a})$ can be identified with its covariant divergence relative to the ρ metric whence

$$\begin{aligned} \tilde{\nabla}_\ell(\rho)[\mu_g g^{a\ell}\lambda_{,a}] &= \partial_\ell[\mu_g g^{a\ell}\lambda_{,a}] = \mu_g(\tilde{\nabla}_\ell(g)\tilde{\nabla}^\ell(g)\lambda) = \mu_g\Delta_g\lambda \\ &= \sqrt{1 + \frac{1}{2}\left|\frac{2\tau\zeta}{\mu_\rho}\right|^2}\,\mu_\rho\tilde{\Delta}_\rho\lambda + 2\tau\rho^{ab}\zeta^\ell_b\left[\tilde{\nabla}_\ell(\rho)\tilde{\nabla}_a(\rho)\lambda\right] \end{aligned} \tag{7.16}$$

where we have used Eq. (6.20) to simplify the ρ-divergence of the right hand side of Eq. (7.16) above. Rewriting this result as

$$\frac{\mu_g}{\mu_\rho\sqrt{1 + \frac{1}{2}\left|\frac{2\tau\zeta}{\mu_\rho}\right|^2}}\Delta_g\lambda = \left\{\rho^{\ell m} + \frac{\frac{2\tau\zeta^{\ell m}}{\mu_\rho}}{\sqrt{1 + \frac{1}{2}\left|\frac{2\tau\zeta}{\mu_\rho}\right|^2}}\right\}\tilde{\nabla}_\ell(\rho)\tilde{\nabla}_m(\rho)\lambda \tag{7.17}$$

which can be further re-expressed using the formula

$$\frac{\tau^2\mu_g}{\mu_\rho} = 1 + \sqrt{1 + \frac{1}{2}\left|\frac{2\tau\zeta}{\mu_\rho}\right|^2}, \tag{7.18}$$

which follows from Eq. (6.16), one thus finally has

$$\left(\frac{1 + \sqrt{1 + \frac{1}{2}\left|\frac{2\tau\zeta}{\mu_\rho}\right|^2}}{\sqrt{1 + \frac{1}{2}\left|\frac{2\tau\zeta}{\mu_\rho}\right|^2}}\right)\left(\frac{1}{\tau^2}\Delta_g\lambda\right) = \tilde{g}^{\ell m}\tilde{\nabla}_\ell(\rho)\tilde{\nabla}_m(\rho)\lambda \tag{7.19}$$

as a formula relating these basic elliptic operators.

8. Existence of solutions to the $\Lambda^\dagger$ equation

To establish the existence of solutions to the $\Lambda^\dagger$ equation, we first need to specify more precisely the function spaces for which the equation is defined. We shall then apply the standard method of continuity to show that for any fixed metric ρ and TT tensor density ζ_a^{TTb} there exists a unique solution $\Lambda^\dagger$ for all τ in the interval $[0, -\infty)$.

Let $H^s(\Sigma)$ designate the Sobolev space of square integrable functions on Σ having square integrable (distributional) derivatives up to order s and let $\mathcal{M}^s(\Sigma)$ represent the corresponding space of H^s-Riemannian metrics on Σ. In the following, we shall assume that $\rho \in \mathcal{M}^s(\Sigma)$ for $s > 4$ and, as usual, also require that ${}^{(2)}R(\rho) = -1$ on Σ. For any such ρ let $\mathcal{S}_d^{TT_{s-1}}(\Sigma)$ designate the (finite dimensional) space of H^{s-1} tensor densities of type $(1,1)$ which, relative to the chosen ρ, are symmetric (i.e., satisfy $\rho^{ab}\zeta_b^{TTc} = \rho^{cb}\zeta_b^{TTa}$), transverse and traceless on Σ. In addition, consider for the same chosen $s > 4$, functions $\Lambda^\dagger \in H^{s+1}(\Sigma)$. The Schauder ring property of H^s maps in 2 dimensions then guarantees that the map

$$\mathcal{F} : \mathcal{S}_d^{TT_{s-1}}(\Sigma) \times H^{s+1}(\Sigma) \to H^{s-1}(\Sigma) \tag{8.1}$$

$$\left(\zeta^{TT}, \Lambda^\dagger\right) \mapsto \tilde{\nabla}_\ell(\rho)\tilde{\nabla}^\ell(\rho)\Lambda^\dagger - \Lambda^\dagger + \sqrt{1 + \frac{1}{2}\left|\frac{2\tau\zeta}{\mu_\rho}\right|^2}$$

where, as above,

$$\frac{2\tau\zeta_b^\ell}{\mu_\rho} = \frac{2\tau\zeta_b^{TT\ell}}{\mu_\rho} + 2\tilde{\nabla}_b(\rho)\tilde{\nabla}^\ell(\rho)\Lambda^\dagger - \delta_b^\ell\left(\tilde{\nabla}_m(\rho)\tilde{\nabla}^m(\rho)\Lambda^\dagger\right) \tag{8.2}$$

is a smooth (i.e., C^∞) map between the indicated function spaces. Here, as before, τ is a real constant in the range $[0, -\infty)$.

In Sec. 7 we showed that the equation $\mathcal{F}(\zeta^{TT}, \Lambda^\dagger) = 0$ has the unique solution $\Lambda^\dagger = 1$ when $\tau\zeta^{TT} = 0$, on Σ and that the equation was elliptic at an arbitrary configuration which is sufficiently smooth (the latter being here guaranteed by our Hilbert space assumptions for ρ, ζ^{TT} and $\Lambda^\dagger$). We want to appeal to the (Banach space version of the) implicit function theorem to show that a solution $\Lambda^\dagger$ to $\mathcal{F}(\zeta^{TT}, \Lambda^\dagger) = 0$ is implicitly determined in terms of $\tau\zeta^{TT}$ on some neighborhood of any given solution. We already know that any such solution must be unique so we need only check that the Frechet derivative of $\mathcal{F}$ with respect to its second argument $\Lambda^\dagger$ defines (at any "background" configuration $(\tau\zeta^{TT}, \Lambda^\dagger) \in \mathcal{S}_d^{TT_{s-1}}(\Sigma) \times H^{s+1}(\Sigma)$) an isomorphism of the function spaces $H^{s+1}(\Sigma)$ and $H^{s-1}(\Sigma)$, i.e., that

$$D_2\mathcal{F}(\zeta^{TT}, \Lambda^\dagger) \cdot \delta\Lambda^\dagger = \sigma \in H^{s-1}(\Sigma) \tag{8.3}$$

is uniquely solvable for $\delta\Lambda^\dagger \in H^{s+1}(\Sigma)$ for arbitrary $\sigma \in H^{s-1}(\Sigma)$. Here $D_2\mathcal{F}$ is given by the first variation formula (7.13). At a background for which $\tau\zeta^{TT} = 0$ (so that $\Lambda^\dagger = 1$) this equation reduces to

$$\rho^{\ell m}\tilde{\nabla}_\ell(\rho)\tilde{\nabla}_m(\rho)\delta\Lambda^\dagger - \delta\Lambda^\dagger = \sigma \tag{8.4}$$

but, for any H^{s+1} metric ρ, the operator $\Delta_\rho - 1$ provides a well-known isomorphism of $H^{s+1}(\Sigma)$ and $H^{s-1}(\Sigma)$ via the Fredholm alternative.

When $\tau \neq 0$, we can combine Eqs. (7.13), (7.14) and (7.19) to re-express the linearized equation (8.3) as

$$\frac{\left(1+\sqrt{1+\frac{1}{2}\left|\frac{2\tau\zeta}{\mu_\rho}\right|^2}\right)^2}{\sqrt{1+\frac{1}{2}\left|\frac{2\tau\zeta}{\mu_\rho}\right|^2}}\frac{1}{\tau^2}\Delta_g\delta\Lambda^\dagger - \delta\Lambda^\dagger = \sigma \tag{8.5}$$

where g is the H^{s-1} metric defined (at the background solution) by Eqs. (6.16) and (8.2). The factor involving the square roots lies in $H^{s-1}(\Sigma)$ and so does not cause a serious problem. Multiplying the equation by its inverse merely converts the source σ to another element of $H^{s-1}(\Sigma)$ and replaces the coefficient of $\delta\Lambda^\dagger$ by a strictly positive coefficient that is sufficiently smooth for the application of the standard elliptic theory argument.

A difficulty seems to arise however, though the fact that the g metric only lies in H^{s-1} and thus that its Christoffel symbols generically only lie in $H^{s-2}(\Sigma)$ and not in $H^{s-1}(\Sigma)$, as was true of the ρ metric. This seems to interfere with the desired H^{s+1}-smoothness of $\delta\Lambda^\dagger$ (when the source function is taken to lie in H^{s-1}). Fortunately, however, our setup ensures that the identity map from (g, Σ) to (ρ, Σ) is harmonic and thus that

$$g^{ab}(\Gamma^c_{ab}(g) - \tilde{\Gamma}^c_{ab}(\rho)) = 0. \tag{8.6}$$

Using this equation to re-express the g-Laplacian on functions we get that

$$\Delta_g\delta\Lambda^\dagger = g^{ab}\tilde{\nabla}_a(\rho)\tilde{\nabla}_b(\rho)\delta\Lambda^\dagger \tag{8.7}$$

and hence that Δ_g maps H^{s+1} to H^{s-1} as desired with no fatal extra loss of derivatives.

It follows from the standard Fredholm argument again that $D_2\mathcal{F}(\zeta^{TT}, \Lambda^\dagger)$ yields the needed isomorphism between H^{s-1} and H^{s+1} and hence that the equation $\mathcal{F}(\zeta^{TT}, \Lambda^\dagger) = 0$ uniquely and implicitly determines $\Lambda^\dagger$ as a smooth functional of ζ^{TT} on some neighborhood of any particular solution. Since we know the unique solution when $\tau = 0$, we deduce that, for any chosen $\zeta^{TT} \in S_d^{TT_{s-1}}(\Sigma)$, there exists a $\tau_0 \in (0, -\infty)$ such that a unique solution $\Lambda^\dagger$ exists for all $\tau \in [0, \tau_0)$.

To show directly that every solution extends to the full interval $\tau \in [0, -\infty)$ we need to establish that the $H^{s+1}(\Sigma)$ -norm of $\Lambda^\dagger$ cannot blow up until τ exhausts this interval. Ideally, this result should follow from estimates derived directly from the Monge-Ampêre equation itself and indeed the derivation of such estimates is part of the aim of an ongoing project with S.-T. Yau which seeks to sharply characterize the asymptotics of solutions in the limit as $\tau \searrow -\infty$ (i.e., at their big bang initial singularities). We shall describe this project more fully in the concluding section below, but to complete the present argument, shall here instead give an elementary proof (based upon a slight refinement of the ideas from Ref. **[6]**) that all the solution curves (of the reduced Hamiltonian system) do indeed exhaust the interval $\tau \in [0, -\infty)$.

The reduced Hamiltonian system under study is a globally smooth system of first order ordinary differential equations defined on $T^*\mathcal{T}(\Sigma) \times \mathbf{R}^+$ (when expressed in terms of the time variable $T = -\frac{1}{\tau} \in \mathbf{R}^+$). From Eq. (5.5) and the properties of the function $\mathcal{S}$ it follows that the momenta, $p_\alpha(T) = \frac{\partial \mathcal{S}}{\partial q^\alpha}(q^\alpha(T), \rho, T)$, remain well-defined (and hence, through Eq. (4.12) yield a correspondingly well-defined $p^{TTab}(x^c, T)$) so long as the base curve, expressed in coordinates through functions $\{q^\alpha(T)\}$, persists as a curve in $\mathcal{T}(\Sigma)$. From the smooth character of the associated differential equations (i.e., Hamilton's equations) the premature breakdown of such a solution could only occur if the base curve runs "off-the-edge" of Teichmüller space before $\tau \searrow -\infty$ (or, equivalently, before $T \searrow 0$). We shall exclude such a breakdown by exploiting known properties of the Dirichlet energy (with the target metric ρ held fixed) as a proper function (i.e., an exhaustion function) on $\mathcal{T}(\Sigma)$ and by deriving estimates which show that this Dirichlet energy cannot blow up along a solution curve until $\tau \searrow -\infty$. These same estimates will also show that the Dirichlet energy tends to its (unique) infimum in the opposite limit (as $\tau \nearrow 0$ or $T \nearrow \infty$) and, again from known properties of this energy function, that this implies that the "moving," conformal metric γ_T tends to the target metric ρ as $T \to \infty$, a result which also follows from the method of continuity argument.

First of all note that, using Eqs. (2.4) and (2.5), Eq. (4.15) can be re-expressed as

$$\tau^2 = -\Delta_g N + N\left[\left|k^{TT}\right|_g^2 + \frac{1}{2}\tau^2\right]. \tag{8.8}$$

Using the inequality given by (2.44), that $\left|k^{TT}\right|_g^2 < \frac{\tau^2}{2}$, one easily derives the maximum principle bounds for the lapse function N:

$$1 < N \leq 2 \quad \text{on } \Sigma. \tag{8.9}$$

Furthermore, the integral of Eq. (8.8) over Σ yields the formula

$$\tau^2 \int_\Sigma d\mu_g (2 - N) = 2\int_\Sigma d\mu_g N \left|k^{TT}\right|_g^2. \tag{8.10}$$

Now, from Eqs. (3.6) and (3.7), taking $T = -\frac{1}{\tau}$ as before, we find that

$$(8.11)\quad \frac{\partial}{\partial T}\int_\Sigma d\mu_g\left(k_c^{TT\,d}k_d^{TT\,c}\right) = \frac{\partial}{\partial T}\left\{\frac{1}{2}\int_\Sigma \tau^2 d\mu_g + \int_\Sigma d\mu_g\,{}^{(2)}R_{(g)}\right\}$$
$$= \tau^3\int_\Sigma d\mu_g\left(1-\frac{N}{2}\right) = \tau\int_\Sigma d\mu_g N\left|k^{TT}\right|_g^2$$
$$= -\frac{1}{T}\int_\Sigma d\mu_g N\left(k_c^{TT\,d}\,k_d^{TT\,c}\right)$$

where, in intermediate steps, we have used the ADM field equation

$$(8.12)\qquad g_{ab,T} = \frac{2N}{\mu_g}(\pi_{ab} - g_{ab}tr_g\pi) + (\mathcal{L}_{(2)_X}g)_{ab}$$
$$= -2Nk_{ab} + (\mathcal{L}_{(2)_X}g)_{ab}$$

to evaluate $\partial_T\mu_g = \frac{1}{2}\mu_g g^{ab}\partial_T g_{ab}$.

In view of the upper and lower bounds on N given by (8.9), one easily derives from Eq. (8.11) the following bounds on

$$(8.13)\qquad F := \int_\Sigma d\mu_g(k_c^{TT\,d}\,k_d^{TT\,c}).$$

Either

(i) $F(T) = 0\,\forall\, T\in[0,\infty)$,

or

(ii) $F(T_0)\left(\frac{T_0}{T_1}\right)^2 \le F(T_1) < F(T_0)\frac{T_0}{T_1}$

$\forall\, T_0, T_1$ such that $T_0 < T_1$ with $T_0, T_1 \in (0,\infty)$.

Thus, unless $F(T) = 0$ identically, one has

$$\frac{\text{const}}{T^2} \le F(T) < \frac{\text{const}}{T} \quad \text{as } T\to\infty$$

and

$$\frac{\text{const}}{T} < F(T) \le \frac{\text{const}}{T^2} \quad \text{as } T\to 0$$

and thus that the Dirichlet energy function (c.f., Eqs. (3.6) and (3.7))

$$(8.14)\qquad \mathcal{A}(\gamma_T,\rho,Id) = 2\int_\Sigma d\mu_g k_c^{TT\,d}\;k_d^{TT\,c} - \int_\Sigma d\mu_g\,{}^{(2)}R(g)$$
$$= 2F(T) - \int_\Sigma d\mu_g\,{}^{(2)}R(g)$$

cannot blow up until $T\to 0$ but definitely does blow up in this limit unless $F(T) = 0$ identically (which corresponds to the trivial solution $\gamma_T = \rho\;\forall\; T\in [0,\infty)$). Furthermore, all solutions have the property that

$$(8.15)\qquad \mathcal{A}(\gamma_T,\rho,Id) \underset{T\longrightarrow\infty}{\longrightarrow} -\int_\Sigma d\mu_g\,{}^{(2)}R(g)$$

so that this energy asymptotes to its global infimum in the indicated limit. From the aforementioned well-known properties of this function (c.f., the discussion in Ref. [**14**]), one sees that all nontrivial solution curves run "off-the-edge" of Teichmüller space precisely as $T \searrow 0$ and that every solution curve (including the trivial ones with $\gamma_T = \rho$) has the property that $\gamma_T \underset{T\to\infty}{\longrightarrow} \rho$ since the (two-point) Dirichlet energy achieves its global infimum precisely at coincident points.

This latter result can also be recovered from the method of continuity argument given above. We showed therein that, for fixed $\rho \in \mathcal{M}^s(\Sigma), s > 4$ and arbitrary $\zeta^{TT} \in \mathcal{S}_d^{TT_{s-1}}(\Sigma)$ the Monge-Ampêre equation yielded $\Lambda^\dagger \in H^{s+1}(\Sigma)$ as a smooth functional of $\tau\zeta^{TT}$ for τ in some interval of the form $[0, \tau_0), \tau_0 < 0$. Furthermore, the limiting value as $\tau \nearrow 0$ was always given by the unique solution, $\Lambda^\dagger = 1$, of this equation when $\tau\zeta^{TT} = 0$. Now, recalling Eqs, (6.16), (6.17) and (6.31) we see that the "rescaled" metric $g^*_{ab} := \tau^2 g_{ab}$ satisfies

$$g^{*a\ell} = \frac{\left[2\tau\frac{\rho^{ab}}{\mu_\rho}\zeta^\ell_b + \sqrt{1 + \frac{2\tau^2|\zeta|}{(\mu_\rho)^2}}\rho^{a\ell}\right]}{\left(1 + \sqrt{\left(1 + \frac{2\tau^2|\zeta|^2}{(\mu_\rho)^2}\right)}\right)} \tag{8.16}$$

and has a well-defined limit as $\tau \nearrow 0$ given by

$$g^*_{ab} \underset{\tau\nearrow 0}{\longrightarrow} 2\rho_{ab} \tag{8.17}$$

since $\tau\zeta^b_a \to 0$ in that limit. Since γ_{ab} is obtained from g_{ab} (or equivalently from g^*_{ab}) by uniformization, it follows that $\gamma_T \underset{T\to\infty}{\longrightarrow} \rho$ in this limit.

9. Einstein solution curves and ray structures on Teichmüller space

In the previous section we saw that, for any fixed metric ρ having ${}^{(2)}R(\rho) = -1$, the TT symmetric tensors relative to ρ (i.e., the tensors $\frac{\rho_{ab}}{\mu_\rho}\zeta_c^{TT_b}$ formed from the mixed TT densities $\zeta_a^{TT_b}$ used there) determine solution curves to the reduced vacuum Einstein equations in CMCSH (constant-mean-curvature-spatially-harmonic) gauge. These curves fall naturally into one parameter families generated by the scale invariance of Einstein's vacuum field equations – replacing ζ^{TT} by $\lambda\zeta^{TT}$, where λ is a constant greater than zero, yields a rescaled solution for which the slice originally having mean curvature τ now has mean curvature τ/λ. To obtain all possible vacuum solutions one must, in addition, allow ρ to vary over a model of Teichmüller space for the given surface Σ, e.g., over a global cross-section for the trivial $\mathcal{D}_0(\Sigma)$-bundle $\mathcal{M}_{-1}(\Sigma) \to \mathcal{M}_{-1}(\Sigma)/\mathcal{D}_0(\Sigma) \approx \mathcal{T}(\Sigma)$ where $\mathcal{M}_{-1}(\Sigma)$ represents the space of metrics γ with ${}^{(2)}R(\gamma) = -1$, $\mathcal{D}_0(\Sigma)$ designates the group of diffeomorphisms of Σ isotopic to the identity and $\mathcal{T}(\Sigma)$ is the abstract

Teichmüller space of Σ. One can construct such cross sections in at least two distinct ways using harmonic maps as discussed for example by Tromba in [**14**, Sec. (3.4)]. In one such construction (due to Earle and Eells [**16**]) the section consists of all metrics $\rho \in \mathcal{M}_{-1}(\Sigma)$ such that the identity map from a fixed domain (γ, Σ) to (ρ, Σ) is harmonic. In a complementary section discussed by Tromba one holds the target (ρ, Σ) fixed and considers all $\gamma \in \mathcal{M}_{-1}(\Sigma)$ such that again the identity from (γ, Σ) to (ρ, Σ) is harmonic. Our result shows that in effect ρ and its conjugate variable ζ^{TT} represent asymptotic "Cauchy data" which, when prescribed in the limit $\tau \nearrow 0$, determine solution curves uniquely via the resolution of the $\Lambda^{\dagger}$ equation discussed above.

We know from the work in [**6**] that every such solution curve extends to the full interval $\tau \in [0, -\infty)$ which represents cosmological expansion from a big bang singularity at $\tau \to -\infty$ to the limit of infinite area as $\tau \nearrow 0$. Each corresponding curve of metrics g_τ can be smoothly and uniquely uniformized to yield a curve γ_τ which lies in a particular Tromba-section of $\mathcal{M}_{-1}(\Sigma)$, namely the submanifold consisting of all metrics γ such that $Id : (\gamma, \Sigma) \to (\rho, \Sigma)$ is harmonic for the appropriately chosen asymptotic conformal metric ρ. The results in [**6**] show that, except for the trivial, fixed point solutions generated by $\zeta^{TT} = 0$, every solution curve runs "off-the-edge" of Teichmüller space as $\tau \to -\infty$. In this section, we want to consider the families of such uniformized solution curves which all have the same asymptotic limit ρ (i.e., the curves determined by the asymptotic data (ρ, ζ^{TT}) for arbitrary ζ^{TT}). The aim is to show that such families (one for each choice of ρ) define so-called "ray structures" on Teichmüller space which are distinct from but somewhat complementary to the ray structures defined by Wolf [**8**]. In a similar spirit, we want to show how one can use the TT tensors at ρ to define a global coordinate system for Teichmüller space that is complementary to the one defined by Wolf (wherein one used TT tensors relative to the domain metric of a harmonic mapping rather than at the target as we do).

Thus we fix $\rho \in \mathcal{M}_{-1}(\Sigma)$ and focus on that subset of solutions determined by data $\{\rho, \zeta^{TT}\}$ where ζ^{TT} is transverse-traceless and symmetric with respect to ρ. For the moment, let us exclude the trivial solution corresponding to $\zeta^{TT} = 0$ and look only at the non-fixed-points. We claim that no two distinct solutions, aside from those obtained by rescaling any given one, ever intersect (except asymptotically where they all tend to ρ). To see this, suppose an intersection did exist at some (uniformized) metric γ. By rescaling we can always arrange that the intersection point for each of the two curves corresponds to the same value of the mean curvature τ. The complementary TT tensors $\overset{(1)}{k}{}^{TT}_{ab}$ and $\overset{(2)}{k}{}^{TT}_{ab}$ which, together with the common γ_{ab}, make up the two sets of conformal Cauchy data at mean curvature time τ for these solutions must be distinct (i.e., have $\overset{(1)}{k}{}^{TT} \neq \overset{(2)}{k}{}^{TT}$) since otherwise, by uniqueness of solutions of the reduced Einstein equations in CMCSH gauge (an elementary ODE uniqueness result in $2+1$ dimensions) they would yield identical solutions contrary to assumption. However our setup would then show that Wolf's basic equation (c.f., Eq. (2.28) above)

$$\rho_{ab} = \left\{ \left(e^{-4\lambda} \gamma^{de} \gamma^{cf} k^{TT}_{ce} k^{TT}_{df} + \frac{1}{2}\tau^2 \right) e^{2\lambda} \gamma_{ab} + 2\tau k^{TT}_{ab} \right\} \tag{9.1}$$

(where λ is uniquely determined by the solution of the Lichnerowicz equation (2.23)) is then satisfied for fixed $\{\rho, \gamma, \tau\}$ by two distinct choices for k^{TT}, namely $\overset{(1)}{k}{}^{TT}$ and $\overset{(2)}{k}{}^{TT}$. But this is impossible by Wolf's result that shows that the TT tensors k^{TT} relative to a fixed metric γ define a global coordinate chart for the Earle-Eells section of $\mathcal{M}_{-1}(\Sigma)$ representing Teichmüller space $\mathcal{T}(\Sigma)$. In other words, a given target metric ρ satisfying Eq. (9.1) corresponds to a unique k^{TT} (note that Wolf, who was not considering Einstein's equations, has effectively taken $\tau = -1$ in our formulation).

By a similar argument we can show that every metric γ in a Tromba-section based on ρ is attained by a solution curve (unique up to scaling if $\gamma \neq \rho$) which is asymptotic to ρ. The result is trivial if $\gamma = \rho$ since one only need take the fixed point solution so assume that $\gamma \neq \rho$ but that γ lies in the Tromba-section based on ρ (i.e., all $\gamma \in \mathcal{M}_{-1}(\Sigma)$ such that $Id : (\gamma, \Sigma) \to (\rho, \Sigma)$ is harmonic). By Wolf's result (after choosing say $\tau = -1$ for convenience to eliminate the scaling freedom) there exists, for the chosen γ and ρ, a unique k^{TT} satisfying Eq. (9.1) (with λ determined uniquely by the Lichnerowicz equation (2.23)). However, our arguments from Sec. 8 have shown that this conformal Cauchy data generates a solution whose asymptotic conformal metric is ρ. To compute the complementary asymptotic data ζ^{TT} for the corresponding solution curve we can appeal to Eqs. (6.15) and (6.21) which show that ζ^{TTb}_a is simply the transverse-traceless summand (relative to the ρ metric decomposition here!) of the density

$$\zeta^b_a = \mu_g g^{bc} k^{TT}_{ac} = \mu_\gamma \gamma^{bc} k^{TT}_{ac} \tag{9.2}$$

and further (in view of Eqs. (6.14)–(6.16) and recalling that $\mu_g g^{ab} = \mu_\gamma \gamma^{ab}$) that this projection is independent of the value of τ at which $(\gamma_{ab}, k^{TT}_{ab})$ are evaluated (and hence of the arbitrary choice $\tau = -1$ made for convenience above).

10. Lagrangian foliations of $T^*\mathcal{T}(\Sigma)$

Various Lagrangian foliations of the cotangent bundle of Teichmüller space have been discussed in the mathematics literature [**17, 18**]. On the other hand, Hamilton-Jacobi theory is closely connected to the construction of Lagrangian submanifolds or Lagrangian foliations of the phase spaces of suitable Hamiltonian systems and we are here in the optimal circumstances of having a globally defined, complete solution to the Hamilton-Jacobi equation for the reduced Einstein equations in 2+1 dimensions. This allows us not only to define two distinct (one parameter families of) Lagrangian foliations of $T^*\mathcal{T}(\Sigma)$ but also to give a simple geometrical interpretation of the leaves of these foliations in terms of the ray structures discussed in the previous section.

As shown in Sec. 5 the Dirichlet energy function $\mathcal{A}$, after subtraction of the Gauss-Bonnet invariant and rescaling by the factor $\frac{1}{\tau}$, yields a global solution of the Hamilton-Jacobi equation for reduced $2+1$ gravity for any choice of the (uniformized) target metric ρ. By virtue of the $\mathcal{D}_0(\Sigma)$ covariance of the formalism, there is no essential loss of generality involved in constraining ρ to lie in a model for Teichmüller space (e.g., in an Earle-Eells section of the bundle $\mathcal{M}_{-1}(\Sigma) \to \frac{\mathcal{M}_{-1}(\Sigma)}{\mathcal{D}_0(\Sigma)} \approx \mathcal{T}(\Sigma)$) which can be globally coordinatized (since $\mathcal{T}(\Sigma) \approx \mathbf{R}^{6\,\mathrm{genus}(\Sigma)-6}$) by a single chart $\{Q^\alpha \mid \alpha = 1, 2, \ldots, 6\,\mathrm{genus}(\Sigma) - 6\}$. Since the Dirichlet energy construction is conformally invariant with respect to the metric on the domain, we can also, without loss of generality, think of the domain metric $\gamma \in \mathcal{M}_{-1}(\Sigma)$ as constrained to lie in a model for $\mathcal{T}(\Sigma)$ which is globally coordinatized by another single chart $\{q^\alpha \mid \alpha = 1, 2, \ldots, 6\,\mathrm{genus}(\Sigma) - 6\}$. Our Hamilton-Jacobi function $\mathcal{S}$ can thus be regarded as expressible as a globally smooth real-valued function defined on $\mathcal{T}(\Sigma) \times \mathcal{T}(\Sigma) \times \mathbf{R}^-$ and written in coordinates in the form $\mathcal{S}(q^\alpha, Q^\alpha, \tau)$, where

$$\mathcal{S}(q, Q, \tau) = \frac{\mathcal{A}(q, Q)}{\tau} - \frac{1}{\tau} \int_\Sigma d\mu_\gamma^{(2)} R(\gamma) \tag{10.1}$$

or, in terms of the "Newtonian" time $T = -\frac{1}{\tau}$ which ranges over $(0, +\infty)$, as $\mathcal{S}(q, Q, T) = -T \left[\mathcal{A}(q, Q) - \int_\Sigma d\mu_\gamma^{(2)} R(\gamma)\right]$. It satisfies the reduced Hamilton-Jacobi equation

$$-\frac{\partial \mathcal{S}}{\partial T} = H_{\mathrm{reduced}} \left(q^\alpha, \frac{\partial \mathcal{S}}{\partial q^\alpha}, T\right) \tag{10.2}$$

$\forall\ \{q^\alpha, Q^\alpha, T\}$ ranging over (the global charts for) $\mathcal{T}(\Sigma) \times \mathcal{T}(\Sigma) \times \mathbf{R}^+$.

In the usual way, we can introduce a global chart for $T^*\mathcal{T}(\Sigma)$ with bundle coordinates $\{(q^\alpha, p_\alpha) \mid \alpha = 1, 2, \ldots 6\,\mathrm{genus}(\Sigma) - 6\}$ in terms of which the canonical symplectic form is expressible as $\omega = \sum_\alpha dq^\alpha \wedge dp_\alpha$. By graphing the gradient of $\mathcal{S}(q, Q, T)$, for fixed $\{Q^\alpha, T\}$ one gets a Lagrangian submanifold of $T^*\mathcal{T}(\Sigma)$ expressible in coordinates as $\{(q^\alpha, p_\alpha = \frac{\partial \mathcal{S}}{\partial q^\alpha})\}$. This clearly has the (maximal) dimension, namely $6\,\mathrm{genus}(\Sigma) - 6$, and is Lagrangian since the pullback of the symplectic form to this submanifold vanishes by virtue of the induced formula for the differentials dp_α, i.e.,

$$dp_\alpha = \sum_\beta \frac{\partial^2 \mathcal{S}}{\partial q^\alpha \partial q^\beta} dq^\beta \tag{10.3}$$

which yields $\sum_{\alpha\beta} \frac{\partial^2 \mathcal{S}}{\partial q^\alpha \partial q^\beta} dq^\alpha \wedge dq^\beta \equiv 0$ by symmetry of $\frac{\partial^2 \mathcal{S}}{\partial q^\alpha \partial q^\beta}$. By allowing the $\{Q^\alpha\}$'s to vary, still holding T fixed, one generates a family of such leaves which, taken all together, define a Lagrangian foliation of $T^*\mathcal{T}(\Sigma)$.

To see this, first note that the points on any given leaf $\{(q^\alpha, p_\alpha = \frac{\partial \mathcal{S}}{\partial q^\alpha}) \mid (Q^\beta, T)\text{fixed}\}$ represent the initial data sets (at time T) for all those solution curves which tend asymptotically to the conformal geometry represented by $\{Q\}$ as $T \to \infty$. Thus no two distinct leaves can intersect since a point of intersection would have to correspond to a solution curve having two distinct asymptotic conformal geometries. Furthermore, there is a (necessarily unique) leaf through any point of $T^*\mathcal{T}(\Sigma)$ since, thanks to Wolf's results, there is, for any base point $\{q^\alpha\}$ a unique covector $p_\alpha dq^\alpha$ (usually represented as a TT tensor or quadratic differential at a representative domain metric $\gamma \in \mathcal{M}_{-1}(\Sigma)$) such that the map given by Eq. (9.1) yields any desired target $\{Q^\alpha\}$. In fact, by Wolf's diffeomorphism result, there is a bijective correspondence between the target points $\{Q^\alpha\}$ and the covectors $p_\alpha dq^\alpha$ at $\{q^\alpha\}$ and in our Hamilton-Jacobi formulation the latter are realized as gradients of $\mathcal{S}$ via $p_\alpha dq^\alpha = \frac{\partial \mathcal{S}}{\partial q^\alpha}(q, Q, T) dq^\alpha$. Thus every point $\{q^\alpha, p_\alpha\}$ in phase space $T^*\mathcal{T}(\Sigma)$ is realized as a pair $\{q^\alpha, \frac{\partial \mathcal{S}}{\partial q^\alpha}(q, Q, T)\}$ when, for any fixed $T \in (0, \infty)$, the pair (q^α, Q^α) ranges over a global chart for $\mathcal{T}(\Sigma) \times \mathcal{T}(\Sigma)$.

A complimentary Lagrangian foliation is defined by graphs of the form $\{(Q^\alpha, P_\alpha = \frac{\partial \mathcal{S}}{\partial Q^\alpha}(q, Q, T))\}$ where again (q^α, Q^α) ranges over a (globally defined) chart for $\mathcal{T}(\Sigma) \times \mathcal{T}(\Sigma)$ at fixed $T > 0$. A particular leaf in this case (corresponding to fixed $\{q^\alpha\}$) consists of all asymptotic data points $\{Q^\alpha, P_\alpha\}$ belonging to solution curves which passed through $\{q^\alpha\}$ at time $T > 0$. Leaves cannot intersect by virtue of our bijectivity result from Sec. 9 which shows that for a fixed asymptotic geometry $\{Q^\alpha\}$ there is a bijective correspondence between the asymptotic covectors $P_\alpha dQ^\alpha$ (usually expressed in terms of TT tensors, c.f., Eq. (5.9)–(5.13)) and the points $\{q^\alpha\}$ exhausting Teichmüller space. This correspondence insures that every point $\{(Q^\alpha, P_\alpha)\}$ in (a global chart for) $T^*\mathcal{T}(\Sigma)$ is achieved as, in addition, $\{Q^\alpha\}$ is allowed to range over its chart for $\mathcal{T}(\Sigma)$.

In the above, T was held fixed. If it is allowed to vary, then the foliations "evolve" in a smooth way that is directly related to the scale invariance of Einstein's equations noted above. This follows from the observation that $p_\alpha = \frac{\partial \mathcal{S}}{\partial q^\alpha} = -T \frac{\partial \mathcal{A}}{\partial q^\alpha}$ and $P_\alpha = \frac{\partial \mathcal{S}}{\partial Q^\alpha} = -T \frac{\partial \mathcal{A}}{\partial Q^\alpha}$ wherein T only appears as an overall multiplicative factor which rescales the momenta as it is varied.

Suppose that $\{Q^\alpha(p)\}$ and $\{q^\alpha(p)\}$ label the same point p in abstract Teichmüller space. Then it is not difficult to see that $\frac{\partial \mathcal{S}}{\partial q^\alpha}(q(p), Q(p), T) = \frac{\partial \mathcal{S}}{\partial Q^\alpha}(q(p), Q(p), T) = 0$. This corresponds to the statement that the only solution curve passing through a given point in Teichmüller space at time $T \in (0, \infty)$ and also asymptoting to that point as $T \to \infty$ is in fact the trivial fixed point solution corresponding to the chosen point.

Concluding remarks

Solutions to the vacuum field equations on $\Sigma \times \mathbf{R}$ can each be used, by simply taking a product with the circle, to generate corresponding vacuum solutions to the $3+1$ dimensional Einstein equations on $\Sigma \times S^1 \times \mathbf{R}$. Each of these $3+1$ dimensional Lorentz manifolds is still flat and spatially compact (with space sections diffeomorphic to $\Sigma \times S^1$) and has by construction a spacelike Killing field tangent to the circular factors. As such it provides a very specialized example of vacuum solutions definable on S^1-bundles over $\Sigma \times \mathbf{R}$ which have a Killing symmetry imposed along the circular fibers of the bundle in question. The most general such ($U(1)$-isometric) solution can be regarded (upon adopting a Kaluza-Klein viewpoint and projecting fields down to the base) as a non-vacuum solution to the $2+1$ dimensional field equations defined on $\Sigma \times \mathbf{R}$ with a certain very specific type of matter source. Perhaps the most elegant form of these reduced field equations is that wherein the $2+1$ dimensional metric is coupled to a wave map field with target space given by the Poincaré half-plane. If trivial bundles are considered this wave map can be polarized (i.,e. restricted so that its image lies along a geodesic in the half-plane) so that the wave map equations reduce to a wave equation but in the general case of non-trivial bundles over $\Sigma \times \mathbf{R}$ such a polarization is incompatible with the topology of the bundle [**19, 20, 21**].

The global study of such $U(1)$-invariant vacuum solutions to the $3+1$ dimensional field equations is an important and still largely open problem in general relativity. Some significant progress on it has been made by Y. Choquet-Bruhat and the present author by assuming Cauchy data which is sufficiently small (as a non-linear perturbation of the trivial, constant wave map solutions) and evolving only in the direction of cosmological expansion [**19, 20, 21**], a device which sidesteps complications produced by the big bang singularities. However, some independent work by these same authors together with J. Isenberg has exploited so-called Fuchsian methods to deal rigorously with the singularities themselves at least when the solutions are "half-polarized" in a well-defined sense [**22, 23**]. The generic non-polarized solution is however, anticipated not to be amenable to this kind of Fuchsian analysis and instead to exhibit a singularity of oscillatory type [**24**].

It seems plausible that the formation of black holes in these $U(1)$-isometric models is suppressed by the symmetry imposed and thus that there is no singular behavior expected for the cosmologically expanding direction at all. In other words, one should have large data global existence for all solutions to this problem. Of course, it is extremely unlikely that one could prove such a result without first learning how to treat $2+1$ dimensional wave maps globally on a background Lorentz manifold. In the present problem the Lorentzian metric is not a background but instead a functional of the evolving wave map and the Teichmüller parameters. Indeed even the (polarized) special case of a wave equation is highly non-linear because of this metrical dependence upon the evolving wave and, at present, only the small data results mentioned above are known.

Another problem related to that considered in the present article is that of constructing CMC foliations of *flat*, higher dimensional, spatially compact Lorentz manifolds. For any compact hyperbolic manifold $(\mathbf{H}^n/\Gamma, h)$, where h has constant negative curvature as a Riemannian metric, one can construct the trivial, flat Lorentz cone spacetime on $\mathbf{H}^n/\Gamma \times \mathbf{R}$ for any $n \geq 2$ which is naturally foliated by CMC slices. For $n \geq 3$, Mostow rigidity forbids any "obvious" deformation of the metrics on $\mathbf{H}^n/\Gamma \times \mathbf{R}$ which preserves flatness since there is no corresponding Teichmüller space of hyperbolic metrics in these cases, but nevertheless, for some choices of $\mathbf{H}^n/\Gamma$ there can exist non-trivial moduli spaces of flat Lorentzian metrics on $\mathbf{H}^n/\Gamma \times \mathbf{R}$ and, for these, it would be of interest to construct CMC foliations. This problem has already been dealt with using indirect methods by L. Andersson who exploited earlier results by G. Mess and others on the properties of such spaces [**25**]. However, it might also be possible to attack this problem directly with methods that parallel, to some extent, those developed in the present paper. In particular, harmonicity of the Gauss map for a CMC slice in such a flat, higher dimensional spacetime will continue hold and the analogue of Eq. (6.3) above is easy to derive. Thus it is of interest to see whether one could derive an elliptic equation (or system) which plays the role for these higher dimensional problems that our Monge-Ampêre equation for $\Lambda^\dagger$ plays here. One would expect the role of holomorphic quadratic differentials to be played by the traceless Codazzi tensors of a fixed hyperbolic metric h on $\mathbf{H}^n/\Gamma$.

Finally, let us return to the problem alluded to at the end of Sec. 8, namely the derivation of estimates for solutions of the Monge-Ampêre equation sufficiently sharp so as to characterize their asymptotics as $\tau \searrow -\infty$ (which limit corresponds to the big bang singularities in these models (c.f. Ref. [**6**])). This problem is currently under study with S.-T. Yau.

As in the case of Wolf's ray structures the natural conjecture here would seem to be that the ray structures introduced in Sec. 9, representing families of Einstein solution curves, have limits, as $\tau \searrow -\infty$, at Thurston boundary points of $\mathcal{T}(\Sigma)$. More precisely, the idea is that, for any fixed ρ, the collection of non-trivial solution curves emerging (at $\tau = 0$) from this (arbitrary) interior point of $\mathcal{T}(\Sigma)$ effectively attach the Thurston boundary to $\mathcal{T}(\Sigma)$ in the sense that each solution ray limits to a boundary point and every such boundary point is the limit of a unique ray. That this should be true is, to a large extent, already known from indirect, barrier-estimate arguments due to L. Andersson [**26**] which in turn are based upon the fundamental work of R. Benedetti and E. Guadagnini [**27**]. The latter authors use a "cosmic time" slicing (essentially a Gaussian normal slicing but having the big bang singularity itself as a $t = 0$ level "surface") and explicitly construct a (dense) subset of the full solution space by suitably "cutting and pasting" negatively curved Friedman-Roberton-Walker and flat Kazner metrics along certain leaves of geodesic laminations of Σ. This representation of the solutions induces different families of curves in $\mathcal{T}(\Sigma)$ from those we get (since generically their slicings are neither smooth nor CMC whereas ours are) but they show that their solution curves in $\mathcal{T}(\Sigma)$ do indeed limit to Thurston boundary points. Andersson's barrier arguments are designed to control the asymptotics of CMC slicings relative to the cosmic time ones near the big bang singularities and thereby to show that Thurston boundary points are attained, in the limit, by the former as well as the latter. Our Monge-Ampêre analysis yields, in principle, a direct characterization of all CMC sliced solutions (and not only the "simplicial" ones dealt with by Andersson) and thus affords the possibility of computing their (CMC-sliced) singularity structures more explicitly.

Acknowledgements

I am grateful to Shing-Tung Yau, Lars Andersson, Arthur Fischer, Raymond Puzio, Michael Wolf, Thibault Damour, Yvonne Bruhat, Richard Schoen, Jeanette Nelson, Tullio Regge, James Hartle and Anthony Tromba for numerous conversations over the years on 2 + 1-dimensional Einstein gravity and Teichmüller theory which greatly clarified the issues discussed in this article. In addition, I am also grateful for the hospitality and support of the Albert Einstein Institute (Golm, Germany), The Erwin Schrödinger Institute (Vienna, Austria), the Institut des Hautes Études Scientifiques (Bures-sur-Yvette, France), The Isaac Newton Institute (Cambridge, UK), the Kavli Institute for Theoretical Physics (Santa Barbara, California), Caltech University (Pasadena, California), Stanford University and the American Institute of Mathematics (Palo Alto, California) where portions of this research were carried out. This research was supported by the NSF grant PHY-0354391 to Yale University.

References

[1] E. Witten, 2+1 *Dimensional Gravity as an Exactly Soluble System*, Nucl. Phys. **B311** (1988), 46–78.

[2] V. Moncrief, *Reduction of the Einstein equations in* 2+1 *dimensions to a Hamiltonian system over Teichmüller space*, J. Math. Phys. **20** (1989), 2907–2914 .

[3] S. Carlip, *Quantum Gravity in* 2 + 1 *Dimensions (Cambridge Monographs on Mathematical Physics)*, Cambridge University Press (1998).

[4] L. Andersson and V. Moncrief, *Elliptic-Hyperbolic Systems and the Einstein Equations*, Ann. Henri Poincaré **4** (2003), 1–34 .

[5] A. Fischer and V. Moncrief, *Hamiltonian reduction, the Einstein flow, and collapse of* 3*-manifolds*, Nucl. Phys. B (Proc. Suppl.) **88** (2000), 83–102.

[6] L. Andersson, V. Moncrief and A. Tromba, *On the global evolution problem in* 2 + 1 *gravity*, J. Geom. Phys. **23** (1997), 191–205.

[7] R. Puzio, *The Gauss map and* 2 + 1 *gravity*, Class. Quantum Grav. **11** (1994), 2667–2675.

[8] M. Wolf, *The Teichmüller Theory of Harmonic Maps*, J. Diff, Geom. **29** (1989), 449–479.

[9] B. Tabak, *A Geometric Characterization of Harmonic Diffeomorphisms Between Surfaces*, Math. Ann. **270** (1985), 147–157.

[10] L. M. Sibner and R. J. Sibner, *A non-linear Hodge-deRham theorem*, Acta. Math. **125** (1970), 57–73.

[11] E. A. Ruh and J. Vilms, *The tension field of the Gauss map*, Trans. Am. Math. Soc. **149** (1970), 569–573.

[12] V. Moncrief, *Reduction of Einstein's Equations for Vacuum Space-Times with Space-like U(1) Isometry Groups*, Ann. Phys. (NY) **167** (1986), 118–142. See appendix herein for an application of the sub and super solutions method.

[13] J. L. Kasdan, *Some applications of Partial Differential Equations to Problems in Geometry*, Surveys in Geometry (copyright by J. L. Kasdan 1983). Some of this material appears in the appendix of Einstein Manifolds by A. Besse (Springer, 2002)

[14] A. J. Tromba, *Teichmüller Theory in Riemannian Geometry (Lectures in Mathematics)* Birkhaüser 1992. See especially section 3.4 of this reference for a discussion of the relevant global cross-sections of $\mathcal{M}_{-1}(\Sigma)$.

[15] M. H. Protter and H. F. Weinberger, *Maximum Principles in Differential Equations*, Springer (1984). See section 3 of chapter 2 for a discussion of the Hopf (strong) maximum principle.

[16] C. Earle and J. Eells, *A fibre bundle description of Teichmüller theory*, J. Diff. Geom. **3** (1969), 19–43.

[17] M. Wolf, *Measured Foliations and Harmonic Maps of Surfaces*, J. Diff. Geom. **49** (1998), 437–467.

[18] J. Hubbard and H. Masur, *Quadratic Differentials and Foliations*, Acta Math. **142** (1979), 221–274.

[19] Y. Choquet-Bruhat and V. Moncrief, *Future global in time Einsteinian spacetimes with $U(1)$ isometry group*, Ann. Henri Poincaré **2** (2001), 1007–1064.

[20] ———, *Non-linear stability of Einsteinian spacetimes with $U(1)$ isometry group* in "Partial Differential Equations and Mathematical Physics" in honor of J. Leray, Kajitani and Vaillant, eds. Birkhaüser (2003).

[21] Y. Choquet-Bruhat, *Future Complete $U(1)$ Symmetric Einsteinian Spacetimes, the Unpolarized Case* in "The Einstein Equations and the Large Scale Behavior of Gravitational Fields, P. T. Chrusciel and H. Friedrich, eds. Birkhaüser (2004).

[22] Y. Choquet-Bruhat, J. Isenberg and V. Moncrief, *Topologically general $U(1)$ symmetric Einstein spacetimes with AVTD behavior*, Nuovo Cim. **119B** (2004), 625–638.

[23] J. Isenberg and V. Moncrief, *Asymptotic Behavior of Polarized and Half-Polarized $U(1)$ Symmetric Vacuum Spacetimes*, Class. Quantum Grav. **19** (2002), 5361–5386.

[24] B. K. Berger and V. Moncrief, *Signature for local Mixmaster dynamics in $U(1)$ symmetric cosmologies*, Phys. Rev. **D 62** (2000), 123501. See also Phys. Rev. **D 62**, 023509 by the same authors.

[25] L. Andersson, *Constant mean curvature foliations of flat spacetimes*, Comm. Anal. Geom. **10** (2002), 1094-1115; arXiv: math. DG/0110245.

[26] ———, *Constant mean curvature folitaions of simplicial flat spacetimes*; arXiv: math. DG/0307338 v1 (2003).

[27] R. Benedetti and E. Guadagnini, *Cosmological Time in $(2+1)$-Gravity*, Nucl. Phys. **B613** (2001), 330–352 arXiv: gr-qc/0003055 v2 (2001).

Department of Mathematics and Department of Physics, Yale University, New Haven, Connecticut USA

Monotonicity and Li-Yau-Hamilton Inequalities

Lei Ni

ABSTRACT. This is a survey article for *Surveys in Differential Geometry* series on the subject of Li-Yau-Hamilton type differential inequalities and related monotonicity formulae.

1. Introduction

The purpose of this essay is to give an expository account on various sharp estimates of Li-Yau-Hamilton type for solutions to geometric evolution equations and their relation to various important monotonicity formulae in the subject.

The Li-Yau-Hamilton type estimates in the geometric evolution equations are originally called differential Harnack estimates since they imply the celebrated Harnack estimates for the parabolic equations originated with the work of Moser [**Mo**]. In the seminar paper [**LY**] P. Li and S.-T. Yau first proved a gradient estimate for positive solutions to the heat equation via the maximum principle and derived a sharp form of Harnack estimate by integrating the proven gradient estimate on space-time paths. Even though its proof has its root in the corresponding earlier works on geometric elliptic equations (cf. [**ChY1, ChY2, Y**], etc.), this gradient estimate differs from its elliptic analogue fundamentally by its sharpness as well as its broader impact to the study of nonlinear geometric evolution equations. Later on, the similar technique was employed by R. Hamilton in the study of Ricci flow [**H2, H3**], as well as the hypersurface mean curvature flow [**H4**]. Based on this history and its great impact towards the study of geometric evolution equations, in [**NT**], the class of sharp differential inequalities which yield a Harnack type estimate by the path-integration, started to be called Li-Yau-Hamilton type estimates. Soon after Hamilton's work the corresponding results for the Kähler-Ricci flow and Gauss curvature flow (Yamabe flow) were proved by H.-D. Cao [**Co**] and B. Chow [**Ch1, Ch2**] respectively. Later

The author was supported in part by NSF Grants and an Alfred P. Sloan Fellowship, USA.

in [**An1**], the result was established for a general class of hypersurface flows by B. Andrews.

On the other hand, Hamilton also generalized Li-Yau's original estimate for the linear heat equation to a matrix form on a class of Riemannian manifolds with nonnegative sectional curvature [**H5**]. More importantly he derived several important monotonicity formulae out of his matrix estimate [**H6**].

In the last decade there were a few further developments along this direction, [**CC1, CC2, CH, NT, CN, N3, N4**]. Many applications in geometry have also been discovered. The most spectacular one is the Li-Yau-Hamilton type estimate for the conjugate heat equation (coupled with the Ricci flow) and its related entropy monotonicity formula discovered by G. Perelman [**P**] (see also [**N5**]). Perelman's inequality and its relation with the *reduced volume* monotonicity suggest a profound connection between the monotonicity formulae and Li-Yau-Hamilton type estimates. It is one of our purposes here to convey this connection.

Due to its expository nature, in this paper we shall only include a complete proof of a statement in the cases either the stated result appears in the first time, or we are compelled by the simplicity of the original arguments. In the rest of the cases, we shall just give a outline or the key steps.

Here is how we organize this paper. In Section 2 we discuss various Li-Yau-Hamilton inequalities on geometric evolution equations. In Section 3 we discuss how various monotonicity formulae can be derived out of the Li-Yau-Hamilton inequalities. In Section 4 we discuss how monotonicity formulae in turn suggest new Li-Yau-Hamilton type inequalities. The order of the appearance of the results may not necessarily follow the chronological order in which the results were proved. Instead we present them in the order which we feel is the most natural and logical. Due to the lack of expertise and the limited time allowed for writing this article we have to omit the applications of Li-Yau-Hamilton inequality altogether, as well as other various important topics, which we shall list in the section of comments for interested parties. Through out the paper n denotes the dimension of a Riemannian manifold M^n and m denotes the complex dimension of a Kähler manifold M^m $(n=2m)$.

2. Li-Yau-Hamilton type inequalities

The most miraculous result is Hamilton's matrix Li-Yau-Hamilton (which we shall abbreviate as LYH later) for Ricci flow of metrics with a bounded nonnegative curvature operator. However its proof as well as the formulation is motivated by the corresponding simpler consideration on linear heat equations. We also found that it is more suggestive to start with the simpler case since its understanding often sheds lights on the more technically complicated nonlinear setting. Hence we start with the solutions to the linear heat equation.

2.1. Linear heat equation. Through out this subsection, let (M, g) be a fixed Riemannian manifold. The main concern is on the positive solution to

$$\left(\frac{\partial}{\partial t} - \Delta\right) u(x,t) = 0. \tag{2.1}$$

A. Li-Yau's gradient estimate for the linear heat equation. In [**LY**], the following result was proved.

THEOREM 2.1. *Let (M, g) be a complete Riemannian manifold. Assume that on the ball $B(o, 2R)$, $\mathrm{Ric}(M) \geq -k$. Then for any $\alpha > 1$, we have that*

$$\sup_{B(o,R)} \left(\frac{|\nabla u|^2}{u^2} - \alpha \frac{u_t}{u}\right) \leq \frac{C\alpha^2}{R^2}\left(\frac{\alpha^2}{\alpha^2 - 1} + \sqrt{k}R\right) + \frac{n\alpha^2 k}{2(\alpha - 1)} + \frac{n\alpha^2}{2t}. \tag{2.2}$$

If (M, g) has nonnegative Ricci, letting $R \to \infty$, (2.2) gives the clean estimate (a Hamilton-Jacobi inequality):

$$\frac{|\nabla u|^2}{u^2} - \frac{u_t}{u} \leq \frac{n}{2t} \tag{2.3}$$

This estimate is sharp in the sense that the equality satisfied for some (x_0, t_0) implies that (M, g) is isometric to $\mathbb{R}^n$ [**N5**]. It can also be easily checked that if u is the fundamental solution on $\mathbb{R}^n$ given by the formula $\frac{1}{(4\pi t)^{\frac{n}{2}}} \exp(-\frac{|x|^2}{4t})$, then the equality holds in (2.3). It was observed by Li-Yau that by integration over the path jointing (x_1, t_1) to (x_2, t_2) with $t_2 > t_1$, (2.3) gives

$$u(x_1, t_1) \leq \left(\frac{t_2}{t_1}\right)^{\frac{n}{2}} u(x_2, t_2) \exp\left(\frac{r^2(x_1, x_2)}{4(t_2 - t_1)}\right)$$

which is a sharp form of Harnack estimates for parabolic equations.

Despite the fundamental importance of Theorem 2.1, its proof is elementary via the maximum principle for parabolic equations by computing $\left(\frac{\partial}{\partial t} - \Delta\right) F$ with $F = \frac{|\nabla u|^2}{u^2} - \alpha\frac{u}{u_t}$. The localized estimate (2.2) is possible due to a term of $-F^2$ appearing in the resulting estimation of $\left(\frac{\partial}{\partial t} - \Delta\right) F$. The Ricci curvature comes into play due to the commutation of the differentiations as in the regular Bochner formula [**SY**]. In the special case that (M, g) has nonnegative Ricci curvature, the proof of (2.3) is based on the following computation. Let $Q := u\left(\Delta \log u + \frac{n}{2t}\right)$. Then

$$\left(\frac{\partial}{\partial t} - \Delta\right) Q = \frac{2}{u}|\Upsilon_{ij}|^2 - \frac{2}{t}Q + \frac{2}{u}R_{ij}\nabla_i u \nabla_j u \tag{2.4}$$

where

$$\Upsilon_{ij} = \nabla_i \nabla_j u + \frac{u}{2t} g_{ij} - \frac{u_i u_j}{u} = u \left(-\nabla_i \nabla_j f + \frac{1}{2t} g_{ij} \right).$$

with f defined by $u = \frac{e^{-f}}{(4\pi t)^{n/2}}$. ($Q = g^{ij}\Upsilon_{ij}$.)

B. The differential estimates on the fundamental solution. In [**N2**], motivated by the work of [**P**] the following result was proved.

THEOREM 2.2. *Let (M, g) be a complete Riemannian manifold with non-negative Ricci curvature. Let $u(x,t) = H(x,t;y,o)$ be the positive heat kernel. Then*

$$t(2\Delta f - |\nabla f|^2) + f - n \leq 0 \tag{2.5}$$

where $u = \frac{e^{-f}}{(4\pi t)^{\frac{n}{2}}}$.

The estimate (2.5) is a Li-Yau-Hamilton type since combining with the heat equation $2|\nabla f|^2 - 2\Delta f + 2f_t + \frac{n}{t} = 0$ we have a Hamilton-Jacobi inequality

$$|\nabla f|^2 + 2f_t + \frac{f}{t} \leq 0. \tag{2.6}$$

For (x_2, t_2) and (x_1, t_1) with $t_2 > t_1$ we have that

$$\begin{aligned}
\sqrt{t_2} f(x_2,t_2) - \sqrt{t_1} f(x_1,t_1) &= \int_{t_1}^{t_2} \frac{d}{dt}\left(\sqrt{t} f(\gamma(t),t)\right) dt \\
&= \int_{t_1}^{t_2} \sqrt{t}\left(f_t + \frac{f}{2t} + \langle \nabla f, \gamma'(t)\rangle\right) dt \\
&\leq \frac{1}{2}\int_{t_1}^{t_2} \sqrt{t}|\gamma'(t)|^2\, dt
\end{aligned}$$

for any path joining from x_1 to x_2. This gives a Harnack type estimate:

$$\sqrt{t_2} f(x_2,t_2) - \sqrt{t_1} f(x_1,t_1) \leq \inf_{\gamma} \frac{1}{2}\int_{t_1}^{t_2} \sqrt{t}|\gamma'(t)|^2\, dt = \frac{r^2(x_1,x_2)}{4(\sqrt{t_2}-\sqrt{t_1})}. \tag{2.7}$$

Since $\lim_{t\to 0} \sqrt{t} f(o,t) \leq 0$ from that $\lim_{t\to 0} u(x,t) = \lim_{t\to 0} \frac{e^{-\frac{\sqrt{t}f(x,t)}{\sqrt{t}}}}{(4\pi t)^{\frac{n}{2}}} = \delta_o(x)$, we have that

$$f(x_2,t_2) \leq \frac{r^2(o,x_2)}{4t_2}. \tag{2.8}$$

This is equivalent to the heat kernel comparison theorem of Cheeger and Yau [**CY**].

The curious reader may ask why the estimate (2.5) is not proved for arbitrary positive solutions. The simple answer is that it is no longer true in such generality. This can be seen for those solutions which are smooth on $M \times [0, T)$. The proof of the result requires both the point-wise computation as Theorem 2.1 (precisely (2.4)) as well as an entropy formula together with some integral estimates. Hence we postpone a more detailed account to a later section. However, there is a geometric consideration which indicates that (2.5) is a natural one and is related to (2.3). It was known that

$$\lim_{t\to 0} 4tf = r^2(o, x)$$

if u is the heat kernel. Rewriting (2.3) as

$$t(2\Delta f) - n \leq 0$$

one therefore deduces the Laplacian comparison theorem:

$$\Delta r^2 \leq 2n.$$

Define $\hat{L}(x,t) := 4tf(x,t)$. Then (2.3) amounts to $\Delta \hat{L} \leq 2n$. Therefore one can view it as a generalized/space-time version Laplacian comparison theorem. Writing (2.5) in terms of $\hat{L}$ we have that

$$\Delta \hat{L} + \hat{L}_t \leq 2n. \tag{2.9}$$

In this sense, (2.5) is a space-time Laplacian comparison theorem. The sharpness of (2.5) follows from that the equality holds for some (x,t) with $t > 0$ if and only if M is isometric to $\mathbb{R}^n$ [**N5**].

C. The matrix LYH inequalities. We found that keeping in mind the connection between the LYH type estimate and the comparison theorem on distance functions is beneficial. The following result of Hamilton [**H5**] corresponds to the Hessian comparison theorem.

THEOREM 2.3. *Assume that (M, g) is a complete Riemannian manifold with nonnegative sectional curvature and parallel Ricci curvature. Then*

$$\nabla_i \nabla_j \log u + \frac{1}{2t} g_{ij} \geq 0. \tag{2.10}$$

Noting the trace of (2.10) is just (2.3). The extra assumption that (M, g) has parallel Ricci is quite restrictive, which essentially means that (M, g) is Einstein. When (M, g) is a Kähler manifold this assumption can be dropped and the nonnegativity of the sectional curvature can be relaxed to the nonnegativity of the bisectional curvature. This was observed in [**CN**].

THEOREM 2.4. *Let (M, g) be a complete Kähler manifold with nonnegative holomorphic bisectional curvature. Then*

$$\nabla_i \nabla_{\bar{j}} \log u + \frac{1}{t} g_{i\bar{j}} \geq 0. \tag{2.11}$$

Both (2.10) and (2.11) yield the Hessian comparison theorem for the distance function as illustrated before. In the next section we shall show how they can suggest estimates for the Ricci/Kähler-Ricci flow.

The proof is as usual via the maximum principle. This time one needs the tensor maximum principle of Hamilton developed in [**H1**]. The rest is on computing

$$\left(\frac{\partial}{\partial t}-\Delta\right)\left(\nabla_i\nabla_j \log u+\frac{1}{2t}g_{ij}\right)$$

and grouping the resulting terms.

D. LYH inequality on $(1,1)$ forms. The above results are all on the positive solution to the heat equation. The curious reader may ask if there is any such result for solutions to a parabolic system. There is one indeed [**N3**]. But we have to restrict ourselves to the complex Kähler manifolds.

Let $h(x,t) = \sqrt{-1}h_{\alpha\bar{\beta}}dz^{\alpha}\wedge dz^{\bar{\beta}}$ be a real $(1,1)$ form satisfying the Lichnerowicz-Laplacian heat equation:

$$\left(\frac{\partial}{\partial t}-\Delta\right)h_{\gamma\bar{\delta}} = R_{\beta\bar{\alpha}\gamma\bar{\delta}}h_{\alpha\bar{\beta}}-\frac{1}{2}\left(R_{\gamma\bar{s}}h_{s\bar{\delta}}+R_{s\bar{\delta}}h_{\gamma\bar{s}}\right). \tag{2.12}$$

We assume that $h_{\alpha\bar{\beta}}(x,t)$ is semi-positive definite (denoted briefly as $h_{\alpha\bar{\beta}}(x,t)\geq 0$) and that M has nonnegative bisectional curvature. For any $(1,\,0)$ vector field V we define

$$\begin{aligned} Z_h(x,t) &= \frac{1}{2}\left(g^{\alpha\bar{\beta}}\nabla_{\bar{\beta}}\operatorname{div}(h)_{\alpha}+g^{\gamma\bar{\delta}}\nabla_{\gamma}\operatorname{div}(h)_{\bar{\delta}}\right)+g^{\alpha\bar{\beta}}\operatorname{div}(h)_{\alpha}V_{\bar{\beta}} \\ &\quad +g^{\gamma\bar{\delta}}\operatorname{div}(h)_{\bar{\delta}}V_{\gamma}+g^{\alpha\bar{\beta}}g^{\gamma\bar{\delta}}h_{\alpha\bar{\delta}}V_{\bar{\beta}}V_{\gamma}+\frac{H}{t}. \end{aligned} \tag{2.13}$$

Here

$$\operatorname{div}(h)_{\alpha} = g^{\gamma\bar{\delta}}\nabla_{\gamma}h_{\alpha\bar{\delta}},\quad \operatorname{div}(h)_{\bar{\delta}} = g^{\alpha\bar{\beta}}\nabla_{\bar{\beta}}h_{\alpha\bar{\delta}}$$

and

$$H = g^{\alpha\bar{\beta}}h_{\alpha\bar{\beta}}.$$

In the context where the meaning is clear we drop the subscript h in Z_h.

THEOREM 2.5. *Let M be a complete Kähler manifold with nonnegative holomorphic bisectional curvature. Let $h_{\alpha\bar{\beta}}(x,t)\geq 0$ be a symmetric $(1,1)$ tensor satisfying (1.1) on $M\times(0,T)$. Assume that for any $\epsilon'>0$,*

$$\int_{\epsilon'}^{T}\int_M e^{-ar^2(x)}\|h\|^2\,dv\,dt<\infty. \tag{2.14}$$

Then

$$Z(x,t) \geq 0, \tag{2.15}$$

for any $(1,0)$ *vector* V. *If* $Z(x_0,t_0)=0$ *for some point* (x_0,t_0) *with* $t_0>0$ *and* $h_{\alpha\bar\beta}(x,t)>0$, *then* M *is flat.*

The proof is via the maximum principle as usual. Instead of explaining on that we choose to formulate the result in more conventional way for the people who are more familiar with the complex geometric notations. Recall the Hodge-Kodaira Laplacian

$$\Delta'' = \bar\partial\bar\partial^* + \bar\partial^*\bar\partial.$$

The standard Kodaira-Bochner formula gives

$$-\Delta'' h = \Delta_L h \doteqdot \Delta h_{\alpha\bar\beta} + R_{\beta\bar\alpha\gamma\bar\delta}h_{\alpha\bar\beta} - \frac{1}{2}\left(R_{\gamma\bar s}h_{s\bar\delta} + R_{s\bar\delta}h_{\gamma\bar s}\right),$$
$$\partial^* h = -\operatorname{div}(h)_{\bar\delta}dz^{\bar\delta}, \quad \bar\partial^* h = \operatorname{div}(h)_\alpha dz^\alpha$$

and

$$g^{\alpha\bar\beta}\nabla_{\bar\beta}\operatorname{div}(h)_\alpha = -\partial^*\bar\partial^* h, \quad g^{\alpha\bar\beta}\nabla_\alpha \operatorname{div}(h)_{\bar\beta} = \bar\partial^*\partial^* h.$$

Hence the result concludes that for $h \geq 0$ satisfying the heat equation

$$\left(\frac{\partial}{\partial t} + \Delta''\right) h = 0 \tag{2.16}$$

the scalar quantity

$$Z_h = \frac{1}{2}\left(-\partial^*\bar\partial^* + \bar\partial^*\partial^*\right)h + \bar\partial^* h(V) - \partial^* h(\bar V) + h(V,\bar V) + \frac{\Lambda h}{t} \geq 0 \tag{2.17}$$

for any vector field V. Here Λh is the standard contraction of h by the Kähler form.

When h is the trace of the curvature form of a Hermitian vector bundle satisfying the Hermitian-Einstein flow, Theorem 2.5 gives a Li-Yau-Hamilton estimate for Hermitian-Einstein flow on nonnegative curved manifolds. Please see [**N3**] for details on this.

2.2. Ricci/Kähler-Ricci flow. For most discussion in this subsection, $(M,g(t))$ is a solution to Ricci/Kähler Ricci flow. The most important result for Ricci flow is Hamilton's [**H3**] matrix LYH inequality.

A. Matrix LYH inequalities on curvature. The following fundamental result for Ricci flow was proved by Hamilton [**H3**] (see also [**H2**]).

THEOREM 2.6. *If* $(M^n, g(t))$, $t\in[0,T)$, *is a solution to the Ricci flow with nonnegative curvature operator, namely* $R_{ijkl}U_{ij}U_{kl}\geq 0$ *for all 2-forms, and if* $(M^n, g(t))$ *is either compact or complete noncompact with bounded curvature, then for any* 1*-form* $W \in C^\infty(\Lambda^1 M)$ *and* 2*-form* $U \in C^\infty(\Lambda^2 M)$ *we have*

$$Z(U,W) \doteqdot M_{ij}W_iW_j + 2P_{pij}U_{pi}W_j + R_{ipjq}U_{ip}U_{jq} \geq 0. \tag{2.18}$$

Here

$$P_{kij} \doteqdot \nabla_k R_{ij} - \nabla_i R_{kj} \tag{2.19}$$

and the symmetric 2*-tensor* M *is defined by*

$$M_{ij} \doteqdot \Delta R_{ij} - \frac{1}{2}\nabla_i\nabla_j R + 2R_{ikjl}R_{kl} - R_{ip}R_{pj} + \frac{1}{2t}R_{ij}. \tag{2.20}$$

The original paper [**H3**] is still the best place to read the proof of this result. The expression is suggested by playing with gradient expanding soliton equations. This again is well explained in [**H3**]. As in Li-Yau's case, the expression of the estimate is obtained by the computation of the special solutions (grouping various gradient terms of the heat kernel on the Euclidean space $\mathbb{R}^n$ yields the LYH expression for the linear heat equation, grouping the soliton equation and its various covariant differentiations on an gradient expanding soliton yields the matrix LYH expression for the Ricci flow). The LYH expression usually vanishes identically on the special solutions. However it is highly nontrivial, if not a miracle, to come up as complicated an expression as (2.18) and conclude that it has a sign for the general solutions/spaces.

The Kähler analogue of the above result was proved by Cao [**Co**]. Define (2.21)

$$\tilde{Y}(X)_{\alpha\bar{\beta}} \doteqdot \frac{\partial}{\partial t}R_{\alpha\bar{\beta}} + R_{\alpha\bar{\gamma}}R_{\gamma\bar{\beta}} + \nabla_\gamma R_{\alpha\bar{\beta}}X^\gamma + \nabla_{\bar{\gamma}}R_{\alpha\bar{\beta}}X^{\bar{\gamma}} + R_{\alpha\bar{\beta}\gamma\bar{\delta}}X^\gamma X^{\bar{\delta}} + \frac{R_{\alpha\bar{\beta}}}{t}$$

for any $(1,0)$-vector $X = X^\gamma \frac{\partial}{\partial z^\gamma}$ and where $X^{\bar{\gamma}} \doteqdot \overline{X^\gamma}$.

THEOREM 2.7. *If* $(\mathcal{M}^n, g(t))$ *is a complete solution to the Kähler–Ricci flow with bounded nonnegative bisectional curvature, then*

$$\left(\tilde{Y}(X)_{\alpha\bar{\beta}}\right) \geq 0 \tag{2.22}$$

for any $(1,0)$*-vector* X.

B. The linear trace LYH inequalities. Recall that the Lichnerowicz Laplacian acting on symmetric 2-tensors is given by

$$\Delta_L h_{ij} = \Delta h_{ij} + 2R_{ikjl}h_{kl} - R_{ik}h_{jk} - R_{jk}h_{ik}.$$

The following result was proved by Chow and Hamilton [**CH**].

THEOREM 2.8. *Suppose* $(M^n, g(t))$, $t \in [0,T)$, *is a complete solution of Ricci flow with bounded nonnegative curvature operator. Assume that* $h(x,t)$ *with* $h(0) \geq 0$, *is a solution to*

$$\frac{\partial}{\partial t} h = \Delta_L h \tag{2.23}$$

and $|h(t)|$ *is bounded. Then* $h(t) \geq 0$ *for* $t \in [0,T)$ *and for any vector* X *we have*

$$Z(X) \doteqdot \nabla_i \nabla_j h_{ij} + R_{ij} h_{ij} + 2(\nabla_j h_{ij}) X_i + h_{ij} X_i X_j + \frac{H}{2t} \geq 0, \tag{2.24}$$

where $H \doteqdot g^{ij} h_{ij}$.

If h_{ij} is the Ricci tensor, one can check that (2.23) holds. Hence the above theorem implies the trace form of (2.18) (choosing $U = X \wedge W$ and tracing the variable W). Also (2.23) is satisfied by the variational tensors of a family of solutions to Ricci flow in a certain sense. Namely, (2.23) is a linearization of the Ricci flow. This explains the term *linear trace.*

In [**NT**], with the motivation of study the Liouville property of plurisubharmonic functions, the Kähler analogue of the above was proved. Let $(\mathcal{M}^n, g(t))$, $t \in [0,T)$, be a complete noncompact solution of the Kähler–Ricci flow with bounded nonnegative bisectional curvature. Let $h = \sqrt{-1} h_{\alpha\bar{\beta}} dz^\alpha \wedge dz^{\bar{\beta}}$ be real $(1,1)$ form satisfying (2.16). Define

$$\begin{aligned} Z(h,V) \doteqdot & \frac{1}{2} g^{\alpha\bar{\beta}} \left(\nabla_{\bar{\beta}} \operatorname{div}(h)_\alpha + \nabla_\alpha \operatorname{div}(h)_{\bar{\beta}} \right) + R_{\alpha\bar{\beta}} h_{\beta\bar{\alpha}} \\ & + g^{\alpha\bar{\beta}} \left(\operatorname{div}(h)_\alpha V_{\bar{\beta}} + \operatorname{div}(h)_{\bar{\beta}} V_\alpha \right) + h_{\alpha\bar{\beta}} V_{\bar{\beta}} V_{\bar{\alpha}} + \frac{H}{t}, \end{aligned}$$

where V is a vector field of type $(1,0)$, H and $\operatorname{div}(h)_\alpha$ are as before.

THEOREM 2.9. *Suppose that* $(\mathcal{M}^n, g(t))$, $t \in [0,T)$, *is a complete solution of the Kähler–Ricci flow with bounded nonnegative bisectional curvature and* $h \geq 0$ *satisfying (2.16) and (2.14). Then*

$$Z(h,V) \geq 0 \tag{2.25}$$

on $\mathcal{M} \times [0,T)$ *for any vector field* V *of type* $(1,0)$.

As before, expressed in terms operators $\bar{\partial}^*$ and ∂^*, the result asserts

$$Z = \frac{1}{2}\left(-\partial^* \bar{\partial}^* + \bar{\partial}^* \partial^*\right) h + \bar{\partial}^* h(V) - \partial^* h(\bar{V}) + \operatorname{Ric}(h) + h(V, \bar{V}) + \frac{\Lambda h}{t} \geq 0. \tag{2.26}$$

Here $\operatorname{Ric}(h)$ is the contraction of h by the Ricci form.

C. A matrix LYH inequality for forward conjugate heat equation. Now we consider the *forward conjugate heat equation:*

$$\left(\frac{\partial}{\partial t} - \Delta\right) u(x,t) = R(x,t)u(x,t). \tag{2.27}$$

Here $R(x,t)$ is the scalar curvature. The following result was proved in [**N4**].

THEOREM 2.10. *Let $(M, g(t))$ be a solution to Kähler-Ricci flow defined on $M \times [0,T]$ (for some $T > 0$) with nonnegative bisectional curvature. In the case that M is complete noncompact, assume further that the bisectional curvature is bounded on $M \times [0,T]$. Let u be a positive solution to (2.27). Then*

$$u_{\alpha\bar{\beta}} + \frac{u}{t} g_{\alpha\bar{\beta}} + uR_{\alpha\bar{\beta}} + u_\alpha V_{\bar{\beta}} + u_{\bar{\beta}} V_\alpha + uV_\alpha V_{\bar{\beta}} \geq 0 \tag{2.28}$$

for any $(1,0)$ vector field V.

By choosing the minimizing vector V in (2.28) we have

$$\nabla_\alpha \nabla_{\bar{\beta}} \log u + R_{\alpha\bar{\beta}} + \frac{1}{t} g_{\alpha\bar{\beta}} \geq 0. \tag{2.29}$$

When $m = 1$, namely M is a Riemann sphere, we have the following estimate which is very similar to Li-Yau's (2.3)

$$\Delta \log u + R + \frac{1}{t} \geq 0. \tag{2.30}$$

This result was first proved by Chow and Hamilton [**CH**].

The result was discovered by the interpolation consideration which shall be explained next.

D. Interpolations. The following estimates was proved by Chow [**Ch3**], which links the Li-Yau estimate to the linear trace estimate (2.30) when M is the Riemann sphere with positive curvature.

PROPOSITION 2.11. *Given $\varepsilon > 0$, if $(M^2, g(t))$ is a solution to the ε-Ricci flow*

$$\frac{\partial}{\partial t} g_{ij} = -2\varepsilon R_{ij} = -\varepsilon R g_{ij} \tag{2.31}$$

on a closed surface with $R > 0$ and if u is a positive solution to

$$\frac{\partial}{\partial t} u = \Delta u + \varepsilon R u, \tag{2.32}$$

then

$$\frac{\partial}{\partial t} \log u - |\nabla \log u|^2 + \frac{1}{t} = \Delta \log u + \varepsilon R + \frac{1}{t} \geq 0. \tag{2.33}$$

There are two different high dimensional generalizations of the above result. They are all in the category of Kähler manifolds. The first one proved in [**N4**] is straightforward and connects Theorem 2.10 with Theorem 2.4 (more precisely, connects (2.11) with (2.29).

For any $\epsilon > 0$, we consider the *ϵ-Kähler-Ricci flow*:

$$\frac{\partial}{\partial t} g_{\alpha\bar{\beta}}(x,t) = -\epsilon R_{\alpha\bar{\beta}}(x,t). \tag{2.34}$$

Consider the positive solution u to the parabolic equation:

$$\left(\frac{\partial}{\partial t} - \Delta\right) u(x,t) = \epsilon R(x,t) u(x,t). \tag{2.35}$$

We shall call (2.35) *forward conjugate heat equation*, since it is the adjoint of the *backward heat equation* $\left(\frac{\partial}{\partial t} + \Delta\right) v = 0$.

THEOREM 2.12. *Assume that the complete solution $(M, g(t))$ (defined on $M \times [0,T]$ for some $T > 0$) to (2.34) has nonnegative bisectional curvature. In the case that M is noncompact, assume further that the bisectional curvature of $g(t)$ is uniformly bounded on $M \times [0,T]$. Let u be a positive solution to (2.35). Then*

$$\nabla_\alpha \nabla_{\bar{\beta}} \log u + \epsilon R_{\alpha\bar{\beta}} + \frac{u}{t} g_{\alpha\bar{\beta}} \geq 0. \tag{2.36}$$

It is obvious that Theorem 2.12 generalizes (2.33). The second one proved in [**N3**], connects Theorem 2.5 with Theorem 2.9 (more precisely (2.17) and (2.26)).

THEOREM 2.13. *Let $(M, g(t))$ be a solution to (2.34) with bounded nonnegative bisectional curvature. Let $h(x,t) \geq 0$ be a real $(1,1)$ form satisfying (2.16). Then*

$$Z_{h,\epsilon} = \frac{1}{2}\left(-\partial^*\bar{\partial}^* + \bar{\partial}^*\partial^*\right) h + \bar{\partial}^* h(V) - \partial^* h(\bar{V}) + \epsilon \operatorname{Ric}(h) + h(V, \bar{V}) + \frac{\Lambda h}{t} \geq 0. \tag{2.37}$$

Moreover, the equality holds for some $t > 0$ implies that $(M, g(t))$ is an expanding gradient soliton, provided that $h_{\alpha\bar{\beta}}(x,t) > 0$ and M is simply-connected.

It is less obvious that Theorem 2.13 also generalizes (2.33). To see this, first observe that $H = \Lambda h$ satisfies

$$\left(\frac{\partial}{\partial t} - \Delta\right) H = \epsilon \operatorname{Ric}(h) \tag{2.38}$$

which generalizes (2.32). Now we restrict ourselves to the case in which h is closed. Using the Kähler identities

$$\partial\Lambda - \Lambda\partial = \bar{\partial}^*, \quad \bar{\partial}\Lambda - \Lambda\bar{\partial} = -\partial^* \tag{2.39}$$

we have that

$$Z_{h,\epsilon} = \Delta H + \partial H(V) + \bar{\partial} H(\bar{V}) + \epsilon \operatorname{Ric}(h) + h(V, \bar{V}) + \frac{H}{t} \geq 0. \tag{2.40}$$

In the case $m = 1$, this gives the estimate (2.33), noticing that $-\Delta' = -\Delta'' = \Delta$.

Since (2.38) and (2.40) make sense for any (p,p) form $h \geq 0$, it is natural question to ask if (2.40) holds for (p,p) form $h \geq 0$. Theorem 2.12 shows that it is the case for (m,m) forms. We conjecture that it is the case for general nonnegative (p,p) forms.

E. LYH for the conjugate heat equation. Assume that $(M, g(t))$ is a solution to Ricci flow on $M \times [0,T]$. Now we consider the *conjugate heat equation:*

$$\left(\frac{\partial}{\partial \tau} - \Delta + R\right) u(x,\tau) = 0. \tag{2.41}$$

Here $\tau = T - t$. This equation is the adjoint of the heat equation $\left(\frac{\partial}{\partial t} - \Delta\right) v = 0$. In [**P**], the following estimate was discovered by Perelman.

THEOREM 2.14. *Assume that* $u = \frac{e^{-f}}{(4\pi\tau)^{\frac{n}{2}}}$ *is the fundamental solution to (2.41) with* $u(o,0) = \delta_o(x)$*. Then*

$$\tau(2\Delta f - |\nabla f|^2 + R) + f - n \leq 0. \tag{2.42}$$

Note that the result is equivalent to the space time Lapalacian comparison

$$\Delta \hat{L} + \hat{L}_\tau \leq 2n \tag{2.43}$$

with $\hat{L} = 4\tau f$. Comparing the above with Theorem 2.2, the most striking part is that the result holds without any curvature assumption. The detailed proof of Theorem 2.14 was given in [**N5**]. The key equation is the following one discovered by Perelman

$$\left(\frac{\partial}{\partial \tau} - \Delta + R\right) V = -2\tau \left|R_{ij} + \nabla_i \nabla_j f - \frac{1}{2\tau} g_{ij}\right|^2 u \tag{2.44}$$

where

$$V = \left[\tau\left(2\Delta f - |\nabla f|^2 + R\right) + f - n\right] u.$$

The equation (2.44) can be derived from the following simpler one which resembles (2.4)

$$\left(\frac{\partial}{\partial \tau} - \Delta + R\right) v_0 = -2|R_{ij} + f_{ij}|^2 u \tag{2.45}$$

if $v_0 = (2\Delta f - |\nabla f|^2 + R)u$.

Combining (2.42) with (2.41) we have that

$$2f_\tau + |\nabla f|^2 - R + \frac{f}{\tau} \le 0 \tag{2.46}$$

which yields for any $\gamma(\tau)$ joining (x_1, τ_1) with (x_2, τ_2),

$$\begin{aligned}\sqrt{\tau_2} f(x_2, \tau_2) - \sqrt{\tau_1} f(x_1, \tau_1) &= \int_{\tau_1}^{\tau_2} \sqrt{\tau}\left(f_\tau + \frac{f}{2\tau} + \langle \nabla f, \gamma' \rangle\right) d\tau \\ &\le \frac{1}{2}\int_{\tau_1}^{\tau_2} \sqrt{\tau}\left(|\gamma'|^2 + R\right) d\tau.\end{aligned}$$

Now observing as before that $\lim_{\tau \to 0} \sqrt{\tau} f(o, \tau) \le 0$, we have that

$$f(x, \bar{\tau}) \le \frac{1}{2\sqrt{\bar{\tau}}} \inf_\gamma \int_0^{\bar{\tau}} \sqrt{\tau}\left(|\gamma'|^2 + R\right) d\tau. \tag{2.47}$$

The right hand side is called *reduced distance*, denoted by $\ell(x, \tau)$. There exists another way of proving (2.47) via the fact that

$$\left(\frac{\partial}{\partial \tau} - \Delta + R\right) \hat{u}(x, \tau) \le 0$$

where $\hat{u}(x, \tau) = \frac{e^{-\ell(x,\tau)}}{(4\pi\tau)^{\frac{n}{2}}}$ satisfying that $\lim_{\tau \to 0} \hat{u}(x, \tau) = \delta_o(x)$. For the purpose of better presentation we work with t instead. Assume that $H(x, t; y, T)$ (with $t \le T$) is the fundamental solution of (2.41) and $h(y, s; x, t)$ (with $s \ge t$) is the fundamental solution to the heat equation. The well-known duality asserts that

$$h(z, s; x, t) = H(x, t; z, s).$$

Now we can check that

$$I(s) = \int_M h(z, s; x, t)\hat{u}(z, s; y, T)\, d\mu_s(z)$$

is monotone increasing in s. Here $\hat{u}(z, s; y, T)$ corresponds to the *reduced distance* starting from (y, T). Hence

$$\begin{aligned} H(x, t; y, T) = h(y, T; x, t) &= \lim_{s \to T} I(s) \\ &\ge \lim_{s \to t} I(s) = \hat{u}(x, t; y, T)\end{aligned}$$

from which (2.47) follows easily.

2.3. Hypersurface flows in $\mathbb{R}^{n+1}$. The LYH estimates for the hypersurface flow in $\mathbb{R}^{n+1}$ were first proved for mean curvature flow by Hamilton [**H4**] and Gauss curvature flow by Chow [**Ch1**]. Here we present the work of Andrews on much more general class of hypersurface flows, which is a lot less computational than the original works of Hamilton and Chow. Our presentation follows that of [**An1, An2**].

First we collect some formulae and results needed on the hypersurface's curvature flow. In our discussion we assume that $X_0 : M \to \mathbb{R}^{n+1}$ is a immersed closed n-dimensional hypersurface. Consider the 1-family of smooth hypersurfaces $X : M \times [0, T) \to \mathbb{R}^{n+1}$ satisfying

$$\frac{\partial X}{\partial t}(x,t) = -f(x,t)\nu(x,t) \tag{2.48}$$

where $f(x,t)$ is a smooth function and $\nu(x,t)$ is the outer unit normal. We assume that $M_t = X(M,t)$ are compact convex hypersurfaces. The following are well-known.

Let g_{ij} be the induced metric, $d\mu$ be the induced measure. Then

$$\frac{\partial}{\partial t} g_{ij} = -2fh_{ij}; \tag{2.49}$$

$$\frac{\partial}{\partial t} d\mu = -fHd\mu \tag{2.50}$$

$$\frac{\partial}{\partial t} h_{ij} = \nabla_i \nabla_j f - fh_{ik}h_{jl}g^{kl}. \tag{2.51}$$

Usually, we write that $f(x,t) = F(W(x,t))$, where $W = \{h^i_j\}$ is the Weingartan map. Then

$$\frac{\partial}{\partial t} W = g^* \left(\mathrm{Hess}_\nabla F\right) + FW^2 \tag{2.52}$$

$$\frac{\partial}{\partial t} H = \mathrm{tr}\left(g^* \left(\mathrm{Hess}_\nabla F\right)\right) + F|A|^2 \tag{2.53}$$

$$\frac{\partial}{\partial t} \nu = \nabla F \tag{2.54}$$

$$\frac{\partial}{\partial t} F = \dot{F}\left(g^* \left(\mathrm{Hess}_\nabla F\right)\right) + F\dot{F}(W^2). \tag{2.55}$$

A. The support function. The support function is very useful for the convex hypersurface flow. Especially because it can be used to reparametrize the surface via the Gauss map and greatly reduce the computation in deriving the LYH inequalities for the hypersurfaces flow.

Let $s(z) : \mathbb{S}^n \to \mathbb{R}$ be the support function of M (precisely, $X : M \to \mathbb{R}^{n+1}$). It can be defined by

$$s(z) = \langle z, X(\nu^{-1}(z))\rangle$$

where $\nu(x) = \nu(X(x))$, namely the normal of the image. We may recover the immersion after the reparametrization by

$$\bar{X}(z) = s(z)z + \bar{\nabla} s$$

where $\bar{\nabla}$ is the standard connection of $\mathbb{S}^n$. For simplicity *when there is no confusion* we just write $X(z)$. Geometrically it is very clear that

$$X(z) = s(z)z + a(z)$$

for some vector $a(z)$, tangent to the sphere (normal to z). Just differentiate we have that

$$\begin{aligned} dX(u) &= \langle \bar{\nabla} s, u\rangle z + su + D_u a \\ &= \langle \bar{\nabla} s, u\rangle z + su + \bar{\nabla}_u a - \bar{g}(u, a)z. \end{aligned}$$

Here $\bar{g}$ is the metric of the sphere D is the direction derivative of $\mathbb{R}^{n+1}$. Noticing that $dX(u)$ is tangent to the sphere we have that

$$\langle \bar{\nabla} s, u\rangle = \bar{g}(u, a) \qquad \text{hence} \quad a = \bar{\nabla} s.$$

Namely we have the equation

$$X(z) = s(z)z + \bar{\nabla} s. \tag{2.56}$$

More precisely we have shown that $X(x) = \bar{X}(\nu^{-1}(x)) = s(z)z + \bar{\nabla} s$. If we extend s to $\mathbb{R}^{n+1}$ homogenously as a degree 1 function we have that $D_z s = s$. Hence

$$Ds = s(z)z + \bar{\nabla} s$$

which implies that

$$X(z) = Ds. \tag{2.57}$$

This gives the immersion from the support function.

For the parametrization via Gauss map as above, it changes the computation on M to computation with respect to the fixed standard metric of $\mathbb{S}^n$.

LEMMA 2.15.

$$W^{-1} = \bar{g}^* \left(\bar{\nabla}\bar{\nabla} s + s\,\mathrm{id}\right). \tag{2.58}$$

PROOF. Observe that $\nu(X(z)) = z$. Hence we have that $W^{-1} = dX$. Now using (2.56) we just compute

$$dX(u) = \bar{\nabla}_u \bar{\nabla} s + su$$

and we have the result. □

The following lemma shows the relation on the time derivatives with respect to different parametrizations.

LEMMA 2.16. *Let $Q(x,t)$ and $\bar{Q}(z,t)$ be the quantities related by $Q(x,t) = \bar{Q}(z(x,t),t)$. Then*

$$\frac{\partial Q}{\partial t} = \frac{\partial \bar{Q}}{\partial t} + h^{-1}(\nabla F, \nabla Q). \tag{2.59}$$

Here h^{-1} stands for the inverse of the second fundamental form.

PROOF. Using the connection $X(x,t) = \bar{X}(z(x,t),t)$ we have that

$$\begin{aligned}-F(x,t)\nu(x,t) &= \frac{\partial}{\partial t}X(x,t)\\ &= \frac{\partial}{\partial t}\bar{X}(z,t) + d\bar{X}\left(\frac{\partial z}{\partial t}\right).\end{aligned}$$

Recalling $\bar{X}(z,t) = s(z,t)z + \bar{\nabla}s$ we have that

$$\frac{\partial}{\partial t}\bar{X}(z,t) = \left(\frac{\partial}{\partial t}s\right)z + \bar{\nabla}\left(\frac{\partial}{\partial t}s\right).$$

By comparing the normal and tangential components we have that

$$\begin{aligned}\frac{\partial}{\partial t}s &= -F = -\Phi(z,t)\\ \frac{\partial z}{\partial t} &= -(d\bar{X})^{-1}\left(\bar{\nabla}\left(\frac{\partial}{\partial t}s\right)\right) = (d\bar{X})^{-1}(\bar{\nabla}F) = (d\bar{X})^{-1}(\bar{\nabla}\Phi).\end{aligned}$$

Here when F is viewed as a function on $\mathbb{S}^n$ we write as $\Phi(z,t)$. Then

$$\frac{\partial}{\partial t}s = -\Phi \tag{2.60}$$

$$\frac{\partial z}{\partial t} = (d\bar{X})^{-1}(\bar{\nabla}\Phi). \tag{2.61}$$

Now we have that

$$\begin{aligned}\frac{\partial Q}{\partial t} &= \frac{\partial \bar{Q}}{\partial t} + \left\langle \bar{\nabla}\bar{Q}, \frac{\partial z}{\partial t}\right\rangle\\ &= \frac{\partial \bar{Q}}{\partial t} + \langle \bar{\nabla}\bar{Q}, (d\bar{X})^{-1}(\bar{\nabla}\Phi)\rangle.\end{aligned}$$

To get what we need we use the following observations. First we have also (from (2.54))

$$\frac{\partial z}{\partial t} = dX(\nabla F).$$

Hence

$$\left\langle \bar{\nabla}\bar{Q}, \frac{\partial z}{\partial t}\right\rangle = \langle \bar{\nabla}\bar{Q}, dX(\nabla F)\rangle = g(\nabla F, dX^{-1}(\bar{\nabla}\bar{Q})).$$

On the other hand for any tangent vector u we have that

$$\begin{aligned}g(\nabla Q, u) &= uQ = dQ(u)\\ &= d\bar{Q}(d\nu(dX(u)))\\ &= \langle d\nu(dX(u)), \bar{\nabla}Q\rangle\\ &= g(W(u), dX^{-1}(\bar{\nabla}\bar{Q})).\end{aligned}$$

Let $uW^{-1}(\nabla F)$ we have that

$$g(\nabla Q, W^{-1}(\nabla F)) = \left\langle \bar{\nabla}\bar{Q}, \frac{\partial z}{\partial t} \right\rangle.$$

This proves the claim. □

When $Q = F$ (then $\bar{Q}(z,t) = \Phi(z,t)$, we have that

$$\frac{\partial}{\partial t}F - h^{-1}(\nabla F, \nabla F) = \frac{\partial}{\partial t}\Phi. \tag{2.62}$$

Note the left hand side often appears in the differential Harnack inequality.

It turns out that it is better that we consider $\Phi(X) = F(X^{-1})$. In terms of the support function we examine the equation (2.60) for several flows.

Mean curvature flow: $F(W) = \sum \lambda_i$, where W is the Weingartan map, hence $\Phi(A) = \sum \frac{1}{\mu_i}$, where μ_i are eigenvalues of $A = W^{-1} = \bar{g}^*\mathrm{Hess}_{\bar{\nabla}} s + s\,\mathrm{id}$. Then the mean curvature flow can be expressed as

$$\frac{\partial s}{\partial t} = -\operatorname{tr}\left((\bar{g}^*\mathrm{Hess}_{\bar{\nabla}} s + s\,\mathrm{id})^{-1}\right). \tag{2.63}$$

Gauss curvature flow: $F(W) = \Pi\lambda_i$, where W is the Weingartan map, hence $\Phi(A) = \Pi\frac{1}{\mu_i}$, where μ_i are eigenvalues of $A = W^{-1} = \bar{g}^*\mathrm{Hess}_{\bar{\nabla}} s + s\,\mathrm{id}$. Then the Gauss curvature flow can be expressed as

$$\frac{\partial s}{\partial t} = -\frac{1}{\det(\bar{g}^*\mathrm{Hess}_{\bar{\nabla}} s + s\,\mathrm{id})} \tag{2.64}$$

which was the form more studied from PDE point of view.

Harmonic mean curvature flow: $F(W) = \frac{1}{\sum \frac{1}{\lambda_i}}$, where W is the Weingartan map, hence $\Phi(A) = \frac{1}{\sum \mu_i}$, where μ_i are eigenvalues of $A = W^{-1} = \bar{g}^*\mathrm{Hess}_{\bar{\nabla}} s + s\,\mathrm{id}$. Then the harmonic mean curvature flow can be expressed as

$$\frac{\partial s}{\partial t} = -\frac{1}{\bar{\Delta} s + ns}. \tag{2.65}$$

The $\frac{1}{H}$-curvature flow: $F(W) = -\frac{1}{\sum \lambda_i}$, where W is the Weingartan map, hence $\Phi(A) = -\frac{1}{\sum \frac{1}{\mu_i}}$, where μ_i are eigenvalues of $A = W^{-1} = \bar{g}^*\mathrm{Hess}_{\bar{\nabla}} s + s\,\mathrm{id}$. Then the $\frac{1}{H}$-curvature flow mean curvature flow can be expressed as

$$\frac{\partial s}{\partial t} = \frac{1}{\operatorname{tr}\left((\bar{g}^*\mathrm{Hess}_{\bar{\nabla}} s + s\,\mathrm{id})^{-1}\right)}. \tag{2.66}$$

B. The LYH inequality. Using the support function and parametrization via the Gauss map, we have the following result.

THEOREM 2.17. *Assume that $-\Phi$ is α-concave for some $\alpha < 0$. Then*

$$\frac{\partial \Phi}{\partial t} + \frac{\alpha \Phi}{(\alpha - 1)t} \geq 0. \tag{2.67}$$

Using (2.62) we have that

$$\frac{\partial}{\partial t} F - h^{-1}(\nabla F, \nabla F) + \frac{\alpha F}{(\alpha - 1)t} \geq 0.$$

$\Psi : T^*\mathbb{S}^n \otimes T\mathbb{S}^n \to \mathbb{R}^n$ is called α-concave if

$$\ddot{\Psi} \leq \frac{\alpha - 1}{\alpha \Psi} \dot{\Psi} \otimes \dot{\Psi}.$$

We first check the applicability of the theorem to two example curvature flows: the Gauss curvature flow and the mean curvature flow.

Recall the following useful computational lemma.

LEMMA 2.18. *When A is diagonal,*

$$\dot{F}(A) = diag\left(\frac{\partial f(\lambda)}{\partial \lambda_1}, \cdots, \frac{\partial f(\lambda)}{\partial \lambda_n}\right); \tag{2.68}$$

$$\ddot{F}(A)(X, X) = \sum_{p,q} \frac{\partial^2 f}{\partial \lambda_p \lambda_q} X_p^p X_q^q + \sum_{p \neq q} \frac{\frac{\partial f}{\partial \lambda_p} - \frac{\partial f}{\partial \lambda_q}}{\lambda_p - \lambda_q} \left(X_p^q\right)^2. \tag{2.69}$$

Hence if $\frac{\partial f}{\partial \lambda_i} > 0$ at some A, then $\dot{F}$ is positive definite. And if f is convex (concave), F is convex (concave).

Mean curvature flow: $F(W) = \sum \lambda_i$, where W is the Weingartan map, hence $\Phi(A) = \sum \frac{1}{\mu_i}$, where μ_i are eigenvalues of $A = W^{-1} = g^* \mathrm{Hess}_{\overline{\nabla}} s + s\,\mathrm{id}$. $\Psi = -\Phi = -\sum \frac{1}{\mu_i}$. Using the computational lemma we have that

$$\begin{aligned}\ddot{\Psi}(X, X) &= \sum_p \frac{2}{\mu_p^3} |X_p^p|^2 - \sum_{p \neq q} \frac{\mu_p + \mu_q}{\mu_p^2 \mu_q^2} |X_p^q|^2 \\ &= -\sum_{pq} \frac{\mu_p + \mu_q}{\mu_p^2 \mu_q^2} |X_p^q|^2\end{aligned}$$

and

$$\dot{\Psi}(X) = \frac{1}{\mu_p^2} X_p^p$$

$$\frac{(\dot{\Psi}(X))^2}{\Psi} = -\frac{\left(\sum \frac{1}{\mu_p^2} X_p^p\right)\left(\sum \frac{1}{\mu_q^2} X_q^q\right)}{\sum \frac{1}{\mu_i}}.$$

Using $\mu_p + \mu_q \geq 2\sqrt{\mu_p\mu_q}$, we have that

$$\sum_{pq} \frac{\mu_p + \mu_q}{\mu_p^2\mu_q^2}|X_p^q|^2 \geq \sum_{pq} \frac{2}{\mu_p^{1+1/2}\mu_q^{1+1/2}}|X_p^q|^2$$

$$\geq \sum_{p} \frac{2}{\mu_p^{1+1/2}\mu_p^{1+1/2}}|X_p^p|^2$$

$$\geq 2\frac{\left(\sum \frac{1}{\mu_p^2}X_p^p\right)\left(\sum \frac{1}{\mu_q^2}X_q^q\right)}{\sum \frac{1}{\mu_i}}.$$

Hence Ψ is -1-concave for the mean curvature flow and we have Hamilton's

$$\frac{\partial}{\partial t}F - h^{-1}(\nabla F, \nabla F) + \frac{F}{2t} \geq 0.$$

Gauss curvature flow: In fact we consider the more general class $F(W) = (\Pi\lambda_i)^\beta$, where W is the Weingartan map, hence $\Phi(A) = \left(\Pi\frac{1}{\mu_i}\right)^\beta$, where μ_i are eigenvalues of $A = W^{-1} = g^*\mathrm{Hess}_{\overline{\nabla}} s + s\,\mathrm{id}$. $\Psi = -\Phi = -\left(\Pi\frac{1}{\mu_i}\right)^\beta$. Using the computational lemma we have that

$$\ddot{\Psi}(X,X) = \sum_{p\neq q} \frac{\beta^2\Psi}{\mu_p\mu_q}X_p^pX_q^q + \sum_p \frac{(\beta^2+\beta)\Psi}{\mu_p^2}|X_p^p|^2$$

$$+ \beta\Psi\sum_{p\neq q}\frac{1}{\mu_p\mu_q}|X_p^q|^2$$

$$= \sum_{pq} \frac{\beta^2\Psi}{\mu_p\mu_q}X_p^pX_q^q + \sum_{pq}\frac{\beta\Psi}{\mu_p\mu_q}|X_p^q|^2$$

which is negative since $\Psi < 0$, and

$$\dot{\Psi}(X) = \frac{\beta\Psi}{\mu_p}X_p^p$$

$$\frac{(\dot{\Psi}(X))^2}{\Psi} = \beta^2\Psi\left(\sum\frac{1}{\mu_p}X_p^p\right)^2.$$

Letting $Y_p^p = \frac{1}{\mu_p}X_p^p$ and observing

$$\left(\sum Y_p^p\right)^2 \leq n\sum |Y_p^p|^2$$

Hence

$$\begin{aligned}\ddot{\Psi}(X,X) &= \frac{(\dot{\Psi}(X))^2}{\Psi} + \sum_{pq} \frac{\beta\Psi}{\mu_p\mu_q}|X_p^q|^2 \\ &\leq \frac{(\dot{\Psi}(X))^2}{\Psi} + \sum_{p} \frac{\beta\Psi}{\mu_p\mu_p}|X_p^p|^2 \\ &\leq \frac{(\dot{\Psi}(X))^2}{\Psi} + \frac{1}{n\beta}\frac{(\dot{\Psi}(X))^2}{\Psi}.\end{aligned}$$

Hence the β-power Gauss curvature flow is $-n\beta$ concave and we have Chow's

$$\frac{\partial}{\partial t}F - h^{-1}(\nabla F, \nabla F) + \frac{n\beta F}{(1+n\beta)t} \geq 0.$$

Despite the generality of Theorem 2.17 the proof, which shall be given next, seems a lot easier than the ones for the special cases in [**H4, Ch1**].

C. PDE satisfied by the speed function. We shall prove Theorem 2.17 and discuss the $\frac{1}{H}$-flow case, to which Theorem 2.17 does not apply.

Start from (2.60). Take $\mathrm{Hess}_{\bar{\nabla}}$ on both sides and notice the commutativity of the $\frac{\partial}{\partial t}$ and $\mathrm{Hess}_{\bar{\nabla}}$ we have that

$$\frac{\partial}{\partial t}\left(\mathrm{Hess}_{\bar{\nabla}} s + \bar{g} s\right) = \mathrm{Hess}_{\bar{\nabla}} \Psi + \Psi \bar{g}.$$

Then

$$\frac{\partial}{\partial t} A = \bar{g}^* \mathrm{Hess}_{\bar{\nabla}} \Psi + \Psi \,\mathrm{id}\,. \tag{2.70}$$

This enables us to compute the time derivative of the speed function Ψ as follows

$$\begin{aligned}\frac{\partial \Psi}{\partial t} &= \dot{\Psi}(A)\left[\frac{\partial A}{\partial t}\right] \\ &= \dot{\Psi}(A)\left[\bar{g}^* \mathrm{Hess}_{\bar{\nabla}} \Psi + \Psi \,\mathrm{id}\right] \\ &= \mathcal{L}\Psi + \dot{\Psi}(A)\,[\mathrm{id}]\,\Psi. \qquad (2.71)\end{aligned}$$

Here $\dot{\Psi}(A)[B]$ is the previous $\dot{\Psi}(B)$. We write this way since it is more clear to specify that $\dot{\Psi}$ is computed at A. We shall return to the previous notation when *there is no confusion*. $\mathcal{L}\varphi = \dot{\Psi}(A)\left[\bar{g}^* \mathrm{Hess}_{\bar{\nabla}} \varphi\right]$.

For the proof we also need the PDE on $P := \frac{\partial}{\partial t}\Psi$. Let

$$Q = \frac{\partial}{\partial t} A = \bar{g}^* \mathrm{Hess}_{\bar{\nabla}} \Psi + \Psi \,\mathrm{id}\,.$$

Taking one more derivative on (2.71) we have that

$$\begin{aligned}\frac{\partial}{\partial t}P &= \ddot{\Psi}(Q,Q) + \dot{\Psi}(\bar{g}^{*}\mathrm{Hess}_{\overline{\nabla}}\Psi_t + \Psi_t\,\mathrm{id}) \\ &= \mathcal{L}P + \dot{\Psi}(\mathrm{id})P + \ddot{\Psi}(Q,Q).\end{aligned} \tag{2.72}$$

Note that the first equation in (2.71) is equivalent to

$$P = \dot{\Psi}(Q). \tag{2.73}$$

Hence in the case of Ψ being α-concave we have that

$$\frac{\partial}{\partial t}P \le \mathcal{L}P + \dot{\Psi}(\mathrm{id})P + \frac{\alpha-1}{\alpha}\frac{P^2}{\Psi}. \tag{2.74}$$

Let $L = tP + \frac{\alpha}{\alpha-1}\Psi$. Using (2.71) and (2.74) we then have that

$$\begin{aligned}\frac{\partial}{\partial t}L &\le \mathcal{L}(tP) + tP\dot{\Psi}(id) + t\frac{\alpha-1}{\alpha}\frac{P^2}{\Psi} + P \\ &\quad + \frac{\alpha}{\alpha-1}\mathcal{L}\Psi + \frac{\alpha}{\alpha-1}\Psi\dot{\Psi}(id) \\ &= \mathcal{L}L + \dot{\Psi}(\mathrm{id})L + \frac{P}{\Psi}\frac{\alpha-1}{\alpha}L.\end{aligned}$$

Theorem 2.17 follows from the maximum principle and the observation that $L \le 0$ at the initial time (noting that $\Psi < 0$ at $t = 0$).

If $\alpha = 1$, we are in the degenerate case in the definition of the α-concavity. If we also have that $\Psi \ge 0$, as in the $\frac{1}{H}$*-curvature flow* case, we can only conclude from the above proof that both

$$\sup_M \Psi_t \quad \text{and} \quad \sup_M \frac{\partial}{\partial t}(\log\Psi) = \sup_M \frac{\Psi_t}{\Psi}$$

are monotone non-increasing.

The following calculation shows that the $\frac{1}{H}$-curvature flow is indeed 1-concave. First $\Psi = \frac{1}{\sum \frac{1}{\mu_i}}$, hence

$$\frac{\partial \Psi}{\partial \mu_i} = \frac{\Psi^2}{\mu_i^2}$$

and for $i \ne j$

$$\frac{\partial^2 \Psi}{\partial \mu_i \partial \mu_j} = \frac{2\Psi^3}{\mu_i^2\mu_j^2},$$

for $i = j$

$$\frac{\partial^2 \Psi}{\partial \mu_i \partial \mu_j} = \frac{2\Psi^2}{\mu_i^3}\left(\frac{\Psi}{\mu_i} - 1\right).$$

Now compute

$$\begin{aligned}\ddot{\Psi}(X,X) &= \sum_{p\neq q} \frac{2\Psi^3}{\mu_p^2\mu_q^2} X_p^p X_q^q + \sum_p \frac{2\Psi^2}{\mu_p^3}\left(\frac{\Psi}{\mu_p} - 1\right)|X_p^p|^2 \\ &\quad - \sum_{p\neq q} \frac{\Psi^2(\mu_p+\mu_q)}{\mu_p^2\mu_q^2}|X_p^q|^2 \\ &= \sum_{p,q} \frac{2\Psi^3}{\mu_p^2\mu_q^2} X_p^p X_q^q - \sum_{p,q} \frac{\Psi^2(\mu_p+\mu_q)}{\mu_p^2\mu_q^2}|X_p^q|^2.\end{aligned}$$

On the other hand

$$\frac{\left(\dot{\Psi}(X)\right)^2}{\Psi} = \Psi^3\left(\sum \frac{1}{\mu_p^2} X_p^p\right)^2.$$

If we let $Y_p^q = \frac{1}{\mu_p\mu_q} X_p^q$ we have that

$$\begin{aligned}\ddot{\Psi}(X,X) &= 2\Psi^3 \sum Y_p^p Y_q^q - \Psi^2 \sum (\mu_p+\mu_q)\left(Y_p^q\right)^2 \\ &\leq 2\Psi^3\left(\sum Y_p^p\right)^2 - 2\Psi^2 \sum \sqrt{\mu_p\mu_q}(Y_p^q)^2 \\ &\leq 2\Psi^3\left(\sum Y_p^p\right)^2 - 2\Psi^2 \sum \mu_p (Y_p^p)^2 \\ &\leq 0.\end{aligned}$$

Since there exists a situation that all the inequalities hold equality, we conclude that Ψ is only concave.

At last, we examine the equation (2.71) for concrete examples.

Mean curvature flow: $\Psi = -\sum \frac{1}{\mu_i} = -H$. Hence, we have that $\mathcal{L}\varphi = \frac{1}{\mu_i^2}\bar{\nabla}_i\bar{\nabla}_i\varphi = (W^2)^{ij}\bar{\nabla}_i\bar{\nabla}_j\varphi$. Here W^2 is the square of the Weingarten map. We abuse ^{ij} to just mean the (i,j)-th component of W^2. The equation satisfied by the speed function (on sphere $\mathbb{S}^n$) is

$$\frac{\partial}{\partial t}\Psi = (W^2)^{ij}\bar{\nabla}_i\bar{\nabla}_j\Psi + |W|^2\Psi.$$

Gauss curvature flow: $\Psi = -\frac{1}{\Pi\mu_i} = -K$. Hence $\mathcal{L}\varphi = (-\Psi)W^{ij}\bar{\nabla}_i\bar{\nabla}_j\varphi$. The equation satisfied by the speed function (on sphere $\mathbb{S}^n$) is

$$\frac{\partial}{\partial t}\Psi = (-\Psi)(W)^{ij}\bar{\nabla}_i\bar{\nabla}_j\Psi - \Psi^2 H.$$

Harmonic mean curvature flow: $\Psi = -\frac{1}{\sum\mu_i}$, $\mathcal{L}\varphi = \Psi^2\bar{\Delta}\varphi$. The equation satisfied by the speed function is

$$\frac{\partial\Psi}{\partial t} = \Psi^2\bar{\Delta}\Psi + n\Psi^3.$$

The $\frac{1}{H}$*-flow:* $\Psi = \frac{1}{\sum \frac{1}{\mu_i}}$, and $\mathcal{L}\varphi = \Psi^2 (W^2)^{ij} \bar{\nabla}_i \bar{\nabla}_j \varphi$. The equation satisfied by the speed function is

$$\frac{\partial \Psi}{\partial t} = \Psi^2 (W^2)^{ij} \bar{\nabla}_i \bar{\nabla}_j \Psi + \Psi^3 |W|^2.$$

3. Implied monotonicity

The LYH inequality often has some immediate consequence on the monotonicity. For instance (2.2) implies that $t^{\frac{n}{2}} u(x,t)$ is monotone increasing in t. In the case that h is closed $(1,1)$ form, (2.17) concludes that

$$\frac{\partial}{\partial t} H + \partial H(V) + \bar{\partial} H(\bar{V}) + h(V, \bar{V}) + \frac{H}{t} \geq 0$$

which in turn says that $tH(x,t)$ is monotone increasing. This also applies to (2.40). Similar conclusions can be drawn from (2.18), (2.22), (2.24) etc. In this section we shall discuss those less obvious monotonicities, mostly in the integral form, implied by the LYH inequalities. They can be roughly divided into three different classes. The first type is obtained by applying the matrix LYH to submanifolds, or high order symmetric functions in stead of the trace. The second type is derived using a consideration from thermodynamics. This includes Perelman's entropy formula. The third type, which contains Hamilton's entropy formula for Ricci flow on surfaces and the Gauss curvature flow, is inferred from the long time existence and an ODE consideration. For the later two types, not the result but the proof of the LYH inequalities is needed.

3.1. Linear heat equation.

Again, we present the linear heat equation due to its simplicity. Most monotonicity derived for the linear heat equation has its nonlinear analogue for Ricci flow. This is somewhat striking.

A. From the matrix LYH inequalities. It was observed first by Hamilton [**H6**] that Theorem 2.3 implies several previously known monotonicity formulae, including Huisken's monotonicity formula for the mean curvature flow in $\mathbb{R}^n$, and Struwe's monotonicity formula for the harmonic maps from Euclidean domains.

Let (M, g) be as in Theorem 2.3. Let $T > 0$ and $\tau = T - t$. Now let $k(x, \tau)$ be the fundamental solution to the *backward heat equation*:

$$\left(\frac{\partial}{\partial \tau} - \Delta \right) k(x, \tau) = 0.$$

Mean curvature flow: Let $\mathcal{V}$ be the family of submanifolds evolved by the mean curvature flow. More precisely, let $X(\cdot, t) : V^v \to M$ be a family of embeddings (v is the dimension of V) satisfying (2.48) with f being the mean curvature $H(x,t)$. Let $\vec{H} = H\nu$ be the mean curvature vector with ν being a unit normal. We use D to denote the differentiation with respect to the metric and connection of M. We use $i, j, k, \cdots$ for basis of vectors tangent to V and $\alpha, \beta, \gamma, \cdots$ for the basis of vectors normal to V. (More precisely $X(x,t)$ satisfies $\frac{\partial}{\partial t} X(x,t) = -\vec{H}(x,t)$ with $\vec{H} = (-\sum_i D_{e_i} e_i)^{\perp}$.) Let $H^\alpha = g^{ij} h^\alpha_{ij}$ where h^α_{ij} is the second fundamental form. We also have the following equations.

$$\frac{\partial}{\partial t} g_{ij} = -2 g_{\alpha\beta} H^\alpha h^\beta_{ij}, \tag{3.1}$$

$$\frac{\partial}{\partial t} d\mu = -|H|^2 d\mu, \tag{3.2}$$

$$\Delta_M k = \Delta_V k + g^{\alpha\beta} D_\alpha D_\beta k + \langle \vec{H}, Dk \rangle. \tag{3.3}$$

With respect to the evolving metrics, there exists the heat operator on V. It is easy to compute that for u defined on M, the following is the *conjugate heat equation:*

$$\Delta_V u - u_\tau - \langle \vec{H}, Du \rangle - |H|^2 u = 0. \tag{3.4}$$

By Theorem 2.3, we have that

$$\Delta_V u - u_\tau - \langle \vec{H}, Du \rangle - |H|^2 u = -\tau^{\frac{n-v}{2}} \left| \vec{H} + D^\perp \log k \right|^2 k.$$

for $u = \tau^{\frac{n-v}{2}} k$. It immediately implies the following result.

THEOREM 3.1. *Let*

$$I(t) = \tau^{\frac{n-v}{2}} \int_V k(x,\tau)\, d\mu_t.$$

Then

$$\frac{d}{dt} I(t) = -\tau^{\frac{n-v}{2}} \int k \left| \vec{H} + D^\perp \log k \right|^2 d\mu_t. \tag{3.5}$$

Harmonic map heat flow: Let $F(\cdot, t)$ be a family of maps from M into another Riemannian manifold N satisfying that harmonic map heat equation

$$\frac{\partial}{\partial t} F = \Delta F$$

where locally

$$(\Delta F)^\alpha = g^{ij} \left(\frac{\partial^2 F}{\partial x^i \partial x^j} - \Gamma^k_{ij} \frac{\partial F^\alpha}{\partial x^k} + \bar{\Gamma}^\alpha_{\beta\gamma}(F) \frac{\partial F^\beta}{\partial x^i} \frac{\partial F^\gamma}{\partial x^j} \right).$$

THEOREM 3.2. *Let F be a solution to the harmonic map heat flow and let*

$$I(t) = \tau \int_M |DF|^2 k \, d\mu.$$

The following identity holds on any M:

(3.6)

$$\frac{d}{dt} I(t) = -2\tau \int_M \left(\left| \Delta F + \frac{Dk}{k} \cdot DF \right|^2 + \left(D_i D_j \log k + \frac{1}{2\tau} g_{ij} \right) D_i F D_j F \right) k.$$

In particular if (M, g) is as in Theorem 2.3 then $I(t)$ is non-increasing in t.

The proof is direct computation and integration by parts. The key identities are

$$\int_M \Delta k |DF|^2 = -2 \int_M D_j k D_j D_i F^\alpha D_i F^\alpha,$$

$$\int_M D_i D_j k D_i F^\alpha D_j F^\alpha = -\int_M \Delta F^\alpha D_j F^\alpha D_j k + \frac{1}{2} \int_M \Delta k |DF|^2.$$

Yang-Mills flow: This is similar to the harmonic map heat flow. Now the equation is on a family of connections $A = (A^\alpha_{j\beta})$ on a vector bundle E. Let $F_A = F^\alpha_{ij\beta}$ be the curvature of A. The Yang-Mills equation is

$$\frac{\partial}{\partial t} A = \operatorname{div} F_A \tag{3.7}$$

where $\operatorname{div} F^\alpha_{k\beta} = g^{ij} D_i F^\alpha_{jk\beta}$.

THEOREM 3.3. *Let*

$$I(t) = \tau^2 \int_M |F|^2 k.$$

Then

$$\begin{aligned} \frac{d}{dt} I(t) = -4\tau^2 \int_M \Bigg(& \left(D_i D_j \log k + \frac{1}{2\tau} g_{ij} \right) F^\alpha_{ik\beta} F^\alpha_{ik\beta} \\ & + \left| \operatorname{div} F^\alpha_{j\beta} + D_i \log k F^\alpha_{ij\beta} \right|^2 \Bigg) k. \end{aligned} \tag{3.8}$$

Kähler case: Applying Theorem 2.4 and Theorem 2.5 one can obtain results similar as the above. Due to the fact that the estimate now is on complex Hessian only, the results are on holomorphic objects. The first result is in the same line as Theorem 3.1, but with the opposite monotonicity since we apply (2.4) to the heat kernel instead of the fundamental solution to the backward heat equation.

THEOREM 3.4. *Let M be a complete Kähler manifold with nonnegative bisectional curvature. Let $H(x, t; y, 0)$ be the fundamental solution of the* *heat equation. Let $\mathcal{V} \subset M$ be any complex subvariety of dimension s. Let $K_{\mathcal{V}}(x,t;y,0)$ be the fundamental solution of heat equation on $\mathcal{V}$. Then*

(i)

$$K_{\mathcal{V}}(x,t;y,0) \le (\pi t)^{m-s} H(x,t;y,0), \text{ for any } x,\ y \in \mathcal{V}. \tag{3.9}$$

If the equality holds, then $\mathcal{V}$ is totally geodesic. Furthermore if $\tilde{M}$ is the universal cover of M with covering map π and $\tilde{\mathcal{V}} = \pi^{-1}(\mathcal{V})$, then $\tilde{M} = \tilde{M}_1 \times \mathbb{C}^k$ for some Kähler manifold $\tilde{M}_1$ which does not contain any Euclidean factors, with $k \ge m-s$. Moreover $\tilde{\mathcal{V}} = \tilde{M}_1 \times \mathbb{C}^l$ with $l < k$.

(ii)

$$\frac{d}{dt}\int_{\mathcal{V}} (\pi t)^{m-s} H(x,t;y,0)\, dA_{\mathcal{V}}(y) \ge 0, \text{ for any } x \in M. \tag{3.10}$$

Similarly, if the equality holds for some $x \in M$ at some positive time t, then $\tilde{M} = \tilde{M}_1 \times \mathbb{C}^k$ with $k \ge m-s$.

The result has applications in obtaining the sharp dimension estimates on the space of holomorphic functions with polynomial growth [**N3**].

Theorem 2.4 also implies the monotonicity of the weighted energy of a holomorphic mapping from M (into any Kähler manifolds), as well as the monotonicity of the weighted energy for Hermitian-Einstein flow on any holomorphic vector bundle over M.

Let F be a holomorphic mapping from M (into, say, another Kähler manifold N), then we have that

$$\frac{d}{dt}\left((T-t)\int_M |\partial F|^2 k(x,t;x_0,T)\, d\mu\right) \le 0 \tag{3.11}$$

where $|\partial F|^2 = g^{\alpha\bar{\beta}} h_{i\bar{j}} F^i_\alpha F^{\bar{j}}_{\bar{\beta}}$, $k(x,t;x_0,T)$ is the fundamental solution to the backward heat equation satisfying $(\frac{\partial}{\partial t} + \Delta)k(x,t;x_0,T) = \delta_{(x_0,T)}(x,t)$. Dually we also have that

$$\frac{d}{dt}\left(t\int_M |\partial F|^2 H(x,t;x_0,0)\, d\mu\right) \ge 0 \tag{3.12}$$

where $H(x,t;x_0,0)$ is the fundamental solution of the heat equation centered at $(x_0,0)$.

Let $H(t)$ be a family of metrics of a holomorphic vector bundle E satisfying the Hermitian-Einstein equation

$$\frac{\partial h}{\partial t} h^{-1} = -\Lambda F + \lambda \mathrm{I} \tag{3.13}$$

where $h^\beta_\alpha(t) = H(t)_{\alpha\bar{\gamma}} H(0)^{\bar{\gamma}\beta}$, a Hermitian symmetric morphism of E. Then

$$\frac{d}{dt}\left((T-t)^2\int_M |F|^2 k(x,t;x_0,T)\, d\mu\right) \le 0 \tag{3.14}$$

where $|F|^2 = |F_{i\bar{j}\alpha^\beta}|^2$.

B. Thermodynamical consideration. In the thermodynamics the concept of the entropy is used to characterize the equilibrium states. In this setting, there is entropy function S which depends on state variable energy E as well as other parameters. The following are assumed: The entropy function S and energy E satisfy

1) $\frac{\partial S}{\partial E} > 0$ (hence define $\frac{1}{\tau} \doteqdot \frac{\partial S}{\partial E}$, with τ being interpreted as the temperature);
2) S is concave in E;
3) S is positively homogenous of degree 1.

Written as functions of S, there is *free energy* $F \doteqdot E - TS$.

In statistical mechanics, the entropy was used to measure the uncertainty. Here the equilibrium is characterized as the distribution maximizing the entropy(=uncertainty). Let $(M,\ d\mu)$ be a manifold with measure. Let $H : (M,\ d\mu) \to \mathbb{R}$ be the Hamiltonian. We have the following definitions.

Partition function: $Z := \int_M e^{-\beta H}\, d\mu$;
Temperature: $\tau := \frac{1}{\beta}$;
Energy: $E := -\frac{\partial}{\partial \beta} \log Z \geq 0$;
Entropy: $S := \beta E + \log Z$;
Free energy: $F := -\frac{1}{\beta} \log Z$.

The following result recovers the aspects of the classical equilibrium thermodynamics.

THEOREM 3.5.

$$E = \langle H \rangle, \text{ with respect to the canonical distribution } \sigma = \frac{1}{Z} e^{-\beta H},$$

$$S = S(\sigma) = -\int_M \sigma \log \sigma \, d\mu,$$

$$\frac{\partial S}{\partial E} = \frac{1}{\tau}, \qquad F = E - S\tau, \qquad \frac{\partial F}{\partial \tau} = -S.$$

Define the *heat capacity*

$$C_V = \frac{\partial E}{\partial \tau} = -\beta^2 \frac{\partial E}{\partial \beta}.$$

COROLLARY 3.6. $C_V \geq 0$ *and*

$$C_V = \beta^2 \frac{\partial^2}{\partial \beta^2} \log Z,$$

$$\frac{\partial^2 S}{\partial E^2} = \left(\frac{\partial E}{\partial \beta} \right)^{-1} \leq 0,$$

$$\frac{\partial S}{\partial \beta} = \beta \frac{\partial E}{\partial \beta} \leq 0.$$

This consideration suggests the following entropy formula. First after (2.4) (replacing t by τ) denote that $c_v = \frac{2}{u}|\Upsilon_{ij}|^2 + \frac{2}{u}R_{ij}\nabla_i u \nabla_j u$. Then

$$\frac{\partial}{\partial \tau}\int_M Q\, d\mu = \int_M c_v\, d\mu - \frac{2}{\tau}\int_M Q\, d\mu.$$

Equivalently,

$$\frac{\partial}{\partial \tau}\left(\tau^2 \int_M Q\, d\mu\right) = C_V$$

where $C_V = \tau^2 \int_M c_v\, d\mu$. This suggests that $\tau^2 \int_M Q$ should be the energy E. Normalizing $\int_M u\, d\mu = 1$, then $\log Z$ (essentially it is the anti-derivative of E in terms of $\frac{1}{\tau}$) can be computed.

$$\log Z = \int_M u \log u + \frac{n}{2}\log(4\pi\tau) + \frac{n}{2}.$$

This gives the entropy

$$S = \int_M (-\tau\Delta f - f + n) u\, d\mu = -\int_M (\tau|\nabla f|^2 + f - n) u\, d\mu.$$

The following is a direct consequence of Corollary 3.6.

THEOREM 3.7. *Let $\mathcal{W} = -S$. Then*

$$\frac{d\mathcal{W}}{d\tau} = -2\tau \int_M \left(\left|\nabla_i \nabla_j f - \frac{1}{2\tau} g_{ij}\right|^2 + R_{ij} f_i f_j\right) u\, d\mu. \tag{3.15}$$

In particular, if M has nonnegative Ricci curvature, $\mathcal{W}(f,\tau)$ is monotone decreasing along the heat equation.

3.2. Ricci flow. The most important monotonicity formula for the Ricci flow is the entropy monotonicity discovered by Perelman [**P**]. However, before this spectacular result, several monotonicity results have been obtained by Hamilton, mostly for surfaces, including his entropy and isoperimetric constant monotonicity.

A. Hamilton's entropy for surfaces. In [**H2**], Hamilton discovered an entropy monotonicity for the Ricci flow on the Riemann sphere with positive curvature. It is derived out of the proof for the LYH inequality for the surface case and the long time existence for the normalized flow: $\frac{\partial}{\partial t} g_{ij} = (r - R) g_{ij}$, where $r = \frac{\int R\, d\mu_t}{A(M_t)}$ is the average of R. It is easy to see that the normalized flow preserves the area $A(M_t)$. Also it is not hard to show that the normalized flow has a long time solution [**H2**].

THEOREM 3.8. *Let $I(t) = \int_M R \log R$. Then $I(t)$ is monotone non-increasing in t.*

Direct computation, using $\left(\frac{\partial}{\partial t} - \Delta\right) R = R^2 - rR$, shows that

$$\frac{d}{dt} I(t) = \int_M (\Delta R) \log R + R^2 - Rr = \int_M (\Delta \log R + R - r)\, R.$$

Let $Q = R(\Delta \log R + R - r)$. Similar as (2.4) we have that

$$\left(\frac{\partial}{\partial t} - \Delta\right) Q = 2\left|\nabla_i \nabla_j \log R + \frac{1}{2}(R-r)g_{ij}\right|^2 R + rQ + Q(R-r) \tag{3.16}$$

which implies Hamilton's LYH inequality for surfaces by applying maximum principle. Integrating the above one has that

$$\frac{d}{dt}\int_M Q\, d\mu = \int_M 2\left|\nabla_i \nabla_j \log R + \frac{1}{2}(R-r)g_{ij}\right|^2 R + r\int_M Q.$$

Using

$$\left(\int_M Q\right)^2 \leq \left(\int_M \left(\frac{Q}{R}\right)^2 R\right)\int_M R \leq 8\pi \int_M 2\left|\nabla_i \nabla_j \log R + \frac{1}{2}(R-r)g_{ij}\right|^2 R$$

the above implies that if denote $Z = \int_M Q\, d\mu = I'$,

$$Z' \geq \frac{1}{8\pi} Z^2 + rZ.$$

By ODE comparison we conclude that $Z \leq 0$, otherwise Z has to blow up at some finite time.

It is easy to see from the proof that if $Z = 0$ for some t_0, (M, g) is a gradient shrinking soliton.

B. From matrix LYH inequalities. The first result is a consequence of Cao's inequality (2.22). A simple consequence is the following trace inequality:

$$\frac{\partial R}{\partial t} + X_s \nabla_{\bar{s}} R + X_{\bar{s}} \nabla_s R + R_{i\bar{j}} X_{\bar{i}} X_j + \frac{R}{t} \geq 0. \tag{3.17}$$

In particular, we have that

$$\frac{\partial R}{\partial t} - \frac{|\nabla R|^2}{R} + \frac{R}{t} \geq 0. \tag{3.18}$$

In [**Co**], the following result was derived on $S_n(x,t) := \frac{\det(R_{i\bar{j}})}{\det(g_{i\bar{j}})}$, the n-th symmetric function of Ric with respect to the Kähler metric ω,

$$\frac{\partial S_n}{\partial t} - \frac{|\nabla S_n|^2}{nS_n} + \frac{nS_n}{t} \geq 0. \tag{3.19}$$

We shall derive the similar results for general symmetric functions of Ric with respect to ω.

First we have to set some notations. Here we follow [**HS**] (see also [**An1**]). Let A be a symmetric (Hermitian symmetric) matrix. Denote by λ_i its eigenvalues.

$$S_k(\lambda) := \sum_{\alpha_1<\alpha_2<\cdots<\alpha_k} \lambda_{\alpha_1}\lambda_{\alpha_2}\cdots\lambda_{\alpha_k}.$$

The symmetric functions $S_k(\lambda)$ can also be viewed as functions of matrix A (when so we write as $S_k(A)$). Define the map

$$A^{(k)}(e_{\alpha_1}\otimes\cdots e_{\alpha_k}) := \frac{1}{k!}\sum_{\sigma\in S(k)}(-1)^{\operatorname{sgn}(\sigma)}A(e_{\sigma(\alpha_1)})\otimes\cdots A(e_{\sigma(\alpha_k)}).$$

Then we have that $S_k(\lambda) = \operatorname{tr} A^{(k)}$. Now it is easy to show that at the points where A is diagonal we have that

$$\left(\frac{\partial S_k(A)}{\partial a_{ij}}\right) = \operatorname{diag}\left(\frac{\partial S_k(\lambda)}{\partial\lambda_1},\cdots,\frac{\partial S_k(\lambda)}{\partial\lambda_n}\right). \tag{3.20}$$

It is also easy to see that

$$S_k(\operatorname{Ric}) = \frac{n!}{k!(n-k)!}\frac{\operatorname{Ric}^k\wedge\omega^{n-k}}{\omega^n}.$$

Let $S_{k,i}(\lambda)$ be the sum of terms of $S_k(\lambda)$ not containing the factor λ_i. We shall need the following identities, which are proved in [**HS**].

LEMMA 3.9.

$$\frac{\partial S_k(\lambda)}{\partial\lambda_i}(\lambda) = S_{k-1,i}(\lambda), \tag{3.21}$$

$$S_k(\lambda) = S_{k,i}(\lambda) + \lambda_i S_{k-1,i}(\lambda), \tag{3.22}$$

$$\sum_{i=1}^n S_{k,i}(\lambda) = (n-k)S_k(\lambda), \tag{3.23}$$

$$\sum_{i=1}^n \lambda_i S_{k-1,i}(\lambda) = kS_k(\lambda), \tag{3.24}$$

$$\sum_{i=1}^n \lambda_i^2 S_{k-1,i}(\lambda) = S_1(\lambda)S_k(\lambda) - (k+1)S_{k+1}(\lambda). \tag{3.25}$$

Armed with the above, we shall first compute $\frac{\partial}{\partial t}(\log S_k(\operatorname{Ric}))$. (In the following we shall simply denote $S_k(\operatorname{Ric})$ by $S_k(x,t)$, S_k, or $S_k(\lambda)$, where λ_i are eigenvalues of Ric with respect to ω).

LEMMA 3.10. *With respect to a unitary frame such that* Ric *is diagonal we have*

(3.26)
$$\frac{\partial}{\partial t}(\log S_k(\lambda)) = \frac{1}{S_k(\lambda)}\left(\sum_{i=1}^{n} S_{k-1,i}(\lambda)\frac{\partial}{\partial t}R_{i\bar{i}} - (k+1)S_{k+1}(\lambda)\right) + S_1(\lambda)$$

$$\nabla_s(\log S_k(\lambda)) = \frac{\sum_i \nabla_s R_{i\bar{i}} S_{k-1,i}(\lambda)}{S_k(\lambda)}. \tag{3.27}$$

PROOF. First by the direct calculation we have that

$$\frac{\partial}{\partial t}(\log S_k(\lambda)) = \frac{k\operatorname{Ric}^{k-1}\wedge\frac{\partial}{\partial t}\operatorname{Ric}\wedge\omega^{n-k} - (n-k)\operatorname{Ric}^{k+1}\wedge\omega^{n-k-1}}{\operatorname{Ric}^k\wedge\omega^{n-k}} + S_1(\lambda).$$

Direct calculation shows that

$$k\operatorname{Ric}^{k-1}\wedge\frac{\partial}{\partial t}\operatorname{Ric}\wedge\omega^{n-k} = \frac{k!(n-k)!}{n!}\sum_{i=1}^{n}\left(\frac{\partial}{\partial t}R_{i\bar{i}}\right)S_{k-1,i}(\lambda)\omega^n$$

and

$$(n-k)\operatorname{Ric}^{k+1}\wedge\omega^{n-k-1} = \frac{(k+1)!(n-k)!}{n!}S_{k+1}(\lambda)\omega^n.$$

Putting them together we prove (3.26). The proof of (3.27) follows from (3.20) and (3.21). □

THEOREM 3.11. *Under the assumption that $(M, g(t))$ is a solution to the Kähler-Ricci flow with bounded nonnegative bisectional curvature, we have that*

$$\frac{\partial S_k}{\partial t} - \frac{1}{k}\frac{|\nabla S_k|^2}{S_k} + \frac{k}{t}S_k \geq 0. \tag{3.28}$$

PROOF. Choose a unitary frame so that Ric is diagonal. By the above lemma and (2.22) we have that

$$\begin{aligned}
&\frac{\partial}{\partial t}(\log S_k(\lambda))\\
&\geq \frac{1}{S_k(\lambda)}\sum_{i=1}^{n} S_{k-1,i}(\lambda)\left(-\lambda_i^2 - \frac{\lambda_i}{t} - X_s\nabla_{\bar{s}}R_{i\bar{i}} - X_{\bar{s}}\nabla_s R_{i\bar{i}} - R_{i\bar{i}s\bar{t}}X_{\bar{s}}X_t\right)\\
&\quad - \frac{(k+1)S_{k+1}(\lambda)}{S_k(\lambda)} + S_1(\lambda).
\end{aligned}$$

Using (3.25), (3.24) and (3.27) we have that

$$\begin{aligned}\frac{\partial}{\partial t}\left(\log S_k(\lambda)\right) \geq& \frac{1}{S_k(\lambda)}\left(-S_1(\lambda)S_k(\lambda)+(k+1)S_{k+1}(\lambda)-\frac{kS_k(\lambda)}{t}\right)\\ &- X_s\nabla_{\bar{s}}\log S_k(\lambda) - X_{\bar{s}}\nabla_s \log S_k\\ &- \frac{1}{S_k(\lambda)}\sum_{i=1}^{n} S_{k-1,i}(\lambda)R_{i\bar{i}s\bar{t}}X_{\bar{s}}X_t - \frac{(k+1)S_{k+1}(\lambda)}{S_k(\lambda)} + S_1(\lambda)\\ =& -\frac{k}{t} - X_s\nabla_{\bar{s}}\log S_k(\lambda) - X_{\bar{s}}\nabla_s\log S_k(\lambda)\\ &- \frac{1}{S_k(\lambda)}\sum_{i=1}^{n} S_{k-1,i}(\lambda)R_{i\bar{i}s\bar{t}}X_{\bar{s}}X_t. \end{aligned} \tag{3.29}$$

Notice that

$$R_{i\bar{i}s\bar{t}}X_{\bar{s}}X_t = R\left(i,\bar{i},X,\bar{X}\right) = |X|^2 R\left(i,\bar{i},\frac{X}{|X|},\frac{\bar{X}}{|X|}\right) \leq |X|^2 R_{i\bar{i}}.$$

Hence

$$\frac{\partial}{\partial t}\log S_k(\lambda) \geq -\frac{k}{t} - X_s\nabla_{\bar{s}}\log S_k(\lambda) - X_{\bar{s}}\nabla_s\log S_k(\lambda) - k|X|^2. \tag{3.30}$$

Here we have used (3.24) again. The claimed result follows by choosing $X = -\frac{1}{k}\nabla \log S_k(\lambda)$. ☐

The theorem in particular implies that $t^k S_k(x,t)$ is monotone non-decreasing. As in the linear case, one can also apply the matrix LYH inequality (2.29) to subvarieties of M.

THEOREM 3.12. *Let M be a complete Kähler manifold with bounded non-negative bisectional curvature. Let $H(x,t;y,0)$ be a fundamental solution to the forward conjugate heat equation on M. Let $\mathcal{V}$ be a complex subvariety of M of dimension s. Let $K_{\mathcal{V}}(x,t;y,0)$ be the fundamental solution to the restricted forward conjugate heat equation (with respect to the induced metrics) on $\mathcal{V}$. Then we have* (3.9) *and* (3.10). *Moreover, the equality (for positive t), in either cases, implies that the universal cover (of M) $\tilde{M}$ has the splitting $\tilde{M} = \tilde{M}_1 \times \mathbb{E}^k$, where $\mathbb{E}^k$ is a gradient expanding Kähler-Ricci soliton of dimension $k \geq m - s$.*

Remark 3.13. One can think of (3.10) as a dual version of Perelman's monotonicity of the reduced volume since the reduced volume in the Section 7 of [**P**] is, in a sense, a 'weighted volume' of M (with weight being the fundamental solution (to the *backward conjugate heat equation*) of a 'potentially infinity dimensional manifold' restricted to M, as explained in Section 6 of [**P**]), while here the monotonicity is on the 'weighted volume' of complex submanifolds with weight being the fundamental solution (of the *forward conjugate heat equation*) of M restricted to the submanifold. The reduced volume monotonicity of Perelman has important applications in the study of Ricci flow. We expect that (3.10) will have some applications in understanding the relation between Kähler-Ricci flow and the complex geometry of analytic subvarieties.

C. Perelman's entropy. Perelman's entropy and its monotonicity can be obtained from the thermodynamical consideration as before from (2.45). Recall that $(M, g(t))$ is a solution to the Ricci flow on $M \times [0, T]$. Let $\tau = T - t$, and consider the positive solution u to the conjugate heat equation

$$\frac{\partial u}{\partial \tau} - \Delta u + Ru = 0.$$

Assume $\int_M u = 1$, Integrating (2.45) we have that

$$\frac{\partial}{\partial \tau} \int_M v_0 = -2 \int_M |R_{ij} + f_{ij}|^2 \, d\mu.$$

Using ODE, it is easy to see that $\int_M Q \, d\mu \geq 0$ if $Q = \frac{n}{2\tau} u - v_0$.) Then

$$\frac{\partial}{\partial \tau} \int_M Q = \int_M c_v - \frac{2}{\tau} \int_M Q$$

with

$$c_v = 2|R_{ij} + f_{ij} - \frac{1}{2\tau} g_{ij}|^2 u.$$

Exactly as the earlier linear case we let

$$E = \tau^2 \int_M Q \, d\mu$$

be the energy and let τ be the temperature, then

$$\log Z = \int_M u \log u + \frac{n}{2} \log(4\pi\tau) + \frac{n}{2}$$

obtained as the anti-derivative of E.

Now Corollary 3.6 implies the following entropy formula.

THEOREM 3.14. *Let*

$$\mathcal{W}(g, u, \tau) := \int_M \big(\tau(|\nabla f|^2 + R) + f - n\big)\, u \, d\mu.$$

Then

$$\frac{d\mathcal{W}}{d\tau} = -2\tau \int_M \left| \nabla_i \nabla_j f + R_{ij} - \frac{1}{2\tau} g_{ij} \right|^2 u \, d\mu \leq 0. \tag{3.31}$$

3.3. Interpolation and localization. The interpolation we mean here is the one between the LYH inequality (2.29) and Perelman's entropy monotonicity (3.31), which we found a little mysterious. This seems to only work for Kähler cases.

Consider the Kähler-Ricci flow:

$$\frac{\partial}{\partial \tau} g_{\alpha\bar{\beta}} = \epsilon R_{\alpha\bar{\beta}} \tag{3.32}$$

where ϵ is a parameter and the *conjugate heat equation*:

$$\left(\frac{\partial}{\partial \tau} - \Delta + \epsilon \mathcal{R}\right) u(x,\tau) = 0. \tag{3.33}$$

When $\epsilon < 0$, (3.32) is a forward Ricci flow equation and (3.33) becomes the *forward conjugate heat equation.* The equations look different from those in Section 2 since in this section the case of $\epsilon < 0$ corresponds to the forward Ricci flow and the case of $\epsilon > 0$ corresponds to the backward Ricci flow. For example $\epsilon = 1$ is exactly the setting for Perelman's entropy and energy monotonicity. Notice that (3.33) becomes the *backward conjugate heat equation* for $\epsilon = 1$. For the positive solution $u(x,\tau)$ we define the $(1,1)$ tensor $Z_{\alpha\bar{\beta}}$ by

$$Z_{\alpha\bar{\beta}} = -(\log u)_{\alpha\bar{\beta}} + \epsilon R_{\alpha\bar{\beta}}.$$

Let Δ_L denote the Lichnerowicz Laplacian on $(1,1)$ tensors. Computation yields

$$\begin{aligned}
&\left(\frac{\partial}{\partial \tau} - \Delta_L\right) Z_{\alpha\bar{\beta}} \\
&\quad = -\epsilon^2 \left(\Delta R_{\alpha\bar{\beta}} + R_{\alpha\bar{\beta}\gamma\bar{\delta}} R_{\bar{\gamma}\delta} + \nabla_\gamma R_{\alpha\bar{\beta}} \left(\frac{1}{\epsilon}\nabla_{\bar{\gamma}} \log u\right)\right. \\
&\qquad \left. + \nabla_{\bar{\gamma}} R_{\alpha\bar{\beta}} \left(\frac{1}{\epsilon}\nabla_{\gamma} \log u\right) + R_{\alpha\bar{\beta}\gamma\bar{\delta}} \left(\frac{1}{\epsilon}\nabla_{\bar{\gamma}} \log u\right)\left(\frac{1}{\epsilon}\nabla_{\delta} \log u\right)\right) \\
&\qquad - (\log u)_{\alpha\gamma}(\log u)_{\bar{\gamma}\bar{\beta}} + \nabla_\gamma (Z_{\alpha\bar{\beta}})\nabla_{\bar{\gamma}} \log u + \nabla_{\bar{\gamma}}(Z_{\alpha\bar{\beta}})\nabla_\gamma \log u \\
&\qquad + \frac{1}{2} Z_{\alpha\bar{\gamma}}\left(\epsilon R_{\gamma\bar{\beta}} + (\log u)_{\gamma\bar{\beta}}\right) + \frac{1}{2}\left(\epsilon R_{\alpha\bar{\gamma}} + (\log u)_{\alpha\bar{\gamma}}\right) Z_{\gamma\bar{\beta}}.
\end{aligned} \tag{3.34}$$

Let

$$\tilde{Z}_{\alpha\bar{\beta}} = Z_{\alpha\bar{\beta}} - \frac{1}{\tau} g_{\alpha\bar{\beta}}$$

and

$$\begin{aligned}
Y_{\alpha\bar{\beta}} = {}& \Delta R_{\alpha\bar{\beta}} + R_{\alpha\bar{\beta}\gamma\bar{\delta}} R_{\bar{\gamma}\delta} + \nabla_\gamma R_{\alpha\bar{\beta}} \left(\frac{1}{\epsilon}\nabla_{\bar{\gamma}} \log u\right) + \nabla_{\bar{\gamma}} R_{\alpha\bar{\beta}} \left(\frac{1}{\epsilon}\nabla_{\gamma} \log u\right) \\
& + R_{\alpha\bar{\beta}\gamma\bar{\delta}} \left(\frac{1}{\epsilon}\nabla_{\bar{\gamma}} \log u\right)\left(\frac{1}{\epsilon}\nabla_{\delta} \log u\right).
\end{aligned}$$

Notice that $\tilde{Y}_{\alpha\bar{\beta}}$ in Theorem 2.7 is related to $Y_{\alpha\bar{\beta}}$ by $\tilde{Y}_{\alpha\bar{\beta}} = Y_{\alpha\bar{\beta}} - \frac{R_{\alpha\bar{\beta}}}{\epsilon\tau}$. From (3.34), we can derive the equation for $\tilde{Z}_{\alpha\bar{\beta}}$ as follows. In terms of $\tilde{Z}$ we have that

$$\begin{aligned}\left(\frac{\partial}{\partial\tau} - \Delta_L\right)\tilde{Z}_{\alpha\bar{\beta}} &= \left(\frac{\partial}{\partial\tau} - \Delta_L\right)Z_{\alpha\bar{\beta}} + \frac{1}{\tau^2}g_{\alpha\bar{\beta}} - \frac{1}{\tau}\epsilon R_{\alpha\bar{\beta}}\\ &= -\epsilon^2\tilde{Y}_{\alpha\bar{\beta}} - (\log u)_{\alpha\gamma}(\log u)_{\bar{\gamma}\bar{\beta}}\\ &\quad + \nabla_\gamma(\tilde{Z}_{\alpha\bar{\beta}})\nabla_{\bar{\gamma}}\log u + \nabla_{\bar{\gamma}}(\tilde{Z}_{\alpha\bar{\beta}})\nabla_\gamma \log u\\ &\quad + \frac{1}{2}\tilde{Z}_{\alpha\bar{\gamma}}\left(\epsilon R_{\gamma\bar{\beta}} + (\log u)_{\gamma\bar{\beta}} - \frac{1}{\tau}g_{\gamma\bar{\beta}}\right)\\ &\quad + \frac{1}{2}\left(\epsilon R_{\alpha\bar{\gamma}} + (\log u)_{\alpha\bar{\gamma}} - \frac{1}{\tau}g_{\alpha\bar{\gamma}}\right)\tilde{Z}_{\gamma\bar{\beta}}.\end{aligned} \tag{3.35}$$

For $\epsilon < 0$, applying Theorem 2.7 we know that $\epsilon^2 Y_{\alpha\bar{\beta}} - \frac{\epsilon}{\tau}R_{\alpha\bar{\beta}} \geq 0$ under the assumption that M is a complete Kähler manifold with bounded nonnegative holomorphic bisectional curvature. Hence the tensor maximum principle and (3.35) imply that $\tilde{Z}_{\alpha\bar{\beta}} \leq 0$.

Let $f = -\log u$, $Z = g^{\alpha\bar{\beta}}Z_{\alpha\bar{\beta}}$. Tracing (3.34) gives

$$\begin{aligned}\left(\frac{\partial}{\partial\tau} - \Delta\right)Z &= -\epsilon R_{\bar{\alpha}\beta}Z_{\alpha\bar{\beta}} - \epsilon^2 g^{\alpha\bar{\beta}}Y_{\alpha\bar{\beta}} - (f)_{\alpha\gamma}(f)_{\bar{\gamma}\bar{\alpha}}\\ &\quad - \nabla_\gamma Z\nabla_{\bar{\gamma}}f - \nabla_{\bar{\gamma}}Z\nabla_\gamma f + Z_{\alpha\bar{\beta}}(\epsilon R_{\beta\bar{\alpha}} - f_{\beta\bar{\alpha}})\end{aligned} \tag{3.36}$$

and

$$\begin{aligned}g^{\alpha\bar{\beta}}Y_{\alpha\bar{\beta}} &= \Delta\mathcal{R} + R_{\alpha\bar{\beta}}R_{\bar{\alpha}\beta} - \nabla_\gamma\mathcal{R}\left(\frac{1}{\epsilon}\nabla_{\bar{\gamma}}f\right)\\ &\quad - \left(\frac{1}{\epsilon}\nabla_\gamma f\right)\nabla_{\bar{\gamma}}\mathcal{R} + R_{\alpha\bar{\beta}}\left(\frac{1}{\epsilon}f_{\bar{\alpha}}\right)\left(\frac{1}{\epsilon}f_\beta\right).\end{aligned}$$

For $\epsilon = 1$, integration by parts as before gives

$$\frac{d}{d\tau}\int_M \tilde{Z}u\,d\mu_\tau = -\int_M\left(|\tilde{Z}_{\alpha\bar{\beta}}|^2 + |f_{\alpha\beta}|^2\right)u\,d\mu_\tau - \frac{2}{\tau}\int_M \tilde{Z}u\,d\mu_\tau. \tag{3.37}$$

The above equation is equivalent to Perelman's entropy monotonicity since the quantity Q in Perelman's entropy formula derivation above is nothing but $-\tilde{Z}u$ (for the Kähler case, there exists a factor 2 difference).

Note that for $\epsilon = -1$, the scalar curvature R does satisfies (3.33) in the case $m = 1$. And renormalization of (3.35) gives the previous Hamilton's result (3.16). Hence the matrix computation above also implies the Hamilton's entropy monotonicity, Theorem 3.8. It is interesting to see if (3.35) also gives high dimensional analogue of Hamilton's entropy monotonicity for Kähler-Ricci flow. It still remains a open problem if Hamilton's entropy formula has high dimensional version.

Localizing LYH inequalities and monotonicity formulae is important to the study of the singularities. The localization of LYH inequality is not an easy matter in general. Li-Yau's original work provides such a localized estimate. There also exists recent fundamental work of Hamilton [**H7**] in this direction. Here we present the localization of the monotonicity formulae via the *heat ball/heat sphere*. This consideration can be traced back to the works of [**Fu, Wa**]. It is much related to mean value theorems for harmonic functions and solutions to the heat equation. Here we just include a general formulation for evolving metrics [**EKNT**].

In [**EKNT**], a very general scheme on localizing the monotonicity formulae is developed. It is for any family of metrics evolved by the equation $\frac{\partial}{\partial t} g_{ij} = -2\kappa_{ij}$. The localization is through the so-called 'heat ball'. More precisely for a smooth positive space-time function v, which often is the fundamental solution to the *backward conjugate heat equation* or the 'pseudo backward heat kernel' $\hat{H}(x_0, y, \tau) = \frac{e^{-\frac{r^2(x_0,y)}{4\tau}}}{(4\pi\tau)^{\frac{n}{2}}}$ (or $\frac{e^{-\ell(y,\tau)}}{(4\pi\tau)^{\frac{n}{2}}}$ in the case of Ricci flow), with $\tau = t_0 - t$, one defines the 'heat ball' by $E_r = \{(y,t) | \, v \geq r^{-n};\ t < t_0\}$. For all interesting cases we can check that E_r is compact for small r (cf. [**EKNT**]). Let $\psi_r = \log v + n \log r$. For any 'Li-Yau-Hamilton' quantity $\mathcal{Q}$ we define the local quantity:

$$P(r) := \int_{E_r} \left(|\nabla \psi_r|^2 + \psi_r (\mathrm{tr}_g \kappa) \right) \mathcal{Q} \, d\mu_t \, dt.$$

The finiteness of the integral can be verified by a local gradient estimate. The general form of the theorem, which is proved in Theorem 1 of [**EKNT**], reads as the following.

THEOREM 3.15. *Let* $I(r) = \frac{P(r)}{r^n}$. *Then*

$$I(r_2) - I(r_1) = -\int_{r_1}^{r_2} \frac{n}{r^{n+1}} \int_{E_r} \left[\left(\left(\frac{\partial}{\partial t} + \Delta - tr_g \kappa \right) v \right) \frac{\mathcal{Q}}{v} + \psi_r \left(\frac{\partial}{\partial t} - \Delta \right) \mathcal{Q} \right] d\mu_t \, dt \, dr. \tag{3.38}$$

It gives the monotonicity of $I(r)$ in the cases that $\mathcal{Q} \geq 0$, which is ensured by the LYH estimates in the case we shall consider, and both $\left(\frac{\partial}{\partial t} + \Delta - \mathrm{tr}_g \kappa\right) v$ and $\left(\frac{\partial}{\partial t} - \Delta\right) \mathcal{Q}$ are nonnegative. The nonnegativity of $\left(\frac{\partial}{\partial t} + \Delta - \mathrm{tr}_g \kappa\right) v$ comes for free if we chose v to be the 'pseudo backward heat kernel'. The nonnegativity of $\left(\frac{\partial}{\partial t} - \Delta\right) \mathcal{Q}$ follows from the key computation, which we call the *pre-LYH equation*, in the proof of the corresponding LYH estimate.

There exist certain localizations on the entropy formula (3.31). These localizations are achieved by suitable cut-off functions and are easier than the above consideration via the heat balls. Please see [**N4**] for details.

3.4. Hypersurface flows. For Gauss curvature flow, define.

$$\mathcal{E}(t) \doteqdot \int_M K \log K \, d\mu.$$

Its monotonicity was first established by Ben Chow, using the proof of his LYH inequality for Gauss curvature flow. It turns out that it holds for slightly more general flows with a certain integrability condition. (This nice observation is due to Andrews [**An2**].)

A. The general formulation. Recall the notations Φ, Ψ, P, etc, from the previous section. For the simplicity we assume that the flow speed $\Phi > 0$.

THEOREM 3.16. *Assume that the speed function $\Psi = \Psi(A)$ is α-concave and satisfies that*

$$\bar{\nabla}_i \left(\Psi^{-2} \dot{\Psi}^{ij} \right) = 0. \tag{3.39}$$

Then

$$\frac{d}{dt} \int_{\mathbb{S}^n} \frac{\partial}{\partial t} \log \Phi \, d\sigma \geq \frac{\alpha - 1}{\alpha} \int_{\mathbb{S}^n} \left(\frac{\partial}{\partial t} \log \Phi \right)^2 d\sigma. \tag{3.40}$$

PROOF. The proof follows from the computation in previous sections on the LYH estimate for hypersurface flows. Recall that $P = \Psi_t$ satisfies that

$$\frac{\partial}{\partial t} P = \mathcal{L} P + \dot{\Psi}(\mathrm{id}) P + \ddot{\Psi}(Q, Q)$$

where $Q = \frac{\partial}{\partial t} A = \bar{g}^* \mathrm{Hess}_{\bar{\nabla}} \Psi + \Psi \, \mathrm{id}$. Hence

$$\begin{aligned}
\frac{d}{dt} \int_{\mathbb{S}^n} \frac{\partial}{\partial t} \log \Phi \, d\sigma &= \int_{\mathbb{S}^n} \frac{\frac{\partial}{\partial t} P}{\Psi} - \frac{P \frac{\partial}{\partial t} \Psi}{\Psi^2} \, d\sigma \\
&= \int_{\mathbb{S}^n} \frac{(\dot{\Psi})^{ij} \bar{\nabla}_i \bar{\nabla}_j P + \dot{\Psi}(\mathrm{id}) P}{\Psi} + \frac{\ddot{\Psi}(Q, Q)}{\Psi} - \left(\frac{P}{\Psi} \right)^2 d\sigma \\
&\geq \int_{\mathbb{S}^n} \frac{(\dot{\Psi})^{ij} \bar{\nabla}_i \bar{\nabla}_j P + \dot{\Psi}(\mathrm{id}) P}{\Psi} + \frac{\alpha - 1}{\alpha} \left(\frac{P}{\Psi} \right)^2 - \left(\frac{P}{\Psi} \right)^2 d\sigma
\end{aligned}$$

noting that $\dot{\Psi}(Q) = P$. Now we compute the first term

$$\begin{aligned}
\int_{\mathbb{S}^n} \frac{(\dot{\Psi})^{ij} \bar{\nabla}_i \bar{\nabla}_j P}{\Psi} \, d\sigma &= \int_{\mathbb{S}^n} \Psi \left(\Psi^{-2} (\dot{\Psi})^{ij} \bar{\nabla}_i \bar{\nabla}_j P \right) d\sigma \\
&= - \int_{\mathbb{S}^n} \bar{\nabla}_i \Psi \left(\Psi^{-2} (\dot{\Psi})^{ij} \bar{\nabla}_j P \right) d\sigma \\
&= \int_{\mathbb{S}^n} \left(\bar{\nabla}_j \bar{\nabla}_i \Psi \right) \Psi^{-2} (\dot{\Psi})^{ij} P \, d\sigma.
\end{aligned}$$

Using that $P = (\dot{\Psi})^{ij}(\bar{\nabla}_i\bar{\nabla}_j\Psi + \Psi\bar{g}_{ij})$ we have that

$$\int_{\mathbb{S}^n} \frac{(\dot{\Psi})^{ij}\bar{\nabla}_i\bar{\nabla}_j P + \dot{\Psi}(\mathrm{id})P}{\Psi}\, d\sigma = \int_{\mathbb{S}^n} \frac{P^2}{\Psi^2}\, d\sigma.$$

□

Now we check that the condition (3.39) holds for both the Gauss curvature flow and harmonic mean curvature flow.

Gauss curvature flow: Using the normal coordinate centered at a point, we have that $A = (A_{ij})$ with $A_{ij} = \bar{\nabla}_i\bar{\nabla}_j s + s\bar{g}_{ij}$. Direct calculation shows that

$$\Psi^{-2}(\dot{\Psi})^{ij} = \det(A)(A^{-1})_{ij} = \frac{\partial \det(A)}{\partial A_{ij}}.$$

Hence it suffices to show

$$\bar{\nabla}_i\left(\frac{\partial \det(A)}{\partial A_{ij}}\right) = 0.$$

This follows from two claims below.

1) For any symmetric tensor A satisfying the Codazzi equation $\bar{\nabla}_k A_{ij} = \bar{\nabla}_i A_{jk}$, we have

$$\bar{\nabla}_i\left(\frac{\partial S_k(A)}{\partial A_{ij}}\right) = 0$$

where S_k is the k-th symmetric function.

2) The $A = \left(\bar{\nabla}_i\bar{\nabla}_j s + s\bar{g}_{ij}\right)$ satisfies the Codazzi equation.

The first claim can be checked directly. For the second one, we have the following computation.

$$\begin{aligned}
&\bar{\nabla}_k\left(\bar{\nabla}_i\bar{\nabla}_j s + s\bar{g}_{ij}\right) - \bar{\nabla}_i\left(\bar{\nabla}_k\bar{\nabla}_j s + s\bar{g}_{kj}\right)\\
&= \bar{\nabla}_k\bar{\nabla}_i\bar{\nabla}_j s - \bar{\nabla}_i\bar{\nabla}_k\bar{\nabla}_j s + \bar{\nabla}_k s\bar{g}_{ij} - \bar{\nabla}_i s\bar{g}_{kj}\\
&= -\bar{R}_{ikjp}\bar{\nabla}_p s + \bar{\nabla}_k s\bar{g}_{ij} - \bar{\nabla}_i s\bar{g}_{kj}\\
&= 0
\end{aligned}$$

where we have used the expression for the curvature $\bar{R}_{ikjp} = \bar{g}_{ij}\bar{g}_{kp} - \bar{g}_{ip}\bar{g}_{kj}$.

Harmonic mean curvature flow: Direct calculation shows that $\Psi^{-2}(\dot{\Psi})^{ij} = \bar{g}_{ij}$. Hence $\bar{\nabla}_i(\Psi^{-2}(\dot{\Psi})^{ij})$ is trivially true.

B. The entropy formulae. Now we look further into the Gauss curvature flow case. Noting that $\mathcal{E} = \int_{\mathbb{S}^n} \log K\, d\sigma$, a consequence of the theorem via the Hölder inequality is that

$$\frac{d^2\mathcal{E}(t)}{dt^2} \geq \frac{n+1}{nA(\mathbb{S}^n)}\left(\frac{d\mathcal{E}(t)}{dt}\right)^2.$$

Now consider the normalized flow. Let

$$\tilde{X} = \psi^{-1}(t)X$$

where $\psi = ((n+1)(T-t))^{\frac{1}{n+1}}$. Hence the new equation is

$$\frac{\partial}{\partial \tau}\tilde{X} = -\tilde{K}\nu + \tilde{X} \tag{3.41}$$

where $\tau = \frac{1}{n+1}\log(1-\frac{t}{T})$. It can be checked that the flow preserves the enclosed volume if $V(M_0) = \omega_{n+1}$, the volume of the unit ball. Now the corresponding equation on the support function $\tilde{s}$ is that

$$\frac{\partial}{\partial \tau}\tilde{s} = \tilde{s} - \tilde{K} = \tilde{s} - \frac{1}{\det(\bar{g}^* \mathrm{Hess}_{\bar{\nabla}}\tilde{s} + \tilde{s}\,\mathrm{id})}. \tag{3.42}$$

Let $\tilde{A} = \bar{g}^*\left(\mathrm{Hess}_{\bar{\nabla}}\tilde{s} + \bar{g}\tilde{s}\right)$. Now

$$\frac{\partial}{\partial \tau}\tilde{A} = \tilde{A} + \bar{g}^*\left(\mathrm{Hess}_{\bar{\nabla}}\tilde{\Psi} + \mathrm{id}\,\tilde{\Psi}\right).$$

Similarly $\tilde{P} \doteqdot \frac{\partial}{\partial \tau}\tilde{\Psi} = \dot{\tilde{\Psi}}\left(\frac{\partial}{\partial \tau}\tilde{A}\right)$ satisfies

$$\begin{aligned}
\frac{\partial}{\partial \tau}\tilde{P} &= \mathcal{L}\tilde{P} + \dot{\tilde{\Psi}}(\mathrm{id})\tilde{P} + \mathcal{L}\tilde{\Psi} + \dot{\tilde{\Psi}}(\mathrm{id})\tilde{\Psi} + \dot{\tilde{\Psi}}(\tilde{A}) + \ddot{\tilde{\Psi}}(\tilde{Q},\tilde{Q}) \\
&= \mathcal{L}\tilde{P} + \dot{\tilde{\Psi}}(\mathrm{id})\tilde{P} + \dot{\tilde{\Psi}}\left(\frac{\partial}{\partial \tau}\tilde{A}\right) + \ddot{\tilde{\Psi}}(\tilde{Q},\tilde{Q}) \\
&= \mathcal{L}\tilde{P} + \dot{\tilde{\Psi}}(\mathrm{id})\tilde{P} + P + \ddot{\tilde{\Psi}}(\tilde{Q},\tilde{Q})
\end{aligned}$$

where $\tilde{Q} = \frac{\partial}{\partial \tau}\tilde{A}$. Let $\tilde{\mathcal{E}}(\tau) = \int_{\mathbb{S}^n}\log\tilde{K}\,d\sigma$. Repeating the calculation in the last subsection we have that

$$\frac{d^2\tilde{\mathcal{E}}(\tau)}{d\tau^2} \geq \frac{n+1}{nA(\mathbb{S}^n)}\left(\frac{d\tilde{\mathcal{E}}(\tau)}{d\tau}\right)^2 + \frac{d\tilde{\mathcal{E}}(\tau)}{d\tau} \tag{3.43}$$

which then implies the following result of Chow. Note that

$$\tilde{\mathcal{E}}(\tau) = \int_{\mathbb{S}^n}\log\left(((n+1)(T-t))^{\frac{n}{n+1}}K(x,t)\right)d\sigma.$$

THEOREM 3.17.

$$\frac{d\tilde{\mathcal{E}}(\tau)}{d\tau} \leq 0. \tag{3.44}$$

PROOF. This is derived out of the long time existence result of Tso [**T**]. Otherwise, assume that $\mathcal{F} \doteqdot \frac{d\tilde{\mathcal{E}}}{d\tau} > 0$ (abbreviated as $\tilde{\mathcal{E}}'$) at some τ_0. By (3.43) we have that $\mathcal{F}' \geq \alpha\mathcal{F}^2$ for all $\tau \geq \tau_0$. This implies that $\mathcal{F}$ must blow up at some finite time. A contradiction! □

This derivation of the monotonicity formula via the long time existence and ODE consideration is originated in [**H2**].

It is also worthwhile to look into the special case that the entropy is constant. Tracing the proof we conclude that $\tilde{K}_\tau = 0$ and

$$\tilde{A} = \lambda \bar{g}^* \left(\mathrm{Hess}_{\bar{\nabla}} \tilde{\Psi} + \bar{g}\tilde{\Psi} \right).$$

Using the PDE satisfied by $\tilde{\Psi}$, it is easy to conclude that the above $\lambda = -1$, which implies $\frac{\partial}{\partial \tau}\tilde{A} = 0$. In particular we have that $\mathrm{Hess}_{\bar{\nabla}}(\tilde{K} - \tilde{s}) + (\tilde{K} - \tilde{s})\bar{g} = 0$. This concludes that $\tilde{K} - \tilde{s}$ is the restriction of a linear function. Keeping in mind that the choices of different origin to define the support function cause the support function to differ by a linear function, we essentially have that $\tilde{s} - \tilde{K} = 0$. Namely the equality holds only on a shrinking soliton (steady solution to the normalized flow).

The following dual version of the entropy formula was motivated by the thermodynamic consideration as before. Define the following entropy-like quantities:

$$\overline{\mathcal{E}}(t) = -\int_{\mathbb{S}^n} \log\left(t^{\frac{n}{n+1}} K\right) d\sigma \tag{3.45}$$

$$\begin{aligned} \mathcal{W}(t) &= \frac{d}{dt}\left(t\overline{\mathcal{E}}(t)\right) = t\frac{d}{dt}\overline{\mathcal{E}}(t) + \overline{\mathcal{E}}(t) \\ &= -t\frac{d}{dt}\mathcal{E}(t) - \frac{nA(\mathbb{S}^n)}{n+1} + \overline{\mathcal{E}}(t). \end{aligned} \tag{3.46}$$

By the LYH inequality we have that

$$\frac{d}{dt}\overline{\mathcal{E}}(t) \le 0.$$

Letting $\mathcal{K} = \frac{\partial}{\partial t}\log K$, the above computation also gives

$$\begin{aligned} \frac{d}{dt}\mathcal{W}(t) &= -t\frac{d}{dt}\left(\frac{d\mathcal{E}(t)}{dt}\right) - 2\frac{d\mathcal{E}(t)}{dt} - \frac{nA(\mathbb{S}^n)}{(n+1)t} \\ &\le -\int_{\mathbb{S}^n} \left(\frac{n+1}{n} t\mathcal{K}^2 + 2\mathcal{K} + \frac{n}{(n+1)t} \right) d\sigma \\ &\le -\frac{nt}{n+1}\int_{\mathbb{S}^n} \left(\frac{n+1}{n}\mathcal{K} + \frac{1}{t} \right)^2 d\sigma. \end{aligned} \tag{3.47}$$

The following corollary summarizes the above observations.

Corollary 3.18. *The entropies $\overline{\mathcal{E}}(t)$ and $\mathcal{W}(t)$ are monotone non-increasing.*

Similar entropy formulae for $\overline{\mathcal{E}}(t)$ and $\mathcal{W}(t)$, which can be similarly defined as

$$\overline{\mathcal{E}}(t) = -\int_{\mathbb{S}^n} \log\left(\Phi t^{\frac{\alpha}{\alpha-1}}\right) d\sigma \quad \text{and} \quad \mathcal{W}(t) = \frac{d}{dt}\left(t\overline{\mathcal{E}}(t)\right),$$

can be shown verbatim provided that the assumption of Theorem 3.16 holds, particularly including the harmonic mean curvature flow.

By tracing the equality in either (3.45) or (3.46) we have that

$$\frac{K_t}{K} + \frac{n}{t(n+1)} = 0$$

and

$$-Q = \bar{g}^* \left(\bar{\nabla}_i \bar{\nabla}_j K + K \bar{g}_{ij} \right) = \lambda A$$

for some λ which may depend on space-time point. Using the equation for K one deduces that $\lambda = -\frac{1}{(n+1)t}$. Hence as before we conclude that the equality holds only if $K + \frac{1}{(n+1)t}s = 0$. Namely the solution is an expanding soliton.

4. The other direction

The previous section shows how LYH inequality implies the monotonicity formulae. This process sometimes can be reversed. Here we show that some geometric considerations which lead to the monotonicity of some geometric quantities also suggest the LYH inequalities (2.5) (for the heat equation) and (2.14) (for Ricci flow) respectively. According to Perelman, it is this consideration that leads the discovery of his entropy formula and (2.14).

4.1. Linear heat equation. Recall that the LYH inequality (2.42) implies (2.8), which is equivalent to the following result of Cheeger and Yau, which asserts that on a complete Riemannian manifold with nonnegative Ricci curvature, the heat kernel $H(x, \tau; o, 0)$ (the fundamental solution of the operator $(\frac{\partial}{\partial \tau} - \Delta)$) has the lower estimate

$$H(x, \tau; o, 0) \geq \frac{1}{(4\pi\tau)^{\frac{n}{2}}} \exp\left(-\frac{r^2(o, x)}{4\tau} \right) \tag{4.1}$$

where $r(o, x)$ is the distant function on the manifold. This fact can be derived out of the maximum principle and the differential inequality

$$\left(\frac{\partial}{\partial \tau} - \Delta \right) \left(\frac{1}{(4\pi\tau)^{\frac{n}{2}}} \exp\left(-\frac{r^2(o, x)}{4\tau} \right) \right) \leq 0.$$

Integrating on the manifold M, this differential inequality also implies the monotonicity (monotone non-increasing) of the integral

$$\tilde{V}(o, \tau) := \int_M \frac{1}{(4\pi\tau)^{\frac{n}{2}}} \exp\left(-\frac{r^2(o, y)}{4\tau} \right) \tag{4.2}$$

Fix a point $o \in M$. Let $\gamma(\tau)$ $(0 \le \tau \le \bar{\tau})$ be a curve parameterized by the time variable τ with $\gamma(0) = o$. Here we image that we have a time function τ, with which some parabolic equation is associated. Define the $\mathcal{L}$-length by

$$\mathcal{L}(\gamma)(\bar{\tau}) = \int_0^{\bar{\tau}} \sqrt{\tau}|\gamma'(\tau)|^2 \, d\tau. \tag{4.3}$$

We can define the $\mathcal{L}$-geodesic to be the curve which is the critical point of $\mathcal{L}(\gamma)$. The simple computation shows that the first variation of $\mathcal{L}$ is given by

$$\delta\mathcal{L}(\gamma) = 2\sqrt{\bar{\tau}}\langle Y, X\rangle(\bar{\tau}) - 2\int_0^{\bar{\tau}} \sqrt{\tau}\left(\langle \nabla_X X + \frac{1}{2\tau}X, Y\rangle\right) d\tau, \tag{4.4}$$

where Y is the variational vector field, from which one can write down the $\mathcal{L}$-geodesic equation. It is an easy matter to see that γ is a $\mathcal{L}$-geodesic if and only if $\gamma(\sigma)$ with $\sigma = 2\sqrt{\tau}$ is a geodesic. In other words, a $\mathcal{L}$-geodesic is a geodesic after certain re-parametrization. Here we insist all curves are parameterized by the 'time'-variable τ. One can check that for any $v \in T_oM$ there exists a $\mathcal{L}$-geodesic $\gamma(\tau)$ such $\frac{d}{d\sigma}(\gamma(\sigma))|_{\sigma=0} = v$. Notice that the variable σ scales in the same manner as the distance function on M. So it is more convenient to work with σ.

Following [**P**] we also introduce the ℓ-'distance' function.

$$\ell_o(y, \bar{\tau}) = \frac{1}{2\sqrt{\bar{\tau}}} L_o(y, \bar{\tau}), \quad \text{where } L_o(y, \bar{\tau}) = \inf_{\gamma} \mathcal{L}(\gamma)).$$

Here our ℓ is defined for a fixed background metric. We also omit the subscript o in the context where the meaning is clear. The similar computations as in [**P**], together with the second variation formula from the Riemannian geometry, show that

$$|\nabla \ell|^2 = \frac{1}{\bar{\tau}}\ell \tag{4.5}$$

$$\ell_\tau = -\frac{1}{\bar{\tau}}\ell \tag{4.6}$$

and

$$\Delta \ell \le \frac{n}{2\bar{\tau}} - \frac{1}{\bar{\tau}^{\frac{3}{2}}}\int_0^{\bar{\tau}} \tau^{\frac{3}{2}} \operatorname{Ric}(X, X) \, d\tau \tag{4.7}$$

where $X = \gamma'(\tau)$ with $\gamma(\tau)$, $0 \le \tau \le \bar{\tau}$ being the minimizing $\mathcal{L}$-geodesic joining o to x. Putting (4.5)–(4.7) together, one obtains a new proof of the result of Cheeger-Yau, which asserts that if M has nonnegative Ricci curvature, then

$$-\ell_\tau + \Delta\ell - |\nabla \ell|^2 - \frac{n}{2\tau} \le 0 \tag{4.8}$$

which is equivalent to

$$\left(\frac{\partial}{\partial \tau} - \Delta\right)\left(\frac{e^{-\ell(x,\tau)}}{(4\pi\tau)^{\frac{n}{2}}}\right) \leq 0.$$

Namely $\frac{e^{-\ell(x,\tau)}}{(4\pi\tau)^{\frac{n}{2}}}$ is a sub-solution of the heat equation. In particular,

$$\frac{d}{d\tau}\int_M \frac{e^{-\ell(x,\tau)}}{(4\pi\tau)^{\frac{n}{2}}}\, d\mu \leq 0.$$

Using the above geometric consideration, one can think of the above result of Cheeger-Yau as a parabolic volume comparison with respect to the positive measure $\frac{e^{-\ell(x,\tau)}}{(4\pi\tau)^{\frac{n}{2}}}\, d\mu$. Recall that the well-known Bishop volume comparison states that if M has nonnegative Ricci curvature

$$\frac{d}{dr}\left(\frac{1}{r^{n-1}}\int_{S_{x_0}(r)} dA\right) \leq 0$$

where $S_o(r)$ denotes the boundary of the geodesic ball centered at o with radius r, dA is the induced area measure. The by-now standard relative volume comparison can be formulated in the similar way as above. Let A be a measurable subset of $S^{n-1} \subset T_oM$ one can define $C_A(r)$ to be the collection of vectors rv with $v \in A$ such that the geodesic $\exp_o(sv)$ is minimizing for $s \leq r$, where $\exp_o(\cdot)$ is the exponential map. Then the relative volume comparison theorem asserts that if M has nonnegative Ricci curvature

$$\frac{d}{dr}\left(\frac{1}{r^{n-1}}\int_{\exp(C_A(r))} dA\right) \leq 0.$$

The following is a parabolic version of such a relative volume comparison theorem parallel to Perelman's work on Ricci flow geometry.

PROPOSITION 4.1. *Assume that M has nonnegative Ricci curvature. Then*

$$\frac{d}{d\tau}\int_{\mathcal{L}\exp_{C_A(\tau)}(\tau)} \frac{e^{-\ell(x,\tau)}}{(4\pi\tau)^{\frac{n}{2}}}\, d\mu \leq 0. \tag{4.9}$$

Better comparison between Theorem 4.1 and the classical relative volume comparison can be seen by identifying the space-time $\widetilde{M} = M \times [0, T]$ with the manifold M, the time-slice $M \times \{a\}$ with the geodesic sphere $S_o(a)$. Notice that in [**P**], one does need such a localized version (for Ricci flow) to prove the important no local collapsing result on the finite time solution to Ricci flow.

The other combinations of (4.5)–(4.7) include the Hamilton-Jacobi equations:

$$|\nabla \ell|^2 + 2\ell_\tau + \frac{\ell}{\tau} = 0, \tag{4.10}$$

$$|\nabla \ell|^2 + \ell_\tau = 0 \tag{4.11}$$

and the inequalities

$$\Delta \bar{L} + \bar{L}_\tau \le 2n, \qquad \text{for } \bar{L} = 4\tau\ell, \tag{4.12}$$

$$\tau(2\Delta\ell - |\nabla \ell|^2) + \ell - n \le 0. \tag{4.13}$$

Notice that assuming (4.10), (4.8) and (4.13) imply each other. One way of thinking is that ℓ as a solution to a Hamilton-Jacobi equation (4.10), one has (4.13) and (4.8). On the other hand Theorem 2.2 asserts that for the fundamental solution $u = \frac{e^{-\ell}}{(4\pi\tau)^{\frac{n}{2}}}$ to the heat equation, one has (4.13). Namely, if we insist the equality in (4.8) (hence we look at solutions to the heat equation in stead of the Hamilton-Jacobi equation (4.10)) we still have (4.13). The proof of this observation leads to the entropy formula (3.15). One should also compare (4.10) with (2.6), (4.11) with (2.3), (4.12) with (2.9).

4.2. Ricci flow. We find it most striking that the above geometric consideration has a close analogue for the space-time of the solution to the backward Ricci flow $\frac{\partial}{\partial \tau} g = \mathrm{Ric}(g)$. This is one of the most important contributions of [**P**].

Recall (from(2.47), a consequence of the LYH (2.42)) the definition of *reduced distance*

$$\ell_{x_0, g(\tau)}(x, \bar{\tau}) := \inf_\gamma \frac{1}{2\sqrt{\bar\tau}} \int_0^{\bar{\tau}} \sqrt{\tau}\left(|\gamma'(\tau)|^2 + R\right) d\tau \tag{4.14}$$

for all $\gamma(\tau)$ with $\gamma(0) = x_0$, $\gamma(\bar{\tau}) = x$ (in the right context, we often omit the subscript $x_0, g(\tau)$). The quantity

$$\tilde{V}_{g(\tau)}(x_0, \tau) := \int_M \frac{1}{(4\pi\tau)^{\frac{n}{2}}} \exp\left(-\ell(x,\tau)\right) d\mu(\tau) \tag{4.15}$$

is called the *reduced volume*. Perelman proved that $\tilde{V}_{g(\tau)}(x_0, \tau)$ is monotone non-increasing in τ. It was proved in [**P**] by the first and second variation consideration that

$$\ell_\tau = R - \frac{\ell}{\tau} + \frac{1}{2\tau^{\frac{3}{2}}}K, \tag{4.16}$$

$$|\nabla \ell|^2 = -R + \frac{\ell}{\tau} - \frac{1}{\tau^{\frac{3}{2}}}K, \tag{4.17}$$

$$\Delta \ell \le -R + \frac{n}{2\tau} - \frac{1}{2\tau^{\frac{3}{2}}}K, \tag{4.18}$$

where $K = K(\gamma, \bar{\tau}) = \int_0^{\bar{\tau}} \tau^{\frac{3}{2}} H(X)\, d\tau$, with γ being the minimizing $\mathcal{L}$ geodesic, $X = \gamma'(\tau)$ and

$$H(X) = -R_\tau - \frac{R}{\tau} - 2\langle X, \nabla R\rangle + 2\,\mathrm{Ric}(X, X).$$

Grouping the above in different ways we have that

$$2\ell_\tau + |\nabla \ell|^2 - R + \frac{\ell}{\tau} = 0, \tag{4.19}$$

$$-\ell_\tau + \Delta \ell - |\nabla \ell|^2 + R - \frac{n}{2\tau} \le 0, \tag{4.20}$$

$$\tau\left(2\Delta \ell - |\nabla \ell|^2 + R\right) + \ell - n \le 0, \tag{4.21}$$

as well as

$$\Delta \bar{L} + \bar{L}_\tau \le 2n \tag{4.22}$$

with $\bar{L} = 4\tau\ell$. The monotonicity of the reduced volume follows from (4.20) easily.

Considering the first and second variation of energy functional as (4.14) and applying it to prove the monotonicity of a quantity similar as (4.15) were originated in the seminar work of Li and Yau [**LY**] on the Schrödinger operator. As in the linear heat equation case, there they have to assume that the Ricci curvature is nonnegative together with other conditions on the potential function of the Schrödinger equation. The most remarkable thing is that for the backward Ricci flow space time, the whole thing fits together without assuming anything on the curvature.

Note that assuming (4.19), the inequalities (4.20) and (4.21) imply each other and each of them can be viewed as a preserved inequality for the solution to the Hamilton-Jacobi equation (4.19) (The stressing of the Hamilton-Jacobi equation is attributed to [**H8**] for the Ricci flow case and the much earlier work [**LY**] for the linear Schrödinger equation). According to a conversation with Perelman, by insisting that (4.20) holds equality (namely solving a conjugate heat equation), Perelman discovered the LYH inequality (4.21), namely (2.42), for the fundamental solution. The proof of this leads to the entropy formula (3.31). Later on Hamilton [**H8**] observed that if one insists on equation (4.19), then for the solution to this Hamilton-Jacobi type equation, (4.21) will be preserved. Based on this he also [**H8**] proposed a more general notion of *reduced distance* and claimed some interesting monotonicity and relations between entropy and the reduced volume. We found that this duality is quite interesting and deserves further understanding.

Naturally one would ask if the similar consideration can lead to an entropy-like formula for the hypersurface flow.

4.3. Dual version and the matrix inequality. The same line of thinking as the above can motivate the matrix LYH in Theorem 2.10. This time one considers the reduced geometry of the Ricci flow (instead of the backward Ricci flow). For the sake of the exposition we shall focus on Kähler-Ricci flow.

Let $g(t)$ be a complete solution to Kähler-Ricci flow on $M^m \times [0,T]$ (where $m = \dim_{\mathbb{C}}(M)$ and $n = 2m$). Fix x_0 and let γ be a path $(x(\eta), \eta)$ joining $(x_0, 0)$ to $(x, \bar{t})$. Following [**P**] (see also [**LY, FIN**]) we define

$$\mathcal{L}_+(\gamma) = \int_0^{\bar{t}} \sqrt{t}\left(R + 4|\gamma'(t)|^2\right) dt. \tag{4.23}$$

Let $X = \gamma'(t) = \frac{dz^\alpha(t)}{dt}\frac{\partial}{\partial z^\alpha}$ and let Y be a variational vector field along γ. Here $|\gamma'(t)|^2 = g_{\alpha\bar{\beta}}\frac{dz^\alpha(t)}{dt}\frac{dz^{\bar{\beta}}(t)}{dt}$. Using $\mathcal{L}_+$ as energy we can define the $\mathcal{L}_+$-geodesics and we denote $L_+(y,t)$ to be the length of a shortest geodesics jointing $(x_0, 0)$ to (y, t). We also define

$$\ell_+(x, t; x_0, 0) := \frac{1}{2\sqrt{t}} L_+(x, t).$$

Following the first and second variation calculation of [**P**] (see also [**FIN**]) we have that

$$|\nabla \ell_+|^2 = -R + \frac{\ell_+}{t} + \frac{K}{t^{3/2}}, \tag{4.24}$$

$$\frac{\partial \ell_+}{\partial t} = R - \frac{K}{2t^{3/2}} - \frac{\ell_+}{t}, \tag{4.25}$$

$$\Delta \ell_+ \le R + \frac{n}{2t} - \frac{K}{2t^{3/2}}. \tag{4.26}$$

Here

$$K := \int_0^t \eta^{3/2} H(X)\, d\eta,$$

where $H(X) := \partial R/\partial t + 2\langle \nabla R, X\rangle + 2\langle X, \nabla R\rangle + 4\,\mathrm{Ric}(X,X) + R/t$, is exactly the traced LYH differential Harnack expression in [**Co**] applying to the $(1,0)$ vector field $2X$.

Grouping (4.24)–(4.26) suitably we have the following

$$2\frac{\partial \ell_+}{\partial t} + |\nabla \ell_+|^2 - R + \frac{\ell_+}{t} = 0, \tag{4.27}$$

$$\frac{\partial \ell_+}{\partial t} + \Delta \ell_+ + |\nabla \ell_+|^2 - R - \frac{n}{2t} \le 0, \tag{4.28}$$

$$\left(\frac{\partial}{\partial t} - \Delta\right)(\bar{L}_+ + 2nt) \ge 0 \tag{4.29}$$

$$t(2\Delta \ell_+ + |\nabla \ell_+|^2 - R) - \ell_+ - n \le 0. \tag{4.30}$$

In particular (4.30) suggests an entropy formula, which is dual to Perelman's (3.31) since it holds equality on expanding solitons.

THEOREM 4.2. *Let $(M, g(t))$ be a solution to Ricci flow. Let u be a solution to the conjugate heat equation with $\int_M u = 1$. Write $u = \frac{e^{-f_+}}{(4\pi(t-t_0))^{\frac{n}{2}}}$ for $t > t_0$. Let*

$$\mathcal{W}_+ = \int_M \left[(t - t_0)\left(|\nabla f_+|^2 + R\right) - f_+ + n\right] u d\mu_t.$$

Then

$$\frac{\partial \mathcal{W}_+}{\partial t} = 2(t - t_0) \int_M \left| R_{ij} + \nabla_i \nabla_j f_+ + \frac{g_{ij}}{2(t - t_0)} \right|^2 d\mu. \tag{4.31}$$

Note that the entropy expression is suggested by applying (4.30) to $\hat{u} = \frac{e^{\ell_+(x,t)}}{(4\pi(t-t_0))^{\frac{n}{2}}}$, which is a super-solution of the conjugate heat equation by (4.28).

If we assume that (M, g) has nonnegative bisectional curvature, by regrouping and observing that $K \ge 0$ we have the following result.

THEOREM 4.3. *Let $(M^m, g(t))$ be a complete solution to Kähler-Ricci (Ricci) flow with bounded nonnegative bisectional curvature (curvature operator). Let $H(y, t; x_0, 0)$ be the fundamental solution to forward conjugate heat equation centered at $(x_0, 0)$. Then*

$$\tilde{u}(x, t) := \frac{1}{(\pi t)^m} \exp\left(-\ell_+(x, t; x_0, 0)\right)$$

satisfies

$$\left(\frac{\partial}{\partial t} - \Delta - R\right) \tilde{u}(x, t) \le 0. \tag{4.32}$$

In particular,

$$\tilde{u}(x, t; x_0, 0) \le H(x, t; x_0, 0) \tag{4.33}$$

and

$$\tilde{\theta}_+^{(x_0,0)}(t) := \int_M \tilde{u}(x,t)\, d\mu_t(x)$$

is monotone decreasing. Moreover, the equality in (4.32), *or* (4.33) *implies that* M *is a gradient expanding soliton.*

PROOF. First (4.24)–(4.26) implies that

$$\left(\frac{\partial}{\partial t} - \Delta - R\right)\left(\frac{1}{(\pi t)^m}\exp(-\ell_+(x,t))\right)$$
$$= -\frac{K}{t^{\frac{3}{2}}}\left(\frac{1}{(\pi t)^m}\exp(-\ell_+(x,t))\right) \le 0.$$

Here we have used the fact that $K \ge 0$ under the assumption that M has bounded non-negative bisectional curvature. Also if the equality holds it implies that $K \equiv 0$. This further implies that M is an expanding soliton from the computation in [**FIN**]. In order to prove (4.33) one just needs to apply the maximum principle and notice that $\lim_{t\to 0} \frac{1}{(\pi t)^m}\exp(-\ell_+(x,t)) = \delta_{x_0}(x)$. The equality case follows from the analysis on the equality case in [**FIN**]. □

Moreover we also have the following observation which motivated Theorem 2.10.

PROPOSITION 4.4. *Assume that* $(M, g(t))$ *be a Kähler-Ricci flow with bounded nonnegative bisectional curvature on* $M \times [0, T)$. *Let* $\tilde{u}(x,t)$ *be as in Theorem 4.3. Then*

$$\log(\tilde{u})_{\alpha\bar{\beta}} + R_{\alpha\bar{\beta}} + \frac{1}{t} g_{\alpha\bar{\beta}} \ge 0. \tag{4.34}$$

The equality holds if and only if $(M, g(t))$ *is an expanding Kähler-Ricci solition.*

5. Comments

Due to the limited time allowed and the lack of expertise we could not address in this exposition many important aspects related to LYH inequalities. Among them, the most notable is the space-time consideration of Chow and Chu [**CC1, CC2, CC3**], and the very recent work of Hamilton on the local approximation of the trace LYH inequality for the Ricci flow [**H7**]. The main theme of the space-time consideration is to interpret the LYH expression as the curvature operator of certain *degenerate metric* on the space time $\mathcal{M} = M \times [0, T]$. This geometric consideration sometimes suggests new LYH estimates [**CK, Che**]. A similar space-time construction for the backward Ricci flow was done by Perelman [**P**]. This construction for the backward Ricci flow is related to the reduced distance and monotonicity of the reduced volume.

Concerning the monotonicity, we completely missed the monotonicity of many important energy functionals, such as the Donaldson functional and the Mabuchi energy, constructed in the study of Hermitian metrics on holomorphic vector bundles or Kähler metrics/potentials. They seem not related to LYH type estimates in general.

References

[An1] B. Andrews, *Contraction of convex hypersurfaces in Euclidean space,* Calc. Var. **2**(1994), 151–171.

[An2] B. Andrews, *Harnack inequalities for evolving hypersurfaces,* Math. Z. **217**(1994), 179–197.

[B] S. Bando, *On the classification of three-dimensional compact Kaehler manifolds of nonnegative bisectional curvature,* J. Differential Geom. **19**(1984), no. 2, 283–297.

[Co] H.-D. Cao, *On Harnack inequalities for the Kähler-Ricci flow,* Invent. Math. **109**(1992), 247–263, MR1172691, Zbl 0779.53043.

[CN] H.-D. Cao and L. Ni, *Matrix Li-Yau-Hamilton estimates for heat equation on Kähler manifolds*, Math. Ann. **331**(2005), 795–807, MR2148797, Zbl pre02156245.

[CY] J. Cheeger and S.-T. Yau, *A lower bound for the heat kernel*, Comm. Pure Appl. Math. **34**(1981), no. 4, 465–480, MR0615626, Zbl 0481.35003.

[ChY1] S. Y. Cheng and S. T. Yau, *Differential equations on Riemannian manifolds and their geometric applications*, Comm. Pure Appl. Math. **28**(1975), no. 3, 333–354.

[ChY2] S. Y. Cheng and S. T. Yau, *Maximal space-like hypersurfaces in the Lorentz-Minkowski spaces*, Ann. of Math. (2) **104**(1976), no. 3, 407–419.

[Che] H.-B. Cheng, *A New Li-Yau-Hamilton Estimate for the Ricci Flow,* Comm. Anal. Geom. **14**(2006), 551–564.

[Ch1] B. Chow, *On Harnack's inequality and entropy for the Gaussian curvature flow,* Comm. Pure Appl. Math. **44**(1991), no. 4, 469–483.

[Ch2] B. Chow, *The Yamabe flow on locally conformally flat manifolds with positive Ricci curvature*, Comm. Pure Appl. Math. **45**(1992), no. 8, 1003–1014.

[Ch3] B. Chow, *Interpolating between Li-Yau's and Hamilton's Harnack inequalities on a surface*, J. Partial Differential Equations **11**(1998), no. 2, 137–140, MR1626999, Zbl 0943.58017.

[CC1] B. Chow and S.-C. Chu, *A geometric interpretation of Hamilton's Harnack inequality for the Ricci flow,* Math. Res. Lett. **2**(1995), no. 6, 701–718, MR1362964, Zbl 0856.53030.

[CC2] B. Chow and S.-C. Chu, *A geometric approach to the linear trace Harnack inequality for the Ricci flow,* Math. Res. Lett. **3**(1996), no. 4, 549–568, MR1406020, Zbl 0868.58082.

[CC3] B. Chow and S.-C. Chu, *Space-time formulation of Harnack inequalities for curvature flow of hypersurfaces,* J. Geom. Anal. **11**(2001), 219–231.

[CH] B. Chow and R. Hamilton, *Constrained and linear Harnack inqualities for parabolic equations*, Invent. Math. **129**(1997), 213–238, MR1465325.

[CK] B. Chow and D. Knopf, *New Li-Yau-Hamilton inequalities for the Ricci flow via the space-time approach*, J. Differential Geom. **60**(2002), no. 1, 1–54, MR1924591, Zbl 1048.53026.

[E1] K. Ecker, *A local monotonicity formula for mean curvature flow*, Ann. of Math. (2) **154**(2001), no. 2, 503–525, MR1865979, Zbl 1007.53050.

[E2] K. Ecker, *Regularity Theory for Mean Curvature Flow*, Progress in Nonlinear Differential Equations and their Applications 57. Birkhuser Boston, Inc., Boston, MA, 2004, MR2024995, Zbl 1058.53054.

[EKNT] K. Ecker, D. Knopf, L. Ni and P. Topping, *Local monotonicity and mean value formulas for evolving Riemannian manifolds,* to appear in Crelle's Journal.

[FIN] M. Feldman, T. Ilmanen and L. Ni, *Entropy and reduced distance for Ricci expanders*, J. Geom. Anal. **15**(2005), 49–62, MR2132265, Zbl 1071.53040.

[Fu] W. B. Fulks, *A mean value theorem for heat equation,* Proc Amer. Math Soc. **17**(1966), 6–11.

[H1] R. Hamilton, *Four-manifolds with positive curvature operator,* J. Differenital. Geom. **24**(1986), 153–179.

[H2] R. Hamilton, *The Ricci flow on surfaces,* Mathematics and general relativity (Santa Cruz, CA, 1986), 237–262, Contemp. Math., **71**, Amer. Math. Soc., Providence, RI, 1988.

[H3] R. Hamilton, *The Harnack estimate for the Ricci flow*, J. Differential Geom. **37**(1993), no. 1, 225–243, MR1316556, Zbl 0804.53023.

[H4] R. Hamilton, *Harnack estimate for the mean curvature flow*, J. Differential Geom. **41**(1995), no. 1, 215–226.

[H5] R. Hamilton, *A matrix Harnack estimate for the heat equation*, Comm. Anal. Geom. **1**(1993), no. 1, 113–126, MR1230276, Zbl 0799.53048.

[H6] R. Hamilton, *Monotonicity formulas for parabolic flows on manifolds*, Comm. Anal. Geom. **1**(1993), no. 1, 127–137, MR1230277, Zbl 0779.58037.

[H7] R. Hamilton, *A local approximate Harnack estimate for the Ricci flow,* preprint, 2006.

[H8] R. Hamilton, *Lectures on Perelman's reduced distance and differential Harnack inequality.*

[Hu] G. Huisken, *Asymptotic behavior for singularities of the mean curvature flow,* J. Differential Geom. **31**(1990), no. 1, 285–299.

[HS] G. Huisken and C. Sinestrari, *Convexity estimates for mean curvature flow and singularities of mean convex surfaces*, Acta Math. **183**(1999), 45–70.

[KM] K. Kodaira and J. Morrow, *Complex Manifolds.* Holt, Rinehart and Winston, Inc. 1971.

[LY] P. Li and S.-T. Yau, *On the parabolic kernel of the Schrödinger operator*, Acta Math. **156**(1986), no. 3–4, 153–201, MR0834612, Zbl 0611.58045.

[Mo] J. Moser, *A Harnack inequality for parabolic differential equations*, Comm. Pure Appl. Math. **17**(1964), 101–134, MR0159139, Zbl 0149.06902.

[N1] L. Ni, *Monotonicity and Kähler-Ricci flow*, Contemp. Math. **367**(2005), 149–165, MR2115758.

[N2] L. Ni, *The entropy formula for linear heat equation*, J. Geom. Anal. **14**(2004), 87–100, MR2030576, Zbl 1062.58028; *addenda,* J. Geom. Anal. **14**(2004), 369–374, MR2051693.

[N3] L. Ni, *A monotonicity formula on complete Kähler manifold with nonnegative bisectional curvature*, J. Amer. Math. Soc. **17**(2004), 909–946, MR2083471, Zbl pre02106879.

[N4] L. Ni, *A matrix Li-Yau-Hamilton inequality for Kähler-Ricci flow,* J. Differential Geom. **75**(2007), 303–358.

[N5] L. Ni, *A note on Perelman's LYH type inequality,* Comm. Anal. Geom. **14**(2006), 883–905.

[NT] L. Ni and L.-F. Tam, *Plurisubharmonic functions and the Kähler-Ricci flow*, Amer. J. Math. **125**(2003), 623–654, MR1981036, Zbl 1027.53076.

[P] G. Perelman, *The entropy formula for the Ricci flow and its geometric applications*, arXiv: math.DG/ 0211159.

[SY] R. Schoen and S. T. Yau, *Lectures on differential geometry.* Lecture notes prepared by Wei Yue Ding, Kung Ching Chang [Gong Qing Zhang], Jia Qing Zhong and Yi Chao Xu. Translated from the Chinese by Ding and S. Y. Cheng. Preface translated from the Chinese by Kaising Tso. Conference Proceedings and Lecture Notes in Geometry and Topology, I. International Press, Cambridge, MA, 1994. v+235 pp.

[T] K.-S. Chou (Kaiseng Tso), *Deforming a hypersurface by its Gauss-Kronecker curvature,* Comm. Pure Math. Appl. **38**(1985), 867–882.

[Wa] N. A. Watson, *A theory of temperatures in several variables,* Proc. London Math. Soc. **26**(1973), 385–417.

[Y] S. T. Yau, *Harmonic functions on complete Riemannian manifolds*, Comm. Pure Appl. Math. **28**(1975), 201–228.

DEPARTMENT OF MATHEMATICS, UNIVERSITY OF CALIFORNIA AT SAN DIEGO, LA JOLLA, CA 92093

E-mail address: lni@math.ucsd.edu

Singularities of Mean Curvature Flow and Flow with Surgeries

Carlo Sinestrari

ABSTRACT. We collect in this paper several results on the formation of singularities in the mean curvature flow of hypersurfaces in euclidean space, under various kinds of convexity assumptions. We include some recent estimates for the flow of 2-convex surfaces, i.e. the surfaces where the sum of the two smallest principal curvatures is positive everywhere. Such results enable the construction of a flow with surgeries for these surfaces similar to the one introduced by Hamilton and Perelman for the Ricci flow. The topological applications of the construction are also described.

1. Introduction

Let $F_0 : \mathcal{M} \to \mathbb{R}^{n+1}$ be a smooth immersion of an n-dimensional hypersurface in Euclidean space, $n \geq 1$. The evolution of $\mathcal{M}_0 = F_0(\mathcal{M})$ by mean curvature flow is a one-parameter family of smooth immersions $F : \mathcal{M} \times [0, T[\to \mathbb{R}^{n+1}$ satisfying

$$\frac{\partial F}{\partial t}(p,t) = -H(p,t)\nu(p,t), \qquad p \in \mathcal{M}, t \geq 0, \tag{1.1}$$

$$F(\cdot, 0) = F_0, \tag{1.2}$$

where $H(p,t)$ and $\nu(p,t)$ are the mean curvature and the outer normal respectively at the point $F(p,t)$ of the surface $\mathcal{M}_t = F(\cdot, t)(\mathcal{M})$.

This evolution has been studied by many authors in the last decades. It occurs in some physical models describing interface evolution, and also in the singular limit of some reaction diffusion equations. It can be checked that problem (1.1)–(1.2) is parabolic and possesses a unique solution locally in time. In general, however, global existence cannot be expected, because the curvature can become unbounded in finite time. This is always the case, for instance, if the surface is closed. Intuitively, certain parts of the surface, or the whole surface, shrink with the flow and develop singularities.

It is natural to look for a generalized definition of the flow which allows to continue the evolution after the formation of singularities. The first approach was introduced by Brakke [**7**], who gave a definition of weak solutions using notions from geometric measure theory. Several other notions have been introduced in the following. Among others, we recall the one based on the level set method which was adopted independently by Chen, Giga and Goto [**8**] and by Evans and Spruck [**14**].

A further reason of interest for the mean curvature flow came from Hamilton's paper on the Ricci flow [**19**]. The Ricci flow is the evolution of the metric of an abstract Riemannian manifold governed by the equation

$$\frac{\partial g_{ij}}{\partial t} = -2R_{ij}, \tag{1.3}$$

where R_{ij} is the Ricci tensor associated with the metric. Like the mean curvature flow, the Ricci flow is a parabolic problem where singularities can appear in finite time. Hamilton's result in [**19**] was that any initial metric on a closed three-dimensional manifold with positive Ricci curvature evolving by Ricci flow converges, up to a suitable rescaling, to a metric of constant curvature. This implies that any three-dimensional Riemannian manifold with positive Ricci curvature is diffeomorphic to a quotient of the standard sphere S^3. Such an approach had an antecedent in the work of Eels and Sampson [**13**], who had used the heat flow to deform a map between two fixed Riemannian manifolds. Hamilton's result, which was soon followed by similar ones for dimensions other than three, showed that the study of geometric flows could provide new important results.

In [**24**] Huisken proved a theorem for the mean curvature flow which had strong analogies with Hamilton's one for the Ricci flow. He showed that any closed convex surface shrinks to a point in finite time and converges, after rescaling, to a round sphere. Also, it became clear that the two flows have many similarities. The basic examples of formation of singularities (e.g., the sphere, the neckpinch) are analogous for the two flow. In both flows some important geometric properties remain invariant. For example, positive curvature operator and positive scalar curvature are preserved under Ricci flow, while convexity and positive mean curvature are preserved under mean curvature flow. All these results follow easily by the maximum principle applied to the evolution equation for the corresponding scalar or tensor quantity. For both flows the formation of singularities induces in some case the convergence to a canonical structure (constant curvature). It should be pointed out that there are no precise connections between Ricci flow and mean curvature flow, that is, there is no way to transform one flow into the other. Also, the proofs in [**19**] and [**24**] differ in many substantial steps. Throughout the paper, we will remark further the analogies and the differences between the two flows.

In the last years, the Ricci flow has become well known in the mathematical community after Perelman's proof [**34, 35, 36**] of Thurston's geometrization conjecture, a result which gives a complete classification of closed three-dimensional manifolds, and includes the celebrated Poincaré conjecture. Perelman's proof accomplishes a program started by Hamilton shortly after his above mentioned paper, a program aimed to prove the conjecture using the Ricci flow. A central tool in the proof is the construction of a flow after singularities by means of a surgery procedure: when a singularity occurs, the parts of the manifold with larger curvature are removed and are replaced with more regular ones. The surgery is performed in such a way that one keeps track of the possible changes of the topology. Hamilton conjectured that it was possible to define a flow where, after a finite number of surgeries, the remaining components converge to canonical structures, allowing to classify the possible topologies of the initial manifold. The hardest part in this analysis is to derive suitable estimates for the flow which give a description of the singular profiles good enough to do the surgeries, and which allow to prove that the remaining components eventually converge to the desired canonical structure. Hamilton carried out his program in the simpler case the Ricci flow of four-dimensional manifolds with positive isotropic curvature in [**23**]. However, the case of a general three-manifold needed for geometrization presented some further major difficulties. These were successfully solved by Perelman, who found some new estimates and introduced new techniques which allowed the conjectures to be proven.

The aim of the present paper is to give a survey of some recent results in collaboration with G. Huisken [**27, 28, 29**], which can be regarded as the counterpart for the mean curvature flow of Hamilton's approach to the Ricci flow. Namely, we define a mean curvature flow with surgeries for hypersurfaces satisfying a suitable curvature restriction, which we explain below. Like in the three-dimensional Ricci flow, our flow with surgeries gives rise after a finite number of steps to components which have a canonical structure, and allows a classification of the possible topologies of the initial manifold. The applications of our construction are not as far-reaching as the ones of Ricci flow, but they show that also mean curvature flow can be used to derive nontrivial topological consequences. We hope that in the future the technique can be extended to more general classes of surfaces to obtain a wider range of applications.

Let us briefly describe our surgery procedure. Following Hamilton [**23**], we call a *neck* a subset of our surface which is close, up to rescaling, to a portion of the standard cylinder $S^{n-1} \times \mathbb{R}$ in a suitable topology. Our surgery consists of removing a neck and filling smoothly the two holes with convex regions diffeomorphic to disks. The surface changes its topology in a controlled way, because the surgery acts like the inverse of a connected sum. The aim is to perform surgeries in such a way that the parts of the surface with largest curvature are removed and the flow can be continued for a longer time. To this purpose, one needs to show that it is possible to find a neck whenever a singularity is formed (except for those cases where we can already describe the topology of the whole manifold) and that the curvature decreases after the neck is removed. In order to prove these properties, it is necessary to have a detailed description of the singularities by means of suitable a priori estimates. This preliminary analysis is in fact the most difficult part of the procedure.

At the present stage, a good description of the singularities of mean curvature flow is only available for hypersurfaces satisfying certain curvature restrictions. This is in contrast with the Ricci flow, where in low dimensions ($n = 2, 3$) it has been possible to study the flow of an arbitrary metric. Our surgery construction applies to a suitable class of hypersurfaces, usually called two-convex. A hypersurface is called k-convex, for an integer $k = 1, \ldots, n$, if the sum of the smallest k principal curvatures is nonnegative at each point. Also, we need the hypothesis that the surface has dimension $n \geq 3$. Our main result is the following.

THEOREM 1.1. *Let $\mathcal{M}_0 \subset \mathbb{R}^{n+1}$ be a closed immersed n-dimensional two-convex hypersurface, with $n \geq 3$. Then there is a mean curvature flow with surgeries with initial value $\mathcal{M}_0$ such that, after a finite number of surgeries, the remaining components are diffeomorphic either to S^n or to $S^{n-1} \times S^1$.*

Due to the structure of our surgeries, we can easily deduce that the initial manifold is the connected sum of finitely many components diffeomorphic to S^n or to $S^{n-1} \times S^1$. Thus we obtain the following classification of 2-convex hypersurfaces.

COROLLARY 1.2. *Any smooth closed n-dimensional two-convex immersed surface $\mathcal{M} \subset \mathbb{R}^{n+1}$ with $n \geq 3$ is diffeomorphic either to S^n or to a finite connected sum of $S^{n-1} \times S^1$.*

Another consequence of our construction is the following Schoenflies type theorem for simply connected two-convex surfaces.

COROLLARY 1.3. *Any smooth closed simply connected n-dimensional two-convex embedded surface $\mathcal{M} \subset \mathbb{R}^{n+1}$ with $n \geq 3$ is diffeomorphic to S^n and bounds a region whose closure is diffeomorphic to a smoothly embedded (n+1)-dimensional standard closed ball.*

We recall that the Schoenflies conjecture states that any smooth n-dimensional embedded surface $\mathcal{M} \subset \mathbb{R}^{n+1}$ diffeomorphic to S^n bounds a region whose closure is diffeomorphic to a smoothly embedded $(n+1)$-dimensional standard closed ball. The result has been proved for any dimension except for $n = 4$. Thus we show that there are no counterexamples to the conjecture in the class of 2-convex surfaces. The proof of Corollary 1.3 requires some additional argument (like proving that an embedded hypersurface remains embedded under the flow with surgeries) but it is a quite direct consequence of the main Theorem 1.1.

It is natural to compare the mean curvature flow with surgeries with the generalized solutions mentioned at the beginning. It should be observed that our flow cannot be regarded as a weak solution of the equation, because it introduces an arbitrary modification of the manifold at the surgery time. However, the flow with surgeries is more natural and useful for the geometric and topological applications we are considering. For weak solutions, instead, one does not have such a clear knowledge of the topology of the manifold after the singular time. Let us mention that for the Ricci flow there is no such choice between different approaches, because no notion of weak solution is available there yet.

Let us mention some related results in the previous literature. The structure of k-convex hypersurfaces immersed in Riemannian manifolds has been studied by various authors [**16, 31, 37, 40**]. The results of these papers, however, concern mainly the homotopy type of the surfaces. Let us also mention that a piecewise smooth mean curvature flow was considered in [**2**] for a class of rotationally symmetric hypersurfaces. It was shown there that such surfaces become singular at isolated points and times, splitting into components which become smooth again immediately afterwards, so that no surgery is needed. The analysis depends strongly on the assumption of rotational symmetry. Recently, the construction of a mean curvature flow with surgeries for two-dimensional manifolds, based on independent techniques from [**29**], has been announced in [**9**].

In the following we give an informal exposition of the surgery procedure of [**29**]. The largest part of the paper is devoted to survey the estimates which are necessary for the construction, including the ones of [**24, 27, 28**]. We give here not only the statements of the main results, but also an outline of the proofs, where possible. Due to the informal character of the paper, many technical details will be omitted; in particular, most of the proofs should be intended as sketches. We hope that this will suffice to give the reader some understanding of the main techniques used in this theory.

2. Examples

There are some well known cases where the solution to the mean curvature flow can be written explicitly, or at least admits a detailed qualitative description. We recall briefly the interesting ones for our purposes. A more detailed description can be found in the monograph by Ecker [**11**].

EXAMPLE 2.1. **Homothetically shrinking solutions** These solutions, also called self-similarly shrinking solutions, are those such that $\mathcal{M}_t = \lambda(t)\mathcal{M}_0$ for a suitable homothety factor $\lambda(t)$ with $\lambda(t) < 1$ for $t > 0$. The simplest example is the sphere. It is easily checked that, if $\mathcal{M}_0 = S^n_R(0)$, the sphere of radius R around the origin, then $\mathcal{M}_t = S^n_{r(t)}(0)$, where $r(t) = \sqrt{R^2 - 2nt}$. It follows that the flow exists up to time $T = R^2/2n$, at which the sphere shrinks to a point. A less obvious example is given by a class of immersed curves in the plane, which have been classified by Mullins [**33**], see also [**1**]. Other shrinking solutions are obtained by taking products of these surfaces with flat factors, e.g., the cylinder $S^{n-1} \times \mathbb{R}$. It has been proved by Huisken [**26**] that the ones we have just described are the only homothetically shrinking solutions which are mean convex, i.e., with positive mean curvature everywhere. An interesting example which is not mean convex is the homothetically shrinking torus whose existence was proved by Angenent [**5**].

EXAMPLE 2.2. **Translating graphs** A translating graph is a surface $\mathcal{M}_0$ which is the graph of some function $y = u(x)$, whose evolution by mean curvature flow exists for all times and is such that $\mathcal{M}_t$ is the graph of $y = u(x) + t$. The simplest example is the so called grim reaper, which is the graph of $u(x) = -\ln(\cos x)$, $x \in (-\pi/2, \pi/2)$. It is the only example in one dimension, up to translations. In higher dimensions, there is a unique rotationally symmetric example, defined in the whole space and asymptotic to a paraboloid. The existence of translating graphs which are not rotationally symmetric has been recently shown in [**38**].

EXAMPLE 2.3. **The standard neckpinch** Suppose that $\mathcal{M}_0$ looks like two large balls connected by a cylindrical part (neck) which is very thin, in such a way that the mean curvature there is much larger than in the balls. Then one expects that the radius of the neck goes to zero in a short time while the balls move little from their original position. The existence of surfaces with this property was first proved rigorously by Grayson [**18**]; a simple proof can be found in [**11**]. An explicit example of initial surface is given in [**2**] (see the next example).

In contrast with Example 2.1, the surface here does not become singular everywhere at the singular time, but only in a restricted region. In a case like this it is interesting to define a weak solution after the singular time. One intuitively expects that the surface should divide in two parts, each of them flowing independently afterwards. The idea of the flow with surgeries is to induce this behavior in a controlled way.

EXAMPLE 2.4. **The degenerate neckpinch** This example is given in [**2**]. For a given $\lambda > 0$, let us set

$$\phi_\lambda(x) = \sqrt{(1 - x^2)(x^2 + \lambda)}, \qquad -1 \leq x \leq 1.$$

For any $n \geq 2$, let $\mathcal{M}^\lambda$ be the n-dimensional surface in $\mathbb{R}^{n+1}$ obtained by rotation of the graph of ϕ_λ. The surface $\mathcal{M}^\lambda$ looks like a dumbbell, where the parameter λ measures the width of the central part. Then, the following properties hold:

(a) if λ is large enough, the surface $\mathcal{M}_t^\lambda$ eventually becomes convex and shrinks to a point in finite time;

(b) if λ is small enough, $\mathcal{M}_t^\lambda$ exhibits a neckpinch singularity as in Example 2.3;
(c) there exists $\lambda > 0$ such that $\mathcal{M}_t^\lambda$ shrinks to a point in finite time, has positive mean curvature up to the singular time, but never becomes convex. The maximum of the curvature is attained at the two points where the surface meets the axis of rotation. After rescaling around either of these points, the asymptotic profile of the surface is given by a translating solution of the flow.

The behavior in (c) is called degenerate neckpinch and was first conjectured by Hamilton for the Ricci flow [**22**, §3]. Intuitively speaking, it is a limiting case of the neckpinch where the cylinder in the middle and the balls on the sides shrink at the same time. One can also build the example in an asymmetric way, with only one of the two balls shrinking simultaneously with the neck, while the other one remains nonsingular. A sharp analysis of the singular behavior for a class of rotationally symmetric surfaces exhibiting a degenerate neckpinch has been done in [**6**].

Degenerate neckpinches are more difficult to handle when one defines a flow with surgeries, because it is less clear how to find a cylindrical region where the surgery can be performed. Hamilton's intuition for the Ricci flow was that, although the region with the largest curvature is strictly convex, one can find almost cylindrical regions on the surface by moving away by a suitable distance (the rescaled profile near the singularity has an "asymptotically necklike end", in the terminology of [**23**]). For the mean curvature flow, one can think of the typical shape of a translating solution, which is asymptotic to a paraboloid. If one considers a strip of the paraboloid far from the vertex, it is close enough to a portion of a cylinder for the purposes of surgery.

3. Invariance properties

Let $F : \mathcal{M} \times [0, T[\to \mathbb{R}^{n+1}$ be a solution of mean curvature flow (1.1)–(1.2) with closed, smoothly immersed evolving surfaces $\mathcal{M}_t = F(\cdot, t)(\mathcal{M})$. We denote the induced metric by $g = \{g_{ij}\}$, the surface measure by $d\mu$, the second fundamental form by $A = \{h_{ij}\}$ and the Weingarten operator by $W = \{h^i_j\}$. We then denote by $\lambda_1 \leq \cdots \leq \lambda_n$ the principal curvatures, i.e., the eigenvalues of W, and by $H = \lambda_1 + \cdots + \lambda_n$ the mean curvature. In addition, $|A|^2 = \lambda_1^2 + \cdots + \lambda_n^2$ will denote the squared norm of A. All these quantities depend on $(p, t) \in \mathcal{M} \times [0, T[$ and satisfy the following equations computed in [**24**].

LEMMA 3.1. *If $\mathcal{M}_t$ evolves by mean curvature flow, the associated quantities introduced above satisfy the following equations (here ∇ and Δ denote respectively the covariant derivative and the Laplace-Beltrami operator* *induced by the metric on $\mathcal{M}_t$):*

$$(i)\ \frac{\partial}{\partial t}g_{ij} = -2Hh_{ij} \quad (ii)\ \frac{\partial}{\partial t}\, d\mu = -H^2 d\mu, \quad (iii)\ \frac{\partial}{\partial t}h^i_j = \Delta h^i_j + |A|^2 h^i_j,$$
$$(iv)\ \frac{\partial}{\partial t}H = \Delta H + |A|^2 H, \quad (v)\ \frac{\partial}{\partial t}|A|^2 = \Delta |A|^2 - 2|\nabla A|^2 + 2|A|^4.$$

The mean curvature flow is a parabolic system of PDEs and satisfies a local existence and uniqueness result for smooth solutions under general hypotheses. For our purposes the following statement will suffice (see, e.g., [**12, 15, 17, 24**]).

THEOREM 3.2. *Let $\mathcal{M}_0 = F_0(\mathcal{M})$ be smooth and closed. Then the mean curvature flow* (1.1)–(1.2) *has a unique smooth solution, which is defined in a maximal time interval* $[0, T[$*, where* $0 < T < +\infty$*, and satisfies* $\max_{\mathcal{M}_t} |A|^2 \to \infty$ *as* $t \uparrow T$.

A first step in the analysis of singularities is to observe that several geometric properties are invariant under the flow. The invariance can be usually proved in an elementary way by means of the maximum principle. Let us give some examples.

PROPOSITION 3.3. *Let $\mathcal{M}_t$, $t \in [0,T)$ be a closed hypersurface evolving by mean curvature flow.*

(i) *If $H \geq 0$ on $\mathcal{M}_0$, then $H > 0$ on $\mathcal{M}_t$ for any $t \in (0,T)$.*
(ii) *If $|A|^2 \leq cH^2$ on $\mathcal{M}_0$, then $|A|^2 \leq cH^2$ on $\mathcal{M}_t$ for any $t \in (0,T)$.*

PROOF. Part (i) follows from Lemma 3.1 and the strong maximum principle. To obtain (ii), we compute the evolution equation of $f := |A|^2/H^2$. We obtain, by Lemma 3.1 and a straightforward computation,

$$\frac{\partial f}{\partial t} = \Delta f + \frac{2}{H}\langle \nabla H, \nabla f\rangle - \frac{2}{H^4}|H\nabla_i h_{kl} - \nabla_i H\, h_{kl}|^2. \tag{3.1}$$

Thus, the maximum principle implies that the maximum of f is nonincreasing. □

COROLLARY 3.4. *Let $\mathcal{M}_t$, $t \in [0,T)$ be a closed n-dimensional hypersurface evolving by mean curvature flow.*

(i) *If $H > 0$ on $\mathcal{M}_0$, then there is $\varepsilon_0 > 0$ such that $\varepsilon_0|A|^2 \leq H^2 \leq n|A|^2$ everywhere on $\mathcal{M}_t$ for all $t \in (0,T)$.*
(ii) *If $\mathcal{M}_0$ has positive scalar curvature, then the same holds for $\mathcal{M}_t$ for all $t \in (0,T)$.*

PROOF. To prove the first inequality in (i), it suffices to take $\varepsilon_0 = \min_{\mathcal{M}_0} H^2/|A|^2$ which is attained by compactness, and to apply Proposition 3.3-(ii). Inequality $H^2 \leq n|A|^2$ is an algebraic property which holds in general. Part (ii) also is a consequence of Proposition 3.3-(ii) because positive scalar curvature is equivalent to $H^2/|A|^2 > 1$. □

Corollary 3.4-(ii) is a particular case of a more general property of the elementary symmetric polynomials of the curvatures, as we now proceed to show. We recall that the *elementary symmetric polynomial* of degree k in n variables $\lambda_1, \ldots, \lambda_n$ is defined as

$$S_k = \sum_{1 \le i_1 < i_2 < \cdots < i_k \le n} \lambda_{i_1} \lambda_{i_2} \cdots \lambda_{i_k}$$

for $k = 1, \ldots, n$. In particular, $S_1 = H$, and S_2 is the scalar curvature. It is not difficult to show that

$$\lambda_1 \ge 0, \ldots, \lambda_n \ge 0 \iff S_1 \ge 0, \ldots, S_n \ge 0. \tag{3.2}$$

These polynomials enjoy some remarkable concavity properties, see e.g. [**4**]. The relevant one for our purposes is the following [**28, 30**].

THEOREM 3.5. *Let $\Gamma_k \subset \mathbb{R}^n$ denote the connected component of $S_k > 0$ containing the positive cone. Then $S_l > 0$ on Γ_k for all $l = 1, \ldots, k$ and the quotient S_{k+1}/S_k is concave on Γ_k.*

The above properties remain unchanged if we regard the polynomials S_k as functions of the Weingarten operator, instead of the principal curvatures, because we have the following result, see [**3**, Lemma 2.2] or [**28**, Lemma 2.11].

THEOREM 3.6. *Let $f(\lambda_1, \ldots, \lambda_n)$ be a symmetric convex (concave) function and let $F(A) = f$(eigenvalues of A) for any $n \times n$ symmetric matrix A whose eigenvalues belong to the domain of f. Then F is convex (concave).*

The concavity of the above expressions allows one to apply the maximum principle to obtain invariance properties. This will be clear after deriving the following evolution equation.

PROPOSITION 3.7. *Let $F(h^i_j)$ be a function homogeneous of degree one. Let $\mathcal{M}_t$ be a closed mean convex surface evolving by mean curvature flow such that h^i_j belongs to the domain of F everywhere. Then*

$$\frac{\partial}{\partial t}\frac{F}{H} - \Delta \frac{F}{H} = \frac{2}{H}\left\langle \nabla H, \nabla \frac{F}{H} \right\rangle - \frac{1}{H}\frac{\partial^2 F}{\partial h^i_j \partial h^k_l} \nabla^p h^j_i \nabla_p h^k_l$$

As a consequence, if F is concave (convex), any estimate of the form $F \ge cH$ (resp. $F \le cH$) is preserved.

Proof. A straightforward computation, using Lemma 3.1(iii)–(iv) and Euler's theorem on homogeneous functions, yields

$$\begin{aligned}\frac{\partial}{\partial t}\frac{F}{H} &= \frac{1}{H}\frac{\partial F}{\partial h^i_j}(\Delta h^i_j + |A|^2 h^i_j) - \frac{F}{H^2}(\Delta H + |A|^2 H)\\ &= \Delta\frac{F}{H} + \frac{2}{H}\left\langle \nabla H, \nabla\frac{F}{H}\right\rangle - \frac{1}{H}\frac{\partial^2 F}{\partial h^i_j \partial h^k_l}\nabla^p h^j_i \nabla_p h^k_l. \qquad \square\end{aligned}$$

In particular, the previous proposition can be applied to $F = S_{k+1}/S_k$, provided $S_k \neq 0$. This leads to the following result, which generalizes Corollary 3.4.

Proposition 3.8. *Let $\mathcal{M}_0$ be a closed hypersurface such that $S_k > 0$ everywhere for a given $k \in \{1, \dots, n\}$ and let $\mathcal{M}_t$ be its evolution by mean curvature flow. Then, for any $l = 2, \dots, k$ there exists γ_l such that $S_l \geq \gamma_l H^l > 0$ on $\mathcal{M}_t$ for all $t \in (0,T)$.*

Proof. We first observe that on $\mathcal{M}_0$ the curvatures $(\lambda_1, \dots, \lambda_n)$ belong everywhere to the set Γ_k defined in Theorem 3.5. By the same theorem, we have $S_l > 0$ on $\mathcal{M}_0$ for $l = 1, \dots, k$ and so, by compactness, we have $S_l \geq c_l H S_{l-1}$ for suitable constants $c_l > 0$, for any $l = 2, \dots, k$. We know from Proposition 3.3 that $H > 0$ everywhere on $\mathcal{M}_t$ for $t \in (0,T]$. Then we can consider the quotient $S_2/H^2 = S_2/S_1 H$. It is defined for every t, it is greater than c_2 at time zero, and its minimum is nondecreasing by Proposition 3.7. It follows that $S_2 \geq c_2 H^2$ also for $t \in (0,T)$. We now apply the same procedure to the quotient $S_3/S_2 H$ to conclude that it is greater than c_3 for $t \in (0,T)$, i.e., $S_3 \geq c_3 S_2 H \geq c_3 c_2 H^3$. Repeating the argument a finite number of times yields the conclusion. $\square$

Further invariance properties for the mean curvature flow can be obtained using Hamilton's maximum principle for tensors [**20**, Section 4]. Let us first recall a definition. We say that an immersed surface $\mathcal{M}$ is k-convex, for some $1 \leq k \leq n$, if the sum of the k smallest curvatures is nonnegative at every point of $\mathcal{M}$. In particular, 1-convexity coincides with convexity, while n-convexity means nonnegativity of the mean curvature H, i.e., mean convexity. Then we have the following result.

Proposition 3.9. *If a closed hypersurface $\mathcal{M}_0$ satisfies $\lambda_1 + \cdots + \lambda_k \geq \alpha H$ for some $\alpha \geq 0$ and $1 \leq k \leq n$, then the same holds for its evolution by mean curvature flow $\mathcal{M}_t$. In particular, if $\mathcal{M}_0$ is k-convex, then so is $\mathcal{M}_t$.*

Proof. The result follows from Hamilton's maximum principle for tensors, provided we show that the inequality $\lambda_1 + \cdots + \lambda_k \geq \alpha H$ describes a convex cone in the set of all matrices, and that this cone is invariant under the system of o.d.e.'s $dh^i_j/dt = |A|^2 h^i_j$, which is obtained by dropping the diffusion term in the evolution equation for the Weingarten operator in Lemma 3.1.

If we denote by $W(v_1, v_2)$ the Weingarten operator applied to two tangent vectors v_1, v_2 at any point, we have

$$\lambda_1 + \cdots + \lambda_k = \min\{W(e_1,e_1) + \cdots + W(e_k,e_k) : \langle e_i, e_j\rangle = \delta_{ij} \text{ for all } 1 \le i \le j \le k\}.$$

This shows that $\lambda_1+\cdots+\lambda_k$ is a concave function of the Weingarten operator, being the infimum of a family of linear maps. Therefore the inequality $\lambda_1 + \cdots + \lambda_k \ge \alpha H$ describes a convex cone of matrices. In addition, system $dh^i_j/dt = |A|^2 h^i_j$ changes the Weingarten operator by homotheties, and thus leaves any cone invariant. The conclusion follows. □

In particular, we obtain that convex surfaces remain convex under the flow. Observe that the same property also follows from Proposition 3.8 by taking $k = n$ and keeping into account property (3.2).

4. Convergence to a point of convex surfaces

As we recalled in the introduction, the singular behavior of convex surfaces under the flow is described by the following result.

THEOREM 4.1. *Let $\mathcal{M}_0$ be an n-dimensional closed convex surface embedded in $\mathbb{R}^{n+1}$. Then $\mathcal{M}_t$ shrinks to a point as $t \to T$. In addition, if we choose a suitable rescaling factor $\rho(t)$, then the surfaces $\rho(t)\mathcal{M}_t$ converge to a sphere as $t \to T$.*

PROOF The above theorem was proved by Huisken in [**24**] in the case $n \ge 2$ and by Gage and Hamilton [**17**] when $n = 1$. Although the result in [**24**] is well known, it is worth describing here some of the main ideas in the proof, since they play an important role in the later developments of the theory. Let us set

$$f = \frac{|A|^2}{H^2} - \frac{1}{n}.$$

Then it is easy to check that

$$fH^2 = \sum_{i<j} (\lambda_i - \lambda_j)^2.$$

Thus, f is nonnegative and it measures how much the curvatures differ from each other. It vanishes identically on a surface if and only if the surface is a sphere. This approach is suggested by [**19**, §8], where Hamilton considered a similar function of the eigenvalues of the Ricci tensor.

We have seen in Proposition 3.3 that the maximum of f is nonincreasing. To prove convergence to a sphere one needs some stronger estimate, showing that f tends to zero as the singular time is approached. Following [**19**], one considers the function $f_\sigma = fH^\sigma$ for a suitably small $\sigma > 0$. Observe that f_σ is a homogeneous function of the curvatures of degree $\sigma > 0$; thus, one would expect f_σ to blow up as the singular time T is approached. The next theorem shows instead that it remains bounded, and this is one of the crucial steps in the proof of Theorem 4.1.

THEOREM 4.2. *If $\sigma > 0$ is small enough, the function f_σ is uniformly bounded for $t \in [0, T)$.*

PROOF. Let us first remark that a similar result holds for the analogous function considered in [**19**] for the Ricci flow. However, the method of proof is quite different. In fact, the result of [**19**] follows from an application of the maximum principle. In our case, instead, the additional factor H^σ induces the presence of a positive zero order term in the evolution equation for f_σ that cannot be directly compensated by the other terms. More precisely, one finds

$$\frac{\partial f_\sigma}{\partial t} \leq \Delta f_\sigma + \frac{2(1-\sigma)}{H}\langle \nabla H, \nabla f_\sigma \rangle - \frac{2}{H^{4-\sigma}}|H\nabla_i h_{kl} - h_{kl}\nabla_i H| + \sigma |A|^2 f_\sigma. \tag{4.1}$$

Thus, a more elaborate procedure is needed to estimate f_σ. Let us first state a useful lower bound for the gradient term in the above inequality. One can prove that on convex surfaces (and in fact under more general hypotheses) there exists c such that

$$|H\nabla_i h_{kl} - h_{kl}\nabla_i H| \geq cH^2 |\nabla H|^2 \tag{4.2}$$

(see [**24**, Lemma 2.3]). We now integrate the inequality on the manifold and try to estimate the L^p norm of f_σ. After integrating by parts we obtain

$$\begin{aligned} \frac{d}{dt}\int_{\mathcal{M}_t} f_\sigma^p d\mu \leq &-\frac{p(p-1)}{2}\int_{\mathcal{M}_t} f_\sigma^{p-2}|\nabla f_\sigma|^2 d\mu - \frac{p}{c}\int_{\mathcal{M}_t} \frac{f_\sigma^{p-1}}{H^{2-\sigma}}|\nabla H|^2 d\mu \\ &+ p\sigma \int_{\mathcal{M}_t} |A|^2 f_\sigma^p d\mu. \end{aligned} \tag{4.3}$$

To show that the last term can be compensated by the other two, we need some estimate involving both zero order curvature terms and gradient terms. To this purpose, we recall the identity [**24**, Lemma 2.1]

$$\frac{1}{2}\Delta |A|^2 = \langle h_{ij}, \nabla_i \nabla_j H \rangle + |\nabla A|^2 + Z, \tag{4.4}$$

where $Z = H\sum \lambda_i^3 - \left(\sum \lambda_i^2\right)^2$. Using this equality, one can compute

$$\begin{aligned} \Delta f_\sigma \geq &\frac{2}{H^{2-\sigma}} h^{ij}\nabla_i\nabla_j H + \frac{2}{H^{2-\sigma}} Z \\ &- \frac{2(1-\sigma)}{H}\langle \nabla H, \nabla f_\sigma \rangle + \left(\frac{2}{nH^{1-\sigma}} - \frac{2-\sigma}{H} f_\sigma \right) \Delta H. \end{aligned}$$

After integrating this inequality on $\mathcal{M}_t$ and performing some standard computations we obtain that, for all $\eta > 0$

$$\int \frac{1}{H^{2-\sigma}} f_\sigma^{p-1} Z d\mu \leq (2\eta p + 5) \int \frac{1}{H^{2-\sigma}} f_\sigma^{p-1} |\nabla H|^2 d\mu + \eta^{-1}(p-1) \int f_\sigma^{p-2} |\nabla f_\sigma|^2 d\mu. \tag{4.5}$$

On the other hand, it can be shown [**24**, Lemma 2.3(i)] that on a uniformly convex surface, say $h_{ij} \geq \varepsilon H g_{ij}$, we have

$$Z \geq n\varepsilon^2 H^{4-\sigma} f_\sigma > n\varepsilon^2 |A|^2 H^{2-\sigma} f_\sigma. \tag{4.6}$$

Thus we can combine estimates (4.3) and (4.5) with an appropriate choice of η to show that, for p suitably large and for σ suitably small the L^p norm of f_σ is decreasing in time.

This property is the starting point for a Stampacchia iteration procedure to obtain that the L^∞ norm of f_σ is bounded. The proof also relies on the Michael-Simon Sobolev inequality [**32**]. For the details, see [**24**, §5].

Several steps remain to complete the proof of Theorem 4.1. Roughly speaking, the above result shows that, at the points where the curvature becomes unbounded, the Weingarten operator approaches the one of a sphere. One then needs to show that the curvature becomes unbounded in the whole surface when the singular time is approached. The main steps are a gradient estimate for the mean curvature and an application of Myers' theorem, see [**24**]. □

5. Convexity estimates for mean convex surfaces

We shall now consider the formation of singularities for surfaces which are mean convex, that is, with positive mean curvature everywhere. As we have seen in Theorem 3.3, this property is preserved by the mean curvature flow. For the study of singularities, mean convexity is a significant generalization of convexity. For instance, it is enough general to allow for the neckpinch behavior described in Section 2; in particular, mean convex surfaces do not necessarily shrink to a point at the singular time. A fundamental result in the analysis of singularities of mean convex surfaces is the following estimate on the elementary symmetric polynomials of the curvatures, proved in [**28**].

THEOREM 5.1. *Let $\mathcal{M}_0 \subset \mathbb{R}^{n+1}$ be a closed mean convex immersed hypersurface and let $\mathcal{M}_t$, $t \in [0, T)$ be its evolution by mean curvature flow. Then, for any $\eta > 0$ there exists $C = C(\eta, \mathcal{M}_0)$ such that $S_k \geq -\eta H^k - C$ for any $k = 2, \ldots, n$ on $\mathcal{M}_t$ for any $t \in [0, T)$.*

Such an estimate easily implies the following one, which has a more immediate interpretation.

THEOREM 5.2. *Under the same hypotheses of the previous theorem, for any $\eta > 0$ there exists $C = C(\eta, \mathcal{M}_0)$ such that $\lambda_1 \geq -\eta H - C$ on $\mathcal{M}_t$ for any $t \in [0, T)$.*

The interest of the above estimate lies in the fact that η can be chosen arbitrarily small and C is a constant not depending on the curvatures. Thus we see that, roughly speaking, the negative curvatures become negligible with respect to the others when the singular time is approached. This implies that the surface becomes asymptotically convex near a singularity. For this reason we call the estimates of the theorems above *convexity estimates.* Let us observe that the result of Theorem 5.2 is very similar to a well-known estimate in the Ricci flow, usually called *Hamilton-Ivey estimate* [**22**, Theorem 24.4]. In contrast to our result, Hamilton-Ivey estimate holds for arbitrary manifolds, but only in the three dimensional case. Observe that Theorem 5.2 cannot be valid for general surfaces even in low dimensions, because the property is violated in Angenent's example of self-similar shrinking torus.

In this section we illustrate the main steps in the proof of Theorem 5.1. For a better understanding of the technique, it is useful to consider first the following weaker result.

THEOREM 5.3. *Let $\mathcal{M}_0$ be a closed hypersurface such that, for some $1 \leq k \leq n-1$, we have $S_1 > 0, \ldots, S_k > 0$ everywhere on $\mathcal{M}_0$. Then, for any $\eta > 0$ there exists $C = C(\eta, \mathcal{M}_0)$ such that $S_{k+1} \geq -\eta H^{k+1} - C$ everywhere on $\mathcal{M}_t$ for any $t \in [0, T)$.*

PROOF. The case $k = 1$ of this theorem (which can be treated in a more explicit way) was proved in [**27**]. The general case can be regarded as a simplified version of the main theorem in [**28**]. The strategy of proof is similar to the one of Theorem 4.2. By our assumptions and by Proposition 3.8, we have $S_k > c_k H^k$ everywhere on $\mathcal{M}_t$ and so the quotient $Q_{k+1} := S_{k+1}/S_k$ is well defined. Let us consider the function

$$f = f_{\sigma,\eta} = \frac{-Q_{k+1} - \eta H}{H^{1-\sigma}}.$$

where $\sigma, \eta > 0$. A straightforward calculation yields the evolution equation

$$\frac{\partial f}{\partial t} = \Delta f + \frac{2(1-\sigma)}{H}\langle \nabla H, \nabla f \rangle - \frac{\sigma(1-\sigma)}{H^2} f |\nabla H|^2 + \frac{1}{H^{1-\sigma}} \frac{\partial^2 Q_{k+1}}{\partial h^i_j \partial h^p_q} \nabla^m h^i_j \nabla_m h^p_q + \sigma |A|^2 f. \tag{5.1}$$

The function $f_{\eta,\sigma}$ will play a similar role to f_σ in the proof of Theorem 4.1. Actually, in the case $k = 1$ the two functions essentially coincide. Even in this case, however, the proof needs to be modified, because we no longer have the convexity assumption that was crucial in some of the estimates there.

In the case $\sigma = 0$, the maximum principle applied to (5.1) and the concavity of Q_{k+1} give a bound on $f_{\sigma,\eta}$. However, the interesting case for us is when $\sigma > 0$. Namely, we need to show that for any $\eta > 0$ there exists $\sigma > 0$ such that $f_{\sigma,\eta}$ is bounded. In fact, proving that $f_{\sigma,\eta} < C$ for some $C > 0$ implies that

$$S_{k+1} \geq S_k(-\eta H - CH^{1-\sigma}) \geq -2\eta H^{k+1} - C'$$

for a suitable $C' = C'(C, \sigma, \eta)$. Since $\eta > 0$ is arbitrary, this proves Theorem 5.3.

To prove the boundedness of f we are going to apply the same technique of Theorem 4.2, consisting of L^p estimates on f followed by an iteration technique. Note that, in contrast to that theorem, the function f here is not positive everywhere: it is negative, for instance, at all convex points of the surface. It will be convenient to prove estimates on f_+^p, where f_+ is the positive part of f. In this way, we only have to consider the points of $\mathcal{M}_t$ where $f > 0$, that is, where $Q_{k+1} \leq -\eta H$.

As a first step, we need to show that the gradient term in the second line of (5.1) is not only non positive, but it has some coercivity, i.e. we need, an analogue of inequality (4.2). This is not an obvious property, because the function Q_{k+1} is homogeneous of degree one, and thus it is not strictly concave. However, analyzing carefully the properties of Q_{k+1} and exploiting the symmetries given by the Codazzi equations we obtain [**28**, Theorems 2.5 and 2.14] that for any $\eta > 0$ there exists $c > 0$ such that

$$\frac{\partial^2 Q_{k+1}}{\partial h^i_j \partial h^p_q} \nabla^m h^i_j \nabla_m h^p_q \leq -c \frac{|\nabla A|^2}{|A|}$$

at all points where $Q_{k+1} < -\eta H$. This allows to obtain from (5.1)

$$(5.2)\quad \frac{d}{dt} \int_{\mathcal{M}_t} f_+^p \, d\mu \leq -\frac{1}{2} p(p-1) \int_{\mathcal{M}_t} f_+^{p-2} |\nabla f|^2 \, d\mu - Cp \int_{\mathcal{M}_t} \frac{f_+^{p-1}}{H^{2-\sigma}} |\nabla A|^2 + 2p\sigma \int_{\mathcal{M}_t} H^2 f_+^p \, d\mu$$

for a suitable $C = C(\eta)$. In the above arguments it is important that we can restrict our attention to the points where $Q_{k+1} < -\eta H$. This allows us to avoid the points where $Q_{k+1} = 0$ and the function Q_{k+1} has weaker concavity properties. Observe, for instance, that $S_{k+1} = 0$ at all $(\lambda_1, \ldots, \lambda_n)$ with $\lambda_1 = \lambda_2 = \cdots = \lambda_{n-k} = 0$ and the other entries are arbitrary. This shows that the hessian of Q_{k+1} has a large kernel at such points.

Now we need an inequality which allows to estimate zero order terms by first order terms, analogue to (4.5). To this purpose, we look for a suitable identity involving derivatives of the curvature together with zero order curvature terms. It turns out that it is convenient to consider the quantity

$$\frac{\partial S_k}{\partial h_{ij}} \nabla_i \nabla_j S_{k+1}.$$

A long but elementary computation, which uses the commutator identities for covariant derivatives and various properties of the elementary symmetric polynomials [**28**, Lemma 2.15], yields the identity

$$\begin{aligned}\frac{\partial S_k}{\partial h_{ij}}\nabla_i\nabla_j S_{k+1} = {} & \frac{\partial S_k}{\partial h_{ij}}\frac{\partial^2 S_{k+1}}{\partial h_{lm}\partial h_{pq}}\nabla_i h_{lm}\nabla_j h_{pq} + \frac{\partial S_k}{\partial h_{ij}}\frac{\partial S_{k+1}}{\partial h_{lm}}\nabla_l\nabla_m h_{ij} \\ & - HS_kS_{k+1} + (k+1)S_{k+1}^2 + k[(k+1)S_{k+1}^2 \\ & - (k+2)S_kS_{k+2}].\end{aligned}$$

Now we have $(k+1)S_{k+1}^2 - (k+2)S_kS_{k+2} \geq 0$ by a classical property of symmetric polynomials called *Newton's inequality*. We deduce that, at all points where $Q_{k+1} < -\eta H$, we have

$$\begin{aligned}\eta H^2 S_k^2 & < -HS_k^2 Q_{k+1} = -HS_kS_{k+1} \\ & < \frac{\partial S_k}{\partial h_{ij}}\nabla_i\nabla_j S_{k+1} - \frac{\partial S_k}{\partial h_{ij}}\frac{\partial^2 S_{k+1}}{\partial h_{lm}\partial h_{pq}}\nabla_i h_{lm}\nabla_j h_{pq} \\ & \quad - \frac{\partial S_k}{\partial h_{ij}}\frac{\partial S_{k+1}}{\partial h_{lm}}\nabla_l\nabla_m h_{ij}.\end{aligned}$$

Let us simply denote by (RHS) the right hand side of this inequality. We have

$$\int_{\mathcal{M}_t} H^2 f_+^p \, d\mu \leq c_k^{-2}\int_{\mathcal{M}_t} H^2 S_k^2 H^{-2k} f_+^p \, d\mu \leq c_k^{-2}\eta^{-1}\int_{\mathcal{M}_t} (RHS)\cdot H^{-2k} f_+^p \, d\mu.$$

where the last integral no longer contains zero order curvature terms. A careful computation involving integration by parts and properties of the polynomial S_k [**28**, Proposition 3.6] then shows that the negative gradient terms in (5.2) can compensate the positive term for a suitable choice of the constants. In this way we prove that, for any $\eta > 0$ and p large enough, there exists $\sigma > 0$ such that the L^p norm of $(f_{\sigma,\eta})_+$ is nonincreasing. This allows to apply the same iteration procedure as in the proof of Theorem 4.2 to conclude that $f_{\sigma,\eta}$ is bounded from above for a suitable $\sigma > 0$, and this proves the theorem. $\square$

The above statement contains the strong assumption of the positivity of S_l for $l = 1, \ldots, k$. To generalize the technique to the case where we only have $H > 0$, we define a suitable perturbation of the second fundamental form. For given $\varepsilon, D > 0$, we define $b_{ij} = b_{ij;\varepsilon,D}$ as follows

$$b_{ij} = h_{ij} + (\varepsilon H + D)g_{ij}.$$

We denote by $\tilde{S}_k$ the symmetric polynomials computed with respect to b_{ij} instead of h_{ij} (we do not write explicitly the dependence on ε, D for simplicity of notation). The interest of this definition is shown by the next result [**28**, Lemmas 2.8, 2.11].

PROPOSITION 5.4. *Given a mean convex hypersurface and* $k \in \{2, \ldots, n\}$, *the following properties hold.*

(i) *Suppose that for any* $\eta > 0$ *there exists* C *such that* $S_l = -\eta H^l - C$, *for* $l = 2, \ldots, k$. *Then, for any* $\varepsilon > 0$ *there exists* D *such that* $\tilde{S}_l > 0$ *for* $l = 2, \ldots, k$.

(ii) *Conversely, suppose that for any* $\varepsilon > 0, \eta > 0$ *there exist* D, C *such that* $\tilde{S}_l > -\eta\tilde{H} - C$ *for* $l = 2, \ldots, k$. *Then, for any* $\eta > 0$ *there exists* K *(in general larger than* C*) such that* $S_l = -\eta H^l - K$, *for* $l = 2, \ldots, k$.

PROOF OF THEOREM 5.1. The perturbation described above allows us to apply an induction procedure similar to the one of Proposition 3.8. We first apply Theorem 5.3 for $k = 1$ and obtain that for any $\eta > 0$ there exists $C > 0$ such that $S_2 \geq -\eta H^2 - C$. By Proposition 5.4, the perturbed polynomial $\tilde{S}_2$ is positive, and so we can consider the quotient $\tilde{S}_3/\tilde{S}_2$. Suppose that the proof of Theorem 5.3 can be carried through also for the perturbed polynomials. Then we obtain that for any $\eta > 0$ there exists C such that $\tilde{S}_3 \geq -\eta\tilde{H}^3 - C$. But then the unperturbed polynomial S_3 satisfies the same estimate, although with a larger constant, by part (ii) of Proposition 5.4. This shows that the procedure can be iterated to show that all polynomials S_k up to $k = n$ satisfy the desired estimate.

The difficult part is to check that the proof of Theorem 5.3 indeed applies also to the perturbed polynomials. The perturbation induces the presence of several additional terms in the equation, some of which require a sharp estimation. The computations become more involved, but it turns out that the same procedure works. Roughly speaking, the additional terms due to ε can be made arbitrarily small by choosing ε close enough to zero, while the ones containing D are negligible because they are of lower order. □

As it is customary in many nonlinear PDEs, it is possible to study the singular behavior of surfaces evolving by mean curvature flow by rescaling techniques. The property of rescalings are described in the references [**26, 27, 28**] and are not strictly needed for the results described in the remainder of the paper. However, we recall them briefly here since they are useful to have a better insight of the surgery procedure.

In the rescaling procedure one dilates in space and time the flow around the points of a sequence along which the curvature becomes unbounded. The dilations are such that the rescaled flows satisfy local uniform curvature bounds and so we have convergence of a subsequence to a smooth limiting mean curvature flow. A precise description of the procedure is given in [**27**], see also [**22**, §16]. In the rescaling procedure the constant term in the convexity estimates of Theorem 5.1 disappears and we obtain that any limiting flow satisfies $\lambda_1 \geq -\eta H$ for arbitrary $\eta > 0$, that is, $\lambda_1 \geq 0$.

Thus we have

COROLLARY 5.5. *Any limit obtained by rescaling near a singularity a mean convex surface evolving by mean curvature flow is convex (not necessarily strictly).*

The above corollary has been also obtained by White [**39**] by completely different techniques. His approach also applies to weak solutions. The following result gives a classification of the possible limits obtained by rescaling near a singularity in the mean convex case.

THEOREM 5.6. *Let $\mathcal{M}_t \subset \mathbb{R}^{n+1}$ be a mean curvature flow of closed mean convex hypersurfaces. Then there is a sequence of rescaled flows near the singular time converging to one of the following flows:*

(i) *a product of the form $S_t^{n-k} \times \mathbb{R}^k$, for some $0 \leq k \leq n-1$ where S_t^{n-k} is an $(n-k)$-dimensional shrinking sphere;*
(ii) *a flow of the form $G_t \times \mathbb{R}^{n-1}$, where G_t is a homothetically shrinking curve in the plane;*
(iii) *a flow of the form $\Gamma_t^{n-k} \times \mathbb{R}^k$, for some $0 \leq k \leq n-1$, where Γ_t^{n-k} is an $(n-k)$-dimensional strictly convex translating solution to the flow.*

The above result is proved in [**26**, Theorem 5.1] and [**28**, Theorem 4.1]. In addition to the convexity estimates, this classification relies on two other important results, namely Huisken's *monotonicity formula* [**25**] and Hamilton's *differential Harnack inequality* [**21**].

6. Cylindrical and gradient estimates for two-convex surfaces

From now on we consider mean curvature flow of hypersurfaces which have dimension $n \geq 3$ and are uniformly 2-convex, that is, satisfy $\lambda_1 + \lambda_2 \geq \alpha H$ everywhere for some $\alpha > 0$. As we have seen in Proposition 3.9, this property is preserved by the flow. The motivation for considering 2-convex surfaces can be intuitively understood in view of the classification of the possible profiles in Theorem 5.6. If our evolving surfaces are uniformly 2-convex, then so is any limit of rescaled flows. This restricts the number of possibilities in Theorem 5.6, since the only uniformly two-convex limits are the sphere S^n, the cylinder $S^{n-1} \times \mathbb{R}$ and the n-dimensional translating solutions Γ_t^n. All these profiles are compatible with the surgery procedure we are willing to define. In fact, if the limit of the rescalings is the sphere S^n, then the original manifold should be diffeomorphic to a sphere, and no surgery is needed since the topology is known. If the limit is a cylinder, then the manifold should possess a cylindrical region where we can do surgery. If it is an n-dimensional translating solution Γ_t^n, we also expect to find a cylindrical region, as explained at the end of §2. This discussion should be regarded only as a heuristic motivation, since the information provided by

Theorem 5.6 is too weak for the purposes of a flow with surgeries. Our actual proof will be independent of Theorem 5.6 and in fact will not use rescaling techniques, except at one stage (Theorem 9.1). We begin with the following result.

THEOREM 6.1. *Let $\mathcal{M}_t$, with $t \in [0,T)$, be a closed 2-convex solution of mean curvature flow. Then, for any $\eta > 0$ there exists a constant C_η such that*

$$|\lambda_1| \leq \eta H \Longrightarrow |\lambda_j - \lambda_k| \leq c\eta H + C_\eta, \qquad j,k > 1$$

everywhere on $\mathcal{M}_t$, for $t \in [0,T)$, where c only depends on n.

We call the above result a *cylindrical estimate* because it shows that, at a point where H is large and λ_1/H is small, the Weingarten operator is close to the one of a cylinder, since it has all eigenvalues close to each other except for λ_1 which is small. Such a property is an important tool for the detection of the cylindrical regions where the surgeries will be performed.

To derive this estimate, we consider again the quotient $|A|^2/H^2$ which was used in the proof of Theorem 4.1. On a cylinder $\mathbb{R} \times S^{n-1}$ we have $|A|^2/H^2 \equiv 1/(n-1)$. The converse does not hold, that is, if we have $|A|^2/H^2 = 1/(n-1)$ at one point, this does not imply that the curvatures are a multiple of the ones of a cylinder. However, if $|A|^2/H^2 = 1/(n-1)$ and in addition $\lambda_1 = 0$, then necessarily $\lambda_2 = \cdots = \lambda_n$. In fact, we have the identity

$$|A|^2 - \frac{1}{n-1}H^2 = \frac{1}{n-1}\left(\sum_{1<i<j}(\lambda_i - \lambda_j)^2 + \lambda_1(n\lambda_1 - 2H)\right). \tag{6.1}$$

In view of this equality, the estimate of Theorem 6.1 is an immediate consequence of the next result [**29**].

THEOREM 6.2. *Let $\mathcal{M}_t$, $t \in [0,T)$, be a closed 2-convex solution of mean curvature flow. Then, for any $\eta > 0$ there exists a constant $C_\eta > 0$ such that*

$$|A|^2 - \frac{H^2}{n-1} \leq \eta H^2 + C_\eta$$

on $\mathcal{M}_t$ for any $t \in [0,T)$.

PROOF. Let us consider, for $\eta \in \mathbb{R}$ and $\sigma \in [0,2]$, the function

$$f_{\sigma,\eta} = \frac{|A|^2 - \left(\frac{1}{n-1} + \eta\right)H^2}{H^{2-\sigma}}. \tag{6.2}$$

Such a function is very similar to f_σ considered in the proof of Theorem 4.1, and in fact it satisfies the same inequality (4.1). However, in this case we do not have a bound from below for Z analogous to (4.6). In fact, Z can be negative on nonconvex surfaces. A typical example is when $\lambda_1 < 0$ and $\lambda_2 = ... = \lambda_n > 0$; then $Z < 0$, even if $|\lambda_1|$ is small compared to the other curvatures.

However, using also the convexity estimate of Theorem 5.2, we can show [**29**, Lemma 5.2] that there exists a constant $\gamma_1 > 0$ with the following property: for any $\delta > 0$ there exists K_δ such that

$$Z \geq \gamma_1 H^2 \left(|A|^2 - \frac{1}{n-1} H^2 - \delta H^2 \right) - K_\delta H^3 \tag{6.3}$$

on $\mathcal{M}_t$ for any $t > 0$. As in the proof of Theorem 5.3, we will estimate the L^p norms of the positive part $(f_{\sigma,\eta})_+$. In this way, we only need to consider the points where the positive part is nonzero, i.e., $|A|^2 - \frac{H^2}{n-1} \geq \eta H^2$. Thus, if we choose $\delta = \eta/2$ in (6.3) the first term is positive and the only negative contribution to the right hand side is the last term which has lower order. It turns out that this is enough to apply the usual iteration technique of the previous theorems and obtain an upper bound for $f_{\sigma,\eta}$, see [**29**, Theorem 5.3]. Such a bound easily implies the estimate of Theorem 6.2. □

Observe that similar results have been obtained by Hamilton for the Ricci flow of arbitrary three-dimensional manifolds in [**22**, Theorem 24.7] and of four-dimensional manifolds with positive isotropic curvature in [**23**, Theorem B3.3].

We next describe an estimate for the gradient of the curvature for our evolving surfaces. With respect to the gradient estimates for mean curvature flow already available in the literature, e.g., [**10, 12**], the estimate here does not depend on the maximum of the curvature in some neighborhood of the point under consideration. To prove this result we need to assume that the surfaces are 2-convex and that their dimension is at least three.

THEOREM 6.3. *Let $\mathcal{M}_t$, $t \in [0,T)$, be a closed n-dimensional 2-convex solution of mean curvature flow, with dimension $n \geq 3$. Then there is a constant $\gamma_2 = \gamma_2(n)$ and a constant $\gamma_3 = \gamma_3(n, \mathcal{M}_0)$ such that the flow satisfies the uniform estimate*

$$|\nabla A|^2 \leq \gamma_2 |A|^4 + \gamma_3 \tag{6.4}$$

for every $t \in [0,T)$.

PROOF. The result is obtained by applying the maximum principle to a suitable function we are going to introduce. An important tool in the proof is the inequality [**24**, Lemma 2.1], valid for any immersed hypersurface,

$$|\nabla A|^2 \geq \frac{3}{n+2} |\nabla H|^2. \tag{6.5}$$

Observe that $\frac{3}{n+2} > \frac{1}{n-1}$ if $n \geq 3$. Let us set

$$\kappa_n = \frac{1}{2}\left(\frac{3}{n+2} - \frac{1}{n-1}\right).$$

By Theorem 6.2 there exists $C_0 > 0$ such that

$$\left(\frac{1}{n-1} + \kappa_n\right) H^2 - |A|^2 + C_0 \geq 0.$$

Let us set

$$g_1 = \left(\frac{1}{n-1} + \kappa_n\right) H^2 - |A|^2 + 2C_0, \qquad g_2 = \frac{3}{n+2}H^2 - |A|^2 + 2C_0.$$

Then we have $g_2 > g_1 \geq C_0$, and so $g_i - 2C_0 = 2(g_i - C_0) - g_i \geq -g_i$ for $i = 1, 2$. Using the evolution equations for $|A|^2, H^2$ (see Lemma 3.1) and inequality (6.5) we find

(6.6)
$$\begin{aligned}\frac{\partial}{\partial t}g_1 - \Delta g_1 &= -2\left(\left(\frac{1}{n-1} + \kappa_n\right)|\nabla H|^2 - |\nabla A|^2\right) + 2|A|^2(g_1 - 2C_0)\\ &\geq 2\left(1 - \frac{n+2}{3}\left(\frac{1}{n-1} + \kappa_n\right)\right)|\nabla A|^2 - 2|A|^2 g_1\\ &= 2\kappa_n\frac{n+2}{3}|\nabla A|^2 - 2|A|^2 g_1.\end{aligned}$$

Similarly

(6.7)
$$\frac{\partial}{\partial t}g_2 - \Delta g_2 = -2\left(\frac{3}{n+2}|\nabla H|^2 - |\nabla A|^2\right) + 2|A|^2(g_2 - 2C_0) \geq -2|A|^2 g_2.$$

In addition (see [**24**, Theorem 7.1])

$$\frac{\partial}{\partial t}|\nabla A|^2 - \Delta|\nabla A|^2 \leq -2|\nabla^2 A|^2 + c_n|A|^2|\nabla A|^2, \tag{6.8}$$

for a constant c_n depending only on n. Using the above relations one obtains, after a straightforward computation, the following inequality for the quotient $|\nabla A|^2/g_1 g_2$:

$$\begin{aligned}&\frac{\partial}{\partial t}\left(\frac{|\nabla A|^2}{g_1 g_2}\right) - \Delta\left(\frac{|\nabla A|^2}{g_1 g_2}\right) - \frac{2}{g_2}\left\langle \nabla g_2, \frac{|\nabla A|^2}{g_1 g_2}\right\rangle\\ &\leq \frac{|\nabla A|^2\,|A|^2}{g_1 g_2}\left((c_n + 4) - 2\kappa_n^2\frac{n+2}{3n}\frac{|\nabla A|^2}{g_1 g_2}\right).\end{aligned}$$

The maximum principle then implies that $|\nabla A|^2 \leq c_1 g_1 g_2$, where c_1 only depends on n and on the initial data. Using the definition of g_1, g_2, this easily yields our assertion. □

Once the estimate for $|\nabla A|$ is obtained, it is easy to obtain similar estimates for the higher order derivatives, as well as the time derivatives. In particular, we have

$$|\partial_t A| \le c_1|A|^3 + c_2. \tag{6.9}$$

Let us note that at this stage no analogous direct a priori estimates for the derivatives of the curvature are known for Ricci flow, since the corresponding estimates obtained by Perelman [**34, 35**] are derived via contradiction arguments.

7. The surgery procedure

We now describe in more detail how we are going to perform our surgeries. As explained in the introduction, the surgery consists of removing an almost cylindrical region (called a neck) and replacing it by two convex caps. It is important that the procedure does not alter the validity of the estimates proved in the previous sections. To check this, it is necessary to

- give a precise definition of neck, specifying the notion of "being close to a cylinder" in a quantitative way;
- give an explicit expression of the surface after the surgery;
- show that the estimates of the previous sections remain valid after the surgeries with the same constants.

These steps are carefully carried out by Hamilton [**23**] in the case of the Ricci flow. We have followed his approach in many parts, with some modifications suggested by our framework of immersed surfaces. Since the complete definitions are lengthy, we give a simplified exposition by omitting most of the technical details.

Hamilton [**23**] gives different notions of necks in the case where $\mathcal{M}$ is an abstract Riemannian manifold. For a given $\varepsilon > 0$, he defines *ε-geometric necks* as diffeomorphism $N : S^{n-1} \times [a,b] \to \mathcal{M}$ such that the standard metric on the cylinder and the pull-back of the metric on $\mathcal{M}$ to the cylinder are ε-close, up to a homothety. By "ε-close" we mean that the norm of the difference of the two metric tensors (measured with respect to the standard metric of the cylinder) is everywhere less than ε. He then defines *ε-curvature necks* as regions of $\mathcal{M}$ where the curvature operator is ε-close at every point, up to a homothety, to the curvature operator of a standard cylinder. The first notion is useful when one wants to define the surgery procedure. The latter one is useful to prove the existence of necks when a singularity is approached. Clearly, a geometric neck is also a curvature neck. Conversely, Hamilton proves that a curvature neck is locally a geometric neck, that is, that one can detect a neck from the curvature alone.

In our context where $\mathcal{M}$ is an immersed manifold, it is natural to consider notions of neck which also take into account the extrinsic curvature. We say that a geometric neck is a *hypersurface neck* if the Weingarten operator of the cylinder is ε-close at every point to the one induced by the parametrization.

A useful feature of Hamilton's definition of geometric neck (which also holds for hypersurface necks) is that the condition of being close to a cylinder is only local. In this way, the axis or the radius of the cylinder is allowed to vary, provided they do it slow enough so that every point has a neighborhood of given radius sufficiently close to a cylinder. For instance, a sufficiently thin torus is an ε-hypersurface neck. Also, we can have a long neck with quite different radii in the different parts. For instance, it may have in the central part a radius smaller by a factor, say, 10 than the radius at the ends. This is the kind of necks we want to remove by surgery, because in this way we reduce the curvature of the surface in that region by the same factor. Another important property enjoyed by geometric necks is that, if two of them overlap, then their union is again a geometric neck. Thus, every ε-neck has a maximal extension.

Let us now describe explicitly our surgery procedure. Suppose that we have an ε-hypersurface neck $N : S^{n-1} \times [a, b] \to \mathcal{M}$, with ε small enough and $b - a$ large enough. We denote by $(\omega, z) \in S^{n-1} \times [a, b]$ the coordinates in the neck. Let us choose $\tau > 0$ suitably small and $\Lambda, B > 0$ suitably large. We are going to replace smoothly the image of $S^{n-1} \times [a, b]$ under N by two appropriate regions diffeomorphic to disks. We only describe the region attached to the first end $z = a$, since the other one is symmetric. First let us denote by $\bar{C}_a : S^{n-1} \times \mathbb{R} \to \mathbb{R}^{n+1}$ the straight cylinder best approximating $\mathcal{M}$ at the cross section $z = a$. Then the standard surgery with parameters τ, Λ, B is performed as follows.

(i) In the region corresponding to $z \in [a, a + 2\Lambda]$, we bend the surface inwards replacing the original parametrization N by

$$\tilde{N}(\omega, z) := N(\omega, z) - \tau \exp\left(-\frac{B}{z - a}\right) \nu(\omega, z).$$

It can be shown that, if the parameter ε measuring the quality of the neck is small enough, and if the surgery parameters τ, Λ, B are chosen appropriately, then the deformed surface is strictly convex in the part with $z \in [a + \Lambda, a + 2\Lambda]$.

(ii) To blend the resulting surface into an axially symmetric one we choose a fixed smooth transition function $\varphi : [0, 3\Lambda] \to \mathbb{R}^+$ with $\varphi = 1$ on $[0, \Lambda]$, $\varphi = 0$ on $[2\Lambda, 3\Lambda]$ with $\varphi' \leq 0$. We then define, for $z \in [a + \Lambda, a + 2\Lambda]$

$$\hat{N}(\omega, z) := \varphi(z)\tilde{N}(\omega, z) + (1 - \varphi(z))\tilde{C}_a(\omega, z). \tag{7.1}$$

where $\tilde{C}_a$ is obtained from the cylinder $\bar{C}_a$ by applying the bending defined in (i).

(iii) We finally modify the radius of $\tilde{C}_a$ for $z \in [a + 2\Lambda, a + 3\Lambda]$ in such a way that it tends to 0 as $z \to a + 3\Lambda$ and that $\tilde{C}_a(S^{n-1} \times [a + 2\Lambda, a + 3\Lambda])$ is a smooth axially symmetric convex cap. We do not need to write an explicit expression here because knowing that the cap is convex and independent of the original surface is sufficient for the estimates.

It is now possible to prove that, if ε is small enough and if the surgery parameters are chosen appropriately, the estimates of Theorems 5.1, 6.1 and 6.3 remain valid for a flow with surgeries. Such a property is not surprising: bending inwards an almost cylindrical region and closing it with a convex cap should not decrease the convexity of the surface, nor produce parts with large gradients of the curvature or affect the cylindrical estimates. However, the proofs are far from being trivial and require careful computations, see [**29**]. Actually, in [**29**], the theorems recalled above are proved directly in the case of a flow with surgeries, to show more clearly that they hold also in this case.

8. Neck detection

After having described the surgery procedure, we have to show that it allows one to define a flow after the singularities until the surface is split into components with known topology. As a first step, we need results ensuring that, as the singular time is approached, either we can find a neck on our surface or we can tell that the surface is convex so that its topology is known. We discuss such results in this section.

We first introduce some notation. Given $p \in \mathcal{M}$, $t, r, \theta > 0$, with $\theta \leq t$, the *backward parabolic neighborhood* centered at (p,t) is the set

$$\mathcal{P}(p,t,r,\theta) = \{(q,s) \ : \ q \in d_t(p,q) \leq r,\ s \in [t-\theta, t]\}, \tag{8.1}$$

where d_t denotes the distance on $\mathcal{M}$ at time t. If we consider a flow with surgeries, the above set may be not well defined. In fact, the r-neighborhood of p at time t may intersect a region which has been inserted with a surgery at some time between $t-\theta$ and t. If this happens, we say that the backward parabolic neighborhood *contains surgeries.*

The next result is an essential tool to prove the existence of necks before a singularity. We shall call it in the following the *neck detection theorem.*

THEOREM 8.1. *Let $\mathcal{M}_t$, $t \in [0,T[$ be a mean curvature flow with surgeries starting from an immersed manifold $\mathcal{M}_0$ which is closed, two-convex and with dimension $n \geq 3$. Let $\varepsilon, \theta, L > 0$ be given. Then we can find η_0, H_0 with the following property. Suppose that $p_0 \in \mathcal{M}$ and $t_0 \in [0,T[$ are such that*

(ND1) $H(p_0,t_0) \geq H_0, \qquad \dfrac{\lambda_1(p_0,t_0)}{H(p_0,t_0)} \leq \eta_0,$

(ND2) *the neighborhood $\mathcal{P}\left(p_0, t_0, \frac{L}{H(p_0,t_0)}, \frac{\theta}{H^2(p_0,t_0)}\right)$ does not contain surgeries.*

Then, for any $t \in [t_0 - \theta/H^2(p_0, t_0), t_0]$, the ball centered at p_0 of radius $L/2H(p_0, t_0)$ is contained in an ε-hypersurface neck. The constants η_0, H_0 only depend on $\mathcal{M}_0$ and on ε, L, θ.

PROOF. We use a contradiction argument based on a rescaling procedure like the ones which are often used by Hamilton and Perelman for the Ricci flow [**23, 34**]. Let us assume that the assertion is false. Then we can find a sequence (p_n, t_n) such that $H(p_n, t_n) \to +\infty$, $\limsup \lambda_1(p_n, t_n)/H(p_n, t_n) \leq 0$, the parabolic neighborhoods do not contain surgeries, but the points do not lie on an ε-neck. A contradiction will be proved if we show that a subsequence of the parabolic neighbourhoods (after rescaling) converge to the flow of a portion of the standard cylinder; in fact, this will imply that they satisfy the conclusion of the theorem for n large enough.

We first perform a parabolic rescaling of the neighborhoods by a factor $H(p_n, t_n)$ and then translate space and time so that (p_n, t_n) becomes $(0, 0)$; in such a way, they all become flows defined in the time interval $[-\theta, 0]$ and satisfying $H(0, 0) = 1$.

To obtain compactness of a sequence of flows we need uniform curvature bounds. We exploit our gradient estimates (6.4) and (6.9) to obtain that $H \leq 2$ in a possibly smaller parabolic neighborhood around $(0, 0)$ for every element of the sequence. In this smaller neighborhood we have therefore convergence of a subsequence to a limit flow $\tilde{\mathcal{M}}_t$. When we pass to the limit, the constant terms in the estimates of Theorem 5.2 and 6.1 disappear. Therefore the limit flow $\tilde{\mathcal{M}}_t$ is convex and satisfies $\tilde{\lambda}_1(0, 0) = 0$. Hence, it is not strictly convex at the final time. By Hamilton's strong maximum principle for tensors, it must satisfy $\tilde{\lambda}_1(0, 0) = 0$ everywhere. By the cylindrical estimates, the other curvatures coincide at each point. Then it is easy to show that the flow is a portion of a shrinking cylinder.

So far we have only proved convergence of a smaller neighborhood around $(0, 0)$. However, since we have proved that $H \leq 1$ everywhere in this neighborhood, we can apply again the gradient estimates to find that $H \leq 2$ in a larger neighborhood, and prove convergence to a cylinder there too. After a finite number of iterations, we prove convergence to a cylinder of the whole original neighborhoods. □

The next result deals with the case of a point where the curvature is large, but λ_1/H is not small. We give the statement for a stationary surface, since the property is not related to mean curvature flow.

THEOREM 8.2. *Let $\mathcal{M}$ be an immersed hypersurface satisfying the gradient estimate (6.4). Let η_0, H_0 be given. Then there exists $\gamma_0 > 1$ with the following property. Let $p \in \mathcal{M}$ be any point such that $H(p) \geq \gamma_0 H_0$ and $\lambda_1(p) > \eta_0 H(p)$. Then either $\mathcal{M}$ is closed and convex or there exists a point $p' \in \mathcal{M}$ such that*

$$\lambda_1(p') \leq \eta_0 H(p'), \qquad H(p') \geq H_0$$

and such that $H(q) \geq H_0$ at all points q with $d(p, q) \leq d(p, p')$.

PROOF. We first use the gradient estimate to show that the curvature cannot decay too fast as we move away from p. Then an elementary computation shows that if $\lambda_1 \geq H$ in a large enough ball around p, then $\mathcal{M}$ must be convex. It follows that the only other possibility is the existence of a point p' not too far from p and with curvature not too much smaller than $H(p)$, which has the claimed properties (see [**29**, Theorem 7.14] for the details). □

The two previous theorems can be combined to prove the existence of necks before the first singular time. In fact, let η_0, H_0 be the values given from Theorem 8.1 for some choices of ε, θ, L, and let γ_0 be the value associated to η_0, H_0 by Theorem 8.2. Then let us pick a time t_0 close to the singular time so that there exists a point p_0 such that $H(p_0, t_0) \geq \gamma_0 H_0$. If $\lambda_1(p_0, t_0) \leq \eta_0 H(p_0, t_0)$ we can directly apply Theorem 8.1 to conclude that p_0 lies on a neck. Otherwise we apply Theorem 8.2 to find another point p_1 where Theorem 8.1 can be applied. If no such point exists, then Theorem 8.2 implies that our surface is convex and therefore diffeomorphic to a sphere; we do not need to continue the flow any longer. After the first surgery the argument is no longer so direct, because we have to ensure that hypothesis (ND2) in Theorem 8.1 is satisfied. We will see in the next section how we can deal with this difficulty.

9. The surgery algorithm

In this final section we provide an algorithm which determines at which time and place the surgeries are to be performed, and we show that the flow with surgeries generated by this algorithm terminates after a finite number of steps. We will fix three values $H_1 < H_2 < H_3$ suitably large. The flow defined by our algorithm will satisfy the following properties:

- the surgeries are performed at times T_i such that $\max H(\cdot, T_i) = H_3$;
- after the surgeries are performed, we have $\max H(\cdot, T_i+) \leq H_2$;
- the regions introduced with the surgeries satisfy $H_1/2 \leq H \leq 2H_1$.

A flow with these properties necessarily terminates after a finite number of steps. This can be seen considering the decrease of area of the surface. The area is decreasing during the smooth evolution, by Lemma 3.1-(ii). Since the surgeries are performed on necks with approximately fixed curvature, each of them decreases the area at least by a given fixed amount. Therefore, there can be only a finite number of them.

In order to define a flow with the above properties, it is fundamental to show that we can use surgeries to decrease the curvature of our surface by a fixed factor. This will be a direct consequence of the next result.

THEOREM 9.1. *Let $\mathcal{M}_0$ be a 2-convex closed hypersurface of dimension $n \geq 3$. It is possible to define η_1, H_1 with the following properties. Suppose that $\mathcal{M}_t$, with $t \in [0, t_0]$, is a mean curvature flow with surgeries starting* *from $\mathcal{M}_0$. Suppose that all the regions inserted in the surgeries have curvature less than $2H_1$. Let p_0 be such that*

$$(9.1) \qquad H(p_0,t_0) \geq 10H_1, \qquad \lambda_1(p_0,t_0) \leq \eta_1 H(p_0,t_0).$$

Then (p_0,t_0) lies on an ε_0-hypersurface neck $\mathcal{N}_0$, which either covers the whole component of $\mathcal{M}_{t_0}$ including p_0, or can be continued in each direction until one of the two following properties hold:

(i) *the mean curvature has decreased to H_1, or*
(ii) *the neck ends with a convex cap.*

PROOF. The precise proof of this statement is one of the longest and most technical results in [**29**]. We explain here in an intuitive way some of the arguments employed for this result.

Let us first consider the case of the first surgery time. We choose η_1, H_1 as in the neck detection Theorem 8.1. Then we have that (p_0,t_0) lies on an ε_0-hypersurface neck, because condition (ND2) is trivially satisfied. Now we extend the neck in both directions in a maximal way. A first possibility is that the neck never ends, that is, the two ends meet, showing that the component of the surface containing p_0 is diffeomorphic to a torus $S^{n-1} \times S^1$. Otherwise, the neck ends somewhere. In this case, we deduce that the points in the final part of the neck do not satisfy hypothesis (ND1) of the neck detection theorem. One possibility is that the curvature is no longer large; then we have proved case (i) of the theorem. The other possibility is that λ_1 is no longer small; if this happens on a large enough region, then the surface must close as a convex cap, as in case (ii). To show this rigorously, a delicate argument is needed, see [**29**, Theorem 8.2]; in particular, it is necessary to make a more restrictive choice of the parameters η_1, H_1 than the one needed to apply Theorem 8.1 at the beginning.

Let us now complete the argument to include the case where there have already been surgeries. We consider again our starting point (p_0,t_0). It is possible to show [**29**, Lemma 7.2] that, if the parameters have been chosen appropriately, assumption (ND2) holds at (p_0,t_0) because of the gradient estimates. In fact, we are assuming that all regions inserted by the surgeries have curvature less than $2H_1$, while we have $H(p_0,t_0) \geq 10H_1$. Roughly speaking, if (ND2) were violated, there would not be enough time from the last surgery to t_0 to let the curvature increase from $2H_1$ to $10H_1$. It is important in this argument that our gradient estimates are not obtained by arguments employing interior parabolic regularity, and therefore they hold with the same constants regardless how close we are to the surgery times. This shows that also in this case (p_0,t_0) possesses a backward parabolic neighborhood which is surgery-free and we can apply Theorem 8.1 to say that p_0 lies on a neck $\mathcal{N}_0$.

As before, we consider the maximal extension of the neck $\mathcal{N}_0$ and we argue that where the neck ends one of the assumptions of Theorem 8.1 must fail. Let us consider the last point p_1 where it is possible to apply

Theorem 8.1. If the points after p_1 violate condition (ND1), then we argue as in the case without surgeries described before. Let us consider instead the case that (ND2) is violated. Observe that the curvature at p_1 may be comparable with H_1 and thus we cannot ensure the validity of (ND2) as in the case of the point p_0. However, we claim that if (ND2) is violated at the points after p_1, we are still able to describe the topology of the region, and to conclude that the neck ends with a convex cap. In fact, if p_1 is the last point where (ND2) holds, this means that the corresponding backward parabolic neighborhood is a neck at every fixed time, and it intersects on the boundary at some previous time t_1 the region inserted by a surgery. By our construction, the region inserted in a surgery starts out cylindrical and bends gradually until it closes with a convex cap. It is then possible to prove that the cylindrical part of the region must coincide with the last part of the neck $\mathcal{N}_0$ [**29**, Lemma 7.12]. This shows that the neck $\mathcal{N}_0$ ends with a convex cap also in this case. □

We now consider the values η_1, H_1 given by Theorem 9.1, and take the associated γ_0 as in Theorem 8.2. We set

$$H_2 = 10\gamma_0 H_1, \qquad H_3 = 10H_2.$$

We then define our surgery algorithm as follows. We stop the flow every time we reach a time T_i such that $H_{max}(T_i) = H_3$. If some connected component of the hypersurface has become convex everywhere, we neglect it. In the remaining components, we operate surgeries in order to remove all points with curvature greater than H_2. To do this, let p be any point such that $H(p, T_i) \geq H_2$.

If $\lambda_1(p, T_i) \leq \eta_1 H(p, T_i)$, we apply Theorem 9.1 to find that p lies on a neck $\mathcal{N}_0$ having one of the behaviors described there. If the neck covers a whole component of the surface, we know that the component is diffeomorphic to $S^{n-1} \times S^1$, and we neglect it. If in both directions of the neck we find points with mean curvature approximately H_1, we perform surgeries to remove the part of the neck in between, which includes the point p. If we find on one side points with curvature H_1 and on the other side a convex cap, then we do surgery only on one side and neglect the rest of the neck together with the cap, since this leaves the topology of the surface unchanged. If the neck ends with a convex cap in both directions, we neglect the component because it is diffeomorphic to a sphere. In all cases, the point p is removed and the possible surgeries are performed on a part of the neck with curvature close to H_1.

The other case is that $\lambda_1(p) > \eta_1 H(p)$ at time T_i. If the component containing p is convex, it can be neglected. Otherwise, by Theorem 8.2, there is another point p' such that $\lambda_1(p') \leq \eta_1 H(p')$ and $H(p') \geq H(p)/\gamma_0 \geq H_2/\gamma_0 = 10H_1$. Then we apply Theorem 9.1 to the point p' as in the former case. Using the fact that $H(q) \geq H(p)/\gamma_0$ at all q such that $d(p,q) \leq d(p,p')$ we can show that the neck containing p' necessarily ends on one side with a convex cap containing p. Hence the procedure described in the previous case removes the point p together with p'.

We then iterate the procedure until every point with curvature greater than H_2 is removed. This requires at most a finite number of surgeries, since each surgery decreases the surface area by a fixed amount. After all such points have been removed, we restart the flow. We repeat the procedure until there are no more components left. In this way we have defined a surgery algorithm with the required properties, and the proof of the main Theorem 1.1 is complete.

References

[1] U. Abresch, J. Langer, *The normalized curve shortening flow and homothetic solutions*, J. Differential Geom. **23** (1986), 175–196.

[2] S.J. Altschuler, S.B. Angenent, Y. Giga, *Mean curvature flow through singularities for surfaces of rotation*, J. Geom. Analysis **5** (1995), 293–358.

[3] B. Andrews, *Contraction of convex hypersurfaces in Euclidean space*, Calc. Var. **2** (1994), 151–171.

[4] B. Andrews, *Pinching estimates and motion of hypersurfaces by curvature functions*, preprint (2004).

[5] S.B. Angenent, *Shrinking doughnuts* in "Nonlinear diffusion equations and their equilibrium states", (1989, Gregynog), Birkhäuser, Boston (1992).

[6] S.B. Angenent, J.J.L. Velazquez, *Degenerate neckpinches in mean curvature flow*, J. Reine Angew. Math. **482** (1997), 15–66.

[7] K.A. Brakke, The motion of a surface by its mean curvature. Princeton University Press, Princeton (1978).

[8] Y.G. Chen, Y. Giga, S. Goto, *Uniqueness and existence of viscosity solutions of generalized mean curvature flow equation*, J. Differential Geom. **33** (1991), 749–786.

[9] T. Colding, B. Kleiner, *Singularity structure in mean curvature flow of mean-convex sets*, Electron. Res. Announc. Amer. Math. Soc. **9** (2003), 121–124 (electronic).

[10] T. Colding, W.P. Minicozzi, *Sharp estimates for mean curvature flow of graphs*, J. Reine Angew. Math. **574** (2004), 187–195.

[11] K. Ecker, Regularity theory for mean curvature flow. Birkhäuser, Boston (2004).

[12] K. Ecker, G. Huisken, *Interior estimates for hypersurfaces moving by mean curvature*, Invent. Math. **105** (1991), 547–569.

[13] J. Eells, J.H. Sampson, *Harmonic mappings of Riemannian manifolds*, Amer. J. Math. **86** (1964), 109–160.

[14] L.C. Evans, J. Spruck, *Motion of level sets by mean curvature, I*, J. Differential Geom. **33** (1991), 635–681.

[15] L.C. Evans, J. Spruck, *Motion of level sets by mean curvature, II*, Trans. Amer. Math. Soc. **330** (1992), 321–332.

[16] A.M. Fraser, *Minimal disks and two-convex hypersurfaces*, Amer. J. Math. **124** (2002), 483–493.

[17] M. Gage, R.S. Hamilton, *The heat equation shrinking convex plane curves*, J. Differential Geom. **23** (1986), 69–96.

[18] M.A. Grayson, *A short note on the evolution of a surface by its mean curvature*, Duke Math. J. **58** (1989), 555–558.

[19] R.S. Hamilton, *Three-manifolds with positive Ricci curvature*, J. Differential Geom. **17** (1982), 255–306.

[20] R.S. Hamilton, *Four-manifolds with positive curvature operator*, J. Differential Geom. **24** (1986), 153–179.

[21] R.S. Hamilton, *The Harnack estimate for the mean curvature flow*, J. Differential Geom. **41** (1995), 215–226.
[22] R.S. Hamilton, *Formation of singularities in the Ricci flow*, Surveys in Diff. Geom. **2** (1995) 7–136, International Press, Boston.
[23] R.S. Hamilton, *Four-manifolds with positive isotropic curvature*, Comm. Anal. Geom. **5** (1997), 1–92.
[24] G. Huisken, *Flow by mean curvature of convex surfaces into spheres*, J. Differential Geom. **20** (1984), 237–266.
[25] G. Huisken, *Asymptotic behavior for singularities of the mean curvature flow*, J. Diff. Geometry **31** (1990), 285–299.
[26] G. Huisken, *Local and global behavior of hypersurfaces moving by mean curvature*, Proceedings of Symposia in Pure Mathematics **54** (1993), 175–191.
[27] G. Huisken, C. Sinestrari, *Mean curvature flow singularities for mean convex surfaces*, Calc. Variations **8** (1999), 1–14.
[28] G. Huisken, C. Sinestrari, *Convexity estimates for mean curvature flow and singularities of mean convex surfaces*, Acta Math., **183** (1999), 45–70.
[29] G. Huisken, C. Sinestrari, *Mean curvature flow with surgeries of two-convex hypersurfaces*, preprint (2006).
[30] M. Marcus, L. Lopes, *Inequalities for symmetric functions and hermitian matrices*, Canad. J. Math. **9** (1957), 305–312.
[31] F. Mercuri, M.H. Noronha, *Low codimensional submanifolds of Euclidean space with nonnegative isotropic curvature*, Trans. Amer. Math. Soc. **348** (1996), 2711–2724.
[32] J.H. Michael, L.M. Simon, *Sobolev and mean value inequalities on generalized submanifolds of* $\mathbb{R}^n$, Comm. Pure Appl. Math. **26** (1973), 361–379.
[33] W.W. Mullins *Two-dimensional motion of idealised grain boundaries*, J. Appl. Phys. **27** (1956), 900–904.
[34] G. Perelman, *The entropy formula for the Ricci flow and its geometric applications*, preprint (2002).
[35] G. Perelman, *Ricci flow with surgery on three-manifolds*, preprint (2003).
[36] G. Perelman, *Finite extinction time for the solutions to the Ricci flow on certain three-manifolds*, preprint (2003).
[37] J.-P. Sha, *p-convex Riemannian manifolds*, Invent. Math. **83** (1986), 437–447.
[38] X.-J. Wang, *Convex solution to the mean curvature flow*, preprint (2004).
[39] B. White, *The nature of singularities in mean curvature flow of mean-convex sets*, J. Amer. Math. Soc. **16** (2002), 123–138.
[40] H. Wu, *Manifolds of partially positive curvature*, Indiana Univ. Math. J. **36** (1987), 525–548.

Dipartimento di Matematica, Università di Roma "Tor Vergata" Via della Ricerca Scientifica, 00133 Roma, Italy
E-mail address: `sinestra@mat.uniroma2.it`

Some Recent Developments in Lagrangian Mean Curvature Flows

Mu-Tao Wang*

ABSTRACT. We review some recent results on the mean curvature flows of Lagrangian submanifolds from the perspective of geometric partial differential equations. These include global existence and convergence results, characterizations of first-time singularities, and constructions of self-similar solutions.

1. Introduction

1.1. The mean curvature flow. A distinguished normal vector field called the *mean curvature vector field* exists on any submanifold of a Riemannian manifold. It is characterized as the unique direction along which the area or the volume of the submanifold would be decreased most effectively. A submanifold is *minimal* if the mean curvature vector vanishes at each point. The mean curvature flow is an evolution process that moves a submanifold by its mean curvature vector field. This turns out to be a nonlinear parabolic system of partial differential equations for the position functions of the submanifold. Brakke [**BR**] pioneered the formulation of weak solutions of the mean curvature flow in the setting of geometric measure theory in the seventies. In the eighties, Huisken and his coauthors [**HU1**] took up the geometric PDE approach to study the mean curvature flow in Riemannian manifolds. There were also level set formulations of viscosity solutions for hypersurfaces by Chen-Giga-Goto [**CGG**] and independently by Evans-Spruck [**ES1**] later. Soon the mean curvature flow joined the rank of most studied geometric evolution equations like the Ricci flow and the harmonic map heat flows. Indeed, in many aspects, its development paralleled that of the Ricci flow.

Suppose M is a m dimensional Riemannian manifold with a Riemannian metric $\langle \cdot, \cdot \rangle$. Let Σ be another n dimensional smooth manifold with $n < m$.

*The author is partially supported by National Science Foundation Grant DMS 0605115 and an Alfred P. Sloan Research Fellowship.

An immersion of Σ in M is given by a mapping $F : \Sigma \to M$ so that the differential dF has full rank at each point of Σ. In terms of local coordinates $x^1 \cdots x^n$ on Σ, this is equivalent to the matrix $g_{ij} = \langle \frac{\partial F}{\partial x^i}, \frac{\partial F}{\partial x^j} \rangle$ being positive definite. Indeed, g_{ij} defines a Riemannian metric which makes the image $F(\Sigma)$ a Riemannian submanifold. The second fundamental form of $F(\Sigma)$ is given by the tensor $A = \left(\nabla_{\frac{\partial F}{\partial x^i}} \frac{\partial F}{\partial x^j}\right)^{\perp}$ where ∇ denotes the Levi-Civita connection of M and $\perp$ denotes the normal part of a vector in the bundle $TM|_{F(\Sigma)}$. The mean curvature vector H is then the trace of the second fundamental form, i.e.,

$$H = g^{ij} \left(\nabla_{\frac{\partial F}{\partial x^i}} \frac{\partial F}{\partial x^j}\right)^{\perp}$$

where g^{ij} is the inverse of g_{ij}.

We recall the volume of $F(\Sigma)$ is calculated by

$$Vol(F(\Sigma)) = \int_{\Sigma} \sqrt{\det g_{ij}} dx^1 \wedge \cdots \wedge dx^n.$$

Suppose $F : \Sigma \times [0, \epsilon) \to M$ is a family of immersion so the variation field $\frac{\partial F}{\partial s}$ is a normal vector field along $F(\Sigma, s)$. The variation of the volume is then given by

$$\frac{d}{ds} Vol(F(\Sigma, s)) = -\int_{\Sigma} \left\langle \frac{\partial F}{\partial s}, H \right\rangle \sqrt{\det g_{ij}} dx^1 \wedge \cdots \wedge dx^n.$$

The mean curvature flow deforms a submanifold in the direction of the mean curvature vector field H. Namely, a family of immersion $F : \Sigma \times [0, T) \to M$ is said to form a mean curvature flow if

$$\left(\frac{\partial F}{\partial t}(x, t)\right)^{\perp} = H(F(x, t)).$$

The flow is a nonlinear weakly parabolic system for F and is invariant under reparametrization of Σ. Indeed, by coupling with a diffeomorphism of Σ, the flow can be made into a normal deformation, i.e., $\frac{\partial F}{\partial t}(x, t)$ is always in the normal direction. One can establish the short-time existence for any smooth compact initial data. For a normal deformation, a simple calculation shows

$$\frac{d}{dt} \sqrt{\det g_{ij}(F(x, t))} = -|H(F(x, t))|^2 \sqrt{\det g_{ij}(F(x, t))},$$

from which it follows that being a immersion is preserved by the mean curvature flow. Integrating this equality gives

$$\frac{d}{dt} Vol(F(\Sigma, t)) = -\int_{\Sigma} |H(F(x, t))|^2 \sqrt{\det g_{ij}(F(x, t))} dx^1 \wedge \cdots \wedge dx^n. \tag{1.1}$$

In the rest of the paper, when there is no confusion, we will not differentiate between an immersed submanifold Σ and its image $F(\Sigma)$. $T\Sigma$ will denote the tangent bundle and $N\Sigma$ will denote the normal bundle of Σ. We shall denote the image $F(\Sigma, t)$ of a mean curvature flow by Σ_t and the volume form $\sqrt{\det g_{ij}(F(x,t))}dx^1 \wedge \cdots \wedge dx^n$ by dv_{Σ_t}.

1.2. Co-dimension one vs. higher co-dimension. The Lagrangian mean curvature flow is a special case of the mean curvature flow in general. They are mostly higher co-dimensional in the sense that $m > n+1$ where m is the dimension of the ambient manifold and n is the dimension of the evolving submanifold. There are abundant results of hypersurface (i.e., $m = n+1$) mean curvatures flows, while relatively little is known in the higher co-dimensional case. Indeed, many techniques and results for hypersurface flows do not generalize to higher co-dimensions. The contrast between the hypersurface and higher co-dimensional can be seen from two points. Firstly, the hypersurface case corresponds to a scalar equation and the maximum principle holds in the following sense. Two embedded hypersurfaces evolving by the mean curvature flow will avoid each other and an embedded hypersurface remains embedded along the mean curvature flow. These are no longer true for higher co-dimensional mean curvature flows. Secondly, the second fundamental form of a hypersurface is a symmetric two tensor and various convexity conditions associated with this tensor play an important role. However, for a higher co-dimensional submanifold, the second fundamental is a symmetric two tensor valued in the normal bundle and there is no natural convexity condition for such a tensor. Of course, both these points are related to the complexity of the normal bundle. The normal bundle of an embedded orientable hypersurface is always trivial, while the normal bundle of a higher co-dimensional submanifold could be highly non-trivial.

The good news is that many great results like Brakke's regularity theorem [**BR**], Hamilton's maximum principle for tensors [**HA**], Huisken's monotonicity formula [**HU2**], and White's regularity theorem [**WH**] are valid in any dimension and co-dimension.

1.3. What is special about being Lagrangian?. A Lagrangian submanifold sits in a symplectic manifold M. Recall a symplectic manifold is a smooth manifold equipped with a closed non-degenerate two-form $\omega(\cdot,\cdot)$. A Lagrangian submanifold Σ is characterized by the vanishing of $\omega|_\Sigma$. To get the volume functional into the context, we consider on M a Riemannian metric $\langle\cdot,\cdot\rangle$ and a compatible almost complex structure J in the sense that $\omega(\cdot,\cdot) = \langle J(\cdot),\cdot\rangle$.

The simplest examples are Lagrangian submanifolds in the complex Euclidean space $\mathbb{C}^n$. Suppose $z^i = x^i + y^i, i = 1 \ldots n$ are complex coordinates on $\mathbb{C}^n$. Let $\omega = \sum_{i=1}^n dx^i \wedge dy^i$ be the standard symplectic form on $\mathbb{C}^n$ and $J\left(\frac{\partial}{\partial x^i}\right) = \frac{\partial}{\partial y^i}$ be the standard almost complex structure. We have

$\omega(X, Y) = \langle JX, Y \rangle$ where $\langle \cdot, \cdot \rangle$ is the standard metric on $\mathbb{C}^n$. Recall a Lagrangian subspace of $\mathbb{C}^n$ is a subspace on which ω restricts to zero. As the form ω is invariant under the unitary group $U(n)$, the set of all Lagrangian subspaces in $\mathbb{C}^n$, so called the Lagrangian Grassmannian, is isomorphic to the homogeneous space $U(n)/SO(n)$. Therefore a Lagrangian submanifold is simply a submanifold whose tangent spaces are all Lagrangian subspaces in $\mathbb{C}^n$.

Two prominent classes of Lagrangian submanifolds are the following.

1) Graphs of symplectomorphisms: In this case, the ambient space is the product of two symplectic manifolds. Take a smooth map $f : (\mathbb{C}^n, \omega_1) \to (\mathbb{C}^n, \omega_2)$ such that $f^*\omega_2 = \omega_1$ and the graph of f in $(\mathbb{C}^n \times \mathbb{C}^n, \omega_1 - \omega_2)$ is such an example.
2) Graphs of one-forms: In this case, the ambient space is the cotangent bundle of a symplectic manifold. Take a smooth map $f : \mathbb{R}^n \to \mathbb{R}^n$ such that $f = \nabla u$ for some scalar function $u : \mathbb{R}^n \to \mathbb{R}$. Identify the first $\mathbb{R}^n$ with the real part of $\mathbb{C}^n$ with coordinates x^i and the second $\mathbb{R}^n$ as the imaginary part with coordinates y^i. The graph of f in $(\mathbb{C}^n = \mathbb{R}^n \oplus \sqrt{-1}\mathbb{R}^n, \omega)$ with $\omega = \Sigma_{i=1}^n dx^i \wedge dy^i$ is such an example.

Two remarkable properties of Lagrangian mean curvature flow makes one speculate that it perhaps behaves in a better way than other mean curvature flows in higher codimension. Firstly, the normal bundle of a Lagrangian submanifold is canonically isometric to the tangent bundle by the almost complex structure $J : T\Sigma \to N\Sigma$. This gives a simpler description of the second fundamental form as a fully symmetric three tensor in $\odot^3 T\Sigma$, the coefficients of which being

$$h_{ijk} = \left\langle \nabla_{\frac{\partial F}{\partial x^i}} \frac{\partial F}{\partial x^j}, J\left(\frac{\partial F}{\partial x^k}\right) \right\rangle.$$

As JH becomes a tangent vector, it is dual to a one-form σ on Σ; they are related by $\sigma(\cdot) = \omega(H, \cdot)$.

Secondly, when the ambient space is Kähler-Einstein, being Lagrangian is a condition that is preserved along the mean curvature flow [**SM1**]. In many occasions, one is tempted to compare the Lagrangian mean curvature flow to the Kähler-Ricci flow. This will be elaborated in the next section when we consider a Calabi-Yau ambient manifold.

1.4. Calabi-Yau case. The Lagrangian mean curvature flow is of particular interest when the ambient space is Calabi-Yau. Let (M, g, ω, J) be a real $2n$ dimensional Kähler-Einstein manifold; i.e., the Ricci form is a multiple of the the Kähler form:

$$Ric = c\omega.$$

Let Σ be an n dimensional Lagrangian submanifold of M. By the Codazzi equation, we have

$$d\sigma = Ric|_\Sigma = c\omega|_\Sigma = 0.$$

Thus σ is a closed one-form and defines a cohomology class $[\sigma]$.

When M is Calabi-Yau with a canonical parallel holomorphic $(n, 0)$ form Ω. By suitable normalization, the restriction of Ω on Σ gives a multi-valued function θ, called the *phase* function. Indeed, $*_\Sigma \Omega = e^{i\theta}$ where $*_\Sigma$ denote the Hodge star operator on Σ. It turns out the mean curvature form is $\sigma = d\theta$. $[\sigma]$ is called the *Maslov class* and can be defined through the Gauss map of Σ.

Lagrangian submanifolds with vanishing mean curvature H are *minimal Lagrangian* submanifolds. When M is Calabi-Yau, a connected minimal Lagrangian has constant phase and is called a *special Lagrangian* which corresponds to a class of calibrated submanifolds first studied by Harvey and Lawson [**HL**]. They also play important roles in the SYZ [**SYZ**] conjecture in Mirror symmetry. Roughly speaking, people expect special Lagrangians to behave much like holomorphic curves. It is thus desirable to have a general method of constructing special (or minimal) Lagrangian submanifolds. Schoen and Wolfson [**SW1**] studied the existence problem by the variational method. Another approach more related to algebraic geometry and symplectic topology has been investigated by Joyce.

Based on the duality between graded Lagrangian submanifolds and stable vector bundles, Thomas and Yau [**TY**] made the following conjecture:

CONJECTURE 1. *Let M be Calabi-Yau and Σ be a compact embedded Lagrangian submanifold with zero Maslov class, then the mean curvature flow of Σ exists for all time and converges smoothly to a special Lagrangian submanifold in the Hamiltonian isotopy class of Σ.*

Suppose Σ_t is the mean curvature flow of Σ. The phase function θ on Σ_t is evolved by the heat equation

$$\frac{d\theta}{dt} = \Delta_{\Sigma_t}\theta, \tag{1.2}$$

where Δ_{Σ_t} is the Laplace operator of the induced metric on Σ_t. Thus being of zero Maslov class is preserved along the flow. This is a rather bold conjecture as it is easy to see that the mean curvature flow of any compact submanifold of the Euclidean space develops finite time singularities. The most common singularity is the so-called *neck-pinching*. Without the assumption on the Maslov class, Schoen and Wolfson [**SW2**] construct example that develop such singularities in finite time. We shall come back to this point in §2.3.

1.5. Overview of the article. In this paper, we review some recent results on the mean curvature flows of Lagrangian submanifolds. The review is by no means comprehensive or complete and is subject to the author's personal preference. Many other interesting works on the Lagrangian mean curvature flow are not discussed in the article, e.g., [**GSSZ**], [**LW**], [**NE2**], and [**PA**].

A fundamental question in geometric flows is under what conditions on the initial data can we prove the global existence and convergence in the smooth category. The mean curvature flow, as a quasi-linear system, forms singularity exactly when the second fundamental form of the submanifold blows up, i.e., Σ_t becomes singular as t approaches T if and only if $\lim_{t\to T}\sup_{\Sigma_t}|A|^2\to\infty$. In §2.1, we discuss global existence and convergence results for special initial data that correspond to the two classes of Lagrangian submanifolds discussed in §1.3.

For more general initial data set, we need to understand singularity formations through the blow-up analysis. Suppose the flow exists on $[0,T)$ and $\sup_{\Sigma_t}|A|^2\to\infty$ as $t\to T$, we perform parabolic blow-up of the solution near T. For simplicity, here we demonstrate the idea in the case when the ambient space is the Euclidean space $\mathbb{R}^N$. This process depends on three parameters t_i(when), y_i(where), and λ_i (how much) and eventually we let $\lambda_i\to\infty$. Take the space time track of the mean curvature flow $\mathfrak{M}=\cup_{t\in[0,T)}\Sigma_t$ in $\mathbb{R}^N\times\mathbb{R}$ and consider the map

$$\mathbb{R}^N\times[0,T)\to\mathbb{R}^N\times[-\lambda_i^2 t_i,\lambda_i^2(T-t_i))$$

by sending $(y,t)\mapsto(\lambda_i(y-y_i),\lambda_i^2(t-t_i))$ and thus $(y_i,t_i)\mapsto(0,0)$.

The image of Σ_t can be described as

$$\Sigma_s^i=\lambda_i\left(\Sigma_{t_i+\frac{s}{\lambda_i^2}}-y_i\right)$$

in which $s=\lambda_i^2(t-t_i)$. It turns out the space-time track $\mathfrak{M}_i=\cup_s\Sigma_s^i$ forms another mean curvature flow by the invariance of the scaling that lives in $[-\lambda_i^2 t_i,\lambda_i^2(T-t_i))$.

In order to obtain a smooth limit, two types of parabolic blow-ups are often used depending on how fast the second fundamental form blows up.

Type I blow-up: Also called a central blow-up where the center $(t_i,y_i)=(T,y_0)$ is fixed and $\lambda_i\to\infty$. When $|A|^2(T-t)$ is bounded, the limit is smooth. Nevertheless, the limit always exists weakly in the sense of geometric measure theory and is an ancient self-similar solution that lives in $(-\infty,0]$. This is the parabolic analogue of a cone.

Type II blow up: The blow-up center (y_i,t_i) is at a point where $|A|^2(T-t)$ almost achieves its maximum. The scale is proportional to $|A|^2$ so we get a smooth limit that often lives on $(-\infty,\infty)$ with uniformly bounded second fundamental form, so called an eternal solution. In §2.2, we discuss the characterization of first time singularity for Lagrangian mean curvature flow under the Type I blow-up procedure.

The singularities are often classified into *type I singularity* or *type II singularity* according to whether $\sup_{\Sigma_t} |A|^2(T-t)$ is bounded or unbounded as $t \to T$ (not to be confused with the type I and type II blow-up procedures). The simplest Lagrangian mean curvature flow reduces to the so-called curve shortening flow on a two-dimensional orientable surface as any curve is Lagrangian. In this case, it can be proved that any type I singularity is a shrinking circle and any type II singularity is a Grim Reaper, both are self-similar solutions. It is thus concievable that the eventual understandings of the singularities will rely on the classifications of self-similar solutions. In §2.3, we discuss the constructions of self-similar solutions in Lagrangian mean curvature flows.

The author would like to thank B. Andrews, K. Ecker, R. Hamilton, M. Haskin, G. Huisken, T. Ilmanen, D. Joyce, Y.-I. Lee, N. C. Leung, A. Neves, K. Smoczyk, M.-P. Tsui, T. Y. H. Wan, and B. White for helpful discussions on this subject.

2. Results

2.1. Global existence and convergence. Given an immersed submanifold Σ_0 of a Riemannian manifold M, we ask when we can find a family of immersions that forms a mean curvature flow Σ_t and when $\Sigma_t \to \Sigma_\infty$ in C^∞ for a smooth immersed submanifold Σ_∞. As was remarked in the overview of the article, this boils down to bounding the second fundamental form for all $t \in [0, \infty)$ and as $t \to \infty$. The one-dimensional curve-shortening flow is a well-studied area and there are many beautiful global existence and convergence results by, e.g., Gage-Hamilton [**GH**] and Grayson[**GR1**][**GR2**]. We refer to the book by Chou-Zhu [**CZ**] for results in this direction.

The next simplest case will be two-dimensional Lagrangian surfaces in a four-dimensional symplectic manifold. We recall that the graph of a symplectomorphism is naturally a Lagrangian submanifold of the product space. In this case, there is the following theorem:

THEOREM 2.1. *Let $(\Sigma^{(1)}, \omega_1)$ and $(\Sigma^{(2)}, \omega_2)$ be two diffeomorphic compact Riemann surfaces of the same constant curvature c. Suppose Σ is the graph of a symplectomorphism $f : \Sigma^{(1)} \to \Sigma^{(2)}$ as a Lagrangian submanifold of $M = (\Sigma^{(1)} \times \Sigma^{(2)}, \omega = \omega_1 - \omega_2)$ and Σ_t is the mean curvature flow with initial surface $\Sigma_0 = \Sigma$. Then Σ_t remains the graph of a symplectomorphism f_t along the mean curvature flow. The flow exists smoothly for all time and Σ_t converges smoothly to a minimal Lagrangian submanifold as $t \to \infty$.*

The assumption on the curvatures of $\Sigma^{(1)}$ and $\Sigma^{(2)}$ makes M a Kähler-Einstein manifold with the product metric. The long time existence for all cases and the smooth convergence for $c > 0$ was proved in [**WA2**]. The smooth convergence for $c \leq 0$ was established in [**WA4**].

Assuming an extra angle condition, Smoczyk [**SM2**] proved the theorem when $c \leq 0$. Smoczyk's proof of the global existence and convergence result is different from that of [**WA2**] and [**WA4**]. Instead of applying blow-up analysis, he proved directly by the maximum principle that the second fundamental form is uniformly bounded independent of time. This gives the global existence and the convergence at the same time.

We remark the existence of such minimal Lagrangian submanifold was proved using variational method by Schoen [**SC**] (see also Lee [**LE**]).

In the following, we briefly describe the proof of the theorem. The proof is divided into three parts.

1) Σ_t remains the graph of a symplectomorphism f_t as long as the flow exists smoothly.

Since M is Kähler-Einstein, Σ_t remains a Lagrangian surface. This can indeed be shown by considering the evolution equation satisfied by the function $*\omega(p) = \omega(e_1, e_2)$ where $\{e_1, e_2\}$ is any oriented orthonormal basis for $T_p\Sigma$. Likewise, we can consider the heat equation satisfied by the function $*\omega_1(p) = \omega_1(e_1, e_2)$. In an orthonormal basis, we can represent the second fundamental form A by $h_{ijk} = \langle \nabla_{e_i} e_j, J(e_k)\rangle$ and the mean curvature vector by $H_k = \langle H, J(e_k)\rangle = \sum_{i=1}^{2} h_{iik}$. It was computed in [**WA2**] that $\eta_t = 2 *_{\Sigma_t} \omega_1$ satisfies

$$\frac{d}{dt}\eta_t = \Delta\eta_t + \eta_t(2|A|^2 - |H|^2) + c\eta_t(1 - \eta_t^2) \tag{2.1}$$

where $|A|^2 = \sum_{i,j,k} h_{ijk}^2$ and $|H|^2 = \sum_k H_k^2$ are the squared norms of the second fundamental form and the mean curvature vector, respectively. As $|H|^2 \leq 2|A|^2$, $\eta_t > 0$ is preserved along the flow by the maximum principle.

Notice that $*\omega_1$ is in fact the Jacobian of $\pi_1|_\Sigma$ where $\pi_1 : \Sigma^{(1)} \times \Sigma^{(2)} \to \Sigma^{(1)}$ is the projection map onto the first factor. By the inverse function theorem, $*\eta_t > 0$ if and only if Σ_t can be locally written a graph over Σ_1. Therefore $\eta_t > 0$ implies Σ_t is the graph of a symplectomorphism.

Indeed, not only $\eta_t > 0$ but by comparing to solutions of the ordinary differential equation $\frac{d}{dt} f = cf(1 - f^2)$, we arrive at

$$\eta_t \geq \frac{\alpha e^c t}{\sqrt{1 + \alpha^2 e^{2ct}}}$$

where α is a constant that satisfies $\frac{\alpha}{\sqrt{1+\alpha^2}} = \min_{\Sigma_0} \eta$. In particular, $\eta_t \to 1$ as $t \to \infty$ uniformly when $c > 0$.

2) Global existence for all finite time

Utilizing the full symmetry of the second fundamental form h_{ijk}, one can show that

$$|H|^2 \leq \frac{4}{3}|A|^2.$$

Therefore,

$$\frac{d}{dt}\eta_t \geq \Delta\eta_t + \frac{2}{3}\eta_t|A|^2 + c\eta_t(1 - \eta_t^2). \tag{2.2}$$

We can apply the type I blow-up procedure to this solution at any space-time point. Equation 2.2 and the positive lower bound of η_t at any finite time will imply the integral of $|A|^2$ vanishes on the type-I blow-up limit. It follows from White's regularity theorem [**WH**] that any such point is a regular point.

3) Convergence at $t = \infty$.

The aim is to bound $|A|^2$ as $t \to \infty$ since Simon's [**SI**] convergence theorem for gradient flows is applicable in this case. Suppose $\sup_{\Sigma_t} |A|^2 \to \infty$, we apply the type II blow-up procedure to the solution at $t = \infty$. Pick a sequence of t_i and point $p_i \in \Sigma_{t_i}$ such that the space-time track $\mathfrak{M}_i$, after shifting (p_i, t_i) to $(0,0)$ and scaling by the factor $|A|(p_i, t_i)$, has uniformly bounded second fundamental form and $|A|(0,0) = 1$.

By compactness, $\mathfrak{M}_i \to \mathfrak{M}_\infty$, which is an eternal solution of the mean curvature flow defined on $(-\infty, \infty)$ with uniformly bounded second fundamental form and $|A|(0,0) = 1$. When $c > 0$, recall $\eta_t \to 1$ as $t \to \infty$, this implies $\eta \equiv 1$ on the limit $\mathfrak{M}_\infty$ and each time slice must be a flat space, contradicting with $|A|(0,0) = 1$.

In cases when $c \leq 0$, η_t no longer converges to 1 as $t \to \infty$ we consider instead the evolution equation for $|H|^2$. It was computed in [**WA4**] that

$$\left(\frac{d}{dt} - \Delta\right)|H|^2 = -2|\nabla H|^2 + 2\sum_{ij}\left(\sum_k H_k h_{kij}\right)^2 + c(2 - \eta^2)|H|^2.$$

Coupling with the equation for η (2.1) and integrating over Σ_t gives

$$\frac{d}{dt}\int_{\Sigma_t} \frac{|H|^2}{\eta} dv_{\Sigma_t} \leq c \int_{\Sigma_t} \frac{|H|^2}{\eta} dv_{\Sigma_t}.$$

This implies $\int_{\Sigma_t} |H|^2 dv_{\Sigma_t} \to 0$ as $t \to \infty$ since $\int_0^\infty \int_{\Sigma_t} |H| dv_{\Sigma_t} dt < \infty$. The limit $\mathfrak{M}_\infty$ obtained earlier has $\int |H|^2 = 0$ and thus each time slice is a minimal area preserving map from $\mathbb{C}$ to $\mathbb{C}$ which must be flat by a result of Ni [**NI**].

In general dimension, Smoczyk and the author [**SW**] proved a general global existence and convergence theorem for Lagrangian graphs in T^{2n}, a flat torus of dimension $2n$.

THEOREM 2.2. *Let Σ be a Lagrangian submanifold in T^{2n}. Suppose Σ is the graph of $f : T^n \to T^n$ and the potential function u of f is convex. Then the mean curvature flow of Σ exists for all time, remains a Lagrangian graph, and converges smoothly to a flat Lagrangian submanifold.*

The flow in terms of the potential u is a fully nonlinear parabolic equation:

$$\frac{du}{dt} = \frac{1}{\sqrt{-1}} \ln \frac{\det(I + \sqrt{-1}D^2 u)}{\sqrt{\det(I + (D^2u)^2)}} \tag{2.3}$$

where D^2u is the Hessian of u. Notice $\frac{\det(I+\sqrt{-1}D^2u)}{\sqrt{\det(I+(D^2u)^2)}}$ is a unit complex number, so the right hand side is always real. This theorem generalizes prior global existence and convergence results in general dimensions in [**WA3**] and [**SM3**].

A important step in the proof is to show the convexity condition $D^2_{ij}u > 0$ is preserved which we describe in the rest of this section. This involves interpreting the convexity condition as the positivity of some symmetric two tensor on Σ_t and compute the parabolic equation with respect to the induced (evolving metric) on Σ_t. It turns out if we denote π_1 (π_2) to be the projection onto the first (second) factor of $T^n \times T^n$. The condition $D^2_{ij}u$ is the same as

$$\langle J\pi_1(X), \pi_2(X)\rangle > 0$$

for any $X \in T\Sigma$. $S(\cdot,\cdot) = \langle J\pi_1(\cdot), \pi_2(\cdot)\rangle$ defines a two-tensor on T^{2n} and the Lagrangian condition implies the restriction of S to any Lagrangian submanifold is a symmetric tensor.

LEMMA 2.1. *This positivity of S is preserved along the mean curvature flow, i.e., $S|_{\Sigma_t} > 0$ for $t > 0$ as long as $S|_{\Sigma_0} > 0$.*

A direct approach is to calculate the evolution of $S|_{\Sigma_t}$ and apply Hamilton's maximal principle for tensors, see equation (3.3) in [**SW**].

Another more systematic approach is to study how the tangent space of Σ_t evolves as this contains the information of $D^2_{ij}u$. Since the tangent space of T^{2n} can be identified with $\mathbb{C}^n$. We may consider the Gauss map of Σ_t given by $\gamma_t : \Sigma_t \to LG(n) = U(n)/SO(n)$, the Lagrangian Grassmannian, by sending a point $p \in \Sigma$ to the tangent space $T_p\Sigma \subset \mathbb{C}^n$. The following theorem is proven in [**WA5**]

THEOREM 2.3. *γ_t is a harmonic map heat flow.*

For an ordinary heat equation $\frac{d}{dt}f = \Delta f$, the conditions $f > 0$ and $f = 0$ are preserved by the maximum principle. For a harmonic map heat flow into a Riemannian manifold, the analogy is that the image of the map will remain in a convex or totally geodesic set.

Since $LG(n)$ is totally geodesic subset of the Grassmannian, this provides an alternate way to show why being Lagrangian is preserved along the mean curvature flow. Also the determinant map from $U(n)/SO(n)$ to $U(1) \equiv S^1$ is totally geodesic. Thus the composition $\gamma_t \circ \det$ is a harmonic map heat flow into S^1. It is easy to see that $\gamma_t \circ \det$ is exactly the phase function θ and this is another way to derive (1.2).

To show $\{L \in LG(n), S|_L > 0\}$ is a convex subset, we study the geodesic equation on $LG(n)$ and the details can be found in [**SW**]. We remark that as being a minimal Lagrangian in $\mathbb{C}^n$ is an invariant property under the symmetry group $U(n)$. The equation of u indeed enjoys more symmetric than a general fully non-linear Hessian equation. This observation provides more equivalent conditions under which the global existence and convergence theorems can be proved (see the last section in [**SW**]).

2.2. Characterization of first-time singularities. In [**WA1**], the author introduced the notion of *almost calibrated* Lagrangian submanifolds in the study of characterizing the first time singularity. Recall for a special Lagrangian, the Lagrangian angle (after a shifting) satisfies $\cos\theta = 1$. A Lagrangian submanifold in a Calabi-Yau manifold is said to be "almost calibrated" if $\cos\theta \geq \epsilon$ for some $\epsilon > 0$. This has proved to be a very useful condition in the study of Lagrangian mean curvature flow.

As $\cos\theta$ satisfies

$$\frac{d}{dt}\cos\theta = \Delta\cos\theta + \cos\theta|H|^2, \tag{2.4}$$

being almost calibrated is another condition that is preserved along the Lagrangian mean curvature flow. The following theorem is proved in [**WA1**].

THEOREM 2.4. *An almost calibrated Lagrangian submanifold does not develop any type I singularity along the mean curvature flow.*

This is established by coupling equation (2.4) with Huisken's monotonicity formula. In particular, no "neck-pinching" will be forming in the Thomas-Yau conjecture if this condition is assumed. To demonstrate the idea, let us pretend the Lagrangian submanifolds are compact and lie in $\mathbb{R}^n$. As long as characterizing finite time singularity is concerned, this does not pose any serious restriction as the ambient curvature will be scaled away under a blow-up procedure. Very mild assumption needs to be imposed at infinity to assure the integration by parts work. Suppose the flow exists on $[0,\infty)$ and consider the backward heat kernel at (y_0, T).

$$\rho_{y_0,T}(y,t) = \frac{1}{(4\pi(T-t))^{\frac{n}{2}}}\exp\left(\frac{-|y-y_0|^2}{4(T-t)}\right) \tag{2.5}$$

Huisken's monotonicity formula implies

$$\frac{d}{dt}\int_{\Sigma_t}\rho_{y_0,T}dv_{\Sigma_t} = -\int_{\Sigma_t}\left|H + \frac{1}{2(T-t)}F^\perp\right|^2\rho_{y_0,T}dv_{\Sigma_t}.$$

Coupling with the equation for $\cos\theta$ (2.4), we obtain

$$\begin{aligned}\frac{d}{dt}\int_{\Sigma_t}\rho_{y_0,T}(1-\cos\theta)dv_{\Sigma_t} = &-\int_{\Sigma_t}\rho_{y_0,T}|H|^2\cos\theta dv_{\Sigma_t}\\ &-\int_{\Sigma_t}\left|H + \frac{1}{2(T-t)}F^\perp\right|^2(1-\cos\theta)dv_{\Sigma_t}.\end{aligned}$$

Both these equations are scaling invariants and continue to hold for the type I blow-up at (y_0, T) (notice that θ is a scaling invariant) and thus

$$\frac{d}{ds}\int_{\Sigma_s^i}\rho = -\int_{\Sigma_s^i}\rho\left|H - \frac{1}{2s}F^\perp\right|^2 \tag{2.6}$$

(2.7)

$$\frac{d}{ds}\int_{\Sigma_s^i}\rho(1-\cos\theta) = -\int_{\Sigma_s^i}\rho|H|^2\cos\theta - \int_{\Sigma_s^i}\rho\left|H - \frac{1}{2s}F^\perp\right|^2(1-\cos\theta)$$

where ρ is the backward heat kernel at $(0,0)$ and $s = -\lambda_i^2(T-t)$. Therefore it is not hard to see that there exists a sequence of rescaled submanifolds on which both $\int |H|^2$ and $\int |H - \frac{1}{2s}F^\perp|^2$ are approaching zero. Thus $H = 0$ and $F^\perp = 0$ weakly on each time slice of the limit. This indicates that each time slice of the limit should be a union of minimal Lagrangian cones. If we assume the singularity is of type I, then the limit flow is smooth and thus must be a flat space. White's regularity theorem implies the point is a smooth point.

Notice that (2.6) implies local area bound and (2.7) implies local bound for the L^2 norm of mean curvatures on Σ_s^i. It follows from compactness theorems in geometric measure theory that the limit is rectifiable and this was carried out by Chen and Li in [**CL**]. As $|H|^2 = |\nabla\theta|^2$, a natural question arises whether the phase θ is a constant on this union of minimal Lagrangian cones. Notice that even a union of special Lagrangian cones may have different phases and hence not necessarily area-minimizing. Chen and Li [**CL**] claimed that the phase function is a constant on the limit by proving a Poincáre inequality for θ. Unfortunately, the proof of Theorem 5.1 in [**CL**] overlooks some technical difficulties. Neves later gives a different proof assuming two extra conditions and using the evolution equation of the Liouville form $\lambda = \sum_{i=1}^n x^i dy^i - y^i dx^i$. We refer to his paper [**NE1**] for the precise statement of the theorem (Theorem B). Neves [**NE1**] was also able to replace the assumption of almost calibrated by zero Maslov class by observing that the equation for $\cos\theta$ can be replaced by

$$\frac{\partial}{\partial t}\theta^2 = \Delta\theta^2 - 2|H|^2.$$

2.3. Constructions of self-similar solutions. A important tool in the study of geometric flows is the blow-up analysis. A blow-up solution of the mean curvature flow sits in the Euclidean space and often enjoys more symmetry. It is important to study these special solutions as singularity models. A mean curvature flow in the Euclidean space is said to be *self-similar* if it is moved by an ambient symmetry. We may consider ansatz of the type

$$F(x,t) = \phi(t)F(x) \tag{2.8}$$

and

$$F(x,t) = F(x) + \psi(t) \tag{2.9}$$

which correspond to scaling symmetry and translating symmetry, respectively. The ansatz coupled with the mean curvature flow equation gives an elliptic equation for $F(x)$. For a solution of the form (2.8), $F(x,t)$ is called an expanding or a shrinking soliton depending on whether $\phi(t)$ is greater or less than one, respectively. A mean curvature flow $F(x,t)$ that satisfies (2.9) is called a translating soliton.

Henri Anciaux constructed examples of Lagrangian shrinking and expanding solitons in [**AN**]. All the examples are based on minimal Legendrian immersions in S^{2n-1} and the solutions are asymptotic to the associated minimal Lagrangian cones.

Yng-Ing Lee and the author [**LWA**] constructed examples of self-similar shrinking and expanding Lagrangian mean curvature flows that are asymptotic to Hamiltonian stationary cones. They were able to glue them together to form weak solutions of the mean curvature flow in the sense of Brakke. In a new preprint of Joyce, Lee and Tsui [**JLT**] constructed new examples of self-similar solutions, in particular translating solitons. [**NT**] gave some characterizations of translating solitons in the two dimensional case.

3. Prospects

There have been several attempts to find counterexamples of the Thomas-Yau conjecture. Other than the examples of Schoen and Wolfson in [**SW2**], Neves [**NE1**] constructed almost calibrated complete non-compact Lagrangian surfaces in $\mathbb{C}^2$ that develop finite time singularities. However, there is still no genuine counterexample to the Thomas-Yau conjecture as it was stated.

It should be noted that Schoen and Wolfson [**SW3**] proved the following existence result of special Lagrangians in a K-3 surface.

THEOREM 3.1. *Let X be a K3 surface with a Calabi-Yau metric. Suppose that $\gamma \in H_2(X;\mathbb{Z})$ is a Lagrangian class that can be represented by an embedded Lagrangian surface. Then γ can be represented by a special Lagrangian surface.*

A mean curvature flow proof of this theorem will confirm the Thomas-Yau conjecture in two dimension. Since it was already shown that there is no type-I singularity, we need to focus on type II singularities in the zero Maslov class or almost calibrated case. A general type-II singularity can be scaled to get an eternal solution with uniformly bounded second fundamental form that exists on $(-\infty, \infty)$. Such a solution of a parabolic equation should be rather special and we hope to say more about it in the near future.

For a general initial data, one is tempted to speculate that, just as in the Ricci flow case, surgeries are necessary in order to continue the flow. It was commented in Perelman's paper [**PE**] that when the surgery scale goes to zero, the solution with surgeries should converge to a "weak solution" of the Ricci flow, a notion that has yet to established. Weak formulations for the mean curvature flow are available. However, as weak solutions are no longer unique, it is necessary to instruct the flow how to continue after singuarities. The examples found in Lee-Wang [**LWA**] and Joyce-Lee-Tsui [**JLT**] start out as shrinking solitons as $t < 0$, approach to Schoen-Wolfson cones as $t \to 0$ and resolve to expanding solitons for $t > 0$. They altogether form a Brakke flow.

References

[AN] H. Anciaux, *Construction of Lagrangian self-similar solutions to the mean curvature flow in* $\mathbb{C}^n$. Geom. Dedicata 120 (2006), 37–48.

[BR] K. A. Brakke, *The motion of a surface by its mean curvature.* Mathematical Notes, 20. Princeton University Press, Princeton, N.J., 1978.

[CGG] Y. G. Chen, Y. Giga, and Y. Goto, *Uniqueness and existence of viscosity solutions of generalized mean curvature flow equations.* J. Differential Geom. 33 (1991), no. 3, 749–786.

[CL] J. Chen and J. Li, *Singularity of mean curvature flow of Lagrangian submanifolds.* Invent. Math. 156 (2004), no. 1, 25–51.

[CZ] K.-S. Chou and X.-P. Zhu, *The curve shortening problem.* Chapman & Hall/CRC, Boca Raton, FL, 2001.

[EH] K. Ecker and G. Huisken, *Mean curvature evolution of entire graphs.* Ann. of Math. (2) 130 (1989), no. 3, 453–471.

[EH2] K. Ecker and G. Huisken, *Interior estimates for hypersurfaces moving by mean curvature.* Invent. Math. 105 (1991), no. 3, 547–569.

[ES1] L. C. Evans and J. Spruck, *Motion of level sets by mean curvature. I.* J. Differential Geom. 33 (1991), no. 3, 635–681.

[GH] M. Gage and R. Hamilton, *The heat equation shrinking convex plane curves.* J. Differential Geom. 23 (1986), no. 1, 69–96.

[GR1] M. Grayson, *The heat equation shrinks embedded plane curves to round points.* J. Differential Geom. 26 (1987), no. 2, 285–314.

[GR2] M. Grayson, *Shortening embedded curves.* Ann. of Math. (2) 129 (1989), no. 1, 71–111.

[GSSZ] K. Groh, M. Schwarz, K. Smoczyk, and K. Zehmisch, *Mean curvature flow of monotone Lagrangian submanifolds.* Math. Z. 257 (2007), no. 2, 295–327.

[HA] R. S. Hamilton, *Four-manifolds with positive curvature operator.* J. Differential Geom. 24 (1986), no. 2, 153–179.

[HA3] R. S. Hamilton, *Harnack estimate for the mean curvature flow.* J. Differential Geom. 41 (1995), no. 1, 215–226.

[HL] R. Harvey and H. B. Lawson, *Calibrated geometries.* Acta Math. 148 (1982), 47–157.

[HU1] G. Huisken, *Flow by mean curvature of convex surfaces into spheres.* J. Differential Geom. 20 (1984), no. 1, 237–266.

[HU2] G. Huisken, *Asymptotic behavior for singularities of the mean curvature flow.* J. Differential Geom. 31 (1990), no. 1, 285–299.

[JO] D. Joyce, *Lectures on special Lagrangian geometry.* Global theory of minimal surfaces, 667–695, Clay Math. Proc., 2, Amer. Math. Soc., Providence, RI, 2005.

[JLT] D. Joyce, Y.-I. Lee, and M.-P. Tsui, *Self-similar solutions and translating solitons for Lagrangian mean curvature flow.* preprint 2007.

[LW] N. C. Leung and T. Y. H. Wan, *Hyper-Lagrangian submanifolds of hyperkhler manifolds and mean curvature flow.* J. Geom. Anal. 17 (2007), no. 2, 343–364.

[LE] Y.-I. Lee, *Lagrangian minimal surfaces in Kähler-Einstein surfaces of negative scalar curvature.* Comm. Anal. Geom. 2 (1994), no. 4, 579–592.

[LWA] Y.-I. Lee and M.-T. Wang, *Hamiltonian stationary self-shrinkers and self-expanders of Lagranian mean curvature flows.* preprint, arXiv:0707.0239.

[NE1] A. Neves, *Singularities of Lagrangian mean curvature flow: zero-Maslov class case.* Invent. Math. 168 (2007), no. 3, 449–484.

[NE2] A. Neves, *Singularities of Lagrangian mean curvature flow: monotone case.* preprint, arXiv:math/0608401.

[NT] A. Neves and G. Tian, *Translating solutions to Lagrangian mean curvature flow.* preprint, arXiv:0711.4341.

[NI] L. Ni, *A Bernstein type theorem for minimal volume preserving maps.* Proc. Amer. Math. Soc. 130 (2002), no. 4, 1207–1210.

[PA] T. Pacini, *Mean curvature flow, orbits, moment maps.* Trans. Amer. Math. Soc. 355 (2003), no. 8, 3343–3357.

[PE] G. Perelman, *Ricci flow with surgery on three-manifolds.* arXiv:math/0303109v1.

[SC] R. Schoen, *The role of harmonic mappings in rigidity and deformation problems.* Complex geometry (Osaka, 1990), 179–200, Lecture Notes in Pure and Appl. Math., 143, Dekker, New York, 1993.

[SW1] R. Schoen and J. Wolfson, *Minimizing area among Lagrangian surfaces: the mapping problem.* J. Differential Geom. 58 (2001), no. 1, 1–86.

[SW2] R. Schoen and J. Wolfson, *Mean curvature flow and lagrangian embeddings.* preprint, 2002.

[SW3] R. Schoen and J. Wolfson, in preparation.

[SI] L. Simon, *Asymptotics for a class of nonlinear evolution equations, with applications to geometric problems.* Ann. of Math. (2) 118 (1983), no. 3, 525–571.

[SM1] K. Smoczyk, *A canonical way to deform a Lagrangian submanifold.* preprint, dg-ga/9605005.

[SM2] K. Smoczyk, *Angle theorems for the Lagrangian mean curvature flow.* Math. Z. 240 (2002), no. 4, 849–883.

[SM3] K. Smoczyk, *Longtime existence of the Lagrangian mean curvature flow.* Calc. Var. Partial Differential Equations 20 (2004), no. 1, 25–46.

[SW] K. Smoczyk and M.-T. Wang, *Mean curvature flows of Lagrangians submanifolds with convex potentials.* J. Differential Geom. 62 (2002), no. 2, 243–257.

[SYZ] A. Strominger, S.-T. Yau, and E. Zaslow, *Mirror symmetry is T-duality.* Nuclear Phys. B 479 (1996), no. 1–2, 243–259.

[TY] R. P. Thomas and S.-T. Yau, *Special Lagrangians, stable bundles and mean curvature flow.* Comm. Anal. Geom. 10 (2002), no. 5, 1075–1113.

[WA1] M.-T. Wang, *Mean Curvature Flow of surfaces in Einstein Four-Manifolds.* J. Differential Geom. 57 (2001), no. 2, 301–338.

[WA2] M.-T. Wang, *Deforming area preserving diffeomorphism of surfaces by mean curvature flow.* Math. Res. Lett. 8 (2001), no. 5–6, 651-662.

[WA3] M.-T. Wang, *Long-time existence and convergence of graphic mean curvature flow in arbitrary codimension.* Invent. Math. 148 (2002), no. 3, 525–543.

[WA4] M.-T. Wang, *A convergence result of the Lagrangian mean curvature flow.* to appear in the Proceedings of the third International Congress of Chinese Mathematicians. arXiv:math/0508354.

[WA5] M.-T. Wang, *Gauss maps of the mean curvature flow.* Math. Res. Lett. 10 (2003), no. 2–3, 287–299.

[WH] B. White, *A local regularity theorem for classical mean curvature flow.* Ann. of Math. (2) 161 (2005), no. 3, 1487–1519.

January 6, 2008

E-mail address: mtwang@math.columbia.edu

Made in United States
Orlando, FL
04 November 2026

90200369R00197